PHYSICS FROM PLANCK SCALE TO ELECTROWEAK SCALE

PROCEEDINGS OF THE US – POLISH WORKSHOP

PHYSICS FROM PLANCK SCALE TO ELECTROWEAK SCALE

Warsaw, Poland 21–24 September 1994

Editors

Pran Nath
(Northeastern Univ., Boston)

Tomasz Taylor
(Northeastern Univ., Boston)

Stefan Pokorski
(Warsaw Univ., Warsaw)

World Scientific
Singapore • New Jersey • London • Hong Kong

Published by

World Scientific Publishing Co. Pte. Ltd.

P O Box 128, Farrer Road, Singapore 9128

USA office: Suite 1B, 1060 Main Street, River Edge, NJ 07661

UK office: 57 Shelton Street, Covent Garden, London WC2H 9HE

PHYSICS FROM PLANCK SCALE TO ELECTROWEAK SCALE
Proceedings of the US–Polish Workshop

ISBN 981-02-2184-3

Printed in Singapore.

PREFACE

There certainly exists physics beyond the Standard Model! Very few, if anybody, would disagree with this supposition. It may well be that such pressing questions as about the rationale for the replication of fermionic families or the origin of fermionic masses have their solution at a much higher energy scale than presently explored experimentally, may be even at the Planck scale. There is at present only one framework, well motivated theoretically, which allows for extrapolation in energy in a concrete and predictive way. This is supersymmetry combined with the idea of grand unification of all elementary forces, realized as GUTs or string models.

This volume contains the lectures given at Joint U.S.-Polish Workshop devoted mainly to supersymmetry, GUTs and strings. The choice of the topic reflects the active involvement of both sides participating in this research. The meeting was a very useful and lively review of the recent progress achieved both in phenomenological and more formal investigations of physics between Planck scale and electroweak scale.

We thank World Scientific for their goodwill and interest in publishing these Proceedings.

The Editors

CONTENTS

WHICH WAY TO THE PLANCK SCALE?

Chairpersons: *S. Raby, C. Bachas*

SUPERSYMMETRY AND COSMOLOGY

Chairperson: *A. White*

SUPERSYMMETRIC GRAND UNIFICATION

Chairperson: *H. Haber*

SUPERSTRING THEORY

Chairpersons: *J. Rosner, T. R. Taylor*

SEARCH FOR LOW ENERGY SUPERSYMMETRY

Chairperson: *J. Bagger*

OVERVIEW OF THE STANDARD MODEL

JONATHAN L. ROSNER

Enrico Fermi Institute and Department of Physics, University of Chicago
5640 S. Ellis Ave., Chicago, IL 60637, USA

ABSTRACT

A brief overview is presented of the standard model of electroweak interactions, including precision tests, the role of the Cabibbo-Kobayashi-Maskawa (CKM) matrix in describing CP violation, windows on intermediate-scale physics, and the nature of the Higgs phenomenon.

1. Introduction

The standard model of electroweak interactions has guided experimental discoveries for more than ten years, starting with the observation of the W and Z bosons and culminating this past year with evidence for the top quark. The new particles' properties have been very close to those anticipated by theory: reassuring to some physicists, frustrating to others. In this brief overview we describe the present state of electroweak theory and indicate ways in which new insights could emerge from experiments within the next decade.

Precise electroweak tests, described in Sec. 2, show promise of shedding light on the Higgs, and restrict our ability to add large numbers of new particles to the theory, as in some models of dynamical electroweak symmetry breaking. A plausible theory of CP violation, based on phases in the Cabibbo-Kobayashi-Maskawa (CKM) matrix, can be checked by many forthcoming experiments on rare K decays and properties of B mesons, as noted in Sec. 3

While electroweak physics is based on physics below the TeV scale, unification of electroweak and strong interactions must take place at scales above 10^{15} GeV, or protons would decay too rapidly. Although many schemes envision a "desert" between 1 TeV and the unification scale, we describe in Sec. 4 several probes of intermediate scales (such as the interesting range $10^9 - 10^{12}$ GeV), including the study of neutrino masses, the investigation of baryogenesis in the early Universe, and the search for light pseudoscalar particles ("axions").

The mechanism of electroweak symmetry breaking is still unknown. Does it proceed via fundamental scalars in the theory, whose masses are protected from large radiative corrections by some mechanism such as supersymmetry, or is it dynamically generated through new interactions which are strong at the TeV scale? These questions are briefly considered in Sec. 5. We summarize in Sec. 6. More details on some of the topics presented here may be found in Ref 1.

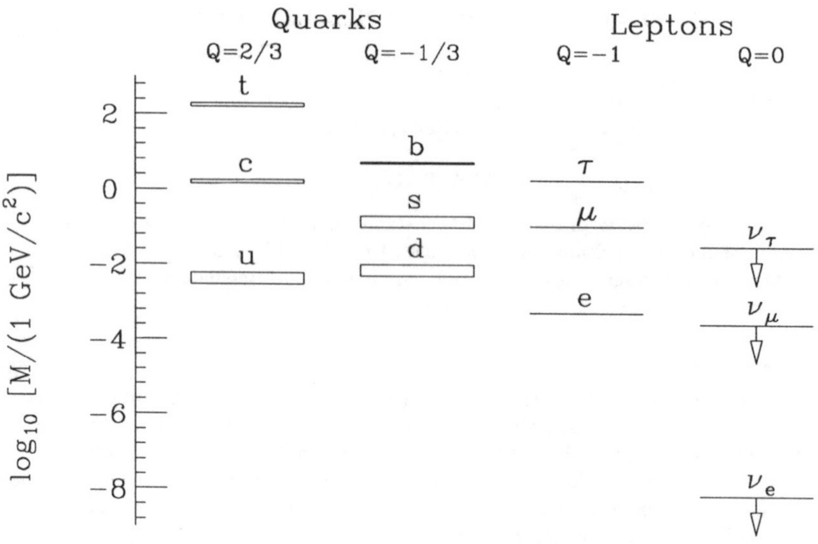

Fig. 1. Masses of quarks and leptons on a logarithmic scale. Widths of bars denote uncertainties in quark masses.

2. Precise electroweak tests

2.1. In celebration of the top quark

In April of this past year the CDF Collaboration at Fermilab presented evidence[2] for a top quark with mass $m_t = 174 \pm 10^{+13}_{-12}$ GeV. The measured cross section is slightly higher than (though consistent with) theoretical expectations based on quantum chromodynamics (QCD). Mechanisms which could give an elevated cross section include the production of more than one flavor of new quark and schemes involving dynamical symmetry breaking.[3,4] The mass lies in the middle of the range anticipated on the basis of electroweak radiative corrections (to be discussed in this section) and analyses of the CKM matrix (to be discussed in the next). But why is the top quark so heavy?

2.2. Quark and lepton masses

The top quark is the last quark to fit into a set of three families of quarks and leptons, whose masses are shown in Fig. 1:

$$\begin{pmatrix} u \\ d \end{pmatrix} \; ; \; \begin{pmatrix} c \\ s \end{pmatrix} \; ; \; \begin{pmatrix} t \\ b \end{pmatrix} \; ; \tag{1}$$

$$\begin{pmatrix} \nu_e \\ e \end{pmatrix} \; ; \; \begin{pmatrix} \nu_\mu \\ \mu \end{pmatrix} \; ; \; \begin{pmatrix} \nu_\tau \\ \tau \end{pmatrix} \; . \tag{2}$$

Only the ν_τ has not yet been directly observed. The fractional error on the top mass is less than that for the light quarks u, d, and s, whose masses are obscured by QCD effects. In logarithmic terms, the t mass seems not all that anomalous, but we have no theory for the pattern in Fig. 1.

2.3. Electroweak unification

The singular four-fermion beta-decay interaction does not permit calculations to higher order. In a theory where this interaction is the low-energy limit of massive charged particle (W) exchange, however, higher-order calculations make sense.[5] The predicted W and a massive neutral particle, the Z, entailed by the simplest version of the theory, were both discovered in 1983. The theory also predicted new weak charge-preserving interactions mediated by Z exchange, first seen in neutrino interactions a decade earlier. The theory has an SU(2) $\times$ U(1) symmetry, broken by the masses of the W, Z, and fermions.

The low-energy limits of W and Z exchange are described by

$$\frac{G_F}{\sqrt{2}} = \frac{g^2}{8M_W^2} \quad , \quad \frac{G_F}{\sqrt{2}}\rho = \frac{g^2 + g'^2}{8M_Z^2} \quad , \tag{3}$$

where G_F is the Fermi constant, $g = e/\sin\theta$ and $g' = e/\cos\theta$ are SU(2) and U(1) coupling constants, e is the proton charge, and θ is the weak mixing angle. The parameter ρ, which can arise from effects of quark loops on W and Z self-energies, is dominated by the top:[6]

$$\rho \simeq 1 + \frac{3G_F m_t^2}{8\pi^2\sqrt{2}} \quad , \tag{4}$$

Consequently, if we define θ by means of the precise measurement at LEP of M_Z,

$$M_Z^2 = \frac{\pi\alpha}{\sqrt{2}G_F\rho\sin^2\theta\cos^2\theta} \quad , \tag{5}$$

then θ will depend on m_t, and so will

$$M_W^2 = \frac{\pi\alpha}{\sqrt{2}G_F\sin^2\theta} \quad . \tag{6}$$

One must remember that the electric charge becomes stronger at the short distances characterizing W and Z exchanges as a consequence of vacuum polarization effects in calculating M_W and M_Z from these expressions.

2.4. The Higgs boson

The replacement of the Fermi interaction by massive W exchange does not cure all the problems of the electroweak theory. At high energies, W^+W^- scattering would violate unitarity (i.e., probability conservation) unless a spinless neutral boson (the *Higgs boson*) existed below about 1 TeV.[7] The search for this boson is one motivation for multi-TeV hadron colliders. Searches in $e^+e^- \to$ Higgs boson $+ \ldots$ at LEP have set a lower limit of about 60 GeV on M_H.[8] (Here and elsewhere $c = \hbar = 1$.)

The Higgs boson affects the parameter ρ through loop diagrams contributing to W and Z self-energies. It is convenient to express contributions to ρ in terms of deviations of the top quark and Higgs boson masses from nominal values. For $m_t = 175$ GeV, $M_H = 300$ GeV, the measured value of M_Z leads to a nominal expected value of $\sin^2\theta_0 \equiv x_0 = 0.2320$. Defining a parameter T of order 1 by $\Delta\rho \equiv \alpha T$, we find

$$T \simeq \frac{3}{16\pi \sin^2\theta}\left[\frac{m_t^2 - (175\text{ GeV})^2}{M_W^2}\right] - \frac{3}{8\pi\cos^2\theta}\ln\frac{M_H}{300\text{ GeV}} \quad . \tag{7}$$

This expression is quadratic in m_t, but only logarithmic in M_H. The weak mixing angle θ, the W mass, and other electroweak observables depend on m_t and M_W.

2.5. S and T parameters

The weak charge-changing and neutral-current interactions are probed under a number of different conditions, corresponding to different values of momentum transfer. For example, muon decay occurs at momentum transfers small with respect to M_W, while the decay of a Z into fermion-antifermion pairs imparts a momentum of nearly $M_Z/2$ to each member of the pair. Although coupling constants and masses on the right-hand sides of (3) vary fairly rapidly with momentum transfer, their quotients vary much less rapidly, as first pointed out by Veltman.[6] Small "oblique" corrections,[9,10] logarithmic in m_t and M_H, arise from contributions of new particles to the photon, W, and Z propagators. Other (smaller) "direct" radiative corrections are important in calcuating actual values of observables.

We may then replace (3) by

$$\frac{G_F}{\sqrt{2}} = \frac{g^2}{8M_W^2}\left(1 + \frac{\alpha S_W}{4\sin^2\theta}\right) \quad , \quad \frac{G_F \rho}{\sqrt{2}} = \frac{g^2 + g'^2}{8M_Z^2}\left(1 + \frac{\alpha S_Z}{4\sin^2\theta\cos^2\theta}\right) \quad , \tag{8}$$

where S_W and S_Z are coefficients representing variation with momentum transfer. Together with T, they express a wide variety of electroweak observables in terms of quantities sensitive to new physics.

Expressing the "new physics" effects in terms of deviations from nominal values of top quark and Higgs boson masses, we have the expression for T written above, while contributions of Higgs bosons and of possible new fermions U and D with electromagnetic charges Q_U and Q_D to S_W and S_Z are[11]

$$S_Z = \frac{1}{6\pi}\left[\ln\frac{M_H}{300\text{ GeV}} + \sum N_C\left(1 - 4\overline{Q}\ln\frac{m_U}{m_D}\right)\right] \quad , \tag{9}$$

$$S_W = \frac{1}{6\pi}\left[\ln\frac{M_H}{300\text{ GeV}} + \sum N_C\left(1 - 4Q_D\ln\frac{m_U}{m_D}\right)\right] \quad . \tag{10}$$

The expressions for S_W and S_Z are written for doublets of fermions with N_C colors and $m_U \geq m_D \gg m_Z$, while $\overline{Q} \equiv (Q_U + Q_D)/2$. The sums are taken over all doublets of new fermions. In the limit $m_U = m_D$, one has equal contributions to S_W and S_Z. For a single Higgs boson and a single heavy top quark, Eqs. (9) and (10) become

$$S_Z = \frac{1}{6\pi}\left[\ln\frac{M_H}{300\text{ GeV}} - 2\ln\frac{m_t}{175\text{ GeV}}\right] \quad ; \quad S_W = \frac{1}{6\pi}\left[\ln\frac{M_H}{300\text{ GeV}} + 4\ln\frac{m_t}{175\text{ GeV}}\right] \quad . \tag{11}$$

We shall use these expressions, together with previous ones, to express all electroweak observables as functions of m_t and M_H.

2.6. Electroweak experiments

Recent direct W mass measurements, in GeV, include 79.92 ± 0.39,[12] 80.35 ± 0.37,[13] 80.37 ± 0.23,[14] 79.86 ± 0.26,[15] with average 80.23 ± 0.18.[16] Recent data[17–19] on the ratio $R_\nu \equiv \sigma(\nu N \to \nu + \ldots)/\sigma(\nu N \to \mu^- + \ldots)$ lead to information on ρ^2 times a function of $\sin^2\theta$ roughly equivalent to the constraint $M_W = 80.27 \pm 0.26$ GeV. Measured Z parameters[20] include $M_Z = 91.1888 \pm 0.0044$ GeV, $\Gamma_Z = 2.4974 \pm 0.0038$ GeV, $\sigma_h^0 = 41.49\pm0.12$ nb (the hadron production cross section), and $R_\ell \equiv \Gamma_{\text{hadrons}}/\Gamma_{\text{leptons}} = 20.795\pm0.040$, which may be combined to obtain the Z leptonic width $\Gamma_{\ell\ell}(Z) = 83.96\pm 0.18$ MeV. Leptonic asymmetries include the forward-backward asymmetry parameter $A_{FB}^\ell = 0.0170 \pm 0.0016$, leading to a value $\sin^2\theta_\ell \equiv \sin^2\theta_{\text{eff}} = 0.23107 \pm 0.0090$, and independent determinations of $\sin^2\theta_{\text{eff}} = (1/4)(1-[g_V^\ell/g_A^\ell])$ from the parameters $A_\tau \to \sin^2\theta = 0.2320\pm0.0013$, $A_e \to \sin^2\theta = 0.2330\pm0.0014$. The last three values may be combined to yield $\sin^2\theta = 0.2317\pm0.0007$. We do not use asymmetries as measured in decays of Z to $b\bar{b}$ (which may reflect additional new-physics effects, to $c\bar{c}$ (which are of limited weight because of large errors), or to light quarks (for which interpretations are more model-dependent). This last result differs by about two standard deviations from that based on the left-right asymmetry parameter A_{LR} measured with polarized electrons at SLC[21]: $\sin^2\theta = 0.2294 \pm 0.0010$.

Parity violation in atoms, stemming from the interference of Z and photon exchanges between the electrons and the nucleus, provides further information on electroweak couplings. The most precise constraint at present arises from the measurement of the *weak charge* (the coherent vector coupling of the Z to the nucleus), $Q_W = \rho(Z-N-4Z\sin^2\theta)$, in atomic cesium,[22] with the result $Q_W(\text{Cs}) = -71.04\pm1.58\pm0.88$. The first error is experimental, while the second is theoretical.[23] The prediction[24] $Q_W(\text{Cs}) = -73.20 \pm 0.13$ is insensitive to standard-model parameters[24,25]; discrepancies are good indications of new physics (such as exchange of an extra Z boson).

2.7. Results of fits to electroweak observables

We have performed a fit to the electroweak observables listed in Table 1. The "nominal" values (including[26] $\sin^2\theta_{\text{eff}} = 0.2320$) are calculated for $m_t = 175$ GeV and $M_H = 300$ GeV. We use $\Gamma_{\ell\ell}(Z)$, even though it is a derived quantity, because it has little correlation with other variables in our fit. It is mainly sensitive to the axial-vector coupling g_A^ℓ, while asymmetries are mainly sensitive to g_V^ℓ. We also omit the total width $\Gamma_{\text{tot}}(Z)$ from the fit, since it is highly correlated with $\Gamma_{\ell\ell}(Z)$ and mainly provides information on the value of the strong fine-structure constant α_s. With $\alpha_s = 0.12\pm0.01$, the observed total Z width is consistent with predictions. The partial width $\Gamma(Z \to b\bar{b})$ will be treated separately below.

Each observable in Table 1 specifies a band in the $S-T$ plane with different slope, as seen from the ratios of coefficients of S and T. Parity violation in atomic cesium is sensitive almost entirely to S.[24,25] The impact of $\sin^2\theta_{\text{eff}}$ determinations on S is considerable. The leptonic width of the Z is sensitive primarily to T. The W mass specifies a band of intermediate slope in the $S-T$ plane; here we assume $S_W = S_Z$.

Table 1. Electroweak observables described in fit

Quantity	Experimental value	Nominal value	Experiment/ Nominal	S Coeff.	T Coeff.
Q_W (Cs)	-71.0 ± 1.8 $^{a)}$	-73.2 $^{b)}$	0.970 ± 0.025	-0.80	-0.005
M_W (GeV)	80.24 ± 0.15 $^{c)}$	80.320 $^{d)}$	0.999 ± 0.002	-0.29	0.45
$\Gamma_{\ell\ell}(Z)$ (MeV)	83.96 ± 0.18 $^{e)}$	83.90 $^{f)}$	1.001 ± 0.002	-0.18	0.78
$\sin^2\theta_{\rm eff}$	0.2317 ± 0.0007 $^{f)}$	0.2320	0.999 ± 0.003	0.0036	-0.0026
$\sin^2\theta_{\rm eff}$	0.2294 ± 0.0010 $^{g)}$	0.2320	0.989 ± 0.004	0.0036	-0.0026

$^{a)}$ Weak charge in cesium[22]; $^{b)}$ Calculation[24] incorporating atomic physics corrections[23]
$^{c)}$ Average of direct measurements[16] and indirect information
from neutral/charged current ratio in deep inelastic neutrino scattering[17–19]
$^{d)}$ Including perturbative QCD corrections[26]; $^{e)}$ LEP average as of July, 1994[20]
$^{f)}$ From asymmetries at LEP[20]; $^{g)}$ From left-right asymmetry in annihilations at SLC[21]

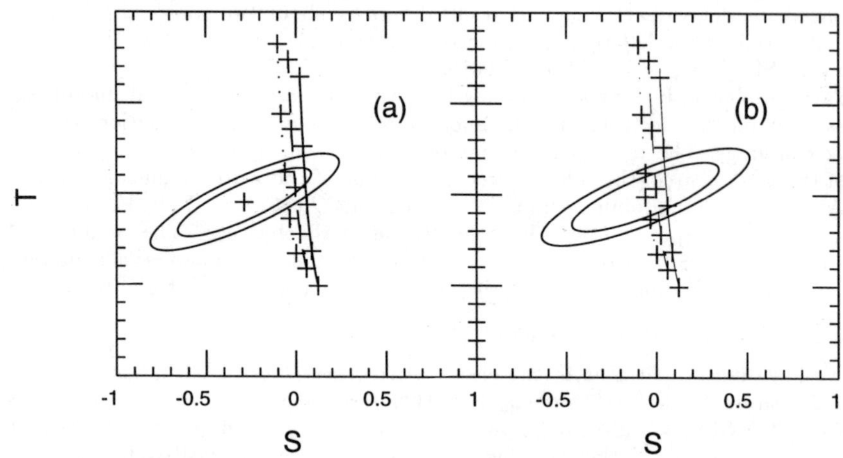

Fig. 2. Allowed ranges of S and T at 68% (inner ellipses) and 90% (outer ellipses) confidence levels. Dotted, dashed, and solid lines correspond to standard model predictions for $M_H =$ 100, 300, 1000 GeV. Tick marks, from bottom to top, denote predictions for $m_t = 100, 140,$ 180, 220, and 260 GeV. (a) Based on data in Table 1; (b) based on data in Table 1 aside from last row.

The resulting constraints on S and T are shown in Figs. 2, both with (a) and without (b) the SLC data. Conclusions about the allowed range of S (and hence about the number of new fermions or other new particles allowed in the theory) are sensitive to the precise value of $\sin^2\theta_{\rm eff}$, though models with large numbers of new fermions (leading to large positive values of S)[9] are excluded.

We have evaluated the overall quality of a fit to the standard model of the data in Table 1, when the constraint $m_t = 174 \pm 17$ GeV is added.[1] The result (see also Fig. 5 of Ref. 1) is a minimum $\chi^2 = 8.3$ for 5 degrees of freedom (d.o.f.) at $M_H \approx 260$ GeV. When the SLC data on $\sin^2\theta$ are omitted, the minimum χ^2 falls to 2.4 at $M_H \approx 700$ GeV for 4 d.o.f. The error in M_H for $\Delta\chi^2 = 1$ corresponds to about a factor of three in either case. Conclusions about the Higgs boson mass clearly are premature!

2.8. The decay $Z \to b\bar{b}$

The measured ratio[27] $R_b \equiv \Gamma(Z \to b\bar{b})/\Gamma(Z \to \text{hadrons})$ lies about 2σ above the standard model prediction.[28] If this discrepancy becomes more significant, there are many possibilities for explaining it within the context of new physics.[29] Two-Higgs-doublet models[28] cannot reduce the discrepancy much without running afoul of other constraints (such as the observed rate[30] for $b \to s\gamma$).

3. The CKM Matrix

The unitary Cabibbo-Kobayashi-Maskawa[31,32] (CKM) matrix describes the weak charge-changing transitions among quarks. Our present understanding of CP violation links the observed effect in the neutral kaon system to a phase in this matrix, whose parameters need to be specified as precisely as possible in order to test the theory. Moreover, theories of quark masses necessarily predict the CKM elements,[33] and may thereby be tested.

3.1. Definitions and magnitudes

We use a convenient parametrization[34] of the matrix:

$$V = \begin{pmatrix} V_{ud} & V_{us} & V_{ub} \\ V_{cd} & V_{cs} & V_{cb} \\ V_{td} & V_{ts} & V_{tb} \end{pmatrix} \approx \begin{bmatrix} 1-\lambda^2/2 & \lambda & A\lambda^3(\rho - i\eta) \\ -\lambda & 1-\lambda^2/2 & A\lambda^2 \\ A\lambda^3(1-\rho-i\eta) & -A\lambda^2 & 1 \end{bmatrix} \quad . \tag{12}$$

The four parameters are measured as follows:

1. The parameter λ is measured by a comparison of strange particle decays with muon decay and nuclear beta decay, leading to $\lambda \approx \sin\theta \approx 0.22$, where θ is the Cabibbo[31] angle.

2. The dominant decays of b-flavored hadrons occur via the element $V_{cb} = A\lambda^2$. The lifetimes of these hadrons and their semileptonic branching ratios then lead to an estimate $A = 0.79 \pm 0.06$.

3. The decays of b-flavored hadrons to charmless final states allow one to measure the magnitude of the element V_{ub} and thus to conclude that $|V_{ub}/V_{cb}| = 0.08 \pm 0.02$ or $\sqrt{\rho^2+\eta^2} = 0.36 \pm 0.09$.

4. The least certain quantity is the phase of V_{ub}: Arg $(V_{ub}^*) = \arctan(\eta/\rho)$. Information on this quantity may be obtained by studying its effect on contributions of higher-order diagrams involving the top quark.

The unitarity of the matrix (up to order λ^3) is implicit in the parametrization; in particular, the relation $V_{ub}^* + V_{td} \simeq A\lambda^3$, when normalized so that its right-hand side is unity, becomes $\rho + i\eta + (1 - \rho - i\eta) = 1$. We shall be concerned with the allowed region in the (ρ, η) plane.

3.2. Indirect information

Two important sources of indirect information on ρ and η are $B^0 - \overline{B}^0$ mixing and CP-violating $K^0 - \overline{K}^0$ mixing.

The presence of "wrong-sign" leptons in semileptonic B decays provided the first evidence for $B^0 - \overline{B}^0$ mixing.[35] Many experiments, including recent observations of time-dependent oscillations,[36] have confirmed the effect, leading to a mixing amplitude $\Delta m/\Gamma = 0.71 \pm 0.07$. The dominant contribution to the mixing is expected to arise from one-loop diagrams ("box graphs") involving internal W and top quark lines, leading to $\Delta m \sim f_B^2 m_t^2 |V_{td}|^2$ (times a slowly varying function of m_t/M_W). Here the "B decay constant," f_B, describes the amplitude for finding a b antiquark and a light quark at the same point in a B meson. Since $|V_{td}| \sim |1 - \rho - i\eta|$, the $B^0 - \overline{B}^0$ mixing amplitude leads to a constraint in the (ρ, η) plane consisting of a circular band with center (1,0). The main contribution to the width of this band is uncertainty in f_B.

A similar set of box diagrams contributes to the parameter ϵ describing CP-violating $K^0 - \overline{K}^0$ mixing. The imaginary part of the mass matrix is proportional to $f_K^2 m_t^2 \mathrm{Im}(V_{td}^2)$ times a slowly varying function of m_t, with a small correction for the charmed quark contribution and an overall factor B_K describing the degree to which the box graphs account for the effect. Since $\mathrm{Im}(V_{td}^2) \sim \eta(1 - \rho)$, the constraint imposed by CP-violating $K^0 - \overline{K}^0$ mixing consists of a hyperbolic band in the (ρ, η) plane with focus at (1,0), whose width is dominated by uncertainty in the magnitude of V_{cb}.[37]

3.3. Constraints on ρ and η

The allowed region[1] in (ρ, η) is shown in Fig. 3. Parameters used, in addition to those mentioned above, include $B_K = 0.8 \pm 0.2$, $f_B = 180 \pm 30$ MeV (in units where $f_\pi = 132$ MeV), $\eta_{QCD} = 0.6 \pm 0.1$ (a correction to the $B - \overline{B}$ mixing diagrams), and $B_B = 1$ for the factor analogous to B_K. The center of the allowed region lies around $\rho \simeq 0$, $\eta \simeq 0.35$. We may ask how our information may be improved.

3.4. Improved tests

We shall concentrate on a few key aspects of B physics, since extensive discussions of processes involving rare kaon decays and the CP-violating parameter ϵ'/ϵ appear elsewhere.[1,33,38]

3.4.1 Decay constant information. The value of f_B affects the interpretation of $B - \bar{B}$ mixing in terms of $|1 - \rho - i\eta|$. One recent lattice gauge theory calculation obtains[39] $f_B = 187 \pm 10 \pm 34 \pm 15$ MeV, $f_{B_s} = 207 \pm 9 \pm 34 \pm 22$ MeV, $f_D = 208 \pm 9 \pm 35 \pm 12$ MeV, $f_{D_s} = 230 \pm 7 \pm 30 \pm 18$ MeV, where the first error is statistical, the second is

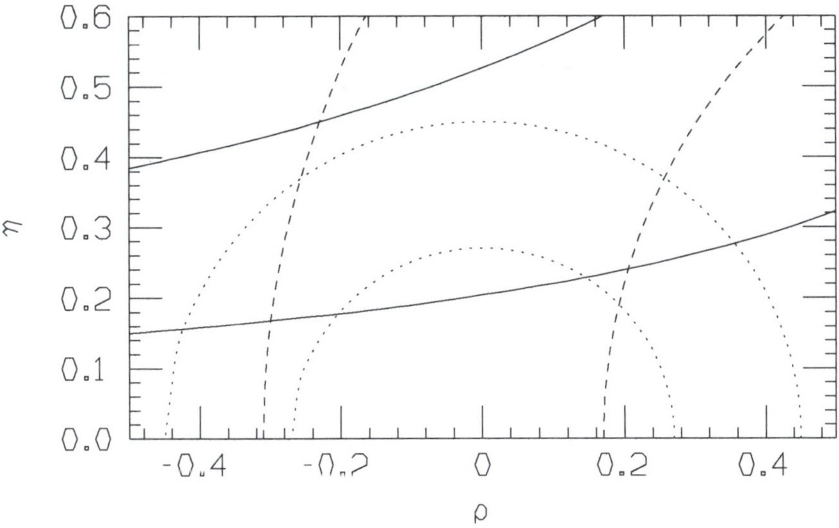

Fig. 3. Region in the (ρ, η) plane allowed by various constraints. Dotted semicircles denote central value and $\pm 1\sigma$ limits implied by $|V_{ub}/V_{cb}| = 0.08 \pm 0.02$. Circular arcs with centers at $(\rho, \eta) = (1, 0)$ denote constraints from $B - \overline{B}$ mixing, while hyperbolae describe region bounded by constraints from CP-violating $K - \overline{K}$ mixing.

associated with fitting and lattice size, and the third arises from scaling. Several other determinations exist,[40] some of which obtain values outside the limits just cited.

The WA75 Collaboration[41] reports a handful of $D_s \to \mu\nu$ candidates, leading to $f_{D_s} = 232 \pm 69$ MeV, while the CLEO Collaboration,[42] based on several dozen such events and an average by Muheim and Stone,[43] obtains $f_{D_s} = 315 \pm 46$ MeV. We average these two values to obtain $f_{D_s} = 289 \pm 38$ MeV. The BES group also has a few candidates for $D_s \to \mu\nu$ or $D_s \to \tau\nu$;[44] the E653 Collaboration has candidates for $D_s \to \mu\nu$ for which results are expected soon.[45]

The lattice determination noted above obtains $f_D/f_{D_s} \simeq 0.9$, while an argument based on the quark model and the observed hyperfine splittings between singlet and triplet charmed nonstrange and strange mesons[46] suggests $f_D/f_{D_s} \simeq 0.8$. Using this last ratio, we obtain $f_D = 231 \pm 31$ MeV, not far below the present upper limit (90% c.l.) of 290 MeV obtained by the Mark III Collaboration[47] on the basis of their search for the decay $D^+ \to \mu^+\nu$. The BES Collaboration should be able to observe this process in events obtained in the reaction $e^+e^- \to \psi(3770) \to D^+D^-$.

Heavy meson decay constants are related to the square of the quark-antiquark wave function at the origin,[48] which can be estimated[46] with the help of spin-dependent isospin mass splittings in K, D, and B systems. The end result of this exercise[1,46] is the result $f_B = 180$ MeV employed above. The error is based on the spread in various theoretical estimates.

3.4.2 B Decays to CP non-eigenstates: pair of light pseudoscalars. A difference between the rates for a process and its charge-conjugate, such as $B^+ \to \pi^+ K^0$ and $B^- \to \pi^- K^0$, signifies CP violation. Under charge conjugation, weak phases change sign, but strong phases do not. In order for a rate difference to appear, there must be both a weak phase difference and a strong phase difference in the channels with isospins $I = 1/2$ and $3/2$. Recently it has been shown that one may be able to measure weak phases via the rates for B decays to pairs of light pseudoscalar mesons *without* having any strong phase differences.[49] The presence of electroweak penguins[50] is one possible obstacle to this program, which is under further investigation.

3.4.3 Decays of neutral B mesons to CP eigenstates: $\pi - B$ correlations. Although produced initially as flavor eigenstates, neutral B mesons can undergo $B^0 - \overline{B}^0$ mixing, leading to time-dependent asymmetries in decays to CP eigenstates like $J/\psi K_S$. Time-integrated decays also can display rate asymmetries, whose interpretation is often independent of final-state effects. For example, the asymmetry in decays of B^0 or $\overline{B}^0$ to $J/\psi K_S$ is equal to $-[x_d/(1+x_d^2)] \sin[\mathrm{Arg}(V_{td}^*)^2]$, where $x_d = (\Delta m/\Gamma)|_d = 0.70 \pm 0.07$ is the mixing parameter mentioned earlier. One has to know the flavor of the neutral B at time of production. One proposed means for "tagging" the B involves its correlation with charged pions produced nearby in phase space.[51] The existence of such a correlation is predicted both by fragmentation and resonance decay pictures.

4. Intermediate mass scales

4.1. Neutrino masses and new mass scales

The pattern of neutrino masses in Fig. 1 is at least as puzzling as the large mass of the top quark. Why are neutrinos so light?

One proposed solution to the apparent suppression of the solar neutrino flux, especially for neutrinos with energies of a few MeV,[52] is based on matter-induced oscillations between electron neutrinos and some other species.[53] An acceptable solution involves a muon neutrino mass of several millielectron volts and a much lighter electron neutrino. A large Majorana mass[54] of $10^9 - 10^{12}$ GeV for the right-handed neutrino, combined with a Dirac mass of order 1 GeV linking the right-handed and left-handed muon neutrino, could give such a mass eigenstate.

If the corresponding tau neutrino Dirac mass is related to the top quark mass (as it is in some grand unified theories), the ν_τ could then have a mass of a few eV. It might well mix with the ν_μ with an angle $\theta \geq m_\mu/m_\tau$, leading to a mixing parameter $\sin^2 2\theta \geq 10^{-2}$.

Present limits[55] from Fermilab Experiment E531 restrict the muon and tau neutrinos to have $\Delta m^2 \leq 1$ eV^2 for large θ and $\sin^2 2\theta \leq$ (a few) $\times 10^{-3}$ for large Δm^2. A wider range of parameters are accessible to CHORUS,[56] an emulsion experiment, and NOMAD,[57] a fine-grained detector, both operating at CERN, which should reach $\sin^2 2\theta \leq$ (a few) $\times 10^{-4}$ for large Δm^2. New short- and long-baseline $\nu_\mu \leftrightarrow \nu_\tau$ oscillation experiments[58,59] are also envisioned for Fermilab, while experiments on atmospheric neutrinos[60] and low-energy reactor neutrinos[61] have recently presented tantalizing hints of oscillation phenomena.[62] A new solar-neutrino experiment, sensitive mainly to high-energy electron neutrinos but eventually to other species as well through neutral-current interactions, will begin operating in a couple of years.[63]

4.2. *Electroweak-strong unification*

The U(1), SU(2), and SU(3) couplings of the electroweak and strong interactions approach one another at high energies,[64] but do not really meet at a single point, in contrast to the predictions of the simplest SU(5) grand unification scheme.[65] One cure[66] for this "astigmatism" is provided by the supersymmetric extension of SU(5), in which the inclusion of superpartners below about 1 TeV modifies the slopes in the relations between α_i^{-1} $(i = 1,2,3)$ and $\ln M^2$ such that the couplings meet slightly above 10^{16} GeV at $\alpha_i^{-1} \approx 26$. Another possibility, however, is to embed SU(5) in an SO(10) model,[67] with SO(10) breaking to $SU(3)_{color}\times SU(2)_R\times SU(2)_L\times U(1)_{B-L}$ around 4×10^{17} GeV and $SU(2)_R\times U(1)_{B-L}$ breaking to U(1)Y at an intermediate mass scale of about 10^{10} GeV. (An early analysis[68] indicated that this intermediate scale could lie anywhere between 10^9 and about 10^{12} GeV.) These possibilities are compared in Fig. 9 of Ref. 1.

4.3. *Baryogenesis*

In 1967 Sakharov[69] identified three key elements of any theory seeking to explain the observed baryon abundance of the Universe, $n_B/n_\gamma \simeq 4 \times 10^{-9}$: (1) C and CP violation; (2) baryon number violation, and (3) a period during which the Universe was out of thermal equilibrium. A toy model is provided by the decay of SU(5) "X" bosons, which couple both to pairs of quarks and to antiquark-lepton pairs. The total rates for X and $\overline{X}$ decays must be the same as a consequence of CPT invariance, but the branching ratios $B(X \to uu)$ and $B(X \to e^+\bar{d})$ can differ from $B(\overline{X} \to \bar{u}\bar{u})$ and $B(\overline{X} \to e^-d)$ as a result of CP violation, leading to a baryon asymmetry once the Universe cools below the energy where these decay processes are balanced by the reverse processes. This example conserves $B - L$, where B is baryon number (1/3 for quarks) and L is lepton number (1 for electrons).

It was pointed out by 't Hooft[70] that the electroweak theory contains an anomaly as a result of nonperturbative effects which conserve $B - L$ but violate $B + L$. If a theory leads to $B - L = 0$ but $B + L \neq 0$ at some primordial temperature T, the anomaly can wipe out any $B + L$ as T sinks below the electroweak scale.[71] Thus, the toy model mentioned above and many others are unsuitable in practice.

One proposed solution is the generation of nonzero $B - L$ at a high temperature, e.g., through the generation of nonzero lepton number L, which is then reprocessed into nonzero baryon number by the 't Hooft anomaly mechanism.[72] The existence of a baryon asymmetry, when combined with information on neutrinos, could provide a window to a new scale of particle physics. Large Majorana masses acquired by right-handed neutrinos would change lepton number by two units and thus would be ideal for generating a lepton asymmetry if Sakharov's other two conditions are met.

The question of baryogenesis is thus shifted onto the leptons: Do neutrinos indeed have masses? If so, what is their "CKM matrix"? Do the properties of heavy Majorana right-handed neutrinos allow any new and interesting natural mechanisms for violating CP at the same scale where lepton number is violated? Majorana masses for right-handed neutrinos naturally violate left-right symmetry and could be closely connected with the violation of P and C in the weak interactions.[73] An open question in this scenario, besides the form of CP violation at the lepton-number-violating scale, is the manner in which CP violation is communicated to the lower mass scale at which we see CKM phases.

4.4. The strong CP problem

The Lagrangian density of QCD in principle can contain a term $(g_3^2\bar{\theta}/32\pi^2)F^a_{\mu\nu}\tilde{F}^{\mu\nu a}$, where $\bar{\theta} \equiv \theta +$ Arg det M, with M the quark mass matrix and θ an angle characterizing the vacuum structure.[74] The neutron electric dipole moment receives a contribution of order $d_n \simeq 10^{-16}\bar{\theta}\ e\cdot$ cm, implying $\bar{\theta} \leq 10^{-9}$. How can we understand the small value of $\bar{\theta}$ without fine-tuning?

Proposed solutions include (1) an unconventional and still controversial interpretation of the vacuum[75] as composed of an incoherent mixture of θ and $-\theta$, in the sense of a density matrix; (2) vanishing of one of the quark masses (e.g., m_u), which probably bends chiral symmetry beyond plausible limits;[76] (3) the promotion[77] of $\bar{\theta}$ to the status of a dynamical variable relaxing to zero, implying the existence[78] of a light Nambu-Goldstone boson, the *axion*. Searches for this particle by means of RF cavities[79] turn out to be uniquely sensitive to the scale of symmetry breaking in the range we have been discussing, $10^9 - 10^{12}$ GeV!

5. Electroweak symmetry breaking

A key question facing the standard model of electroweak interactions is the mechanism for breaking SU(2) × U(1). We discuss two popular alternatives; Nature may turn out to be cleverer than either.

5.1. Fundamental Higgs boson(s)

If there really exists a relatively light fundamental Higgs boson in the context of a grand unified theory, one has to protect its mass from large corrections. Supersymmetry is the popular means for doing so. Then one expects a richer neutral Higgs structure, charged scalar bosons, and superpartners, all below about 1 TeV.

5.2. Strongly interacting Higgs sector

The scattering of longitudinally polarized W and Z bosons violates unitarity above a TeV or two if there does not exist a Higgs boson below this energy.[7] The behavior is similar to that of pion-pion scattering in the non-linear sigma model above a few hundred MeV, where we wouldn't trust the model. Similarly, we mistrust the present version of electroweak theory above a TeV. If the theory has a strongly interacting sector, its $I = J = 0$ boson (like the σ of QCD) may be its least interesting and most elusive feature. The rich spectrum of resonances in QCD is now understood in terms of the interactions of quarks and gluons. We hope that the exploration of the potentially rich TeV physics of the electroweak sector can proceed through the use of both hadron and electron-positron colliders.

6. Summary

Increasingly precise electroweak and top quark measurements show promise of shedding indirect light on the Higgs sector, when present accuracies are improved by about a factor of three. Electron-positron collisions at LEP and SLC, hadron collisions at Fermilab, and even table-top experiments on atomic parity violation all will play key roles.

The Cabibbo-Kobayashi-Maskawa (CKM) matrix is the leading candidate for the source of the observed CP violation in the neutral kaon system, through phases which require the existence of the third family of quarks. The observation of events consistent with a top quark has given encouragement to this scheme, but further confirmation is needed, through the study of CP-violating and other rare kaon decays and through systematic exploration of the properties of hadrons containing b quarks.

Several prospective windows exist on the "intermediate" mass scale of 10^9 – 10^{12} GeV, including the study of neutrino masses, partial unification of interactions, baryogenesis, and axions. Thus, there appears to be at least a lamppost, if not an oasis, in the grand desert between the TeV scale and the unification scale.

Candidates for understanding the Higgs sector include a theory of fundamental scalars with masses protected by supersymmetry and a composite-Higgs theory which could, in principle, be merely the first hints of compositeness on a deeper level. Perhaps even quarks and leptons are composite. I, for one would not be distressed if we were simply uncovering one more layer of the onion in our journey toward the Planck scale.

7. Acknowledgements

I am grateful to Jim Amundson, Aaron Grant, Michael Gronau, Oscar Hernández, Nahmin Horowitz, Mike Kelly, David London, Sheldon Stone, Tatsu Takeuchi, and Mihir Worah for fruitful collaborations on some of the topics mentioned here, to them and to I. Dunietz, M. Goodman, and G. Zapalac for discussions, to the Fermilab Theory Group for hospitality during preparation of this lecture, and to our hosts in Warsaw for providing an enjoyable and informative workshop and a chance to get acquainted with their fine city. This work was supported in part by the United States Department of Energy under Grant No. DE AC02 90ER40560.

References

1. J. L. Rosner, Enrico Fermi Institute Report No. 94-38, August, 1994, presented at DPF 94 Meeting, Albuquerque, NM, Aug. 2 – 6, 1994, Proceedings to be published by World Scientific
2. CDF Collaboration, F. Abe *et al.*, *Phys. Rev.* D **50** (1994) 2966; *Phys. Rev. Lett.* **72** (1994) 225. For earlier limits see S. Abachi *et al.*, *Phys. Rev. Lett.* **72** (1994) 2138
3. E. Eichten and K. Lane, *Phys. Lett.* B **327** (1994) 129
4. C. T. Hill and S. J. Parke, *Phys. Rev.* D **49** (1994) 4454
5. S. L. Glashow, *Nucl. Phys.* **22** (1961) 579; S. Weinberg, *Phys. Rev. Lett.* **19** (1967) 1264; A. Salam, in *Proceedings of the Eighth Nobel Symposium*, ed. by N. Svartholm (Stockholm, Almqvist and Wiksell, and New York, Wiley, 1968) p. 367
6. M. Veltman, *Nucl. Phys.* **B123** (1977) 89
7. M. Veltman, *Acta Physica Polonica* **B8** (1977) 475; *Phys. Lett.* **70B** (1977) 253; B. W. Lee, C. Quigg, and H. B. Thacker, *Phys. Rev. Lett.* **38** (1977) 883; *Phys. Rev.* D **16** (1977) 1519
8. S. Olsen, plenary talk, DPF 94 Meeting (see Ref. 1)

9. M. Peskin and T. Takeuchi, *Phys. Rev. Lett.* **65** (1990) 964; *Phys. Rev.* D **46** (1992) 381, and references therein
10. J. L. Rosner, *Rev. Mod. Phys.* **64** (1992) 1151. In Eqs. (2.18) – (2.21) of that article the expressions for S_W and S_Z should be interchanged.
11. D. C. Kennedy and P. G. Langacker, *Phys. Rev. Lett.* **65** (1990)
12. CDF Collaboration, F. Abe *et al.*, *Phys. Rev. Lett.* **65** (1990) 2243; *Phys. Rev.* D **43** (1991) 2070
13. UA2 Collaboration, J. Alitti *et al.*, *Phys. Lett.* B **276** (1992) 354
14. CDF Collaboration, F. Abe *et al.*, presented by Y. K. Kim at Moriond Conference on Electroweak Interactions and Unified Theories, Meribel, Savoie, France, Mar. 12–19, 1994
15. D0 Collaboration, S. Abachi *et al.*, Fermilab report FERMILAB-CONF-93-396-E, December, 1993, presented by Q. Zhu at Ninth Topical Workshop on Proton-Antiproton Collider Physics, Oct. 18–22, 1993
16. M. Demarteau *et al.* in CDF note CDF/PHYS/CDF/PUBLIC/2552 and D0 note D0NOTE 2115, May, 1994
17. CCFR Collaboration, C. G. Arroyo *et al.*, *Phys. Rev. Lett.* **72** (1994) 3452; R. Bernstein, paper no. 51, DPF 94 Meeting (see Ref. 1)
18. CDHS Collaboration, H. Abramowicz *et al.*, *Phys. Rev. Lett.* **57** (1986) 298; A. Blondel *et al.*, *Zeit. Phys.* C **45** (1990) 361
19. CHARM Collaboration, J. V. Allaby *et al.*, *Phys. Lett.* B **177** (1986) 446; *Zeit. Phys.* C **36** (1987) 611
20. LEP report LEPEWWG/94-02, 12 July 1994, submitted to 27th International Conference on High Energy Physics, Glasgow, Scotland, 21-27 July 1994
21. SLD Collaboration, K. Abe *et al.*, *Phys. Rev. Lett.* **73** (1994) 23
22. M. C. Noecker, B. P. Masterson, and C. E. Wieman, *Phys. Rev. Lett.* **61** (1988) 310
23. S. A. Blundell, J. Sapirstein, and W. R. Johnson, *Phys. Rev.* D **45** (1992) 1602
24. W. J. Marciano and J. L. Rosner, *Phys. Rev. Lett.* **65** (1990) 2963; *ibid.* **68** (1992) 898(E)
25. P. G. H. Sandars, *J. Phys.* B **23** (1990) L655
26. G. DeGrassi, B. A. Kniehl, and A. Sirlin, *Phys. Rev.* D **48** (1993) 3963
27. LEP Electroweak Heavy Flavours Working Group, D. Abbaneeo *et al.*, report LEPHF/94-03, 15 July 1994
28. Aaron K. Grant, Enrico Fermi Institute report EFI 94-24, June 1994, to be published in Phys. Rev. D
29. S. Chivukula, plenary talk, DPF 94 Meeting (see Ref. 1)
30. CLEO Collaboration, B. Barish *et al.*, CLEO-CONF-94-1, July, 1994, presented at 27th International Conference on High Energy Physics (see Ref. 20)
31. N. Cabibbo, *Phys. Rev. Lett.* **10** (1963) 531
32. M. Kobayashi and T. Maskawa, *Prog. Theor. Phys.* **49** (1973) 652
33. J. L. Rosner, in *B Decays*, edited by S. Stone (World Scientific, Singapore, 1994), p. 470
34. L. Wolfenstein, *Phys. Rev. Lett.* **51** (1983) 1945
35. ARGUS Collaboration, H. Albrecht *et al.*, *Phys. Lett.* B **192** (1987) 245
36. V. Sharma, plenary talk, DPF 94 Meeting (see Ref. 1)

37. S. Stone, presented at PASCOS 94 Conference, Syracuse, NY, May 19-24, 1994. Proceedings to be published by World Scientific
38. J. L. Rosner, Enrico Fermi Institute Report No. 94-25, presented at PASCOS 94 Conference (see Ref. 37)
39. C. W. Bernard, J. N. Labrenz, and A. Soni, *Phys. Rev.* D **49** (1994) 2536
40. A. Duncan *et al.*, Fermilab report FERMILAB-PUB-94/164-T, July, 1994 (unpublished); UKQCD Collaboration, R. M. Baxter *et al.*, *Phys. Rev.* D **49** (1994) 1594; P. B. Mackenzie, plenary talk, DPF 94 Meeting (see Ref. 1)
41. WA75 Collaboration, S. Aoki *et al.*, *Prog. Theor. Phys.* **89** (1993) 131
42. CLEO Collaboration, D. Acosta *et al.*, *Phys. Rev.* D **49** (1994) 5690
43. F. Muheim and S. Stone, *Phys. Rev.* D **49** (1994) 3767
44. BES Collaboration. M. Kelsey, paper no. 414, DPF 94 Meeting (see Ref. 1)
45. E653 Collaboration, R. Edelstein and N. W. Reay, private communication
46. J. F. Amundson *et al.*, *Phys. Rev.* D **47** (1993) 3059; J. L. Rosner, *The Fermilab Meeting - DPF 92* (Division of Particles and Fields Meeting, American Physical Society, Fermilab, 10 – 14 November, 1992), ed. by C. H. Albright *et al.* (World Scientific, Singapore, 1993), p. 658
47. Mark III Collaboration, J. Adler *et al.*, *Phys. Rev. Lett.* **60** (1988) 1375; *ibid.* **63** (1989) 1658(E)
48. E. V. Shuryak, *Nucl. Phys.* **B198** (1982) 83
49. M. Gronau, J. L. Rosner, and D. London, *Phys. Rev. Lett.* **73** (1994) 21; M. Gronau, O. F. Hernández, D. London, and J. L. Rosner, *Phys. Rev.* D **50** (1994) 4529; O. F. Hernandez, D. London, M. Gronau, and J. L. Rosner, *Phys. Lett.* B **333** (1994) 500
50. N. G. Deshpande and X.-G. He, University of Oregon report OITS-553, August, 1994 (unpublished)
51. M. Gronau, A. Nippe, and J. L. Rosner, *Phys. Rev.* D **47** (1992) 1988; M. Gronau and J. L. Rosner, in *Proceedings of the Workshop on B Physics at Hadron Accelerators*, Snowmass, Colorado, ed. by P. McBride and C. S. Mishra, Fermilab report FERMILAB-CONF-93/267 (Fermilab, Batavia, IL), p. 701; *Phys. Rev. Lett.* **72** (1994) 195; *Phys. Rev.* D **49** (1994) 254
52. S. Bludman, N. Hata, D. C. Kennedy, and P. Langacker, *Phys. Rev.* D **47** (1993) 2220
53. L. Wolfenstein, *Phys. Rev.* D **17** (1978) 2369; S. P. Mikheev and A. Yu. Smirnov, *Yad. Fiz.* **42** (1985) 1441 [Sov. J. Nucl. Phys. **42** (1985) 913]; *Nuovo Cim.* **9C** (1986) 17; *Usp. Fiz. Nauk* **153** (1987) 3 [Sov. Phys. - Uspekhi **30** (1987) 759]
54. M. Gell-Mann, P. Ramond, and R. Slansky in *Supergravity*, edited by P. van Nieuwenhuizen and D. Z. Freedman (Amsterdam, North-Holland, 1979), p. 315; T. Yanagida *Proceedings of the Workshop on Unified Theory and Baryon Number in the Universe*, edited by O. Sawada and A. Sugamoto (Tsukuba, Japan, National Laboratory for High Energy Physics, 1979)
55. Fermilab E531 Collaboration, N. Ushida *et al.*, *Phys. Rev. Lett.* **57** (1986) 2897
56. CHORUS Collaboration, M. de Jong *et al.*, CERN report CERN-PPE-93-131, July, 1993 (unpublished)
57. NOMAD Collaboration, CERN report CERN-SPSLC-93-19, 1993 (unpublished)
58. K. Kodama *et al.*, Fermilab proposal FERMILAB-PROPOSAL-P-803, October, 1993 (approved Fermilab experiment)

59. W. W. M. Allison *et al.*, Fermilab proposal FERMILAB-PROPOSAL-P-822-UPD, March, 1994 (unpublished)
60. Kamiokande Collaboration, Y. Fukuda *et al.*, *Phys. Lett.* B **335** (1994) 237
61. LSND Collaboration, R. A. Reeder, paper no. 146, DPF 94 Meeting (see Ref. 1)
62. F. Halzen, plenary talk, DPF 94 Meeting (see Ref. 1)
63. E. B. Norman *et al.*, in *The Fermilab Meeting - DPF 92* (Ref. 46), p. 1450
64. H. Georgi, H. Quinn, and S. Weinberg, *Phys. Rev. Lett.* **33** (1974) 451
65. H. Georgi and S. L. Glashow, *Phys. Rev. Lett.* **32** (1974) 438
66. U. Amaldi *et al.*, *Phys. Rev.* D **36** (1987) 1385; U. Amaldi, W. de Boer, and H. Fürstenau, *Phys. Lett.* B **260** (1991) 447; P. Langacker and N. Polonsky, *Phys. Rev.* D **47** (1993) 4028
67. H. Georgi in *Proceedings of the 1974 Williamsburg DPF Meeting*, ed. by C. E. Carlson (New York, AIP, 1975) p. 575; H. Fritzsch and P. Minkowski, *Ann. Phys. (N.Y.)* **93** (1975) 193
68. R. Robinett and J. L. Rosner, *Phys. Rev.* D **25** (1982) 3036
69. A. D. Sakharov, *Pis'ma Zh. Eksp. Teor. Fiz.* **5** (1967) 32 [*JETP Lett.* **5** (1967) 24]
70. G. 't Hooft, *Phys. Rev. Lett.* **37** (1976) 8
71. V. A. Kuzmin, V. A. Rubakov, and M. E. Shaposhnikov, *Phys. Lett.* B **155** (1985) 36; *ibid.* **191** (1987) 171
72. M. Fukugita and T. Yanagida, *Phys. Lett.* B **174** (1986) 45; P. Langacker, R. Peccei, and T. Yanagida, *Mod. Phys. Lett.* A **1** (1986) 541
73. B. Kayser, in *CP Violation,* edited by C. Jarlskog (World Scientific, Singapore, 1989), p. 334
74. R. Peccei, in *CP Violation* (see Ref. 73), p. 503
75. R. G. Sachs, *Phys. Rev. Lett.* **73** (1994) 377
76. H. Leutwyler, Bern Univ. report BUTP-94-8, May, 1994, presented at 2nd IFT Workshop on Yukawa Couplings and the Origins of Mass, Gainesville, FL, 11-13 Feb. 1994, Proceedings to be published by International Press
77. R. D. Peccei and H. R. Quinn, *Phys. Rev. Lett.* **38** (1977) 144; *Phys. Rev.* D **16** (1977) 1791
78. S. Weinberg, *Phys. Rev. Lett.* **40** (1978) 223; F. Wilczek, *Phys. Rev. Lett.* **40** (1978) 279
79. P. Sikivie, in *Perspectives in the Standard Model* (Theoretical Advanced Study Institute in Elementary Particle Physics, Boulder, Colo., June 2 – 28, 1991), edited by R. K. Ellis, C. T. Hill, and J. D. Lykken (World Scientific, River Edge, NJ, 1992), p. 399

Aspects of New Particle Searches

André Sopczak
PPE Division, CERN

Abstract

The high statistics of the combined LEP lineshape data are used to derive constraints on hypothetical extensions of the Minimal Standard Model (MSM). The data comprises about eight million visible Z decays, recorded between 1989 and 1993. This report gives limits for simple tests on models which predict additional Z boson decays or modified Z-couplings. As an application the two-doublet Higgs model is considered. New model-independent limits on branching ratios or masses of pair-produced neutral and charged Higgs bosons are reported. The results of the four LEP experiments are combined in order to constrain a singly-produced Higgs boson, even with strongly suppressed production rate compared to the MSM prediction, in the mass range 0 to 62.8 GeV.

1. Introduction

Severe limits on 'New Physics' beyond the Minimal Standard Model [1] can be obtained from precision measurements of the Z parameters. Any hypothetical Z decay into new particles $Z \to X$ (Fig. 1a), radiative contributions from non-MSM virtual particles (Fig. 1b), or modifications to the MSM Z-couplings (Fig. 1c) are constrained by measurements of the total Z width Γ_Z, the invisible Z width Γ_Z^{inv}, the leptonic widths Γ_Z^{ee}, $\Gamma_Z^{\mu\mu}$, $\Gamma_Z^{\tau\tau}$, or the ratio of the hadronic to leptonic Z decay width R. Thus, constraints on physics beyond the MSM can be expressed as limits on deviations from the MSM Z decay width predictions. In particular, such limits can be used to constrain the existence of Higgs bosons in models with more than one Higgs doublet; charginos, neutralinos and light gluinos in Supersymmetric Models with or without R-parity conservation; additional heavy charged or neutral leptons; or anomalous gauge boson couplings.

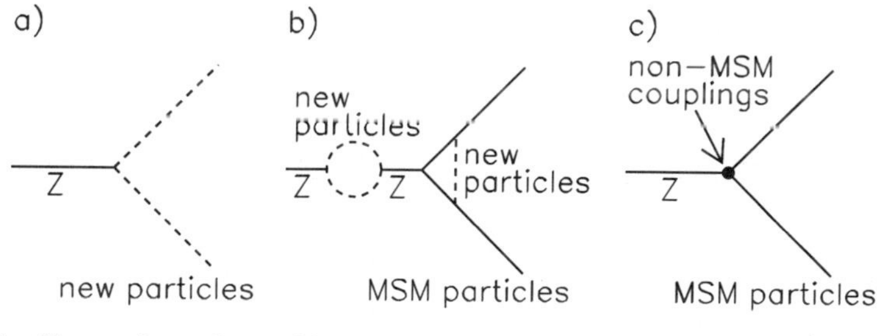

Figure 1: Illustration of possible effects in extensions of the MSM which can be constrained by a comparison of measured Z parameters with MSM predictions.

The present analysis includes 1992 data [2] and preliminary 1993 data collected by the four LEP experiments [3], corresponding to a total of about eight million visible Z decays. The Z parameters are obtained by fitting the lineshape of the Z decay into charged leptons and hadrons. All measurements are in agreement with the MSM predictions. Details of the experimental analysis and similar interpretations as presented here can be found in the corresponding publications of the four LEP experiments [2].

2. Measurement and Theory

Table 1 summarizes the measured values of Γ_Z, Γ_Z^{inv}, Γ_Z^{ee}, $\Gamma_Z^{\mu\mu}$, $\Gamma_Z^{\tau\tau}$, and R, as well as their MSM upper and lower bounds for one-sided 95% CL's. One-sided CL's are used because a new decay would always increase the Z width; they are derived assuming Gaussian errors by extending the 1 σ error to 1.64 σ [4]. The measured values are averages from the four LEP experiments taking into account common systematic errors [3]. Theoretical upper and lower bounds are obtained with an analytical program (ZFITTER version 4.6 [5]) by varying the strong coupling constant α_s, the top quark mass m_t, and the MSM Higgs mass m_h, independently within their one-sided 95% CL limits. The uncertainty in these values constitutes the dominant error on the MSM predictions.

For α_s the world average $\alpha_s(m_Z) = 0.125 \pm 0.005$ [3] is used. Note that this average is based on data from νN experiments, $p\bar{p}$ colliders, the SLD measurement of the left-right asymmetry and the LEP experiments. For m_t the limit implied by the recently reported CDF evidence for the top quark is used, i.e. $m_t = (174 \pm 10^{+13}_{-12})$ GeV [6]. For m_h a combined lower mass limit [7], resulting from the data of the four LEP experiments [8], and a theoretical upper mass bound following from consistency arguments in the MSM [9] is used. Thus, the ranges used for α_s, m_t and m_h are:

$$0.117 < \alpha_s(m_Z) < 0.133, \quad (148 < m_t < 201)\ \text{GeV}, \quad (63.5 < m_h < 1000)\ \text{GeV}.$$

The central values of the MSM predictions are the arithmetic means of the upper and lower bounds.

Parameter	Measurements			Theory (MSM)		
	Mean Value	Lower Bound	Upper Bound	Lower Bound	Upper Bound	Mean Value
Γ_Z	2497.4 ± 3.8	2491.2	2503.6	2480.6	2512.3	2496.5
Γ_Z^{inv}	499.8 ± 3.5	494.1	505.5	499.7	503.4	501.6
Γ_Z^{ee}	83.85 ± 0.21	83.51	84.19	83.56	84.33	83.95
$\Gamma_Z^{\mu\mu}$	83.95 ± 0.30	83.46	84.44	83.56	84.33	83.95
$\Gamma_Z^{\tau\tau}$	84.26 ± 0.34	83.70	84.82	83.37	84.13	83.75
R	20.795 ± 0.040	20.729	20.861	20.692	20.842	20.767

Table 1: Measured Z parameters, MSM predictions, and their lower and upper limits at one-sided 95% CL's. All decay widths are given in MeV.

In order to obtain a conservative limit on non-MSM effects from Γ_Z, one considers the intervals:

$$(\Gamma_Z)^{exp}_{min} - (\Gamma_Z)^{th}_{max} \text{ and } (\Gamma_Z)^{exp}_{max} - (\Gamma_Z)^{th}_{min},$$

where both the experimental and theoretical limits are taken at the one-sided 95% CL. Similar intervals are defined for the other parameters listed in Table 1. If the value of the predicted mean value minus the measured mean value is negative (positive), it is added to the lower (upper) limit. This conservative approach avoids setting tighter constraints than allowed by the agreement between theory and measurement. Otherwise, e.g., a measurement of the central value of the total Z decay width significantly below the MSM expectation would naively lead to a too strong bound on physics processes beyond the MSM. Table 2 summarizes the intervals and differences obtained.

Parameter	Interval	Difference	Sum
$(\delta\Gamma_Z)_{min}$	−21.1	−0.9	−22.0
$(\delta\Gamma_Z)_{max}$	23.0	0	23.0
$(\delta\Gamma_Z^{inv})_{min}$	−9.3	0	−9.3
$(\delta\Gamma_Z^{inv})_{max}$	5.8	1.8	7.6
$(\delta\Gamma_Z^{ee})_{min}$	−0.82	0	−0.82
$(\delta\Gamma_Z^{ee})_{max}$	0.63	0.10	0.73
$(\delta\Gamma_Z^{\mu\mu})_{min}$	−0.87	0	−0.87
$(\delta\Gamma_Z^{\mu\mu})_{max}$	0.88	0	0.88
$(\delta\Gamma_Z^{\tau\tau})_{min}$	−0.43	−0.51	−0.94
$(\delta\Gamma_Z^{\tau\tau})_{max}$	1.45	0	1.45
$(\delta R)_{min}$	−0.113	−0.028	−0.141
$(\delta R)_{max}$	0.169	0	0.169

Table 2: Allowed changes of Γ_Z, Γ_Z^{inv}, Γ_Z^{ee}, $\Gamma_Z^{\mu\mu}$, $\Gamma_Z^{\tau\tau}$, and R due to non-MSM contributions, using twice one-sided 95% CL limits. Max indicates the maximum experimental value minus the minimum theoretical value, and min indicates the minimum experimental value minus the maximum theoretical value. The interval and difference are defined in the text. All decay widths are given in MeV.

Considering the new decay channel $Z \to X$, let the decay ratios of X be defined as $x_j \equiv \Gamma(X \to j)/\Gamma(X \to \text{anything})$, where $j = h, l, i$ for hadrons, leptons and invisible particles, respectively. In this definition, $x_h + x_l + x_i = 1$. Let the hadronic and leptonic branching ratios of the Z be b_h and b_l, respectively. In the definition of R, the hadronic Z decays are summed over all five quark types produced at LEP, while the leptonic Z decay width is given for a massless charged lepton pair assuming lepton universality. Let $\Gamma_Z^X \equiv \Gamma(Z \to X)$, then

1) The limit on Γ_Z^X from Γ_Z is given by:

$$\Gamma_Z^X \leq (\delta\Gamma_Z)_{\max} = 23.0 \text{ MeV}. \tag{1}$$

2) The limit on Γ_Z^X from $\Gamma_Z^{\rm inv}$ is given by:

$$x_i \Gamma_Z^X \leq (\delta\Gamma_Z^{\rm inv})_{\max} = 7.6 \text{ MeV}. \tag{2}$$

3) A contribution from Z $\rightarrow$ X decays would change the ratio $R = b_h/b_l$ by:

$$\delta R = \frac{\Gamma_Z b_h + \Gamma_Z^X x_h}{\Gamma_Z b_l + \frac{1}{3}\Gamma_Z^X x_l} - \frac{b_h}{b_l} \approx R\frac{\Gamma_Z^X}{\Gamma_Z}\left(\frac{x_h}{b_h} - \frac{x_l}{3b_l}\right), \tag{3}$$

an approximation which is valid when $\Gamma_Z^X \ll \Gamma_Z$. For $x_h = 1$ and $x_l = 0$, $(\delta R)_{max}$ leads to $\Gamma_Z^X \leq 14$ MeV. For $x_h = 0$ and $x_l = 1$, $(\delta R)_{min}$ results in $\Gamma_Z^X \leq 1.7$ MeV, however, this limit is weaker than those from Γ_Z^{ee}, $\Gamma_Z^{\mu\mu}$, $\Gamma_Z^{\tau\tau}$. Recently, a lower limit on the gluino mass of 3.8 GeV derived from R-measurements has been reported at 90% CL [10].

Radiative contributions from non-MSM virtual particles or modifications to the MSM Z-couplings are constrained by the upper and lower limits given in Table 2.

3. Discussion and Example

The most stringent limits on deviations from the non-MSM effects on the Z decay widths are summarized in Table 3. Both upper and lower limits are given at one-sided 95% CL. As a consequence, modified MSM Z-couplings or amplitudes of non-MSM radiative corrections are constrained to the interval at 90% CL. The limits on new decay modes obtained from Γ_Z are independent of the decay branching fractions, while the limits from $\Gamma_Z^{\rm inv}$ constrain only invisible Z decay modes. The limits from Γ_Z^{ee}, $\Gamma_Z^{\mu\mu}$, $\Gamma_Z^{\tau\tau}$, and R constrain the corresponding leptonic and hadronic Z-couplings, respectively. The limits for unspecified and invisible decay modes are of most general use. The limits on Γ_Z^{ee} are tighter, since the Zee-coupling contributes both to Z production and decay. One should note that the charged leptonic and hadronic limits are not able to constrain Z decays if the resulting new particles subsequently decay; dedicated searches are necessary for such specific final states. This is due to the precise selection criteria applied for leptonic and hadronic Z decay event topologies. If a model predicts the invisible, charged leptonic and hadronic branching fractions of Z decays, a χ^2-method allows setting tighter constraints.

The present study updates the analyses given in [11] which were based on 1990 and 1991 LEP data. Only slightly tighter limits are obtained by including the 1992 data as they were entirely taken on the Z pole. Including the 1993 data which contain both off and on peak results, limits are significantly improved: the experimental errors are reduced by about a factor two compared to those used in [11]. In this regard, little improvement is expected from the 1994 data as they are taken again on the Z-pole only. For unspecified Z decays, the 1993 improvement of $\Delta\Gamma(Z)$ is mainly due to the increased predicted MSM lower bound on Γ_Z following from the new top mass constraints of the CDF experiment.

Origin	Decay Mode	$\Delta\Gamma(Z)$ (MeV)		$\Delta Br(Z)$ (in %)
Γ_Z	$Z\rightarrow$anything	−22.0	23.0	0.92
Γ_Z^{inv}	$Z\rightarrow$invisible	−9.3	7.6	0.30
Γ_Z^{ee}	$Z\rightarrow e^+e^-$	−0.82	0.73	0.029
$\Gamma_Z^{\mu\mu}$	$Z\rightarrow \mu^+\mu^-$	−0.87	0.88	0.035
$\Gamma_Z^{\tau\tau}$	$Z\rightarrow \tau^+\tau^-$	−0.94	1.45	0.058
R	$Z\rightarrow$hadrons	−12	14	0.56

Table 3: One-sided 95% CL lower and upper limits on $\Delta\Gamma(Z)$ for Z decaying into any, invisible, charged leptonic, and hadronic channel. The corresponding branching ratio upper limits on $\Delta Br(Z)$ are also given.

As an example, a limit on $\cos^2(\beta-\alpha)$ in the general two-doublet Higgs model is derived. The definitions used are: $\tan\beta$ the ratio of the vacuum expectation values of the Higgs doublets and α the mixing angle between the neutral scalar Higgs fields. The Z decay width into neutral Higgs pairs in the general two-doublet Higgs model is given by [12]:

$$\Gamma(Z\rightarrow hA) = \Gamma(Z\rightarrow\nu\bar{\nu})\tfrac{1}{2}\cos^2(\beta-\alpha)\lambda^{3/2}(\frac{m_h^2}{m_Z^2},\frac{m_A^2}{m_Z^2}), \qquad \lambda(a,b) = (1-a-b)^2-4ab, \tag{4}$$

with $\Gamma(Z\rightarrow\nu\bar{\nu})$ derived from a combined Z lineshape fit: $\Gamma(Z\rightarrow\nu\bar{\nu}) = 166.6\pm1.2$ MeV [3]. Without any assumption on the Higgs decay modes, the constraint $\Delta\Gamma(Z)\leq$ 23.0 MeV sets a limit on $\cos^2(\beta-\alpha)$ as a function of m_h and m_A:

$$\cos^2_{max}(\beta-\alpha) = \frac{2\Gamma_Z^X}{\Gamma(Z\rightarrow\nu\bar{\nu})}\lambda^{-3/2}(\frac{m_h^2}{m_Z^2},\frac{m_A^2}{m_Z^2}). \tag{5}$$

Figure 2 shows the excluded $\cos^2(\beta-\alpha)$ range at 95% CL as a function of m_h for $m_A = 20$ GeV. In conjunction with a constraint on $\sin^2(\beta-\alpha)$, derived from the search for the MSM Higgs boson, this limit leads to an exclusion of a large (m_h, m_A) parameter range [7] given in Sec. 6. Further constraints can result from an analysis of the one-loop vertex corrections to the Zbb-coupling involving additional neutral and charged Higgs bosons; such corrections could decrease $\Gamma(Z\rightarrow b\bar{b})$ and thus the hadronic decay width, depending on the unknown parameters of the two-doublet Higgs model [13]. In this case the limit $\Delta\Gamma(Z\rightarrow \text{hadrons}) \gtrsim$ 12 MeV applies.

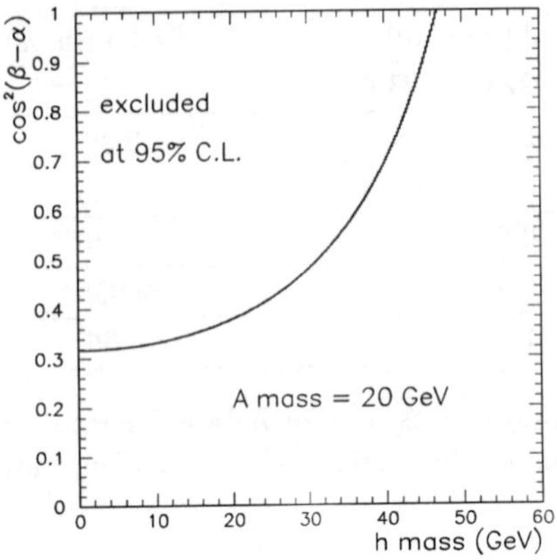

Figure 2: Limit on $\cos^2(\beta - \alpha)$ of the general two-doublet Higgs model as a function of m_h for $m_\mathrm{A} = 20$ GeV. The limit is based on the constraint $\Delta\Gamma(\mathrm{Z} \to \text{anything}) \leq 23.0$ MeV, set by the precision lineshape measurements. No assumptions on the decay branching ratios of the Higgs bosons are made.

4. Combined MSM Higgs Boson Mass Limit and Sensitivity Prospects

The number of expected events is given by each LEP experiment [8], including preliminary results presented at the Glasgow Conference 1994, and shown in Fig. 3 for data corresponding to a total of 7.1 million hadronic Z decays. In first approximation, a combined Higgs mass limit is derived from the summation of the number of expected Higgs events, the calculation of the 95% CL limit taking the background events into account and the correction for the efficiency reduction due to tighter selection cuts with increasing statistics. It is remarkable that as an effect of candidate events the combined Higgs mass limit is for the first time lower than the one previously reported (63.5 GeV [7]). The combined Higgs mass limit is 62.8 GeV. The reduction of four-fermion background will be crucial to increase the sensitivity. This can be achieved by using microvertex b-quark tagging.

The evolution of the Higgs mass limits is shown in Fig. 4. The sensitivity can be extrapolated assuming 50% efficiency in the $\mu^+\mu^-$, $\mathrm{e}^+\mathrm{e}^-$, and $\nu\bar{\nu}$ channels. The combined LEP limit lies below the extrapolated line, since all experiments have tuned the events selection on their own maximum Higgs mass sensitivity. With about 20 million hadronic Z decays a sensitivity of 65 to 70 GeV could be obtained at LEP-I, depending on additional four-fermion background events. In the next phase of the LEP programme, LEP-II, higher masses can be reached as discussed in [14].

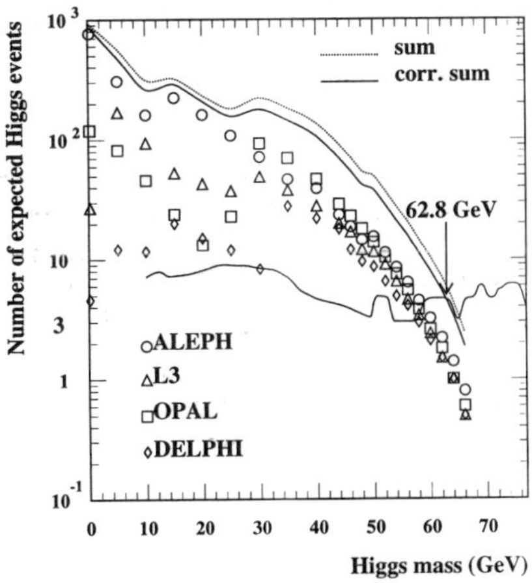

Figure 3: Combined MSM Higgs boson mass limit from ALEPH, DELPHI, L3, and OPAL results.

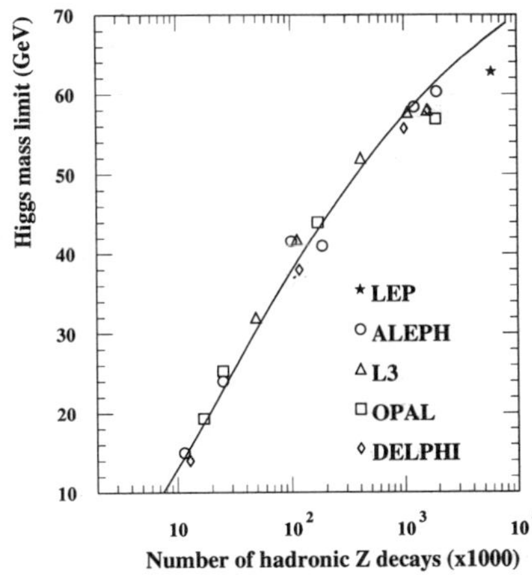

Figure 4: Evolution of MSM Higgs boson mass limits and extrapolation of future sensitivity at LEP-I.

5. Limit on $\sin^2(\beta - \alpha)$

Searches for Higgs boson bremsstrahlung $Z^0 \to Z^{0*}h^0$ in the mass range 0 to about 70 GeV constrain the quantity $\sin^2(\beta - \alpha)$ of the two-doublet Higgs model. Besides the suppression of the Higgs production rate by $\sin^2(\beta - \alpha)$, the Higgs boson couplings to the up- and down-type fermions may be enhanced or suppressed compared to the predictions in the MSM. Assuming that the MSM predictions of the branching fractions of Higgs boson decay remain approximately valid in extensions of the MSM, a limit on $\sin^2(\beta - \alpha)$ is given in Fig. 5. The effects of changes in the selection efficiencies at production thresholds are clearly visible. The further structure in the exclusion contour is due to a few candidate events. Their number is in agreement with the expectations from four-fermion background reactions.

6. Excluded Region in the (m_h, m_A) plane

The limits on the contribution of pair-produced Higgs to Γ_Z give an upper limit on the quantity $\cos^2(\beta - \alpha)$. A mass pair (m_h, m_A) is excluded if the corresponding upper limit on $\sin^2(\beta - \alpha)$ from the bremsstrahlung process is lower than the lower limit coming from the pair-production process. The selection efficiency for possible $h^0 \to A^0A^0$ decays has been evaluated by dedicated searches. The resulting exclusion contour is given in Fig. 6.

7. Constraints on Charged Higgs Bosons of the Two-Doublet Higgs Model

The partial width for Z decay into a charged Higgs pair depends only on the mass of the charged Higgs [12]. Searches for the three relevant processes at LEP $Z^0 \to H^+H^- \to \tau^+\nu\tau^-\nu$, $\tau\nu cs$, $c\bar{s}\bar{c}s$ are performed.

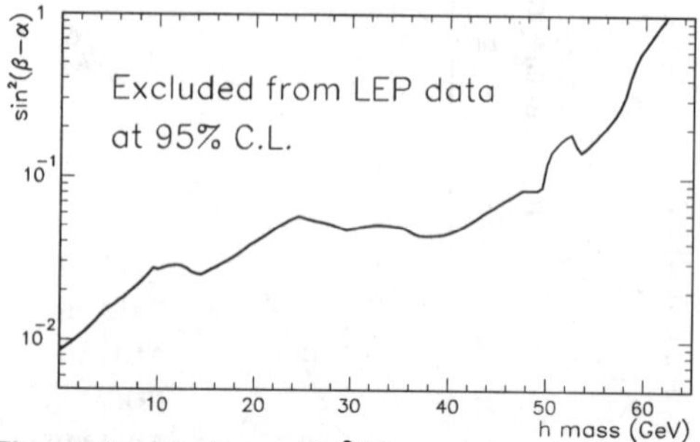

Figure 5: Limit on $\sin^2(\beta - \alpha)$ in the two-doublet Higgs model.

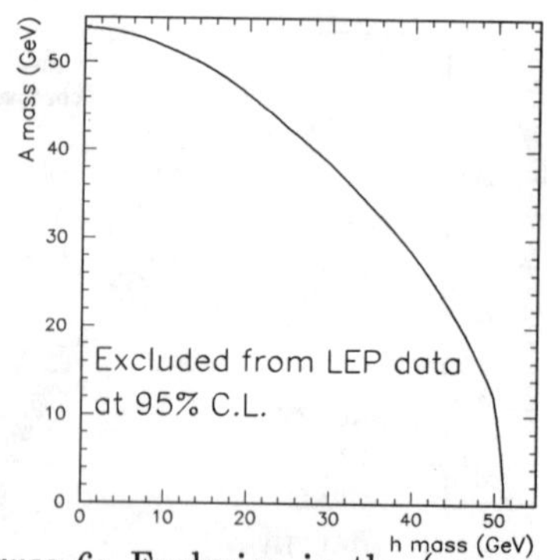

Figure 6: Exclusion in the $(m_{\rm h}, m_{\rm A})$ plane of the two-doublet Higgs model.

The search in the cscs decay channel is the most difficult one due to the irreducible $Z^0 \rightarrow q\bar{q}$ background. The lower charged Higgs mass limit from LEP is about 44 GeV, independently of the Higgs branching ratio. Figure 7 shows two recent results of 95% CL mass limits on charged Higgs bosons as a function of its leptonic (hadronic) branching ratio obtained from the search in the three decay channels [15, 16]. It is remarkable that both experiments find a slightly weaker exclusion contour in the cscs channel for Higgs masses around 40 GeV at L3 and around 37 GeV at DELPHI. This is due to a small excess of data compared to the background expectation.

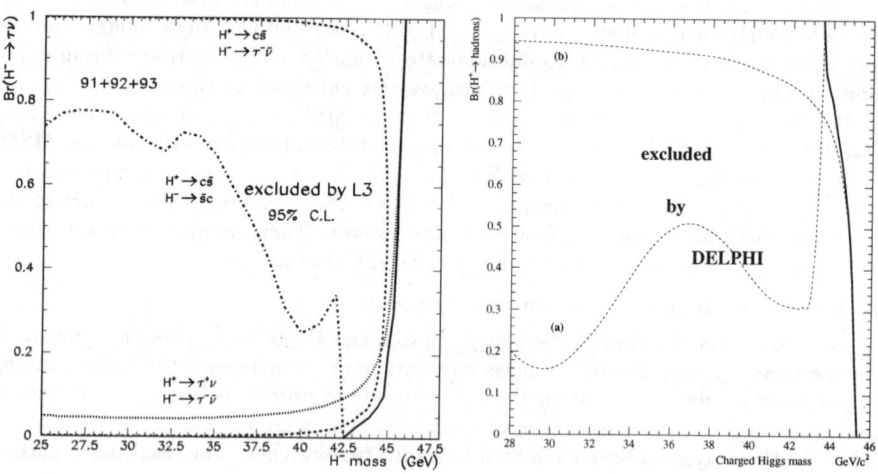

Figure 7: Left plot: preliminary L3; right plot: DELPHI. Excluded regions of pair-produced charged Higgs bosons as a function of the charged Higgs mass and the leptonic (for L3) (hadronic for DELPHI) Higgs decay branching fraction. The thick contour line determines the combined mass limit.

8. Constraints on Neutral Higgs Bosons of the Two-Doublet Higgs Model

Signatures resulting from the following expected Higgs decay modes are investigated: $h^0/A^0 \rightarrow \tau^+\tau^-$, $h^0/A^0 \rightarrow b\bar{b}$, $h^0 \rightarrow A^0A^0$. Searches for 4-jet, 6-jet, $\tau\tau$ 2-jet, and 4-τ signatures are performed. No signal has been observed in any of these channels. Figure 8 shows the excluded regions in the (m_h, m_A) plane for branching ratio limits on: $\Gamma(Z^0 \rightarrow h^0A^0)/\Gamma(Z^0 \rightarrow q\bar{q})$. Detection sensitivities have been largely increased compared to earlier results [17] due to an increase of data by a factor of about 4 and applying lifetime b-quark tagging in addition to semileptonic tagging.

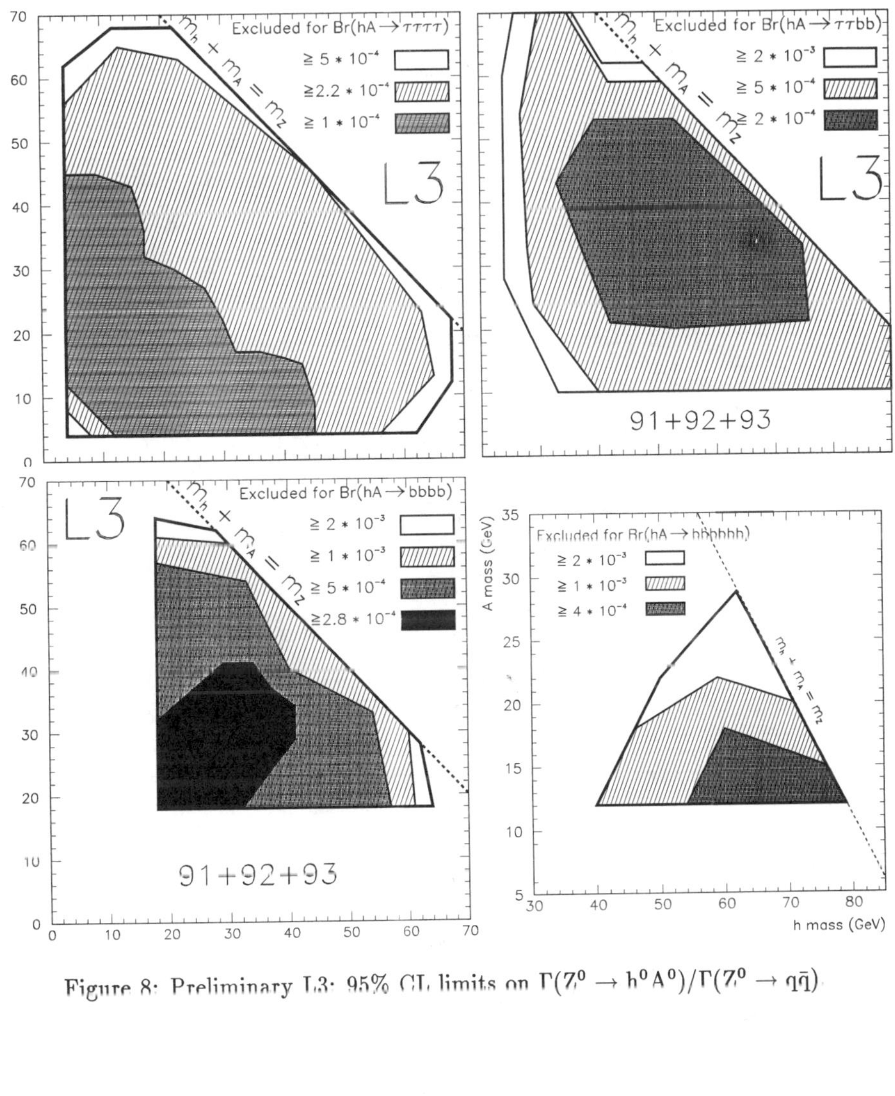

Figure 8: Preliminary L3: 95% CL limits on $\Gamma(Z^0 \rightarrow h^0A^0)/\Gamma(Z^0 \rightarrow q\bar{q})$.

9. Constraints on Neutralinos

Many direct searches for supersymmetric particles have been performed. Excellent new results from the CDF and D0 collaborations of chargino, s-quark, and gluino searches have been reported [18]. A preliminary update of model-independent limits on neutralino production in Z decays is given in Fig. 9 [15].

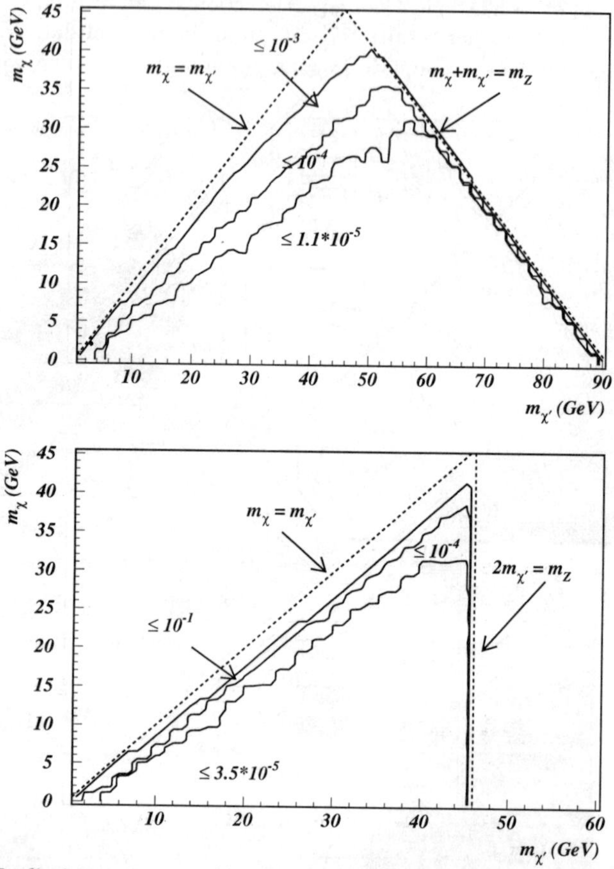

Figure 9: Preliminary L3: 95% CL limit on $Br(Z^0 \to \chi\chi')$ (upper plot) and $Br(Z^0 \to \chi'\chi')$ (lower plot). The dashed lines show the kinematical limits.

10. Conclusions

Stringent limits on extensions of the MSM are presented, and where possible, the results of the four LEP experiments are combined. The largely model-independent presentation is intended for straightforward tests of theoretical models which predict modified branching fractions of Z decays.

Acknowledgments

I would like to thank Sally Dawson, Wolfgang Hollik, Joachim Mnich, Stefan Pokorski, and Janusz Rosiek for fruitful discussions and I express my gratitude to the Institute of Theoretical Physics at the Warsaw University for their hospitality.

References

1. S.L. Glashow, *Nucl. Phys.* **22** (1961) 579;
   A. Salam, *Phys. Rev.* **127** (1962) 331; and in: "Elementary Particle Theory", ed. N. Svartholm (Stockholm, 1968), p. 361;
   S. Weinberg, *Phys. Rev. Lett.* **19** (1967) 1264.
2. ALEPH Collab., D. Buskulic et al., *Z. Phys.* **C 62** (1994) 539;
   DELPHI Collab., P. Abreu et al., *Nucl. Phys.* **B 418** (1994) 403;
   L3 Collab., M. Acciarri et al., *Z. Phys.* **C 62** (1994) 551;
   OPAL Collab. R. Akers et al., *Z. Phys.* **C 61** (1994) 19.
3. The LEP Electroweak Working Group, "Updated Parameters of the Z^0 Resonance from Combined Preliminary Data of the LEP Experiments", Internal Note ALEPH 94-74, DELPHI 94-33, L3 1599 and OPAL TN235;Preprint CERN, PPE 94-187 (1994);
   D. Schaile, Proc. 27th Rochester Conference, Glasgow, Scotland, July 1994, in press *World Scientific*.
4. Particle Data Group, *Phys. Rev.* **D 50** (1994) 1173.
5. D. Bardin *et al.*, *Nucl. Phys.* **B 351** (1991) 1; *Z. Phys.* **C 44** (1989) 493; *Phys. Lett.* **B 255** (1991) 290;
   M. Bilenky and A. Sazonov, JNIR Dubna Preprint E2-89-792 (1989).
6. CDF Collaboration, F. Abe *et al.*, *Phys. Rev.* **D 50** (1994) 2966.
7. A. Sopczak, Proc. XV Int. Warsaw Meeting, Kazimierz, Poland, May 1992, *World Scientific* 1993, p. 45; Proc. XV Autumn School, Lisbon, Portugal, Oct. 1993, Preprint CERN, PPE 94-73 (1994) in press *Nucl. Phys.* **B**, *Proc. Suppl.*
8. ALEPH Collab., D. Buskulic *et al.*, *Phys. Lett.* **B 313** (1993) 299; and Physics Notes 94-28, 94-29, and 94-30, March 1994 (unpublished);
   DELPHI Collab., P. Abreu *et al.*, Preprint CERN, PPE 94-46/Rev (1994) in press *Nucl. Phys.* **B**;
   L3 Collab., O. Adriani *et al.*, *Phys. Lett.* **B 303** (1993) 391;
   OPAL Collab., R. Akers *et al.*, *Phys. Lett.* **B 327** (1994) 397.
9. P. Langacker, Proc. CCAST (World Laboratory) Symposium/Workshop, Beijing, China, May 1990, *Gordon and Breach* 1991, p. 461.
10. M. Schmelling and R.D. St.Denis, *Phys. Lett.* **B 329** (1994) 393.
11. A. Sopczak, Internal L3 Notes 1169, May 1992 and 1372, March 1993 (unpublished).
12. S. Dawson, J.F. Gunion, H.E. Haber and G.L. Kane, "The Physics of the Higgs Bosons: Higgs Hunter's Guide" *Addison Wesley*, Menlo Park, (1989).
13. J. Rosiek, *Phys. Lett.* **B 252** (1990) 135;
    A. Denner, R. Guth, W. Hollik and J.H. Kühn, *Z. Phys.* **C 51** (1991) 695.
14. A. Sopczak, Int. J. Mod. Phys. **A9** (1994) 1747.
15. L3 publication in preparation.
16. DELPHI Collab., P. Abreu *et al.*, Preprint CERN, PPE 94-83 (1994) in press *Z. Phys.* **C**.
17. L3 Collab., O. Adriani *et al.*, *Z. Phys.* **C 57** (1993) 355.
18. K. Sliwa, these proceedings.

Present and future searches with e^+e^- colliders for the neutral MSSM Higgs bosons – the complete 1-loop analysis[1]

Janusz Rosiek
Institut de Fisica Corpuscular, Universitat de Valencia

André Sopczak
PPE Division, CERN

Abstract

New mass regions unexcluded by direct searches are revealed by an analysis of experimental results from LEP1 using full 1-loop diagrammatic calculations of radiative corrections in the MSSM. Simulations of experimental signal efficiencies and full 1-loop calculations are combined to study the sensitivity for neutral Higgs bosons at LEP2 and the NLC. Compared with previous studies based on an Effective Potential Approach, we identify mass regions where the discovery potential depends on the MSSM parameters other than the top and stop masses. We propose our method of interpretation to be adopted by the four LEP experiments for better precision. The possibility of the experimental distinction between the Higgs boson in the MSM and the lightest Higgs boson in the MSSM is discussed for the NLC.

1. Introduction

The search for Higgs particles is one of the most challenging problems of experimental particle physics. Particular attention is given to the search for the Higgs bosons with properties predicted by the Minimal Supersymmetric Standard Model (MSSM). At least one of them, the lighter CP-even scalar h^0, is expected to be relatively light ($m_h < 140$ GeV).

Numerous results of searches for Higgs bosons in the framework of the MSSM are reported from the LEP1 experiments Aleph, Delphi, L3 and Opal [1] (for an overview see [2]). In the next phase of the LEP programme, LEP2, $\sqrt{s} > 2m_W$ is planned to be reached in 1996 [3]. The physics potential of the e^+e^- Next Linear Collider, NLC, running at $\sqrt{s} \geq 500$ GeV, is intensively discussed [4]. A realistic analysis of the MSSM Higgs sector phenomenology have to include radiative corrections [5-11]. Several approaches have been developed to compute them: the Effective Potential Approach (EPA) [6], the Renormalization Group Equations (RGE) approach [7, 8], and the Full 1-loop Diagrammatic Calculations (FDC) in the on-shell renormalization scheme [9, 10]. Previous interpretations in the MSSM of searches for Higgs bosons, performed by the LEP experiments [1], are based on common assumptions:

1. Radiative corrections to the MSSM Higgs boson masses, production and decay rates are considered in the EPA approximation, in its most simplified version. Only the leading part of the contributions arising from the top quark and stop is considered.
2. The MSSM parameter space is strongly constrained by neglecting the dependence on most of the SUSY particle masses and couplings. In the most simplified

[1]Supported in part by a postdoctoral fellowship from Spanish MEC, by grant #PB92-0084, and by Polish Committee for Scientific Research.

version of the EPA, only one free SUSY parameter, $m_{\tilde{t}}$ is considered.

3. In the EPA interpretations $m_{\tilde{t}} > m_t$ is assumed, which is experimentally unjustified.

In two aspects, we improve the interpretations of existing experimental results and perspectives of future searches for the neutral MSSM Higgs bosons. First, we interpret existing LEP1 data and simulations of detector efficiencies expected for LEP2 and the NLC using more accurate theoretical calculations of the production and decay rates of h^0, H^0 and A^0 in the FDC approach, and we compare the results with the EPA predictions. Second, we study the effect of using a larger set of parameters describing the MSSM, taking into account the existing experimental constraints on the SUSY parameters. We discuss three main aspects of the neutral MSSM Higgs boson searches: implications of the existing experimental data from LEP1, perspectives of the searches at LEP2, and at the NLC. A more detailed description of our analysis can be found in [12].

2. Parameter Space of the MSSM

The most general version of the MSSM Lagrangian contains a large number of free parameters. Most of the masses and couplings of the SUSY particles (referred to as SUSY parameters) have small impact on the Higgs sector. Using numerical simulations, we identified which parameters are important for the Higgs boson phenomenology. These parameters have been varied independently:

- (m_h, m_A) or (m_H, m_A) – the investigated Higgs boson mass combinations.
- m_{sq} – the common mass parameter for all squarks. The assumption of the same mass parameters for the three squark generations has a small effect. Results depend mostly on the stop mass parameter and only weakly on the masses of other sfermions.
- m_g – the gaugino mass. We assumed the commonly used GUT relation for the SU(2) and U(1) gaugino masses: $m_{U(1)} = \frac{5}{3}\tan^2\theta_W m_{SU(2)}$, $m_{SU(2)} = m_g$
- μ – the mixing parameter of the Higgs doublets in the superpotential.
- A – the mixing parameter in the sfermion sector. As for m_{sq} only one universal parameter is considered for all generations. The mixing is proportional to Am_{sq}.

In this paper, the top quark mass is fixed to $m_t = 175$ GeV. SUSY parameters are scaned in the ranges given in Table 1.

Parameter	m_{sq} (GeV)	m_g (GeV)	μ (GeV)	A
Range	200—1000	200—1000	−500—500	−1—+1

Table 1: Ranges of SUSY parameters used for independent variation in the study of the MSSM neutral Higgs boson searches.

Tree-level experimental bounds on $\tan\beta$ are assumed to hold approximately. Its value is constrained to $0.5 \leq \tan\beta \leq 50$. The lower bound, based on [15], affects the theoretically allowed regions in the (m_h, m_A) and (m_H, m_A) planes. The variation of the upper bound has no significant effect. As an illustration of the effects of variation of the SUSY parameters, Fig. 1 shows upper limits on m_h as a function of m_A.

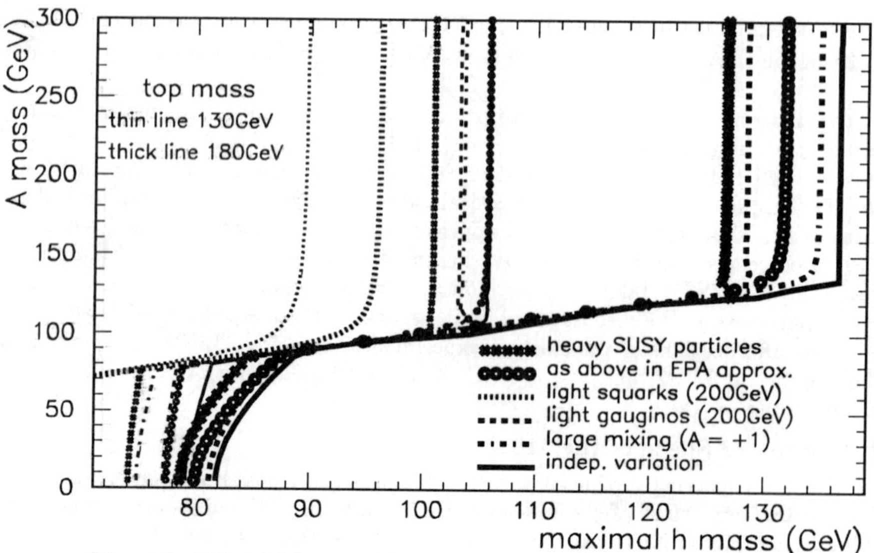

Figure 1: Upper h^0 bounds for various values of SUSY parameters.

3. Excluded Mass Regions at LEP1

In order to derive precise bounds on h^0 and A^0 masses, the limits on the Higgs boson production rates given in [13] have been used. In the mass plane (m_h, m_A), each point with a step size of 1 GeV up to Higgs boson masses of 120 GeV has been analyzed separately. For each mass combination, the production cross sections of the reactions $e^+e^- \to h^0Z^{0\star} \to h^0\overline{f}f$, $e^+e^- \to h^0A^0$, and the branching ratios for h^0 and A^0 decays have been computed. Then, the number of expected Higgs boson events for each investigated final state and each mass bin has been calculated. A given (m_h, m_A) combination is excluded if for *all* SUSY parameter sets the expected number of events in at least one of the channels is excluded at 95% CL.

The left plot of Fig. 2 shows regions in the (m_h, m_A) plane which are excluded by individual channels. The right plot shows the regions that can be excluded by the combined analysis of all channels; this reveals an interesting result. The full scan over the SUSY parameter space (thick solid line) gives, in comparison with the epsilon approximation (dotted line), a substantial additional triangle-shape unexcluded mass range for 45 GeV $< m_A <$ 80 GeV, 25 GeV $< m_h <$ 50 GeV, marked with a bold solid line. The existence of this region can be understood in the following way: in the range $m_h + m_A < m_Z$ the reaction $e^+e^- \to h^0A^0$ is allowed kinematically and both main discovery channels $e^+e^- \to h^0Z^{0\star}$ and $e^+e^- \to h^0A^0$ contribute. If radiative corrections reduce the cross section of one of them below the experimental sensitivity, the complementary cross section will be large enough to exclude this mass combination. The unexcluded triangle begins above the $m_h + m_A = m_Z$ limit. In this range the cross section $e^+e^- \to h^0Z^{0\star}$ can be small for some SUSY parameters. The complementary process is already forbidden kinematically, thus no signal can be

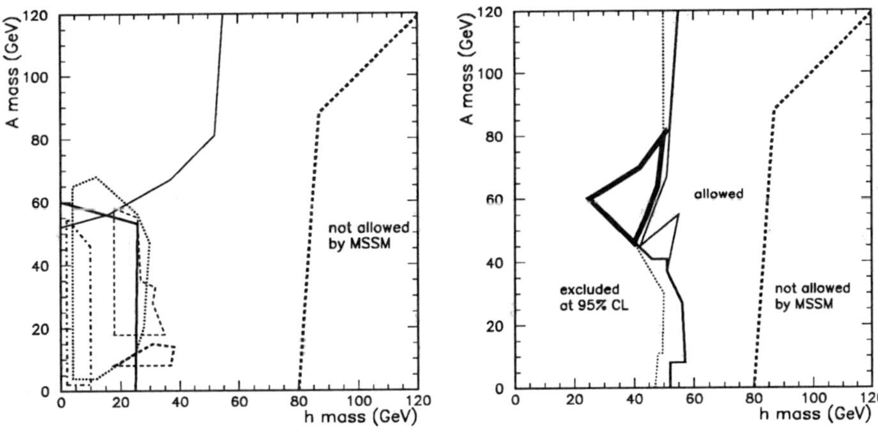

Figure 2: Excluded mass regions in the (m_h,m_A) plane for $\sqrt{s} = m_Z$. Left plot: individually excluded mass regions for various searches: thick solid line: Z^0 lineshape; thin solid line: $Z^0 \rightarrow Z^{0*}h^0$; dotted line: $\tau^+\tau^-bb$; dotted-dashed line: $\tau^+\tau^-\tau^+\tau^-$; thin dashed: $b\bar{b}b\bar{b}$; thick dashed: 6b. Right plot: combined excluded regions: dotted line: EPA (epsilon approximation), thin solid line: FDC result for fixed and heavy SUSY particle masses, thick solid line: FDC results with a full scan over the SUSY parameters, very thick line: new unexcluded region in FDC.

observed. With increasing m_A, the cross section for the $e^+e^- \rightarrow h^0Z^{0*}$ reaction becomes less sensitive to the SUSY parameters and similar to the $e^+e^- \rightarrow H^0_{MSM}Z^{0*}$ cross section, recovering the Higgs mass bound of about 55 GeV. The right plot shows also that the regions obtained in the FDC for fixed and heavy SUSY particle masses (thin solid line) and in the EPA (dotted line) are similar. Thus, the EPA result can approximately be recovered for a specific set of SUSY parameters.

4. LEP2 Discovery Potential

The production reactions relevant for LEP2 $e^+e^- \rightarrow h^0Z^0, h^0A^0$ and $e^+e^- \rightarrow H^0Z^0, H^0A^0$ have been investigated based on sensitivities given in [14]. Four regions are distinguished in the (m_h,m_A) and (m_A,$\tan\beta$) planes:

(A) The sensitivity region where, by direct searches, a Higgs signal cannot escape detection, for any choice of the SUSY parameters from the ranges given in Table 1 and for fixed top quark mass of 175 GeV.

(B) The region where the perspectives of direct searches depend on the SUSY parameters. This means that searches can have sensitivity or not, depending on the specific choice of these parameters.

(C) The non-sensitivity region where no signal can be found independent of the choice of the SUSY parameters.

(D) The theoretically disallowed region where (m_h,m_A) combinations are not allowed in the MSSM for any choice of SUSY parameters and requiring $\tan\beta \geq 0.5$.

The mass regions where at least one CP-even Higgs boson could be discovered at LEP2 for $\sqrt{s}$ = 175, 190 and 210 GeV are shown in the (m_h,m_A) and (m_A,$\tan\beta$)

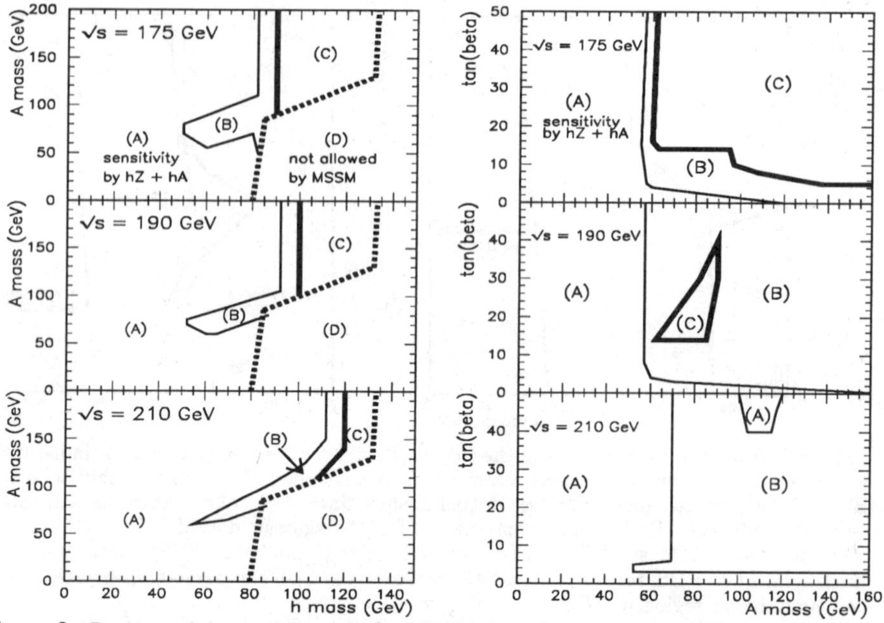

Figure 3: Regions of detectability of h^0 at LEP2 for $\sqrt{s}$ = 175, 190, and 210 GeV with $\mathcal{L} = 500\ pb^{-1}$ in the (m_h,m_A) and (m_A,$\tan\beta$) planes. For the description see text.

planes in Fig. 3. A substantial region (B) reflects an uncertainty in the discovery potential connected with the variation over the SUSY parameters. The effects of increasing center-of-mass energy can clearly be seen.

5. Discovery Potential of the NLC

In the analysis of the physics potential of the NLC we use a center-of-mass energy of $\sqrt{s}$ = 500 GeV and an estimated sensitivity of 10 fb for all $e^+e^- \to h^0Z^0$, H^0Z^0, h^0A^0, H^0A^0 channels [4]. For a total luminosity of the NLC of $\mathcal{L} = 30\ fb^{-1}$, this sensitivity corresponds to a discovery of a signal if more than 300 events are produced (before selection cuts are applied). Under these assumptions the NLC can cover entirely the MSSM parameter space and at least one Higgs boson must be found or the MSSM is ruled out. This conclusion holds for a search for the pair production reactions $e^+e^- \to h^0A^0, H^0A^0$ and for the bremsstrahlung reactions $e^+e^- \to h^0Z^0, H^0Z^0$.

The left plot of Fig. 4 shows (m_h,m_A) regions where the heavier scalar, H^0, could be detected, and the right plot shows that all three scalars could be observed up to $m_A \approx 200$ GeV, region (A). For $m_h \approx 90$ GeV and $m_A = 180$ GeV a small region (B) exists, where the perspectives of the simultaneous h^0, H^0 and A^0 discovery depend on the SUSY parameters. For larger m_A only h^0 could be found, region (C). The important conclusion resulting from the analysis of the production rates of the neutral MSSM Higgs bosons at the NLC is that either h^0 alone, or all three neutral MSSM scalars h^0, H^0 and A^0 could be found simultaneously.

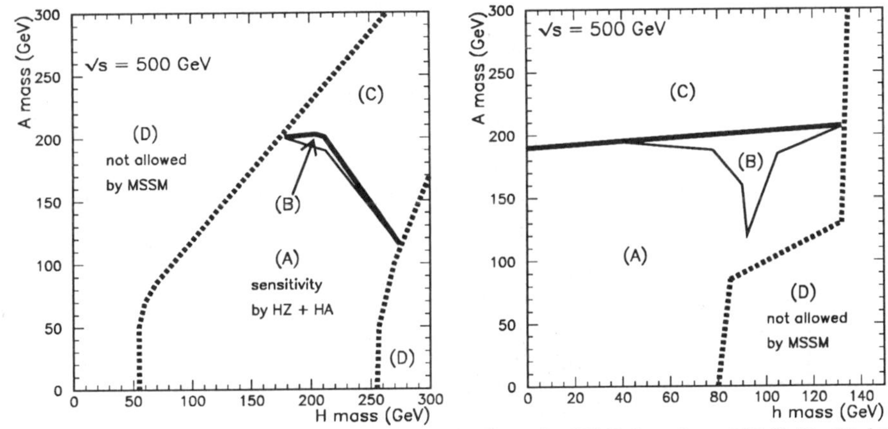

Figure 4: Left plot: regions of detectability of H^0 at the NLC for $\sqrt{s} = 500$ GeV. Right plot: regions of simultaneous detectability of h^0, H^0 and A^0. For the description see text.

6. Distinction between MSSM and MSM Higgs Bosons

Various possibilities exist to identify a Higgs boson unambiguously as a non-minimal one after a first discovery. The observation of two (or more) Higgs scalars would be very interesting, since it immediately excludes the MSM and gives support to multi-doublet Higgs models, with the MSSM as a possible example[2]. Another method of identifying the origin of a scalar is based on measuring its production cross section or decay branching ratios. The left plot of Fig. 5 shows the regions (A,B,C) where the cross section difference between MSM and MSSM Higgs boson bremsstrahlung is larger, depending on the SUSY parameters, and smaller than 20%, respectively. For the branching ratios analysis, the dominant channel of h^0 decay is $h^0 \to b\bar{b}$. Supersymmetric decay modes could allow an effective distinction between MSSM and MSM if they are kinematically open. The total contribution of the other channels such as $h^0 \to \bar{c}c$, $\tau^+\tau^-$, gg, $\gamma\gamma$, $Z^0\gamma$ is less than 20% (10% for $\tan\beta \geq 3$). Deviations of the small branching fractions from the MSM expectations will be difficult to detect[3]. We assume that the MSSM and MSM could be distinguished experimentally if the difference between $BR(h^0 \to \bar{b}b)$ and $BR(H^0_{MSM} \to \bar{b}b)$ is larger than about 8% [19]. Regions where this condition is fulfilled are shown in right plot (they are defined as for the comparison of the cross sections). In the range about $m_A \geq 50$ GeV and $m_h \leq$ 80 GeV, the knowledge of $BR(h^0 \to \bar{b}b)$ at the level of 8% is insufficient to establish the nature of the scalar. In a large region (B), the perspectives of distinction depend on the SUSY parameters. For $m_A \geq 100$ GeV the branching ratios in the MSSM

[2] In this paper, only the neutral MSSM scalars are discussed. For a large part of the MSSM parameter space $H^\pm$ could be discovered, giving rise to unambiguous evidence of physics beyond the MSM. Simulations in the EPA approximation [17] show that experiments at the NLC could detect H^+ with masses up to about 210 GeV, which corresponds to $m_A \approx 200$ GeV.

[3] Recently, a possible distinction of the h^0 from the H^0_{MSM} on the basis of the off-shell decay $h^0 \to WW^*$ has been discussed [18].

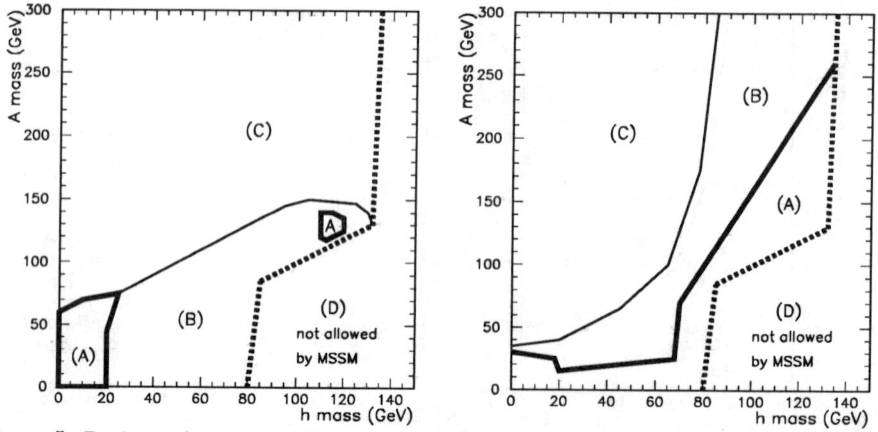

Figure 5: Regions where the MSSM and MSM Higgs boson could be distinguished on the basis of the measurements of the cross section for the $e^+e^- \rightarrow Z^0h^0$ reaction (left plot), and branching ratio for the $h^0 \rightarrow b\overline{b}$ decay (right plot). For the description see text.

and MSM differ typically by about 6%. This means that a very high experimental precision is required to identify the Higgs boson as a MSSM one. It is important to note that although the above comparison is independent (branching ratios) or weakly dependent (cross sections) on the center-of-mass energy, the experimental feasibility depends strongly on the machine energy and total luminosity.

7. Conclusions

Aspects of searches for neutral MSSM Higgs bosons at present and future e^+e^- colliders have been presented. Full 1-loop diagrammatic calculations of radiative corrections to the Higgs particle production and decay rates are applied. The dependence of the results on all important model parameters is investigated. In addition to the stop mass, several other SUSY parameters are varied independently. This variation changes significantly the results compared with the simpler EPA approach. We show that for fixed and heavy SUSY particle masses the results of the EPA can approximately be recovered in the FDC. Nevertheless, differences of the order of few GeV on the investigated Higgs mass bounds also exist in this case. Performing a full scan over the MSSM parameters, the differences become large. For LEP1, FDC gives in comparison with EPA an additional unexcluded region for $m_h \approx 25 - 50$ GeV and $m_A \approx 45 - 80$ GeV. We propose our method of interpretation to be adopted by the four LEP experiments, using higher statistics available now.

For LEP2, detailed experimental simulations for non-minimal Higgs bosons are combined with the improved calculations. Significant effects of the variation of the SUSY parameters on the accessible mass parameter ranges are found. The possibilities of a discovery of the lightest supersymmetric scalar depend strongly on the achievable center-of-mass energy. For $m_A > 120$ GeV, h^0 with mass $m_h < (\sqrt{s}-100)$ GeV could always be found. In the m_h range from $(\sqrt{s}-100)$ GeV up to the kinematical bound, perspectives of discovery depend on the specified set of the SUSY parameters.

At the NLC even at the most unfavourable parameter choice at least one MSSM neutral Higgs boson should be found or the MSSM is ruled out. The NLC has good chances to discover more than one Higgs particle, and most likely either one or all three MSSM neutral scalars could be observed. If only one CP-even scalar is found via the bremsstrahlung process $e^+e^- \to Z^0h^0$, it will be difficult to assess its origin. In a large mass region such a scalar could be distinguished from the MSM Higgs boson on the basis of its fermionic decay branching ratios if the measurements achieve a precision better then about 6%.

References

1. ALEPH Collab., D. Buskulic *et al.*, *Phys. Lett.* **313B** (1993), 312; DELPHI Collab., P. Abreu *et al.*, *Nucl. Phys.* **373B** (1992), 3; L3 Collab., O. Adriani *et al.*, *Z. Phys.* **57C** (1993), 355; OPAL Collab., R. Akers *et al.*, Preprint CERN, PPE/94-104 (1994) in press *Z. Phys.* **C**.
2. E.g. A. Sopczak, Preprint CERN, PPE/94-73 (1994) in press *Nucl. Phys.* **B**, *Proc. Suppl.*
3. C. Wyss, LEPC Meeting, CERN LEPC/94, 6 Feb. 1994.
4. E.g. P. Janot, Proc. Workshop on physics and experiments with linear e^+e^- colliders, Waikoloa, Hawaii, USA, 26–30 April 1993 (World Scientific), p.192; P. Zerwas, Proc. Int. Europhysics Conf. on High Energy Physics, Marseille, France, 22-28 July 1993 (Edition Frontieres), p.184.
5. J. Ellis, G. Ridolfi, F. Zwirner, *Phys. Lett.* **257B** (1991), 83; H.E. Haber, R. Hempfling, *Phys. Rev. Lett.* **66** (1991), 1815; Y. Okada, M. Yamaguchi, T. Yanagida, *Phys. Lett.* **262B** (1991), 54; R. Barbieri, M. Frigeni, M. Caravaglios, *Phys. Lett.* **258B** (1991), 167.
6. J. Ellis, G. Ridolfi and F. Zwirner, *Phys. Lett.* **262B** (1991), 477; R. Barbieri and M. Frigeni, *Phys. Lett.* **258B** (1991), 395; A. Brignole, J. Ellis, G. Ridolfi and F. Zwirner *Phys. Lett.* **271B** (1991), 123.
7. M. Carena, K. Sasaki and C. E. M. Wagner, *Nucl. Phys.* **381B** (1992), 66; H.E. Haber, R. Hempfling, *Phys. Rev.* **D48** (1993), 4280.
8. P.H. Chankowski, S. Pokorski and J. Rosiek, *Phys. Lett.* **281B** (1992), 100.
9. P.H. Chankowski, S. Pokorski and J. Rosiek, *Phys. Lett.* **274B** (1992) 191; **286B** (1992) 307; *Nucl. Phys.* **423B** (1994) 437; **423B** (1994), 497.
10. A. Brignole, *Phys. Lett.* **281B** (1992), 284.
11. Z. Kunszt and F. Zwirner, *Nucl. Phys.* **385B** (1992), 3.
12. J. Rosiek and A. Sopczak, HEP-PH/9409412, Preprint CERN-PPE/94-134 in press *Phys. Lett.* **B**.
13. L3 Collab., O. Adriani *et al.*, *Phys. Lett.* **294B** (1992) 457; *Z. Phys.* **C57** (1993) 355; A. Sopczak, "Search for Non-minimal Higgs Bosons from Z^0 Decays with the L3 Experiment at LEP", Ph.D. Thesis, U. of Cal., San Diego, 1992.
14. A. Sopczak, *Int. J. Mod. Phys.* **A9** (1994), 1747.
15. E.g. B. Grządkowski, and J.F. Gunion *Phys. Lett.* **243B** (1990), 301
16. E.g. A. Djouadi, J. Kalinowski and P.M. Zerwas, *Z. Phys.* **C57** (1993), 569.
17. A. Sopczak, Proc. Workshop on physics and experiments with linear e^+e^- colliders, Waikoloa, Hawaii, USA, 26–30 April 1993 (World Scientific) p.666; Preprint CERN PPE/93-197 in press *Z. Phys.* **C**, Proc. DESY 93-123C.
18. M. Hildreth, T. Barklow, and D. Burke, *Phys. Rev. Lett.* **71** (1993), 3441.
19. M. Hildreth, Proc. of the Workshop on physics and experiments with linear e^+e^- colliders, Waikoloa, Hawaii, USA, 26–30 April 1993 (World Scientific), p.635.

RECENT ELECTROWEAK RESULTS AT LEP

Pascal PERRODO
CERN, PPE Division
1211 Geneva, Switzerland

ABSTRACT

The electroweak results including data taken in 1993 from the four LEP experiments are presented. Improvements in measuring the LEP beam energy by resonant depolarization and in upgrading the luminosity monitors lead to a precise measurement of M_Z and Γ_Z. The $Z \to b\bar{b}$ partial width is measured within 1% as well as the forward-backward charge asymmetry in $b\bar{b}$ events using efficient silicon vertex detectors. Finally the electroweak radiative corrections are measured within 10^{-3}, allowing to set limits on top mass compatible with CDF first measurement.

1. Introduction

In 1993 LEP has performed an energy scan which led to a better measurement of the Z mass and width. The electroweak results of the four LEP experiments including '93 data have been combined by the LEP Electroweak Working Group and are described hereafter. The improvements on Z line shape results have been made possible through a better beam energy measurement by resonant depolarization and the upgrade of the luminosity detectors of the four LEP experiments. On the other hand the high efficiency micro-vertex silicon detectors lead to pure and numerous $b\bar{b}$ samples. So the $Z \to b\bar{b}$ partial width has been measured very accurately as well as the forward-backward charge asymmetry in $b\bar{b}$ events using a jet charge method. Asymmetries for c and s quarks begin to be measured with significant precision. τ physics (asymmetry and polarization) benefits also from higher statistics. Finally the top quark and Higgs boson masses are investigated through the electroweak radiative corrections. The LEP results are compared with other experiments predictions ($p\bar{p}$, neutrino scattering, CDF top mass measurement).

2. LEP performances

2.1. Delivered luminosity

Each LEP experiment has recorded over years an integrated luminosity of about : 1.2 pb^{-1} in 1989, 7 pb^{-1} in 1990, 13 pb^{-1} in 1991, 23 pb^{-1} in 1992, 35 pb^{-1} in 1993 ($\approx$ 60 pb^{-1} has been recorded for the 1994 data). The corresponding numbers of events are given in table 1.

Many progresses on the LEP machine have led to this improvement : since 1992 the quadrupoles and sextupoples are realigned each year, the beam-beam tune has

Table 1. LEP experiments statistics ($\times 10^{-3}$) for Z events into hadronic and leptonic final states.

	ALEPH	DELPHI	L3	OPAL	LEP
$q\bar{q}$: '90-'91	451	356	416	454	1677
'92	680	697	678	733	2788
'93 prel.	653	677	658	653	2641
total	1784	1730	1752	1840	7106
$\ell^+\ell^-$: '90-'91	55	37	40	58	190
'92	82	69	58	88	297
'93 prel.	79	71	62	81	293
total	216	177	160	227	780

been improved, the luminosity lifetime in the machine is now higher than 10 hours and since 1992 the "Pretzel scheme" allows 8 bunches per beam instead of 4 previously.

The choice of the beam energies has been the following : in 1989-1991, a scan at seven values from 88.25 GeV to 94.25 GeV spaced by 1 GeV, in 1992 at 91.2 GeV, in 1993 a scan at three points 89.4, 91.2 and 93.0 GeV. The energy values and the volumes of data taken on and off peak in '93 have been optimized to minimize the error on Γ_Z.

2.2. Beam energy

Improvements on M_Z and Γ_Z precisions have been permitted by the very precise measurement of the beam energy by resonant depolarization. A natural transverse spin polarization builds up in storage rings (Sokolov-Ternov effect[2]). This effect has been observed in all electron storage rings where it has been searched for[3,4,5,6,7,8]. Practically the averaged spin polarization is obtained by measuring the angle of the light coming from a polarized laser (532 nm) backscattered by the beam (Compton scattering).

Resonant depolarization is a well known method which has already been used extensively to measure the masses of hadronic $J=1^{--}$ resonances[3,4]. The spin precession frequency is related in a very precise way to the beam energy through the formula $\nu = (g_e - 2)E_{beam}/2m_e$ (Thomas precession) where m_e is the electron mass and $g_e - 2$ the anomalous magnetic moment for the electron. A magnetic resonance with the spin precession can be driven by a periodic magnetic perturbation generated by a radio-frequency magnet called kicker. By varying the frequency of the perturbation one observes a depolarization depth (see fig. 1) with a minimum located at the beam energy. The overall precision of the measure[9] is 200 keV.

The first energy measurements were done at LEP 1991[7]. Since then the method has been improved and 25 measurements were performed[10] during the 1993 running.

This very precise method allowed to understand many effects affecting the beam energy and correct for them. A correction for magnet temperature has been applied. The now well known terrestrial tide effect[11] (see fig. 2) causing variations of 10 MeV has been taken into account. Finally residual ground motion effects could be under-

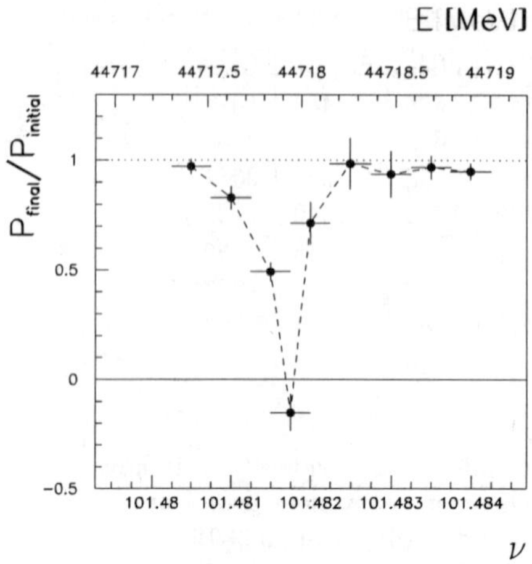

Fig. 1. Width measurement of the artificial depolarizing resonance.

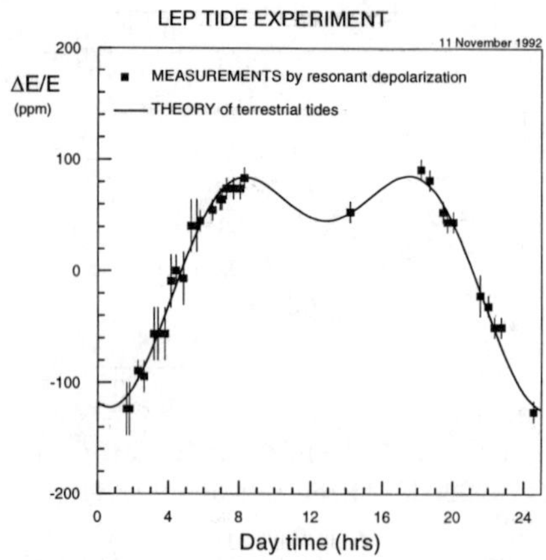

Fig. 2. Measured beam energy variations and the expectation from the tides.

stood by relating them to beam orbit position measurements in the vacuum pipe. They could be correlated with the rainfall and also the level of water in the nearby Geneva lake!

3. Improvements in luminosity measurement

An other major improvement in the years 1992-1993 has been the upgrade of the luminosity monitors for the four LEP experiments. Precise luminosity determination allows more precision for the extraction of the hadronic cross section, $\sigma_{had}^{peak,0}$, and the number of massless neutrino families, N_ν. Luminosity is measured by counting the Bhabha events at low scattering angles. For that purpose two electromagnetic calorimeters are located close to the beam line on each side of the detector. Bhabhas are identified as two back to back electron showers of about the beam energy in coincidence. The integrated Bhabha cross section varies as $(\cos\theta_{min})^{-2}$ where θ_{min} is the minimum angle of the fiducial phase space. The statistical accuracy is up to now better than 10^{-3}. The knowledge of the luminosity monitor geometrical position is crucial to compete with the statistical error. One needs to know the radial distance to the beam line within 25 μm to match the statistical precision quoted above.

ALEPH has upgraded its luminosity monitor in 1992 with a calorimeter[12] where the absorber is made of tungsten plates and the readout are silicon pads designed with a precision better than 10 μm. DELPHI had a silicon telescope from 1992 to 1993 and installed in 1994 a lead-scintillator-silicon sandwich. L3 has added 1993 a silicon tracker[13] in front of the BGO calorimeter to improve the resolution on θ. Finally OPAL built a device[14] similar to the ALEPH ones, operational in 1993.

At present, the theoretical uncertainty is still the largest error on luminosity. It is evaluated by comparing a first order QED calculation (BHLUMI[16]) with additional leading-log second order calculations[18]. Recent work on progress[19] has allowed to quote an error of 10^{-3}. The overall breakdown of the systematic error for the luminosity[15] is given in the table 2.

Table 2. Systematic errors for the ALEPH Si-Cal, OPAL Si-W, L3 Si-tracker+BGO and DELPHI (the numbers from the new DELPHI device are not yet available) luminosity measurements.

source	ALEPH	OPAL	L3	DELPHI
Exp. syst.	0.074%	0.062%	0.114%	-
Simulation stat.	0.060%	0.037%	0.100%	-
TOTAL exp. error	0.095%	0.072%	0.156%	0.28%
Theor. error	.25%			

4. Z line shape results

The input quantities are the hadronic cross section, the three leptonic cross sections and forward-backward asymmetries, measured at various beam energies by the four LEP experiments[21,22,23,24]. Two fits are performed[1,20] using a Breit-Wigner for the Z and taking into account the photon exchange, the $Z-\gamma$ interference term as well as the photon emission in the initial and final states. The first one is a fit with nine parameters, M_Z, Γ_Z, $\sigma_{had}^{peak,0}$, the leptonic ratios $R_e(\equiv \Gamma_{had}/\Gamma_e)$, R_μ, R_τ and the forward-backward asymmetries A_{FB}^e, A_{FB}^μ and A_{FB}^τ. The results demonstrates the universality of the three leptonic families. Then a five parameter fit is performed assuming universality giving M_Z, Γ_Z, $\sigma_{had}^{peak,0}$, R_ℓ and A_{FB}^ℓ. Table 3 shows the results as well as the number of neutrino families with error from theoretical computations, the leptonic width and $sin^2\theta_W^{eff}$ extracted from the leptonic asymmetry A_{FB}^ℓ.

Table 3. Combined line shape parameters from the LEP experiments.

Quantity	LEP average	χ^2/d.o.f.
M_Z (GeV)	$91.1887 \pm 0.0017 \pm 0.0040_{LEP}$	1.6/3
Γ_Z (GeV)	$2.4976 \pm 0.0027 \pm 0.0027_{LEP}$	2.2/3
$\sigma_{had}^{peak,0}$ (nb)	$41.49 \pm 0.05 \pm 0.10_{Theory}$	4.7/3
R_ℓ	20.795 ± 0.040	3.4/3
$A_{FB}^\ell \times 10^4$	$170 \pm 14 \pm 8$	5.5/3
N_ν	$2.988 \pm 0.010 \pm 0.019_{Theory}$	
Γ_ℓ	83.96 ± 0.18 MeV	
$sin^2\theta_W^{eff}$	0.23107 ± 0.00090	

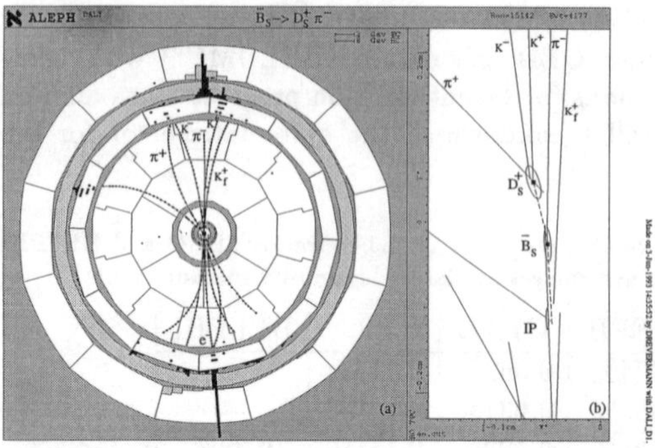

Fig. 3. An example of event with lifetime (with full reconstruction of a B_s decay into D_s in this case).

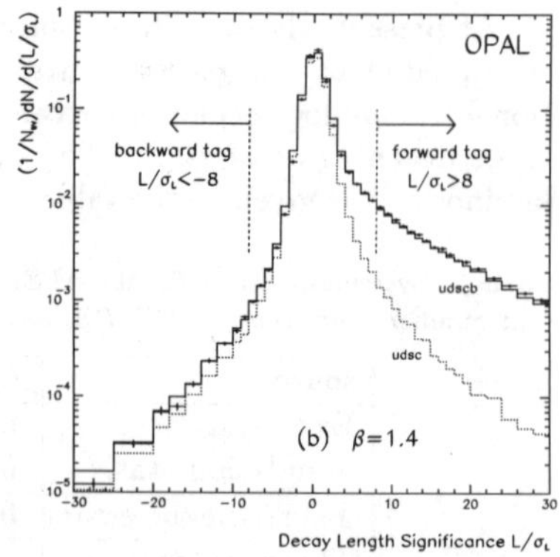

Fig. 4. Decay length significance distribution in OPAL for b and non b events.

5. $\mathbf{b\bar{b}}$ partial width measurement

The $Z \to b\bar{b}$ partial width or $R_b = \Gamma_b/\Gamma_{had}$ is one of the measurable quantities sensitive to the top mass. Several methods have been used : 1) event shape tag, using the fact that tracks from an hemisphere containing a b hadron have an higher invariant mass and are more spherical than light flavours ones ; 2) high p_T lepton tag, based on the semileptonic decays of b hadrons which have on average a higher p_T than those from light quarks ; 3) lifetime tag, exploiting the long B lifetime which leads to many tracks not coming from the Z decay point[25,26,27,28] (see fig. 3). The micro vertex detectors allow to extrapolate the tracks very precisely near the interaction point. One can build an estimator (see fig. 4) which tells what is the probability that an event or an hemisphere contains a bunch of tracks not coming from the primary vertex.

Table 4. R_b measurements using micro-vertex detectors and double tag method. The LEP average includes also results from high p_T lepton and event shape analyses.

Experiment	value	exp. error	modeling error
DELPHI	0.2214	± 0.0020	± 0.0028
ALEPH	0.2187	± 0.0022	± 0.0026
OPAL	0.2171	± 0.0021	± 0.0021
LEP average	0.2192	± 0.0018	

The are two b hadrons per b event which decay almost independently. The lifetime estimator applied on each hemisphere (double tag method) allows **to determine the efficiency and the purity of the b tag from data**, without relying on the b decay tables which are not known with high precision. The most serious background is due to charm events. The results of the LEP experiments have been combined together taking into account correlations and are shown in table 4. It is important to note that R_b is extracted assuming R_c predicted by the Standard Model.

6. Asymmetry measurements

For unpolarized beams the forward-backward charge asymmetry is given by the formula $A^f_{FB} \simeq \frac{3}{4}\mathcal{A}_e\mathcal{A}_f$ where $\mathcal{A}_f \equiv 2g_{Vf}g_{Af}/(g^2_{Vf}+g^2_{Af})$. g_{Vf} and g_{Af} are the vectorial and axial coupling constants in the Standard Model. They can be expressed as $g_{Vf} = I^3_{Lf} - 2Q_f sin^2\theta^{eff}_W$ and $g_{Af} = I^3_{Lf}$ where Q_f is the electric charge and I^3_{Lf} the weak isospin. It is important to note that $\mathcal{A}_f$ is more sensitive to $sin^2\theta^{eff}_W$ for $e^\pm$ than for quarks[29] so the forward-backward asymmetries at LEP measure essentially $sin^2\theta^{eff}_W$ through $\mathcal{A}_e$. Therefore it is convenient to use in the following $sin^2\theta^{eff}_W$ as

$$sin^2\theta^{eff}_W \stackrel{Def}{=} \frac{1}{4}(1-\frac{g_{Ve}}{g_{Ae}}) \tag{1}$$

Several asymmetry measurements are hereafter described. The corresponding values for $sin^2\theta^{eff}_W$ are summarized in table 6.

6.1. τ polarization

The τ polarization can be measured through its decays products. For unpolarized beams it is given by

$$\mathcal{P}_\tau(cos\theta) \simeq -\frac{\mathcal{A}_\tau + \frac{2cos\theta}{1+cos^2\theta}\mathcal{A}_e}{1+\frac{2cos\theta}{1+cos^2\theta}\mathcal{A}_e\mathcal{A}_\tau} \tag{2}$$

where θ is the polar angle. This allows to extract $\mathcal{A}_e$ and $\mathcal{A}_\tau$ separately. As $Z \to \tau^+\tau^-$ partial width is small LEP experiments have developed new analyses[30,31,32,33] using as many channels as possible : $\tau \to \pi\nu_\tau$, $\tau \to e\nu_e\nu_\tau$, $\tau \to \mu\nu_\mu\nu_\tau$, $\tau \to \rho\nu_\tau \to \pi^-\pi^0\nu_\tau$ and $\tau \to a_1\nu_\tau \to \pi^-\pi^+\pi^0\nu_\tau$. The energy spectrum of the charged prong in final states depends upon the τ helicity. For ρ and a_1, where the final state is more complicated, an optimal variable[34,35] was build to extract $\mathcal{P}_\tau$ (see fig. 5). Then $\mathcal{A}_e$ and $\mathcal{A}_\tau$ can be extracted by a fit over $cos\theta$ range. The results show compatibility

Table 5. LEP asymmetries from τ polarization. For $\mathcal{A}_e$ and $\mathcal{A}_\tau$ determination, V-A charged current (i.e. $\xi = -1$) is assumed.

$\mathcal{A}_\tau$	=	0.143 ± 0.010	$sin^2\theta^{eff}_W$	=	0.2320 ± 0.0013
$\mathcal{A}_e$	=	0.135 ± 0.011	$sin^2\theta^{eff}_W$	=	0.2330 ± 0.0014
ξ_ν	=	$-1.006 \pm 0.032 \pm 0.020$ (ALEPH)			

between different decay channels. Moreover $\mathcal{A}_e$ and $\mathcal{A}_\tau$ are found to be compatible ; this assumes maximally parity violating V-A charged current. On the other hand by comparing the two τs of a same event one can also measure the ν_τ helicity, ξ_ν, which is found to be compatible[36,37] with -1. Combined LEP results are shown table 5.

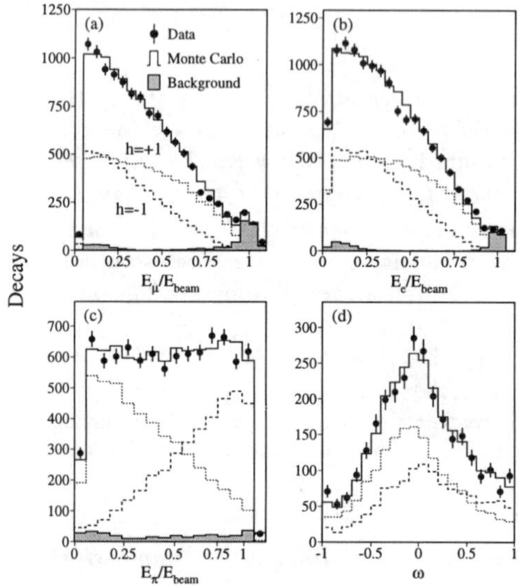

Fig. 5. Extraction of the τ polarizarion in L3 experiment for various channels : a) $\tau \longrightarrow \mu\nu_\mu\nu_\tau$; b) $\tau \longrightarrow e\nu_e\nu_\tau$; c) $\tau \longrightarrow \pi\nu_\tau$; d) $\tau \longrightarrow \rho\nu_\tau$.

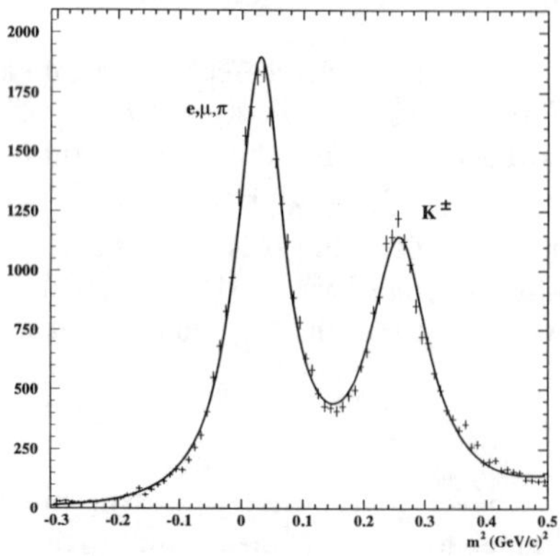

Fig. 6. Charged kaons isolation using DELPHI Cerenkov detector (RICH).

6.2. Quarks asymmetries

It is interesting to measure forward-backward asymmetry for quarks because the asymmetry is higher than for leptons and their statistics is high. But tagging a specific flavor remains difficult, except for b. The measurements performed up to now are : 1) $b\bar{b}$ asymmetry using leptons or lifetime tag ; they have led to precise measurements and have been explained in details[40] ; 2) $c\bar{c}$ asymmetry using fast $D^{*\pm}$ mesons[41,42,43] ; 3) asymmetry measurement with all flavours ; 4) $s\bar{s}$ asymmetry using fast kaons and lambdas.

Measuring the asymmetry of quarks assumes that the final hadrons in one jet keep memory of the original quark charge. This has been suggested by Feynman[38] and proved in neutrino scattering experiments[39]. This allows to define the jet charge Q in one event hemisphere as the mean of the charges of charged tracks, weighted by the projection of their momentum along the event thrust axis to a certain power κ. The difference between the jet charge for the hemisphere pointing into the e^- line direction (Forward) and the other one (Backward), $Q_{FB} = Q_F - Q_B$, can be related to the asymmetries through

$$< Q_{FB} >= \sum_{flavour f} \frac{\Gamma_f}{\Gamma_{had}} \delta_f A_{FB}^f \tag{3}$$

where δ_f is the average difference between the jet charge from the hemisphere containing the quark and the one from the antiquark. The values of δ_f for the five flavours have to be known for this analysis. For this purpose one can build a quantity insensitive to the asymmetry

$$< Q_F Q_B > \simeq -\frac{1}{4} \sum_{flavour f} \frac{\Gamma_f}{\Gamma_{had}} \delta_f^2 \tag{4}$$

which brings a constrain on δ_f.

The inclusive quark asymmetry analysis relies on δ_f from Monte-Carlo hadronisation models except for δ_b. The most important sources of errors are due to strangeness and baryon production in fragmentation. ALEPH[48] and DELPHI[49] have used the jet charge method while OPAL[49] used a different one based on the study of the three most energetic tracks in each hemisphere.

The situation is much simpler in the case of one flavour, let say b. In this case δ_b can be measured from data using $< Q_F Q_B > \simeq -\frac{1}{4}\delta_b^2$. In this case A_{FB}^b can be directly extracted using Eq. (3) and is found[44,45,46] to be $A_{FB}^b = 0.0967 \pm 0.0038$. The combined result (see table 6) is the most precise $sin^2\theta_W^{eff}$ measurement at LEP.

Finally the s quark asymmetry has been investigated by DELPHI using its Cerenkov detector (see fig. 6). High momentum strange particles have a high probability to come from an original s quark. Using also fast Λ, K_S^0 (and using jet charge analysis for this last one) s quark asymmetry has been measured[51] to be $A_{FB}^s = 0.129 \pm 0.027 \pm 0.018$, compatible with the b asymmetry.

Table 6. Summary of $sin^2\theta_W^{eff}$ measurements from LEP and SLD[47].

Method	$sin^2\theta_W^{eff}$	
A_{FB}(leptons)	0.23107 ± 0.00090	
P_τ	0.23200 ± 0.00130	
P_τ^{FB}	0.23300 ± 0.00140	
$A_{FB}^{b\bar{b}}$	0.23268 ± 0.00068	
$A_{FB}^{c\bar{c}}$	0.23100 ± 0.00210	
Q_{FB}	0.23200 ± 0.00160	
LEP average	0.23210 ± 0.00044	χ^2/dof=2.8/5
A_{LR}(SLD)	0.22940 ± 0.00100	
average with SLD	0.23167 ± 0.00040	χ^2/dof=8.9/6

7. Radiative corrections, top quark and Higgs boson masses

The radiative effects at LEP at the Z pole consist of : QED effects (photons emission), loops in Z and γ propagators, vertex diagrams, both sensitive to heavy particles (like top). One can define[20] $\Delta\rho$, Δ_{3Q}, Δr and δ_{vb} by Eq. (1) as :

$$\begin{aligned} M_Z^2 sin^2\theta_W^{eff} cos^2\theta_W^{eff} &= \frac{\pi\alpha(M_Z^2)}{\sqrt{2}G_F(1+\Delta\rho)(1+\Delta_{3Q})} \\ M_W^2(1-\frac{M_W^2}{M_Z^2}) &= \frac{\pi\alpha(M_Z^2)}{\sqrt{2}G_F(1-\Delta r)} \\ \Gamma_\ell &= \frac{G_F M_Z^3}{24\sqrt{2}\pi}(1+\Delta\rho)\left[1+(\frac{g_{V\ell}}{g_{A\ell}})^2\right](1+\frac{3\alpha(M_Z^2)}{4\pi}) \\ \Gamma_b &= \Gamma_d(1+\delta_{vb}) \end{aligned} \quad (5)$$

where G_F is the Fermi constant and $\alpha(M_Z^2)$ is the QED constant evaluated at M_Z^2.

$\Delta\rho$ is related to the isospin breaking in the W and Z propagator loops and varies quadratically with M_t. Δ_{3Q} absorbs the Z mass running $M_Z(q^2)$; it depends on $\ln(M_H)$. Δr reflects corrections to the W propagator. Finally δ_{vb} includes all corrections to the $Z \to b\bar{b}$ vertex[52] and at first order $\delta_{vb} \propto M_t^2$.

The quantities $\Delta\rho$, Δ_{3Q} and δ_{vb} have been determined with LEP data and show good agreement with Standard Model predictions (see table 7). There are 3 unknowns in the Standard model, the strong coupling constant α_s and the two masses M_t and M_H. The value $\alpha_s(M_Z^2) = 0.123 \pm 0.006$ used here comes event shape measurement[53], the top mass is taken to be[54] $M_t = 174 \pm 16$ GeV and $M_H = 300^{+700}_{-240}$ GeV. In order to show how others models take place among these measurements, $N_c = 4$ technicolor predicts $\Delta_{3Q} = -.0300$ which is ruled out by LEP data.

Table 7. Radiative corrections quantities measured by the LEP experiments (assuming $\alpha_s = 0.123 \pm 0.006$) and compared with Standard Model predictions assuming $M_t = 174 \pm 16$ GeV and $M_H = 300^{+700}_{-240}$ GeV.

Quantity	measured value	SM prediction
$\Delta\rho$	0.0033 ± 0.0019	$0.0044^{+0.0020}_{-0.0016}$
Δ_{3Q}	-0.0039 ± 0.0025	$-0.0060^{+0.0008}_{-0.0017}$
δ_v	-0.0097 ± 0.0084	$-0.0175^{+0.0029}_{-0.0027}$

The top mass can be extracted from electroweak LEP data together with the measurement of $sin^2\theta_W^{eff}$=0.2253±0.0047 from neutrino scattering experiments[55,56,57] and the W mass measurement[58] $M_W = 80.23 \pm 0.18$ GeV. The result of the fit is presented for several Higgs masses values in table 8 and shows good agreement with the direct top mass measurement recently performed at Fermilab.

The χ^2 of several fits are now shown (fig. 9) as a function of M_H. On these plots we compare the fit performed with the measured Γ_b (solid line) and without it (dashed line). On fig. 9 a "indirect data" means that this fit is not using the CDF top mass

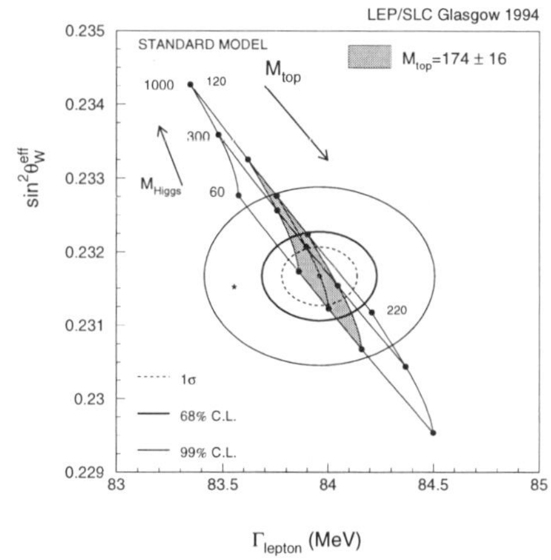

Fig. 7. Standard Model predictions[52] function of M_t and M_{Higgs}. The star represents the Born level calculation.

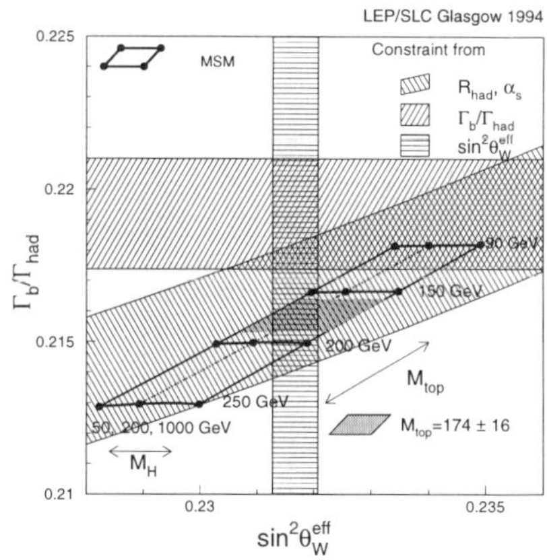

Fig. 8. Standard Model predictions for R_b and $sin^2\theta_W^{eff}$ (function of M_t and M_{Higgs}) compared to data.

direct measurement. With the use of CDF top mass (fig. 9 b) the agreement between theory and data is much better. The data seems to prefer a light Higgs around 90 GeV but the slow χ^2 variations doesn't give significant upper limit for M_H. When adding the direct Higgs mass limit ($M_H > 63$ GeV from LEP experiments) the behaviour of the χ^2 is not changed at high Higgs masses, so top mass doesn't bring any information about Higgs mass.

8. Outlook

In 1994 LEP has delivered about 66 pb^{-1} per experiment at the Z peak. LEP will run 1995 around the Z mass with substantial increase of luminosity expected (bunch trains). Associated with precision energy measurement by resonant depolarization, it seems possible to measure Γ_Z up to 2 MeV. Data all over the world seems to support the Standard model very well. Polarized electron beams is on the other hand a powerful tool to measure $sin^2\theta_W^{eff}$ with very high precision ($\approx .0001$). The top mass needs to be confirmed. It seems possible to measure it at Fermilab in the future with a precision of 5 GeV. In addition LEP II program expects to give the W mass measurement up to 30 MeV. LEP II is the next place for Higgs search before LHC comes. If it is not discovered directly there, all the possible improvements in W mass, Γ_ℓ, $sin^2\theta_W^{eff}$ and M_t will nevertheless be big enough to begin to set a firm limit on Higgs mass, or to give clear clues for new physics.

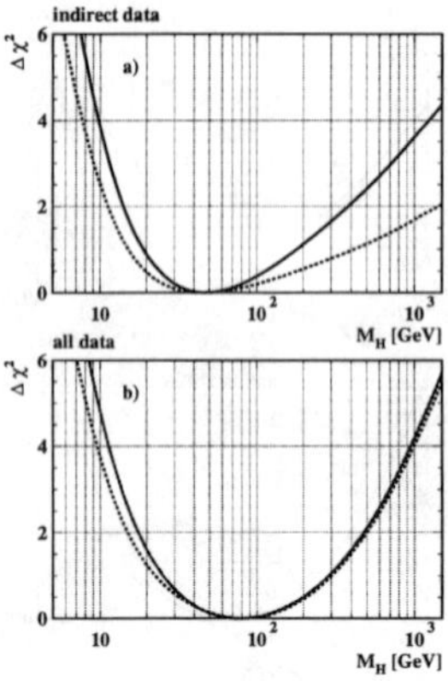

Fig. 9. χ^2 shape from global (M_t, M_H) fit. The data are purely electroweak (see text for explanations).

Table 8. Top mass predictions from LEP data for three Higgs mass values.

$M_t = 159 \pm 12$ GeV $\chi^2 = 13.3$ $M_H = 60$ GeV
$M_t = 178 \pm 11$ GeV $\chi^2 = 15.1$ $M_H = 300$ GeV
$M_t = 196 \pm 10$ GeV $\chi^2 = 16.9$ $M_H = 1000$ GeV

9. Acknowledgements

I am very indebted to Alain Blondel who provided to me the basic material for the talk, a lot of useful explanations and read the manuscript. Bolek Pietrzyk helped me understanding the subtleties of Bhabha physics and simulation. I am also grateful to Dieter Schlatter who helped me outlining my talk and this writing.

10. References

1. The LEP collaborations and the LEP Electroweak working group, Combined Preliminary Data on Z Parameters from the LEP Experiments and constraints on the Standard Model. CERN-PPE/94-187.
2. A.A.Sokolov and I.M.Ternov, Sov. Phys. Doklady,8 (1964)1203.
3. A.A.Zholentz et al, Phys. Lett. 96B (1980) 214.
4. A.S.Artamonov et al, Phys. Lett. 118B (1982) 225.
5. D.P.Barber et al, Phys. Lett. 135B (1984) 498.
6. W.W.McKay et al, Phys. Rev. D29 (1984) 2483.
7. L.Arnaudon et al, Phys. Letters B284 (1992) 431.
8. For reviews, see B.W.Montague, Phys. Rep. 113 (1984) ; A.Blondel CERN-PPE/93-125 (1993) ; M.Böge, DESY 94-087.
9. R.Assmann et al, Accurate determination of the LEP beam Energy by resonant depolarization CERN-SL/94-71(BI),
   Polarization studies at LEP in 1993, CERN-SL/94-08(AP).
10. L.Arnaudon et al, CERN-PPE/92-125 (1992), CERN-SL/92-37(DI) ;
    Phys. Lett. B307 (1993) 187 ;
    L.Arnaudon et al, CERN SL/93-21 (DI), April 1993.

11. G.Fischer and A.Hofmann, in second workshop on LEP performance, J. Poole ed., CERN-SL/92-29 (DI) (1992), 337.
12. B. Bloch, EPS conf. (July 1993) ; ALEPH note 93-149 ; D.Bederède et al, SiCAL - A High precision Luminosity Calorimeter for ALEPH, to appear in Nucl. Inst. Meth.
13. M.Merk, talk given at XXIX rencontres de Moriond, Meribel (1994).
14. OPAL Coll., preliminary result ICHEP 1994, OPAL PN-142.
15. B. Pietrzyk, High precision measurements of the luminosity at LEP, to be published in the Proceedings of the Tennessee International Symposium on radiative Corrections, Gatlinburg, TN, USA, June 27-July 1, 1994.
16. S.Jadach, E.Richter-Was, Z.Was, B.F.L.Ward, Phys. Lett. B268 (1991) 253 ; Comput. Phys. Commun. 70 (1992) 305 ; S.Jadach and B.F.L.Ward, Phys. Rev. D40 (1989) 3582.
17. M.Böhm, A.Denner, W.Hollik, Nucl. Phys. B306 (1988) 687 ; F.A.Berends, R.Kleiss, W.Hollik, Nucl. Phys. B304 (1988) p.712 ; Computer program `BABAMC` by R.Kleiss.
18. W.Beenakker and B.Pietrzyk, Phys. Lett. B296 (1992) 296 ; Phys. Lett. B304 (1993) 366.
19. B.F.L.Ward, S.Jadach, E.Richter-Was and Z.Was, talk given at Glasgow 1994, UTHEP-94-0801.
20. A. Blondel, Precision electroweak physics at LEP, CERN-PPE/94-133.
21. ALEPH Coll., Z. Phys C62 (1994) 539. updated results for ICHEP 1994, GLS0557, GLS0558.
22. DELPHI Coll., Nucl. Phys. B417 (1994) 3 ; Nucl. Phys. B418 (1994) 403 ; ICHEP 1994 GLS0301.
23. L3 Coll., CERN-PPE/94-45 (1994) submitted to Z. Phys C ; Updated results for ICHEP 1994, L3 note 1620.
24. OPAL Coll., Z. Phys. C61 (1994) 19. Updated results for ICHEP 1994, OPAL physics note PN-143.
25. D.Brown, MPI preprint MPI-PhE 93-25 Munich (1993).
26. ALEPH Coll., Phys. Lett. B313 (1993) 535.
27. DELPHI Coll., ICHEP 1994 (Glasgow), GLS0301b, GLS0229.
28. OPAL Coll., CERN-PPE/94-106 (1994).
29. M. Böhm and W. Hollik, CERN 89-08 ("Yellow book" #1), p203.
30. ALEPH Coll., Z. Phys. C 59 (1993) 369 ; update for ICHEP 1994, GLS0551.
31. DELPHI Coll., Z. Phys. C 55 (1992) 555 ; update for ICHEP 1994, GLS0207.
32. L3 Coll., Phys. Rep. 236 (1993).
33. OPAL Coll., Phys. Lett. B266 (1991) 201 ; CERN-PPE/94-120 (1994).
34. A. Rougé, Z. Phys. C48 (1990) 45.
35. M. Davier at al., Phys.Lett. B306 (1993) 411.
36. ALEPH Coll., Phys. Lett. B321(1994) 168. J. Raab (ALEPH Coll.), Talk given at "Tau lepton physics workshop in Montreux" Sept. 1994, to be published in Phys. Lett. B
37. ARGUS Coll., Phys. Lett. B 337 (1994) 383.

38. R.D.Field and R.P.Feynman, Nucl. Phys. B136 (1978) 1.
39. J.P.Berge et al, Nucl. Phys. B184 (1981) 13.
40. R. Clare, Heavy Quark Physics at LEP and SLC, to be published in the Proceedings of the Tennessee International Symposium on radiative Corrections, Gatlinburg, TN, USA, June 27-July 1, 1994.
41. ALEPH Coll., Z. Phys. C62 (1994) 1 ; ICHEP 1994, GLS0562.
42. DELPHI Coll., Z. Phys. C59 (1993) 533 ; ICHEP 1994, GLS0231.
43. OPAL Coll., Z. Phys. C60 (1993) 601 ; OPAL Phys. note PN125.
44. ALEPH Coll., CERN-PPE/94-84 (1994).
45. DELPHI Coll., ICHEP 1994, GLS0230.
46. OPAL Coll., ICHEP 1994, OPAL phys. Note PN127.
47. SLD Coll., Phys. Rev. Lett. 73 (1994) 25.
48. ALEPH Coll., Phys. Lett. B259 (1991) 377 ; ALEPH-Note 93-044.
49. DELPHI Coll., Phys. Lett. 277B (1992) 231.
50. OPAL Coll., Phys. Lett. 294B (1992) 436.
51. DELPHI Coll., ICHEP 1994, GLS0232.
52. A. Blondel and C. Verzegnassi, Phys. Lett. B311 (1993) 346.
53. S. Bethke, Proceedings of the Linear Collider Workshop, Hawaii, April 1993 ; S. Banerjee, α_s measurements from event shape variables using Z data, EPS conference, Marseille, July 22-28, 1993, also L3 note 1460.
54. CDF Coll., Phys. Rev. Lett. 73 (1994) 225.
55. A.Blondel et al, Z. Phys. C45 (1990) 361.
56. J.V.Allaby et al, (CHARM Coll.), Phys. Lett. B177 (1986) 446, Z. Phys. C36 (1987) 611.
57. C.G.Arroyo et al, (CCFR Coll.), Phys. Rev. Lett. 72 (1994) 3452.
58. M.Demarteau et al, Combining W mass measurements, CDF/PHYS/CDF/PUBLIC/2552, D0NOTE 2115 (1994).

Non-Minimal Higgs Sectors: The Decoupling Limit and its Phenomenological Implications

HOWARD E. HABER

Santa Cruz Institute for Particle Physics
University of California, Santa Cruz, CA 95064

Abstract

In models with a non-minimal Higgs sector, a decoupling limit can be defined. In this limit, the masses of all the physical Higgs states are large (compared to the scale of electroweak symmetry breaking) except for one neutral CP-even Higgs scalar, whose properties are indistinguishable from the Higgs boson of the minimal Standard Model. The decoupling limit of the most general CP-conserving two-Higgs doublet model is formulated. Detection of evidence for a non-minimal Higgs sector at future colliders in the decoupling limit may present a formidable challenge for future Higgs searches.

1. Introduction

With the recent "discovery" of the top quark, the only missing piece of the Standard Model is the Higgs boson. The Standard Model posits the existence of one complex weak doublet (with hypercharge $Y = 1$) of elementary scalar (Higgs) fields. The dynamics of these fields is assumed to trigger electroweak symmetry breaking. Three Goldstone bosons are absorbed by the $W^{\pm}$ and Z leaving one CP-even neutral Higgs scalar to be discovered.

Despite the enormous success of the Standard Model, the existing data sheds little light on the spectrum and dynamics of the electroweak symmetry breaking sector. The scalar spectrum could be minimal (as in the Standard Model) or non-minimal, with a rich spectrum of scalar states (including perhaps bound states of higher spin). The dynamics could involve either weakly interacting or strongly interacting forces. Many examples displaying each of the above features have been studied in the literature; comprehensive reviews can be found is refs. 1 and 2. In this paper, I will assume that electroweak symmetry breaking is a result of the dynamics of a weakly-coupled Higgs sector. Such a theory may be technically natural if embedded in a theory of low-energy supersymmetry. However, the results obtained in this work do not necessarily require the existence of low-energy supersymmetry.

In this paper, I shall pose the following questions. Assume that a scalar state (*i.e.*, a candidate for the Higgs boson) is discovered in a future collider experiment. What are the expectations for its properties? Will it resemble the Higgs boson of the minimal Standard Model or will it possess some distinguishing trait? If the properties of this scalar state are difficult to distinguish from the Standard Model Higgs boson, what are the requirements of future collider experiments for detecting the existence (or non-existence) of a non-minimal Higgs sector? Some of these questions have also been addressed by other authors; see *e.g.*, refs. 3 and 4.

These questions would be moot if the experiment that first discovers the lightest Higgs scalar also discovers a second scalar state. For example, in a theory with a light CP-even scalar (h^0) and a light CP-odd scalar (A^0), both states would be produced in e^+e^- collisions via s-channel Z-exchange ($e^+e^- \to Z \to h^0 A^0$) if kinematically allowed. The simultaneous discovery of h^0 and A^0 would clearly indicate the existence of a non-minimal Higgs sector. In this paper, I shall not consider such a scenario. As we shall see, the case where only one light scalar state is initially discovered may present a formidable challenge to unraveling the underlying structure of the electroweak symmetry breaking sector.

For simplicity, I focus in this paper on the (CP-conserving) two-Higgs doublet model. In this model, the physical scalar states consist of a charged Higgs pair ($H^\pm$), two CP-even scalars (h^0 and H^0, with $m_{h^0} \leq m_{H^0}$) and one CP-odd scalar (A^0). The ultimate conclusions of this paper will survive in models with more complicated scalar sectors. Following the discussion above, the working hypothesis of this paper is that h^0, assumed to be the lightest scalar state, will be the first Higgs boson to be discovered. Moreover, the mass gap between h^0 and the heavier scalars is assumed to be sufficiently large so that the initial experiments which can detect h^0 will not have sufficient energy and luminosity to initially discover any of the heavier scalar states.

How will h^0 be discovered and where? Present LEP bounds[5] imply that $m_{h^0} \gtrsim 60$ GeV. This bound will be improved by LEP-II,[6] which will be sensitive to Higgs masses up to roughly $\sqrt{s} - m_Z - 10$ GeV. The LEP search is based on $e^+e^- \to Z \to Zh^0$ where one of the two Z's is on-shell and the other is off-shell. At hadron colliders, an upgraded Tevatron with an integrated luminosity of 10 fb^{-1} can begin to explore the intermediate-mass Higgs regime[7] ($80 \lesssim m_{h^0} \lesssim 130$ GeV). The Higgs search at the LHC will significantly extend the Higgs search to higher masses[8] (although the intermediate mass regime still presents some significant difficulties for the LHC detector collaborations). The dominant mechanism for Higgs production at hadron colliders is via gg-fusion through a top-quark loop. If $m_{h^0} > 2m_Z$, the "gold-plated" detection mode is $h^0 \to ZZ$; each Z subsequently decays leptonically, $Z \to \ell^+\ell^-$ (for $m_{h^0} \gtrsim 130$ GeV, $h^0 \to ZZ^*$, where Z^* is off-shell, provides a viable signature). Other decay modes are required in the case of the intermediate mass Higgs (for recent reviews, see ref. 9). At a future e^+e^- linear collider (NLC), the Higgs mass reach of LEP-II will be extended.[10] In addition, with increasing $\sqrt{s}$, Higgs boson production via W^+W^- fusion begins to be the dominant production process. Finally, one novel possibility is to run the NLC in a $\gamma\gamma$-collider mode. In this mode, Higgs production via $\gamma\gamma$-fusion (which is typically dominated by a W^+W^- and/or a $t\bar{t}$ loop) may be detectable, depending on the particular Higgs final state decay.[11,12] Note that almost all of the Higgs search techniques outlined above involve the h^0ZZ (and in some cases the $h^0W^+W^-$) vertex. In a few cases, it is the $h^0t\bar{t}$ vertex (and possibly the $h^0b\bar{b}$ vertex) that plays the key role. These observations are relevant for the phenomenological considerations of this paper.

In section 2, I review the Higgs sector of the minimal supersymmetric extension of the Standard Model (MSSM). In this context, I discuss why one might expect that h^0 is the lightest scalar whose properties are nearly identical to that of the Standard Model Higgs boson. In section 3, I place the results of section 2 in a more general context. I define the "decoupling limit" of the general two-Higgs doublet model; in this limit, h^0 is indistinguishable from the Standard Model Higgs boson. In section 4, I discuss the phenomenological challenges of the decoupling limit for the Higgs search at future colliders. After briefly mentioning the prospects for non-minimal Higgs detection at the LHC, I consider in more detail the prospects for the discovery of the non-minimal Higgs sector at the NLC. Conclusions are presented in section 5.

2. The Higgs Sector of the MSSM—A Brief Review

In this section, I provide a very brief review of the Higgs sector of the minimal supersymmetric extension of the Standard Model (MSSM).[13] let us consider the MSSM Higgs potential at tree-level which depends on two complex doublet scalar fields H_1 and H_2 of hypercharge -1 and $+1$, respectively :

$$\begin{aligned} \mathcal{V} &= m_{11}^2|H_1|^2 + m_{22}^2|H_2|^2 - m_{12}^2(\epsilon_{ij}H_1^iH_2^j + \text{h.c.}) \\ &\quad + \tfrac{1}{8}(g^2+g'^2)\left(|H_1|^2-|H_2|^2\right)^2 + \tfrac{1}{2}g^2|H_1^*H_2|^2\,, \end{aligned} \tag{2.1}$$

where $m_{ii}^2 \equiv |\mu|^2 + m_i^2$ $(i=1,2)$. The parameters m_1^2, m_2^2 and m_{12}^2 are soft-supersymmetry-breaking parameters, μ is the Higgs superfield mass parameter, and g and g' are the SU(2)$\times$U(1) gauge couplings.

The three parameters m_{11}^2, m_{22}^2 and m_{12}^2 of the Higgs potential can re-expressed in terms of the two Higgs vacuum expectation values, $\langle H_i^0\rangle \equiv v_i/\sqrt{2}$ and one physical Higgs mass. One is free to choose the phases of the Higgs fields such that v_1 and v_2 are positive. Then, m_{12}^2 must be positive, in which case it follows from eq. (2.1) that

$$m_{A^0}^2 = \frac{m_{12}^2}{\sin\beta\cos\beta}\,. \tag{2.2}$$

Note that $m_W^2 = \frac{1}{4}g^2(v_1^2+v_2^2)$, which fixes the magnitude $v_1^2+v_2^2 = (246\ \mathrm{GeV})^2$. This leaves two parameters which determine all the Higgs sector masses and couplings: m_{A^0} and $\tan\beta \equiv v_2/v_1$. The charged Higgs mass is given by

$$m_{H^\pm}^2 = m_W^2 + m_{A^0}^2\,. \tag{2.3}$$

The neutral CP-even Higgs bosons, H^0 and h^0, are obtained by diagonalizing a 2×2 mass matrix. The eigenstates are

$$\begin{aligned} H^0 &= (\sqrt{2}\,\mathrm{Re}\,H_1^0 - v_1)\cos\alpha + (\sqrt{2}\,\mathrm{Re}\,H_2^0 - v_2)\sin\alpha \\ h^0 &= -(\sqrt{2}\,\mathrm{Re}\,H_1^0 - v_1)\sin\alpha + (\sqrt{2}\,\mathrm{Re}\,H_2^0 - v_2)\cos\alpha \end{aligned} \tag{2.4}$$

which defines the CP-even Higgs mixing angle α. The corresponding CP-even Higgs mass eigenvalues are

$$m_{H^0,\,h^0}^2 = \frac{1}{2}\left(m_{A^0}^2 + m_Z^2 \pm \sqrt{(m_{A^0}^2 + m_Z^2)^2 - 4m_Z^2 m_{A^0}^2\cos^2 2\beta}\right)\,, \tag{2.5}$$

where by definition, $m_{h^0} \le m_{H^0}$. Explicit formulae for α can also be derived. Here, I shall note one particularly useful relation

$$\cos^2(\beta-\alpha) = \frac{m_{h^0}^2(m_Z^2 - m_{h^0}^2)}{m_{A^0}^2(m_{H^0}^2 - m_{h^0}^2)}\,. \tag{2.6}$$

Consider the limit where $m_{A^0} \gg m_Z$. Then, from the above formulae,

$$\begin{aligned} m_{h^0}^2 &\simeq m_Z^2\cos^2 2\beta\,, \\ m_{H^0}^2 &\simeq m_{A^0}^2 + m_Z^2\sin^2 2\beta\,, \\ m_{H^\pm}^2 &= m_{A^0}^2 + m_W^2\,, \\ \cos^2(\beta-\alpha) &\simeq \frac{m_Z^4\sin^2 4\beta}{4m_{A^0}^4}\,. \end{aligned} \tag{2.7}$$

Two consequences are immediately apparent. First, $m_{A^0} \simeq m_{H^0} \simeq m_{H^\pm}$, up to corrections of $\mathcal{O}(m_Z^2/m_{A^0})$. Second, $\cos(\beta-\alpha) = 0$ up to corrections of $\mathcal{O}(m_Z^2/m_{A^0}^2)$.

It is known that one-loop radiative corrections have a significant impact on the MSSM Higgs sector.[14] Nevertheless, the conclusions just stated are not modified. For example, in the limit where $m_{A^0} \gg m_Z$ and $m_Z \ll m_t \ll M_{\tilde{t}}$, the one-loop radiatively corrected Higgs squared masses are[15]

$$
\begin{aligned}
m_{h^0}^2 &\simeq m_Z^2 \cos^2 2\beta + \frac{3g^2}{8\pi^2 m_W^2}\left[m_t^4 + \tfrac{1}{2} m_t^2 m_Z^2 \cos 2\beta\right] \ln\left(\frac{M_{\tilde{t}}^2}{m_t^2}\right), \\
m_{H^0}^2 &\simeq m_{A^0}^2 + m_Z^2 \sin^2 2\beta + \frac{3g^2 \cos^2\beta}{8\pi^2 m_W^2}\left[\frac{m_t^4}{\sin^2\beta} - m_t^2 m_Z^2\right] \ln\left(\frac{M_{\tilde{t}}^2}{m_t^2}\right), \\
m_{H^\pm}^2 &\simeq m_{A^0}^2 + m_W^2 + \frac{3g^2}{32\pi^2}\left[\frac{2m_t^2 m_b^2}{m_W^2 \sin^2\beta \cos^2\beta} - \frac{m_t^2}{\sin^2\beta} - \frac{m_b^2}{\cos^2\beta}\right] \ln\left(\frac{M_{\tilde{t}}^2}{m_t^2}\right).
\end{aligned}
\tag{2.8}
$$

The formula for $m_{h^0}^2$ exhibits the importance of the one-loop radiative corrections. The tree-level upper bound, $m_{h^0} \leq m_Z$ is significantly modified by terms that are enhanced for large values of m_t and $M_{\tilde{t}}$. Nevertheless, the numerical value of the radiative corrections to the squared masses is never greater than $\mathcal{O}(m_Z^2)$. Thus, in the limit of $m_{A^0} \gg m_Z$, one again finds that $m_{A^0} \simeq m_{H^0} \simeq m_{H^\pm}$, up to corrections of $\mathcal{O}(m_Z^2/m_{A^0})$. One can also show that $\cos(\beta - \alpha) = \mathcal{O}(m_Z^2/m_{A^0}^2)$ as before.

The phenomenological implications of these results may be discerned by reviewing the coupling strengths of the Higgs bosons to Standard Model particles (gauge bosons, quarks and leptons) in the two-Higgs doublet model. The coupling of h^0 and H^0 to vector boson pairs or vector-scalar boson final states is proportional to either $\sin(\beta - \alpha)$ or $\cos(\beta - \alpha)$ as indicated below.[1,13]

$\cos(\beta - \alpha)$	$\sin(\beta - \alpha)$
$H^0 W^+ W^-$	$h^0 W^+ W^-$
$H^0 ZZ$	$h^0 ZZ$
$ZA^0 h^0$	$ZA^0 H^0$
$W^\pm H^\mp h^0$	$W^\pm H^\mp H^0$
$ZW^\pm H^\mp h^0$	$ZW^\pm H^\mp H^0$
$\gamma W^\pm H^\mp h^0$	$\gamma W^\pm H^\mp H^0$

Note in particular that *all* vertices in the theory that contain at least one vector boson and *exactly one* heavy Higgs boson state (H^0, A^0 or $H^\pm$) is proportional to $\cos(\beta - \alpha)$. This can be understood as a consequence of unitarity sum rules which must be satisfied by the tree-level amplitudes of the theory.[16]

In models with non-minimal Higgs sectors, the Higgs couplings to quarks and leptons are model-dependent. Typically, one imposes constraints on the Higgs-fermion interaction in order to avoid Higgs mediated tree-level flavor changing neutral currents. For example, in the MSSM, one Higgs doublet couples exclusively to down-type fermions and the second Higgs doublet couples exclusively to up-type fermions. In this model, the couplings of the neutral CP-even Higgs bosons to $f\bar{f}$ relative to the Standard Model value are given by (using 3rd family notation)

$$\begin{aligned}
h^0 b\bar{b}: \qquad & -\frac{\sin\alpha}{\cos\beta} = \sin(\beta-\alpha) - \tan\beta\cos(\beta-\alpha)\,, \\
h^0 t\bar{t}: \qquad & \frac{\cos\alpha}{\sin\beta} = \sin(\beta-\alpha) + \cot\beta\cos(\beta-\alpha)\,, \\
H^0 b\bar{b}: \qquad & \frac{\cos\alpha}{\cos\beta} = \cos(\beta-\alpha) + \tan\beta\sin(\beta-\alpha)\,, \\
H^0 t\bar{t}: \qquad & \frac{\sin\alpha}{\sin\beta} = \cos(\beta-\alpha) - \cot\beta\sin(\beta-\alpha)\,.
\end{aligned} \tag{2.9}$$

In contrast to the Higgs couplings to vector bosons, none of the couplings in eq. (2.9) vanish when $\cos(\beta-\alpha)=0$. This is a model-independent feature of the Higgs couplings to fermions. One finds a similar behavior for the Higgs self-couplings. Namely, the various Higgs self-couplings are model-dependent since they depend on the form of the scalar potential. Nevertheless, one can show that the Higgs self-couplings do not vanish when $\cos(\beta-\alpha)=0$.

The significance of $\cos(\beta-\alpha)=0$ is now evident: in this limit, couplings of h^0 to gauge boson pairs and fermion pairs are identical to the couplings of the Higgs boson in the minimal Standard Model*. More precisely, in the limit of $m_{A^0} \gg m_Z$, the effects of the heavy Higgs states ($H^\pm$, H^0 and A^0) decouple, and the low-energy effective scalar sector is indistinguishable from that of the minimal Standard Model.

In the MSSM, the decoupling regime ($m_{A^0} \gg m_Z$) sets in rather quickly once m_{A^0} is taken above m_Z. Although m_{A^0} is a free parameter of the MSSM, its origin is intimately connected to the scale of supersymmetry breaking. From eq. (2.2), we see that $m_{A^0}^2$ is proportional to the soft-supersymmetry breaking parameter m_{12}^2. Generically, one would expect that m_{A^0} is of the same order as a typical soft-supersymmetry-breaking mass parameter. In supergravity based models, $m_{12}^2 \equiv B\mu$, where B is a soft-supersymmetry-breaking parameter. Models of this type with universal soft-supersymmetry-breaking scalar masses at the Planck scale tend to yield values of m_{A^0} that typically lie above m_Z.[17] In such approaches, one would expect the couplings of h^0 to be nearly identical to those of the Standard Model Higgs boson.

$\star$ Likewise, the $h^0h^0h^0$ and $h^0h^0h^0h^0$ couplings also reduce to their Standard Model values when $\cos(\beta-\alpha)=0$.

3. Decoupling Properties of the Two-Higgs Doublet Model[18]

The decoupling properties of the MSSM Higgs sector are not special to supersymmetry. Rather, they are a generic feature of non-minimal Higgs sectors.[19] In this section, I demonstrate this assertion in the case of the most general CP-conserving two-Higgs doublet model. Let Φ_1 and Φ_2 denote two complex $Y = 1$, $\mathrm{SU}(2)_L$ doublet scalar fields[†] The most general gauge invariant scalar potential is given by[‡]

$$\begin{aligned}\mathcal{V} &= m_{11}^2\Phi_1^\dagger\Phi_1 + m_{22}^2\Phi_2^\dagger\Phi_2 - [m_{12}^2\Phi_1^\dagger\Phi_2 + \mathrm{h.c.}] \\ &\quad + \tfrac{1}{2}\lambda_1(\Phi_1^\dagger\Phi_1)^2 + \tfrac{1}{2}\lambda_2(\Phi_2^\dagger\Phi_2)^2 + \lambda_3(\Phi_1^\dagger\Phi_1)(\Phi_2^\dagger\Phi_2) + \lambda_4(\Phi_1^\dagger\Phi_2)(\Phi_2^\dagger\Phi_1) \\ &\quad + \left\{\tfrac{1}{2}\lambda_5(\Phi_1^\dagger\Phi_2)^2 + \left[\lambda_6(\Phi_1^\dagger\Phi_1) + \lambda_7(\Phi_2^\dagger\Phi_2)\right]\Phi_1^\dagger\Phi_2 + \mathrm{h.c.}\right\}\,.\end{aligned} \tag{3.1}$$

In principle, m_{12}^2, λ_5, λ_6 and λ_7 can be complex. However, I shall ignore the possibility of CP-violating effects in the Higgs sector by choosing all coefficients in eq. (3.1) to be real. The scalar fields will develop non-zero vacuum expectation values if the mass matrix m_{ij}^2 has at least one negative eigenvalue. Imposing CP invariance and $\mathrm{U}(1)_{\mathrm{EM}}$ gauge symmetry, the minimum of the potential is

$$\langle\Phi_1\rangle = \frac{1}{\sqrt{2}}\begin{pmatrix}0\\ v_1\end{pmatrix}, \qquad \langle\Phi_2\rangle = \frac{1}{\sqrt{2}}\begin{pmatrix}0\\ v_2\end{pmatrix}, \tag{3.2}$$

where the v_i can be chosen to be real and positive. It is convenient to introduce the following notation:

$$v^2 \equiv v_1^2 + v_2^2 = \frac{4m_W^2}{g^2}, \qquad t_\beta \equiv \tan\beta \equiv \frac{v_2}{v_1}. \tag{3.3}$$

The mass parameters m_{11}^2 and m_{22}^2 can be eliminated by minimizing the scalar potential. The resulting squared masses for the CP-odd and charged Higgs states are

---

† In terms of the $Y = +1$ fields of the previous section, $H_1^i = \epsilon_{ij}\Phi_1^{j\,*}$ and $H_2 = \Phi_2$.

‡ In most discussions of two Higgs doublet models, the terms proportional to λ_6 and λ_7 are absent at tree-level. This can be achieved by imposing a discrete symmetry $\Phi_1 \to -\Phi_1$ on the model. Such a symmetry would also require $m_{12} = 0$ unless we allow a soft violation of this discrete symmetry by dimension-two terms. This latter requirement is sufficient to guarantee the absence of Higgs mediated tree level flavor changing neutral currents.

$$m_{A^0}^2 = \frac{m_{12}^2}{s_\beta c_\beta} - \tfrac{1}{2}v^2(2\lambda_5 + \lambda_6 t_\beta^{-1} + \lambda_7 t_\beta)\,, \tag{3.4}$$

$$m_{H^\pm}^2 = m_{A^0}^2 + \tfrac{1}{2}v^2(\lambda_5 - \lambda_4)\,, \tag{3.5}$$

where $s_\beta \equiv \sin\beta$ and $c_\beta \equiv \cos\beta$. The two CP-even Higgs states mix according to the following squared mass matrix:

$$\mathcal{M}^2 = m_{A^0}^2 \begin{pmatrix} s_\beta^2 & -s_\beta c_\beta \\ -s_\beta c_\beta & c_\beta^2 \end{pmatrix} + v^2 \begin{pmatrix} \mathcal{M}_{11}^2 & \mathcal{M}_{12}^2 \\ \mathcal{M}_{12}^2 & \mathcal{M}_{22}^2 \end{pmatrix}\,, \tag{3.6}$$

where

$$\begin{pmatrix} \mathcal{M}_{11}^2 & \mathcal{M}_{12}^2 \\ \mathcal{M}_{12}^2 & \mathcal{M}_{22}^2 \end{pmatrix} \equiv \begin{pmatrix} \lambda_1 c_\beta^2 + 2\lambda_6 s_\beta c_\beta + \lambda_5 s_\beta^2 & (\lambda_3+\lambda_4)s_\beta c_\beta + \lambda_6 c_\beta^2 + \lambda_7 s_\beta^2 \\ (\lambda_3+\lambda_4)s_\beta c_\beta + \lambda_6 c_\beta^2 + \lambda_7 s_\beta^2 & \lambda_2 s_\beta^2 + 2\lambda_7 s_\beta c_\beta + \lambda_5 c_\beta^2 \end{pmatrix} \tag{3.7}$$

It is convenient to define four squared mass combinations:

$$\begin{aligned} m_L^2 &\equiv \mathcal{M}_{11}^2 \cos^2\beta + \mathcal{M}_{22}^2 \sin^2\beta + \mathcal{M}_{12}^2 \sin 2\beta\,, \\ m_D^2 &\equiv \left(\mathcal{M}_{11}^2 \mathcal{M}_{22}^2 - \mathcal{M}_{12}^4\right)^{1/2}\,, \\ m_T^2 &\equiv \mathcal{M}_{11}^2 + \mathcal{M}_{22}^2\,, \\ m_S^2 &\equiv m_{A^0}^2 + m_T^2\,. \end{aligned} \tag{3.8}$$

In terms of these quantities,

$$m_{H^0,h^0}^2 = \tfrac{1}{2}\left[m_S^2 \pm \sqrt{m_S^4 - 4m_{A^0}^2 m_L^2 - 4m_D^4}\,\right]\,, \tag{3.9}$$

and

$$\cos^2(\beta-\alpha) = \frac{m_L^2 - m_{h^0}^2}{m_{H^0}^2 - m_{h^0}^2}\,. \tag{3.10}$$

Suppose that all the Higgs self-coupling constants λ_i are held fixed such that $\lambda_i \lesssim 1$, while taking $m_{A^0}^2 \gg \lambda_i v^2$. Then the $\mathcal{M}_{ij}^2 \sim \mathcal{O}(v^2)$, and it follows that:

$$m_{h^0} \simeq m_L\,, \qquad m_{H^0} \simeq m_{A^0} \simeq m_{H^\pm}\,, \tag{3.11}$$

and

$$\cos^2(\beta-\alpha) \simeq \frac{m_L^2(m_T^2 - m_L^2) - m_D^4}{m_{A^0}^4}\,. \tag{3.12}$$

Comparing these results with those of eq. (2.7), one sees that the MSSM results are simply a special case of the more general two-Higgs doublet model results just obtained.

The limit $m_{A^0}^2 \gg \lambda_i v^2$ (subject to $\lambda_i \lesssim 1$) will be called the *decoupling limit* of the model. From eq. (3.4), it follows that $m_{12}^2 \gg \lambda_i v^2$ (assuming that neither t_β nor t_β^{-1} is close to 0). This condition clearly depends on the original choice of scalar field basis Φ_1 and Φ_2. For example, I can diagonalize the squared mass terms of the scalar potential [eq. (3.1)] thereby setting $m_{12} = 0$. In the decoupling limit in the new basis, eq. (3.4) would then imply that β must be near 0 or near $\pi/2$. A basis independent characterization of the decoupling limit is simple to formulate. Starting from the scalar potential in an arbitrary basis, form the matrix m_{ij}^2. Denote the eigenvalues of this matrix by m_a^2 and m_b^2 respectively; note that the eigenvalues are real but can be of either sign. By convention, I shall take $|m_a^2| \leq |m_b^2|$. Then, the decoupling limit corresponds to $m_a^2 < 0$, $m_b^2 > 0$ such that $m_b^2 \gg |m_a^2|, v$ (with $\lambda_i \lesssim 1$).

We are now ready to consider the questions posed in section 1. Let us assume that $m_{h^0} \ll m_{H^0}, m_{A^0}, m_{H^\pm}$. From eq. (3.9), it follows that

$$0 < m_{A^0}^2 m_L^2 + m_D^4 \ll m_S^4 . \tag{3.13}$$

Eq. (3.12) implies that in the decoupling limit, $\cos(\beta - \alpha) = \mathcal{O}(m_Z^2/m_{A^0}^2)$, which means that the h^0 couplings to Standard Model particles match precisely those of the Standard Model Higgs boson. However, in the following, I shall *not* impose the decoupling limit. Rather, I shall only require eq. (3.13), which reflects my basic assumption that one scalar state, h^0, is lighter than the other Higgs bosons. Eq. (3.13) is satisfied in one of two cases:

(*i*) m_Z^2, m_L^2, $m_D^4/m_{A^0}^2 \ll m_{A^0}^2$, m_S^2. That is, each term on the left-hand side of eq. (3.13) is separately smaller in magnitude than m_S^4, or

(*ii*) $m_{A^0}^2 m_L^2$ and m_D^4 are both of $\mathcal{O}(m_S^4)$, but due to cancelation of the leading behavior of each term, the sum satisfies eq. (3.13).

In case (i), one finds that

$$m_{h^0}^2 \simeq \frac{m_{A^0}^2 m_L^2}{m_S^2} + \frac{m_D^4}{m_S^2} + \frac{m_{A^0}^4 m_L^4}{m_S^6} + \mathcal{O}\left(\frac{m_L^4}{m_S^4}\right) , \tag{3.14}$$

and $m_{H^0}^2 \sim \mathcal{O}(m_S^2)$. In the same approximation,

$$\cos^2(\beta - \alpha) \simeq \frac{m_L^2}{m_S^2}\left(1 - \frac{m_{A^0}^2}{m_S^2}\right) + \frac{1}{m_S^4}\left[m_L^4\left(\frac{2m_{A^0}^2}{m_S^2} - \frac{3m_{A^0}^4}{m_S^4}\right) - m_D^4\right] . \tag{3.15}$$

The behavior of $\cos(\beta - \alpha)$ depends crucially on how close $m_{A^0}^2/m_S^2$ is to 1. If $m_L^2 \ll m_S^2$ [see eq (3.8)], we recover the results of the decoupling limit [eqs. (3.11) and (3.12)]. On the other hand, if $m_T^2 \sim \mathcal{O}(m_S^2)$, then eq. (3.15) implies that $\cos(\beta-\alpha) \sim \mathcal{O}(m_Z/m_{A^0})$. This is a particular region of the parameter space where some of the λ_i are substantially larger than 1, and yet the h^0 couplings do not significantly deviate from those of the Standard Model. Nevertheless, the onset of decoupling is slower than the $\cos(\beta-\alpha) \sim \mathcal{O}(m_Z^2/m_{A^0}^2)$ behavior found in the decoupling regime. In order to find a parameter regime which exhibits complete non-decoupling, one must consider case (ii) above. In this case, despite the fact that $m_{h^0} \ll m_{H^0}, m_{A^0}, m_{H^\pm}$, one nevertheless finds that $\cos(\beta-\alpha) \sim \mathcal{O}(1)$, which implies that the couplings of h^0 deviate significantly from those of the Standard Model Higgs boson. Although it might appear that case (ii) requires an unnatural cancelation, it is easy to construct a simple model of non-decoupling. Consider a model where: $m_{12}^2 = \lambda_6 = \lambda_7 = 0$, $\lambda_3+\lambda_4+\lambda_5 = 0$, $\lambda_2 \lesssim \mathcal{O}(1)$, and $\lambda_1, \lambda_3, -\lambda_5 \gg 1$. This model yields: $m_{h^0}^2 = \lambda_2 v^2 s_\beta^2$, $m_{H^0}^2 = \lambda_1 v^2 c_\beta^2$, $m_{A^0}^2 = -\lambda_5 v^2$, $m_{H^\pm}^2 = \frac{1}{2}\lambda_3 v^2$, and $\cos^2(\beta-\alpha) = c_\beta^2$. Note that in this model, the heavy Higgs states are not approximately degenerate (as required in the decoupling limit).

4. Phenomenological Challenges of the Decoupling Limit

We have seen that in the decoupling limit, the couplings of h^0 to Standard Model gauge bosons and fermions approach those of the Standard Model Higgs boson. Suppose that a future experiment has already discovered and studied the properties of h^0. What are the requirements of experiments at future colliders for proving the existence or non-existence of a non-minimal Higgs sector? To be specific, let us assume in this section that h^0 has been discovered with couplings approximating those of the Standard Model Higgs boson and $m_{A^0} > 250$ GeV.

At the LHC, the rate for $gg \to A^0$, H^0 and $gb \to H^- t$ provides a substantial number of produced Higgs bosons per year (assuming that the heavy Higgs masses are not too large).[21] Unfortunately, there may not be a viable final state signature. For example, since $\cos^2(\beta-\alpha) \ll 1$, the branching ratio of $H^0 \to ZZ$ is significantly suppressed, so that the gold-plated Standard Model Higgs signature is simply absent. At present, there is no known proven technique for detecting A^0, H^0 and $H^\pm$ signals at the LHC in the decoupling regime of parameter space. An attempt to isolate a Higgs signal in $t\bar{t}$ final states has been discussed in ref. 20. Another method consists of a search for Higgs signals in $t\bar{t}t\bar{t}$, $t\bar{t}b\bar{b}$, and $b\bar{b}b\bar{b}$ final states.[22] These can arise from $gg \to Q\bar{Q}'(H^0, A^0 \text{ or } H^\pm)$, where Q is a heavy quark (b or t), and the Higgs boson subsequently decays into a heavy quark pair. As noted below eq. (2.9), even in the decoupling limit, the couplings of H^0, A^0 and $H^\pm$ to heavy quark pairs are not suppressed. Whether such signals can be extracted from the substantial QCD backgrounds (very efficient b-tagging is one of the many requirements for such a search) remains to be seen.

Let us now turn to e^+e^- colliders. First, consider the process $e^+e^- \to h^0A^0$ (via s-channel Z-exchange). Since the Zh^0A^0 coupling is proportional to $\cos(\beta-\alpha)$, the production rate is suppressed in the decoupling regime. For example, in the MSSM, if $m_{A^0} > m_{h^0}$, then LEP-II will not possess sufficient energy and/or luminosity to directly produce the A^0.[23] Of course, with sufficient energy, one can directly pair-produce the heavy Higgs states via $e^+e^- \to H^+H^-$ and $e^+e^- \to H^0A^0$ without a rate suppression. At the NLC, with $\sqrt{s} = 500$ GeV and 10 fb^{-1} of data, it has been shown[23,24] that no direct signals for A^0, H^0, and $H^\pm$ can be seen if $m_{A^0} \gtrsim 230$ GeV. Although this result was obtained in the MSSM, it also applies to the decoupling regime of more general Higgs sectors. These results seems to imply the following rather general conclusion: *evidence for the non-minimal Higgs sector at the NLC requires a machine with energy $\sqrt{s} > 2m_{A^0}$, sufficient to pair-produce heavy Higgs states.*

Can this conclusion be avoided? There are two possible methods.

1. Produce one heavy Higgs state in association with light states*.
2. Make precision measurements of h^0 couplings to Standard Model particles in order to detect a deviation from the Standard Model expectations.

First, consider production mechanisms which result in a singly produced heavy Higgs state. It was noted in section 2 that whenever a single heavy Higgs state couples directly to a gauge boson plus other particles, the coupling is suppressed by $\cos(\beta-\alpha)$. To avoid this suppression, one must couple the heavy Higgs state to either fermions or scalars. For example, consider $e^+e^- \to Q\bar{Q}'(H^0, A^0, \text{or } H^\pm)$, where $Q = b$ or t. The production rates have been computed by Djouadi *et al.*[25] Unfortunately, the three-body phase space greatly suppresses the rate once the heavy Higgs state is more massive than the Z. In particular, for $m_{A^0} > \sqrt{s}/2$, these processes do not provide viable signatures for the heavy Higgs states. A similar conclusion is obtained when the heavy Higgs state couples to light scalars. Scott Thomas and I have computed the rate for $e^+e^- \to h^0H^0Z$. We assumed that the dominant contribution arises in the s-channel Z-exchange, where the virtual Z^* decays via $Z^* \to Zh^{0*} \to Zh^0H^0$. In the limit where $m_{H^0} \gg m_{h^0}, m_Z$, we obtained[18]

$$\frac{\sigma(e^+e^- \to h^0H^0Z)}{\sigma(e^+e^- \to h^0Z)} \simeq \frac{g^2_{H^0h^0h^0}}{32\pi^2 s^3 m^2_{H^0}}\Bigg\{(s-m^2_{H^0})\left[(s-m^2_{H^0})^2+12sm^2_{H^0}\right] -6m^2_{H^0}s(s+m^2_{H^0})\ln\left(\frac{s}{m^2_{H^0}}\right)\Bigg\}, \tag{4.1}$$

where the $H^0h^0h^0$ coupling in the decoupling limit [*i.e.*, when $\cos(\beta-\alpha)=0$] is

---

* In this context, light states refer to all Standard Model fermions and bosons, with masses of order m_Z or less. Thus, the gauge bosons, h^0, and even the top-quark will be considered among the light states!

given by

$$g_{H^0h^0h^0} = \frac{-3igm_Z}{4\cos\theta_W}\sin 4\beta\,. \tag{4.2}$$

This rate [eq. (4.1)] is also too small for a viable Higgs signal.

Second, consider precision measurements of h^0 branching ratios at the NLC. In a Monte Carlo analysis, Hildreth *et al.*[26] evaluated the anticipated accuracy of h^0 branching ratio measurements at the NLC, assuming $\sqrt{s} = 500$ GeV and a data set of 50 fb^{-1}. For example, taking $m_{h^0} = 120$ GeV, they computed an extrapolated error of $\pm 7\%$ for the 1-σ uncertainty in $BR(h^0 \to b\bar{b})$ and $\pm 14\%$ for $BR(h^0 \to \tau^+\tau^-)$. Other channels yielded substantially larger uncertainties. To see whether these are significant measurements, one can compare these results with the theoretical expectations of the MSSM as a function of m_{A^0} and $\tan\beta$. Hildreth and I have found[27] that deviations in $BR(h^0 \to b\bar{b})$ and $BR(h^0 \to \tau^+\tau^-)$ from the Standard Model can be about 7% for values of m_{A^0} as large as 450 GeV and about 15% for values of m_{A^0} as large as 250 GeV. (See ref. 23 for a related analysis.) These results imply that a precision measurement of $h^0 \to b\bar{b}$ has the potential for detecting the existence of the non-minimal Higgs sector even if the heavier Higgs states cannot be directly detected at the NLC. Of course, one will have to push the precision of this measurement beyond its present expectations, since a 2-σ deviation is not compelling evidence for new physics. Other Higgs decay channels are not competitive.

There is one novel approach which could extend the discovery limits for heavy Higgs bosons at the NLC. A high energy, high luminosity photon beam can be produced by the Compton backscatter of an intense laser beam off a beam of electrons.[28] This provides a mechanism for turning the NLC into a high energy, high luminosity $\gamma\gamma$ collider. All neutral Higgs states couple to $\gamma\gamma$ at one-loop via loops of charged matter.[29] Since the couplings of the heavy Higgs states to fermions and scalars are not suppressed in the decoupling limit, the $\gamma\gamma$ couplings of the heavy Higgs states are also not suppressed relative to the $h^0\gamma\gamma$ coupling. Thus, one can search for the non-minimal Higgs sector at the $\gamma\gamma$ collider by either measuring the $h^0\gamma\gamma$ coupling with sufficient precision or by directly producing A^0 and/or H^0 in $\gamma\gamma$ fusion. In the decoupling regime, the $h^0\gamma\gamma$ coupling approaches the corresponding coupling of the Standard Model Higgs boson[†]. As a result, this is not a viable method for detecting deviations from the Standard Model. Thus, one must focus on $\gamma\gamma \to (A^0, H^0)$. In ref. 11, Gunion and I showed that parameter regimes exist where one could extend the heavy Higgs mass reach above $\sqrt{s}/2$. For example, at a 500 GeV $\gamma\gamma$ collider, a statistically significant A^0 signal in $b\bar{b}$ and Zh^0 final states could be seen above backgrounds if $m_{A^0} < 2m_t$.

† The contributions of supersymmetric loops to the $h^0\gamma\gamma$ amplitude vanish in the limit of large supersymmetric particle masses.

5. Conclusions

In the most general CP-conserving two Higgs doublet model, a decoupling limit can be defined in which the lightest Higgs state is a CP-even neutral Higgs scalar, whose properties approach those of the Standard Model Higgs boson. This result is more general, and applies to non-minimal Higgs sectors that contain the Standard Model Higgs doublet and respect standard phenomenological constraints (such as $m_W = m_Z \cos\theta_W$ at tree-level). In the MSSM, the decoupling limit corresponds to $m_{A^0} \gg m_Z$ (independent of $\tan\beta$). Moreover, the approach to decoupling is rapid once m_{A^0} is larger than m_Z. Thus, over a very large range of MSSM parameter space, the couplings of h^0 to gauge bosons, quarks and leptons are nearly identical to the couplings of the Standard Model Higgs boson.

If the h^0 is discovered with properties approximating those of the Standard Model Higgs boson, then the discovery of the non-minimal Higgs sector will be difficult. At the LHC, A^0, H^0 and $H^\pm$ production rates via gluon-gluon fusion are not suppressed. However, isolating signals of the heavy Higgs states above background presents a formidable challenge. At the NLC, if $\sqrt{s} > 2m_{A^0}$, then pair production of H^+H^- and H^0A^0 is easily detected. However, below pair-production threshold, detection of the non-minimal Higgs sector is problematical. For example, the cross sections for single heavy Higgs boson production (in association with light particles) are too small to be observed. However, experiments at the $\gamma\gamma$ collider may extend the NLC discovery limits of the heavy Higgs states (via $\gamma\gamma$ fusion to H^0 or A^0). Precision measurements of $h^0 \to b\bar{b}$ could provide additional evidence for a non-minimal Higgs sector. However, current experimental expectations at the NLC predict only a 2-σ deviation from Standard Model expectations if m_{A^0} is just beyond the kinematic limit of direct NLC detection.

If h^0 is discovered with distinguishable properties relative to Standard Model predictions, then there are three possibilities. The simplest possibility is that other scalar states of the non-minimal Higgs sector lie close in mass to h^0 and will be discovered via direct production soon after the discovery of h^0. If the other Higgs states are not discovered (and it appropriately strong mass limits are obtained), then two alternative scenarios emerge. If the Higgs parameters lie in the non-decoupling regime (*i.e.*, there exists at least one or more large Higgs self-couplings), then the Higgs sector is probably strongly coupled, which suggests the existence of a rich electroweak symmetry breaking sector at an energy scale near 1 TeV. On the other hand, if the Higgs sector is weakly coupled, then there must exist new non-Standard Model decay channels accessible in h^0 decay. For example, in Majoron models, $h^0 \to JJ$ (where J is the Majoron) provides an invisible decay mode for h^0,[30] which would cause deviations from Standard Model expectations for h^0 branching ratios to gauge bosons, quarks and leptons. A similar possibility exists in some supersymmetric models where the decay $h^0 \to \widetilde{\chi}_1^0\widetilde{\chi}_1^0$ (where $\widetilde{\chi}_1^0$ is the lightest neutralino) can be significant,[31] leading again to deviations in the h^0 branching ratios to Standard Model final states.

Once the first evidence for the Higgs boson is established, it will be crucial to ascertain the underlying dynamics of the electroweak symmetry breaking sector. If the data reveals that the Higgs sector is non-minimal, then we will have an important clue to the structure of the electroweak symmetry breaking sector. However, one must be prepared for the more pessimistic scenario—the discovery of a Higgs boson whose properties are not experimentally distinguishable from the Higgs boson of the minimal Standard Model. This scenario presents a formidable challenge for future collider experiments in their attempt to probe the physics of the electroweak symmetry breaking sector and the nature of TeV-scale physics beyond the Standard Model.

Acknowledgments

This paper describes work performed in a number of separate collaborations with Jack Gunion, Mike Hildreth, Yosef Nir, and Scott Thomas. I am especially grateful for their insights and diligence. In addition, I am pleased to acknowledge George Pocsik and his colleagues at Eötvös University and Stefan Pokorski and his colleagues at the University of Warsaw for their warm hospitality and their labors in providing such a stimulating environment during their workshops. This work was supported in part by the Department of Energy.

References

1. J.F. Gunion, H.E. Haber, G.L. Kane and S. Dawson, *The Higgs Hunter's Guide* (Addison-Wesley, Redwood City, CA, 1990).
2. V.A. Miransky, *Dynamical Symmetry Breaking in Quantum Field Theories* (World Scientific, Singapore, 1993).
3. P.H. Chankowski, S. Pokorski and J. Rosiek, *Phys. Lett.* **B281** (1992) 100.
4. G.L. Kane, in *Perspectives on Higgs Physics*, edited by G.L. Kane (World Scientific, Singapore, 1993) pp. 223–228.
5. L. Montanet *et al.* [Particle Data Group] *Phys. Rev.* **D50** (1994) 1173.
6. A. Sopczak, *Int. J. Mod. Phys.* **A9** (1994) 1747.
7. A. Stange, W. Marciano and S. Willenbrock, *Phys. Rev.* **D49** (1994) 1354; **D50** (1994) 4491; S. Mrenna and G.L. Kane, CALT-68-1938 (1994) [hep-ph 9406337]; J.F. Gunion, UCD-94-24 (1994) [hep-ph 9406405].
8. D. Froidevaux, Z. Kunszt and J. Stirling *et al.*, in *Proceedings of the Large Hadron Collider Workshop*, Aachen 1990, CERN Repoprt 90-10 (1990).
9. S. Dawson, in *Perspectives on Higgs Physics*, edited by G.L. Kane (World Scientific, Singapore, 1993) pp. 129–155; Z. Kunszt, *ibid.* pp. 156–178.

10. V. Barger, K. Cheung, A. Djouadi, B.A. Kniehl, and P. Zerwas, *Phys. Rev.* **D49** (1994) 79.
11. J.F. Gunion and H.E. Haber, *Phys. Rev.* **D48** (1993) 5109.
12. D.L. Borden, D.A. Bauer and D.O. Caldwell, *Phys. Rev.* **D48** (1993) 4018.
13. J.F. Gunion and H.E. Haber, *Nucl. Phys.* **B272** (1986) 1; **B278** (1986) 449 [E: **B402** (1993) 567].
14. For a review of the influence of radiative corrections on the MSSM Higgs sector, see H.E. Haber, in *Perspectives on Higgs Physics*, edited by G.L. Kane (World Scientific, Singapore, 1993) pp. 79–128.
15. See, *e.g.*, H.E. Haber and R. Hempfling, *Phys. Rev.* **D48** (1993) 4280.
16. J.F. Gunion, H.E. Haber, and J. Wudka, *Phys. Rev.* **D43** (1991) 904.
17. See, *e.g.*, M. Carena, M. Olechowski, S. Pokorski, and C.E.M. Wagner, *Nucl. Phys.* **B419** (1994) 213; G.L. Kane, C. Kolda, L. Roszkowski and J.D. Wells, *Phys. Rev.* **D49** (1994) 6173; **D50** (1994) 3498; P. Nath and R. Arnowitt, CERN-TH.7227/94 (1994) [hep-ph 9406403]; W. de Boer, R. Ehret, W. Oberschulte, D.I. Kazakov, IEKP-KA/94-05 [hep-ph 9405342], to be published in *Z. Phys. C*.
18. H.E. Haber and S. Thomas, SCIPP preprint in preparation.
19. H.E. Haber and Y. Nir, *Nucl. Phys.* **B335** (1990) 363.
20. D. Dicus, A. Stange and S. Willenbrock, *Phys. Lett.* **B333** (1994) 126.
21. For a recent review, see J.F. Gunion, in *Perspectives on Higgs Physics*, edited by G.L. Kane (World Scientific, Singapore, 1993) pp. 179–222.
22. J. Dai, J.F. Gunion and R. Vega, *Phys. Rev. Lett.* **71** (1993) 2699; UCD-94-7 (1994) [hep-ph 9403362].
23. P. Janot, in *Physics and Experiments with Linear e^+e^- Colliders*, Workshop Proceedings, Waikoloa, Hawaii, 26–30 April, 1993, edited by F.A. Harris, S.L. Olsen, S. Pakvasa and X. Tata (World Scientific, Singapore, 1993) pp. 192–217.
24. A. Djouadi, J. Kalinowski and P.M. Zerwas, *Z. Phys.* **C57** (1993) 569.
25. A. Djouadi, J. Kalinowski and P.M. Zerwas, *Mod. Phys. Lett.* **A7** (1992) 1765; *Z. Phys.* **C54** (1992) 255.
26. M.D. Hildreth, T.L. Barklow, and D.L. Burke, *Phys. Rev.* **D49** (1994) 3441.
27. H.E. Haber and M.D. Hildreth, in preparation.
28. H.F. Ginzburg, G.L. Kotkin, V.G. Serbo and V.I. Telnov, *Nucl. Inst. and Methods* **205** (1983) 47; H.F. Ginzburg, G.L. Kotkin, S.L. Panfil, V.G. Serbo and V.I. Telnov, *Nucl. Inst. and Methods* **219** (1984) 5.
29. A.I. Vainshtein, M.B. Voloshin, V.I. Zakharov and M. Shifman, *Yad. Fiz.* **30** (1979) 1368 [*Sov. J. Nucl. Phys.* **30** (1979) 711].
30. A.S. Joshipura and J.W.F. Valle, *Nucl. Phys.* **B397** (1993) 105.
31. K. Griest and H.E. Haber, *Phys. Rev.* **D37** (1988) 719.

ARE THERE SOLITONS IN THE TWO-HIGGS STANDARD MODEL?

C.P. BACHAS
Centre de Physique Théorique
Ecole Polytechnique
91128 Palaiseau, France

and

T.N. TOMARAS
Physics Department, University of Crete
and Research Center of Crete
714 09 Heraklion, Greece

ABSTRACT

We present some evidence, based on the analysis of lower-dimensional models, for the possible existence of classically-stable winding solitons in the two-higgs electroweak theory.

The search for stable lumps in the Weinberg-Salam model has a long history. It has revealed a rich structure of classical solutions including the sphaleron [1,2,3,4], deformed sphalerons [5,6,7,8] and vortex strings [9,10,11,12,13]. Such solutions could play a role in understanding (B+L)-violation and structure formation in the early universe, but they are all classically-unstable or/and extended. They have therefore no direct present-day manifestation, contrary to long-lived particles whose relic density could at least in principle be detected.

The existence of particle-like excitations has, on the other hand, been argued for in the context of a strongly- interacting higgs sector [14,15,16,17,18]. The advocated particles can be thought of as technibaryons of an underlying technicolor model. They are described in bosonic language by winding solitons of an effective non-renormalizable lagrangian for the pseudo-goldstone-boson (or technipion) field, much like skyrmions[19] of the effective chiral lagrangian of QCD. This is of course a phenomenological description, since the properties of such hypothetical particles cannot be calculated reliably within a semi-classical expansion. Furthermore, in view of the difficulties facing technicolor models, the possibility of a strongly-interacting higgs sector is not theoretically appealing.

It would be clearly more interesting if classically-stable winding excitations could arise in a *weakly-coupled* scalar sector. To be more precise let us decompose the higgs-doublet field into a real (positive) magnitude and a group-phase: $\Phi = FU$, and consider static configurations with $U(x)$ wrapping N times around the $SU(2)$ manifold. These are potentially unstable for at least three distinct reasons:

(a) because N is not conserved whenever the magnitude F goes through zero; *(b)* because N is not gauge-invariant and can, in particular, be non-vanishing even in a vacuum state; and *(c)* because scalar-field configurations can loose their energy by shrinking to zero size [20]. We refer to these for short as the *radial, gauge* and *scale* instabilities. They can be eliminated formally by *(a)* taking the physical-higgs mass $m_H \to \infty$, *(b)* decoupling the electroweak gauge fields, and *(c)* adding appropriate higher-derivative terms to the action. The question is whether classical stability can be maintained while relaxing the above conditions. This has been investigated numerically in the *minimal* case of one doublet: although one may indeed relax both the weak gauge coupling [16,5] and the higgs mass [21] up to some finite critical values, stability cannot apparently be achieved without the non-renormalizable higher-derivative terms in the action. On the other hand, as we have demonstrated recently, metastable winding solitons do arise in renormalizable models in two [22] and three [23] space-time dimensions. The way this happens is we believe instructive and could guide the search for such semi-classical solitons in four dimensions.

The simplest context in which the *radial* instability is an issue is a two-dimensional model of a complex-scalar field with mexican-hat potential. $V = \frac{1}{4}\lambda(\Phi^*\Phi - v^2)^2$. To find winding solitons we must take space to be periodic with period L *. The condition for classical stability can in this case be derived analytically and reads[22]: $m_H L > \sqrt{5}$, where $m_H = \sqrt{2\lambda}v$. The classically-relevant parameter is thus the radial-higgs mass in units of the soliton size. This follows also by comparing the loss in potential energy to the gain in gradient energy when trying to undo the winding by reducing the magnitude of the scalar. Note that the loop-expansion parameter λL^2, can be taken to zero independently so as to reach a semiclassical limit.

The above winding solitons become unstable classically if we gauge the $U(1)$ symmetry of the model. The *gauge* instability is in fact more severe than in four dimensions, because no energetic barrier opposes the turning-on of a static space-like gauge field, which is necessary to reach a winding-vacuum state. The minimal abelian-higgs model has thus only unstable (sphaleron) solutions [25] †. The situation changes, however, drastically if there are more than one complex scalars. The gauge-invariant relative phases of any two of them cannot in this case wind around non-trivially in a vacuum state. An explicit analysis of this extended abelian-higgs model [22] shows that winding solitons persist down to scalar masses close to the inverse soliton size: gauging and the extra higgs enhance the stability region found in the global model.

The *scale* instability becomes an issue for the first time in three space-time dimensions. To be more precise we consider a real-triplet scalar field $\Phi_a(x)$ $(a = 1, 2, 3)$ with mexican-hat potential : $V = \frac{1}{4}\lambda(\Phi_a\Phi_a - v^2)^2$. The limit $\lambda \to \infty$ corresponds to the $O(3)$ non-linear σ-model. This is known to possess winding solitons, characterized by non-trivial mappings of the two-sphere onto itself, and having arbitrary size [26].

*Alternatively we may add a mass term: $\delta V = -\mu^2 v Re(\Phi)$, that lifts the vacuum degeneracy. Stable winding excitations, which reduce to the sine-Gordon solitons in the $\lambda \to \infty$ limit, can be shown [24] numerically to exist for $\lambda v^2/\mu > 18.8$.

†It was claimed erroneously in [22] that it has no static solutions whatsoever. This is only correct in the $\lambda \to \infty$ limit.

For finite λ on the other hand, or in the presence of a symmetry-breaking potential, Derrick's scaling argument [20] shows that these solitons are unstable to shrinking. One can of course again invoke higher-derivative terms to stabilize the scale[27]. The same result is however in this case achieved by a massive $U(1)$ gauge field with only renormalizable couplings [23]. This can be established by perturbing around the $O(3)$ non-linear σ-model limit, or else by solving numerically the equations of motion.

What do these lower-dimensional solitons teach us? First, they are interesting in their own right, since they correspond to a new class of wall and string defects in renormalizable four-dimensional models. Second, they suggest by analogy that classically-stable winding solitons may exist in a weakly-coupled two-higgs extension of the standard model. These hypothetical solitons would: *(b)* be characterized by the non-trivial winding of the relative phase of the two doublets, and thus be immune to the gauge mode of decay; *(c)* have a scale stabilized by electroweak magnetic fields and hence of order $1/m_W$; and *(a)* hopefully stay stable for higgs masses near m_W and thus compatible with perturbative unitarity ‡. Mathematically the situation is the same as in the hidden-gauge-boson models [28,29] of strong and electroweak interactions, except that the role of the hidden gauge bosons is here played by $W^{\pm}$ and Z themselves. Although previous numerical investigations [28,30] have shown no sign of stable solitons in these contexts, a systematic search is in our opinion necessary in order to settle definitely the issue [31].

Acknowledgements

This research was supported in part by the EEC grants CHRX-CT94-0621 and CHRX-CT93-0340, as well as by the Greek General Secretariat of Research and Technology grant $91E\Delta358$.

References

1. R. Dashen, B. Hasslacher and A. Neveu, *Phys. Rev.* **D10** (1974) 4138 .
2. C. Taubes, *Comm. Math. Phys.* **86** (1982) 257 and 299;
   N.S. Manton, *Phys. Rev.* **D28** (1983) 2019;
   P. Forgacs and Z. Horvath, *Phys. Lett.* **B138** (1984) 397 .
3. F. Klinkhamer and N. Manton, *Phys. Rev.* **D30** (1984) 2212 .
4. J. Boguta, *Phys. Rev. Lett.* **50** (1983) 148;
   J. Burzlaff, *Nucl. Phys.* **B233** (1984) 262 .

‡Though admittedly premature, some other physical properties of such would-be particles are fun to contemplate: being classically stable they could easily have cosmological life times. They would have a mass in the $\sim 10\ TeV$ region, zero charge and dipole moments in their ground state, and geometrical interaction cross sections of order $1/m_W^2$. Assuming maximum production at the electroweak phase transition, a rough estimate of their present abundance shows that they could be candidates for cold dark matter in the universe.

5. G. Eilam, D. Klabucar and A. Stern, *Phys. Rev. Lett.* **56** (1986) 1331;
G. Eilam and A. Stern, *Nucl. Phys.* **B294** (1987) 775 .
6. J. Kunz and Y. Brihaye, *Phys. Lett.* **B216** (1989) 353 and **B249** (1990) 90 .
7. L. Yaffe, *Phys. Rev.* **D40** (1989) 3463 .
8. F. Klinkhamer, *Phys. Lett.* **B236** (1990) 187 .
9. Y. Nambu, *Nucl. Phys.* **B130** (1977) 505 .
10. M.B. Einhorn and R. Savit, *Phys. Lett.* **B77** (1978) 295;
V. Soni, *Phys. Lett.* **B93** (1980) 101;
K. Huang and R. Tipton, *Phys. Rev.* **D23** (1981) 3050 .
11. T. Vachaspati, *Phys. Rev. Lett.* **68** (1992) 1977
and *Nucl. Phys.* **B397** (1993) 648;
T. Vachaspati and M. Barriola, *Phys. Rev. Lett.* **69** (1992) 1867 .
12. M. James, T. Vachaspati and L. Perivolaropoulos, *Phys. Rev.* **D46** (1992) R5232 and *Nucl. Phys.* **B395** (1993) 534 .
13. L. Perivolaropoulos, *Phys. Lett.* **B316** (1993) 528;
M. Earnshaw and M. James, *Phys. Rev.* **D48** (1993) 5818 .
14. J.M. Gipson and H.C.Tze, *Nucl. Phys.* **B183** (1981) 524;
J.M. Gipson, *Nucl. Phys.* **B231** (1984) 365 .
15. E. D'Hoker and E. Fahri, *Nucl. Phys.* **B241** (1984) 104, and
Phys. Lett. **134B** (1984) 86 .
16. V.A. Rubakov, *Nucl. Phys.* **B256** (1985) 509;
J. Ambjorn and V.A. Rubakov, *Nucl. Phys.* **B256** (1985) 434 .
17. R. MacKenzie, F. Wilczek and A. Zee, *Phys. Rev. Lett.* **53** (1984) 2203;
R. MacKenzie, *Mod. Phys. Lett.* **A7** (1992) 293 .
18. J.W. Carlson, *Nucl. Phys.* **B253** (1985) 149, and **B277** (1986) 253 .
19. T.H.R. Skyrme, *Proc. Roy. Soc.* (London), Ser. **A260** (1961) 127 .
20. G.H. Derrick, *J. Math. Phys.* **5** (1964) 1252 .
21. Y. Brihaye, J. Kunz and F. Mousset, *Z. Phys.* **C56** (1992) 231;
J. Baacke, G. Eilam and H. Lange, *Phys. Lett.* **B199** (1987) 234 .
22. C. Bachas and T.N. Tomaras, *Nucl. Phys.* **B428** (1994) 209 .
23. C. Bachas and T.N. Tomaras, CPTH-A323.0794 and Crete-94-14 preprint (September 1994) .
24. K. Pallis, U. of Thessaloniki undergraduate thesis (1994) .
25. Y. Brihaye, S. Giller, P. Kosinski and J. Kunz, *Phys. Lett.* **B293** (1992) 383.
26. A.A. Belavin and A.M. Polyakov, *JETP Lett.* **22** (1975) 245 .
27. R.A. Leese, M. Peyrard and W.J. Zakrzewski, *Nonlinearity* **3** (1990) 773 .
28. Y. Brihaye, J. Kunz and C. Semay, *Phys. Rev.* **D42** (1990) 193;
Phys. Rev. **D44** (1991) 250, and references therein .
29. A.Dobado and J.J. Herrero, *Nucl. Phys.* **B319** (1989) 491, and references therein .
30. B. Kastening, R.D. Peccei and X. Zhang, *Phys. Lett.* **B266** (1991) 413 .
31. C. Bachas, P. Tinyakov and T.N. Tomaras, work in progress .

SUSY SIGNATURES WITH BROKEN R PARITY *

JOSÉ W. F. VALLE †
Instituto de Física Corpuscular - C.S.I.C.,
Departament de Física Teòrica, Universitat de València
46100 Burjassot, València, SPAIN

ABSTRACT

A short review is made of some of the experimental signatures that may be associated to supersymmetric extensions of the standard model involving the violation of R parity. The new signatures may be accessible to experiments performed either at accelerators or at underground installations. The complementarity between these two approaches in the search for signals beyond the standard model occurs in these models since the violation of R parity is correlated with the large mass of the tau neutrino.

1. Introduction

It is agreed that, although extremely successful wherever it has been tested, our present standard $SU(2) \otimes U(1)$ model leaves open many of the fundamental issues of present-day particle physics. The simplest formulation of the model relies on the Higgs mechanism and implies the existence of a fundamental scalar boson. If such an elementary higgs boson exists it is widely believed that some stabilizing principle should be operative at the electroweak scale in order to explain the stability of its mass scale against quantum corrections associated with physics at superhigh energies. Supersymmetry (SUSY) gives the best known principle to do this task, provided it does indeed set in somewhere at a scale $M_{SUSY} \sim 10^3$ GeV. The observed joining of the three gauge coupling strengths as they are evolved from the presently accessible energies up to to a common scale of $\sim 10^{16}$ GeV provides additional support in favour of the existence of a rich supersymmetric spectrum of fundamental states at the weak scale. Unveiling the details of this structure constitutes one of the main goals in the agenda of the next generation of elementary particle colliders.

The prototype of these models is called the minimal supersymmetric standard model (MSSM) [1]. This model realizes SUSY in the presence of a discrete R parity (R_p) symmetry, postulated *ad hoc*. Under this symmetry all standard particles are even while their partners are odd. As a result of this selection rule all SUSY particles must be produced only in pairs, with the lightest of them, called LSP, being stable. It has been suggested as a candidate for the cold dark matter of the universe and several methods of detection at underground installations have been suggested.

---

*Work supported by DGICYT under grant PB92-0084
†E-mail VALLE@flamenco.ific.uv.es or 16444::VALLE

So far all searches for supersymmetric particles have been negative. However, presently accessible energies cover only a small part of the parameter space of supersymmetric theories. One may summarize the present situation as follows. The electrically charged weakly interacting SUSY states, sleptons, charginos as well as SUSY higgs bosons have bounds close to the available beam energy at LEP. There is only a small room for improvement left at LEP1 on the masses of the electrically neutral SUSY particles. As for the strongly interacting SUSY states, gluinos and squarks, their mass bounds come form the Tevatron and there is little room for improvement with the present setup.

So far, however, little is known about the origin of the assumed R parity symmetry, and whether or not it is indeed a necessary requirement to be imposed upon supersymmetric extensions of the standard model. Therefore there is no firm theoretical basis for the so-called minimal supersymmetric standard model, the usual reference model so far considered in most investigations of SUSY phenomenology.

It is indeed of great interest to investigate theories without R parity [2]. There are many ways to break it, either explicitly or spontaneously (RPSUSY models). Here we focus on the case of spontaneous R_p breaking in the $SU(2) \otimes U(1)$ theory. The viability of this possibility has been explicitly demonstrated [3,4]. The basic feature of the models is that the breaking of R-parity is driven by right-handed *isosinglet* sneutrino vacuum expectation values (VEVS) [3], so that the associated Goldstone boson (majoron) is mostly singlet and as a result the Z does not decay by majoron emission, in agreement with LEP observations [5].

Supersymmetric models with spontaneously broken R parity necessarily imply nonvanishing neutrino masses. Here I will focus on the simplest models, where the magnitude of R parity violation is correlated with the nonzero value of the ν_τ mass and, as such, is restricted by a variety of neutrino physics experiments.

In this talk I will illustrate explicitly how one has not yet reached the border of what can be reached with present accelerators in the searches for SUSY particles in spontaneously broken R parity models. I will show how one can have in this case genuine SUSY signals that can be searched for even with present LEP1 sensitivities. Searches for spontaneously broken R parity signatures can also be performed at at LEP2 and hadron supercolliders. Here the most favoured situation is to search not for the single SUSY particle production (although it may be non negligible at the LHC) but rather for the signatures associated with the decay of the LSP.

I will also describe how this realization of the basic supersymmetric picture may affect also the physics of the electroweak breaking sector in a very important way, introducing the possibility of invisibly decaying Higgs bosons.

2. Supersymmetry with Spontaneously broken R parity

If R parity is broken spontaneously it shows up primarily in the couplings of the W and the Z, leading to rare Z decays such as the single production of the charginos

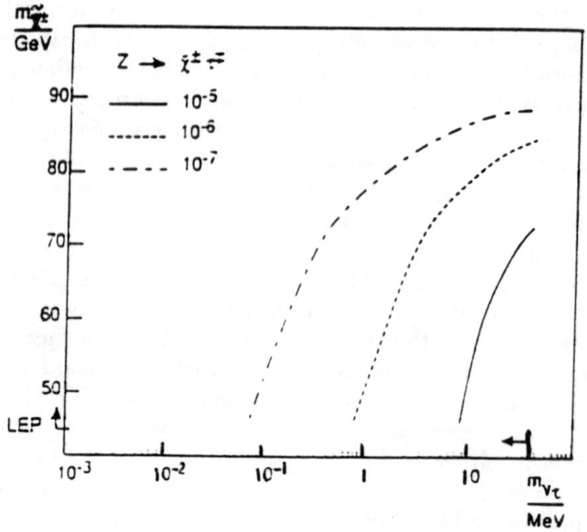

Figure 1: Allowed branching ratios for $Z \to \tilde{\chi}^{\pm}\tau^{\mp}$

and neutralinos [6,7], for example,

$$Z \to \chi^{\pm}\tau^{\mp} \tag{1}$$

where the lightest chargino mass is assumed to be smaller than the Z mass. In the models of interest to us the magnitude of R parity violation is correlated with the nonzero value of the ν_τ mass. Although restricted by many experiments, the R parity violating Z decay branching ratios can be quite sizeable. For example, the $Z \to \chi^{\pm}\tau^{\mp}$ branching ratio can easily exceed 10^{-5}, well within present LEP sensitivities. This is illustrated in Figure 1. Similarly, the lightest neutralino (LSP) could also be singly-produced as [6,7]

$$Z \to \chi^0\nu_\tau \tag{2}$$

Being unstable due to R parity violation, χ^0 is not necessarily an origin of events with missing energy, since some of its decays are into charged particles. Thus the decay $Z \to \chi^0\nu_\tau$ would give rise to zen events, similar to those of the MSSM but where the missing energy is carried by the ν_τ . Another possibility for zen events in these models is the usual neutralino pair production process, where one χ^0 decays visibly and the other invisibly. The corresponding zen-event rates can be larger than in the MSSM.

Although the ν_τ can be quite massive in these models, it is perfectly consistent with cosmology [8] including primordial nucleosynthesis [9], since it decays sufficiently fast by majoron emission [10]. On the other hand, the ν_e and ν_μ have a tiny mass difference [3] which can be chosen to lie in the range where resonant ν_e to ν_μ conversions

Table 1: Allowed branching ratios for rare decays in the RPSUSY model. χ denotes the lightest electrically charged SUSY fermion (chargino) and χ^0 is the lightest neutralino.

channel	strength
$Z \to \chi\tau$	$\lesssim 6 \times 10^{-5}$
$Z \to \chi^0\nu_\tau$	$\lesssim 10^{-4}$
$\tau \to \mu + J$	$\lesssim 10^{-3}$
$\tau \to e + J$	$\lesssim 10^{-4}$

provides an explanation of solar neutrino deficit [11]. Due to this peculiar hierarchical pattern, one can go even further, and regard the rare R parity violating processes as a tool to probe the physics underlying the solar neutrino conversions in this model [12]. Indeed, the rates for such rare decays can be used in order to discriminate between large and small mixing angle MSW solutions to the solar neutrino problem [11]. Typically, in the nonadiabatic region of small mixing one can have larger rare decay branching ratios, as seen in Figure 5 of ref. [12].

It is also possible to find manifestations of R parity violation at the superhigh energies available at hadron supercolliders such as LHC. Either SUSY particles, such as gluinos, are pair produced and in their cascade decays the LSP decays or, alternatively, one violates R parity by singly producing the SUSY states. An example of this latter situation has been discussed in the literature [13]. In this reference one has studied the single production of weakly interacting supersymmetric fermions (charginos and neutralinos) via the Drell Yan mechanism, leading to possibly detectable signatures. More detailed work on this will be desirable, as well as an investigation of the capabilities for probing for RPSUSY signatures in the pair production with unstable LSP. I should emphasize that the allowed LSP decay channels in the spontaneously broken R parity models have been given in the literature [7,13].

Another possible signal of the RPSUSY models based on the simplest $SU(2) \otimes U(1)$ gauge group is rare decays of muons and taus. In this model the spontaneous violation of R parity generates a physical Goldstone boson, called majoron. Its existence is quite consistent with the measurements of the invisible Z decay width at LEP, as it is mostly a singlet under the $SU(2) \otimes U(1)$ gauge symmetry. In this model the lepton number is broken close to the weak scale and can produce a new class of lepton flavour violating decays, such as those with single majoron emission in μ and τ decays. These would be "seen" as bumps in the final lepton energy spectrum, at half of the parent lepton mass in its rest frame. The allowed rates for single majoron emitting μ and τ decays have been determined [14] and are also shown in table 1 to be compatible with present experimental sensitivities [15]. As an illustration, I borrow Figure 2 from ref.

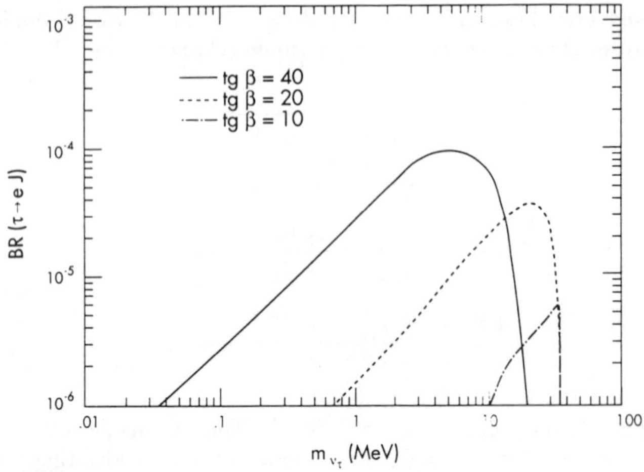

Figure 2: Allowed branching ratios for $\tau \to e + J$ versus m_{ν_τ}

[14]. This example also illustrates how the search for rare decays can be a more sensitive probe of neutrino properties than the more direct searches for neutrino masses, and therefore complementary. Moreover, they are ideally studied at a tau-charm factory [16].

3. Invisibly Decaying Higgs Bosons.

There has been a great effort devoted to the physics associated to the electroweak breaking sector and its possible manifestations at present and future particle colliders. The simplest possibility that the higgs boson does indeed exist as an elementary particle justifies the important role that the study of supersymmetric extensions of the standard model has played in these investigations.

Another quite interesting feature of models with spontaneously broken R parity lies in the electroweak breaking sector. These models also entail the spontaneous violation of an ungauged U(1) lepton number symmetry, thus implying the existence of a physical Goldstone boson, called majoron [‡] As already mentioned above this majoron is consistent with the measurements of the invisible Z decay width at LEP

[‡]Note that the majoron was originally proposed in the framework of the seesaw model of neutrino masses, and required the introduction of a relatively high energy scale associated to the mass of the right-handed neutrinos [17]. Many other attractive theoretical alternatives where lepton number is violated spontaneously at the weak scale or lower have been recently developed [18].

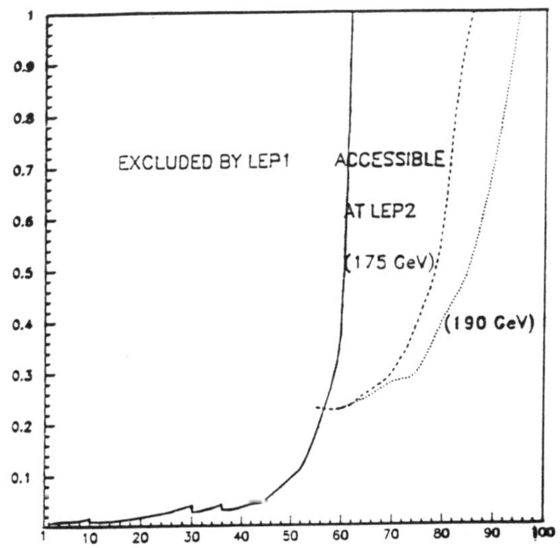

Figure 3: Region in the ϵ^2 vs. m_H that can be excluded by the present LEP1 analyses (solid curve). Also shown are the LEP2 extrapolations (dashed).

due to its mostly singlet nature. As a result it also has very tiny couplings to gauge bosons and to matter. Nevertheless, it can have significant couplings to the Higgs bosons, leading to the striking possibility that these may decay with a substantial branching ratio into the invisible mode [18,19,20]

$$h \to J + J \tag{3}$$

where J denotes the majoron. The presence of this invisible decay channel can affect the corresponding Higgs mass bounds in an important way.

The production and subsequent decay of a Higgs boson which may decay visibly or invisibly involves three independent parameters: its mass M_H, its coupling strength to the Z, normalized by that of the standard model, ϵ^2, and its invisible decay branching ratio. The LEP searches for various exotic channels can be used in order to determine the regions in parameter space that are already ruled out [21]. The exclusion contour in the plane ϵ^2 vs. M_H is shown in Figure 3 [22].

Another mode of production of invisibly decaying Higgs bosons is that in which a CP even Higgs boson is produced at LEP in association with a massive CP odd scalar [23]. Present limits on this production mode are given in Figure 4, taken from ref. [23]. In this plot we have assumed BR $(H \to J\, J) = 100\%$ and a visibly decaying A boson.

The invisible decay of the Higgs boson may also affect the strategies for searches

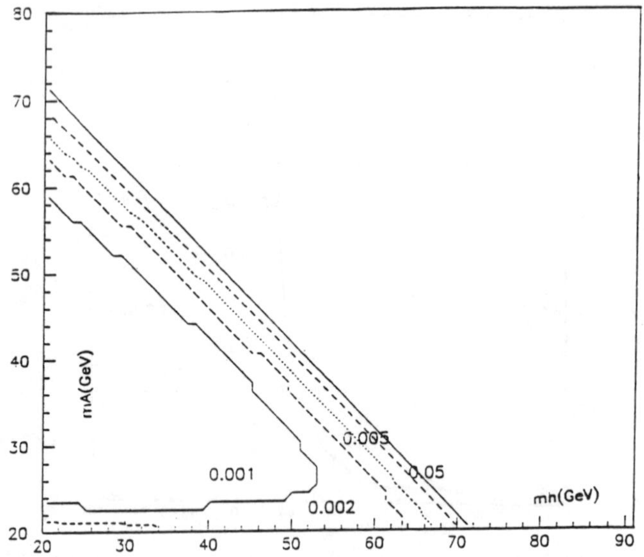

Figure 4: Limits on ϵ_A^2 in the m_A, m_H plane that can be placed by present LEP1 searches based on the $e^+e^- \to H\ A \to J\ Jb\bar{b}$ production channel.

at higher energies. For example, the ranges of parameters that can be covered by LEP2 searches for a total integrated luminosity of 500 pb^{-1} and various centre-of-mass energies have been given in Figure 3. Similar analysis were made for the case of a high energy linear e^+e^- collider (NLC) [24], as well as for the LHC [25].

4. Summary

In summary, there is a wealth of phenomena covering a broad range of energies and of experimental situations that may probe the physics of supersymmetric extensions of the standard model with spontaneously broken R parity. Apart from signatures in the neutrino sector, such neutrinoless double beta decays and possible distortions in beta decay spectra, there are related flavour violating processes accessible at muon and tau factories, as well as novel supersymmetric signals at high energy e^+e^- collisions at LEP, summarized in Table 1. There are also good prospects to observe some of these signatures at the upcoming linear e^+e^- colliders as well as at the LHC. Finally we have discussed the rather peculiar possibility that the Higgs boson may decay dominantly by two majoron emission, leading to missing momentum events. New search strategies are under study to cover all possible signals associated with this possibility in the context of the new high energy colliders.

5. Acknowledgements

I thank my all my collaborators on the subject reported here, as well as the organizers for a very pleasant and exciting meeting.

1. H. Haber and G. Kane, *Phys. Rep.* **117** (1985) 75; H. P. Nilles, *Phys. Rep.* **110** (1984) 1
2. J. W. F. Valle, *Physics at New Accelerators, Looking Beyond the Standard Model*, invited talk at *Neutrino 92 Nucl. Phys. B (Proc. Suppl.)* **31** (1993) 221-232
3. A Masiero, J. W. F. Valle, *Phys. Lett.* **B251** (1990) 273; J. C. Romão, C. A. Santos, and J. W. F. Valle, *Phys. Lett.* **B288** (1992) 311
4. G. Giudice, A. Masiero, M. Pietroni, A. Riotto, *Nucl. Phys.* **B396** (1993) 243; M. Shiraishi, I. Umemura, K. Yamamoto, *Phys. Lett.* **B313** (1993) 89; I. Umemura, K. Yamamoto, *Nucl. Phys.* **B423** (1994) 405
5. J. Steinberger, in *Electroweak Physics Beyond the Standard Model*, ed. J. W. F. Valle and J. Velasco (World Scientific, Singapore, 1992), p. 3.
6. P. Nogueira, J. C. Romão, J. W. F. Valle, *Phys. Lett.* **B251** (1990) 142; R. Barbieri, L. Hall, *Phys. Lett.* **B238** (1990) 86.
7. M. C. Gonzalez-Garcia, J. W. F. Valle, *Nucl. Phys.* **B355** (1991) 330
8. J. W. F. Valle, *Gauge Theories and the Physics of Neutrino Mass*, *Prog. Part. Nucl. Phys.* **26** (1991) 91-171 and references therein.
9. For a review see G. Steigman; proceedings of the *International School on Cosmological Dark Matter*, (World Scientific, 1994), ed. J. W. F. Valle and A. Perez, p. 55
10. J. W. F. Valle, *Phys. Lett.* **B131** (1983) 87; G. Gelmini, J. W. F. Valle, *Phys. Lett.* **B142** (1984) 181; M. C. Gonzalez-Garcia, J. W. F. Valle, *Phys. Lett.* **B216** (1989) 360. A. Joshipura, S. Rindani, PRL-TH/92-10; for an early discussion see J. Schechter and J. W. F. Valle, *Phys. Rev.* **D25** (1982) 774
11. M. Mikheyev, A. Smirnov, *Sov. J. Nucl. Phys.* **42** (1986) 913; L. Wolfenstein, *Phys. Rev.* **D17** (1978) 2369; *ibid.* **D20** (1979) 2634.
12. J. C. Romão and J. W. F. Valle. *Phys. Lett.* **B272** (1991) 436; *Nucl. Phys.* **B381** (1992) 87.
13. M. C. Gonzalez-Garcia, J. C. Romão, J. W. F. Valle, *Nucl. Phys.* **B391** (1993) 100
14. J. C. Romão, N. Rius, J. W. F. Valle, *Nucl. Phys.* **B363** (1991) 369.
15. Particle Data Group, *Phys. Rev.* **D50** (1994) 1173
16. R. Alemany *et al.*, in ECFA/93-151, ed. R. Aleksan, A. Ali, p. 191-211
17. Y. Chikashige, R. Mohapatra, R. Peccei, *Phys. Rev. Lett.* **45** (1980) 1926
18. A. Joshipura and J. W. F. Valle, *Nucl. Phys.* **B397** (1993) 105
19. J. C. Romao, F. de Campos, and J. W. F. Valle, *Phys. Lett.* **B292** (1992) 329.
20. A. S. Joshipura, S. Rindani, *Phys. Rev. Lett.* **69** (1992) 3269; R. Barbieri, and L. Hall, *Nucl. Phys.* **B364** (1991) 27; G. Jungman and M. Luty, *Nucl.*

Phys. **B361** (1991) 24; S. Bertolini, A. Santamaria, *Phys. Lett.* **B213** (1988) 487; E. D. Carlson and L. B. Hall, *Phys. Rev.* **D40** (1989) 3187

21. A. Lopez-Fernandez, J. Romao, F. de Campos and J. W. F. Valle, *Phys. Lett.* **B312** (1993) 240; B. Brahmachari, A. Joshipura, S. Rindani, D. P. Roy, K. Sridhar, *Phys. Rev.* **D48** (1993) 4224.
22. F. de Campos et al., talk at Moriond94, HEP-PH/9405382, to appear in the proceedings.
23. F. de Campos, M. A. Garcia-Jareño, A. Joshipura, D. P. Roy, J. W. F. Valle, *Phys. Lett.* **B336** (1994) 446-456
24. O. Eboli, M. C. Gonzalez-Garcia, A. Lopez-Fernandez, S. Novaes, J. W. F. Valle, *Nucl. Phys.* **B421** (1994) 65
25. J. D. Bjorken, SLAC Report, SLAC-PUB-5673 (1991); J. C. Romao, F. de Campos, L. Diaz-Cruz, and J. W. F. Valle, *Mod. Phys. Lett.* **A9** (1994) 817; J. Gunion, *Phys. Rev. Lett.* **72** (1994) 199; D. Choudhhury, D. P. Roy, *Phys. Lett.* **B322** (1994) 368.

GAUGE-COUPLING UNIFICATION IN THE MSSM: A FOURTH GENERATION BELOW THE TOP?

J. F. GUNION[(a)], DOUGLAS W. McKAY[(b)] and H. POIS[(a)] *

(a) Davis Institute for High Energy Physics,
Department of Physics, U. C. Davis, Davis, CA 95616
(b) University of Kansas, Department of Physics and Astronomy,
Lawrence, Kansas, 66045

ABSTRACT

We explore the possibility of a fourth generation in the gauge-coupling-unified, minimal supersymmetric (MSSM) framework. We find that a sequential fourth generation (with a heavy neutrino ν') is still allowed, provided $\lambda_b(M_U) = \lambda_\tau(M_U)$ Yukawa unification is relaxed. For the theory to remain perturbative up to M_U, the new leptonic generation must lie within reach of LEP-II and the new b', t' quarks must have masses within the reach of the Tevatron, and we comment on t as well as b', t' RGE 'fixed-point' solutions. Experiments at Fermilab are already sensitive to certain mass regions and we comment on direct b' searches and on the $m_{t'} \simeq m_t$ case in light of new CDF data. Discovery may involve novel decay signatures; however, CDF and LEP-II should confirm or exclude an MSSM fourth generation in the near future.

1. Introduction

The experimentally consistent prediction of gauge-coupling unification in the minimal supersymmetric model (MSSM)[1] has led to many detailed investigations of Yukawa and superparticle phenomenology within the MSSM context.[2,3] Virtually all studies have assumed three generations, since broad interest in a possible fourth generation ended abruptly when Mark II and detectors at LEP found that only three light neutrinos are allowed.[4] Here, we argue that by including an $SU(3)\times SU(2)\times U(1)$ gauge singlet, right-handed neutrino ν_R[5] and using the most recent experimental constraints and theoretical inputs, a fourth generation gauge-unified MSSM is in fact, still possible, and briefly explore the relevance to present and future experiments (we identify the fourth generation by the CKM matrix hierarchy: $V_{tb} \simeq V_{t'b'} > V_{t'b} \sim V_{tb'}$).

Even prior to the LEP results, a fourth generation in the gauge-coupling unified MSSM context was disfavored since it spoils the good (given the uncertainty in m_b[6]) prediction for m_b/m_τ,[7] *assuming* that $R_{b/\tau} \equiv \lambda_b(M_U)/\lambda_\tau(M_U) = 1$. However, recent theoretical work indicates that this strict equality may no longer hold given (i) possibly large threshold effects at SUSY breaking and $\mathcal{O}(M_U)$ scales,[8] and/or (ii) because there

*Conference Speaker.

may be *no* symmetry in the model at M_U which leads to an exact equality among the Yukawa couplings; this is the case for several candidate supergravity theories (e.g. flipped $SU(5)$[9]) and for mild extensions of the minimal $SU(5)$ model. For example, $R_{b/\tau} \neq 1$ in general when a **45** Higgs representation[10] is added.

2. 2. Two-Loop Results

We have explored the fourth generation parameter space by solving the two-loop gauge and Yukawa RGE's, and have relaxed the $R_{b/\tau} = 1$ constraint. We neglect the Yukawa coupling effects in the gauge sector. Gauge-coupling unification along with the experimental central values of $\sin^2\theta_W(m_Z) = 0.2324$[11] and $\alpha_{QED}(m_Z) = 1/127.9$ then determines $\alpha_3(m_Z)$; the result (at two-loop) is $\alpha_3(m_Z) = 0.130$. We also obtain $M_U = 5.46 \times 10^{16}$ GeV and $\alpha_U = 0.0936$.[11] In addition, we input values of $\tan\beta$ and the running masses $m_t(m_t)$, $m_b(m_b)$, $m_{\tau'}(m_{\tau'})$, and $m_{\nu'}(m_{\nu'})$. A bound of $\lambda_f < 3$ is imposed in order to guarantee perturbative validity for the RGE's, which proves to be quite constraining.

We find that $\lambda_{\nu'}$ and $\lambda_{\tau'}$ are intrinsically the most susceptible to Landau-pole divergence in evolving up to M_U; $\lambda_{\nu'}$ or $\lambda_{\tau'}$ are typically driven to non-perturbative values unless $m_{\nu'}, m_{\tau'}$ are near their lower bound of $\sim m_Z/2$. Thus, to explore large t, t', b' Yukawas and large $m_{b',t'}$ values, we employ large m_t, take $m_{\tau'} = m_{\nu'} = 50\,\mathrm{GeV}$, and examine the region of perturbatively allowed solutions in the $m_{b'} - m_{t'}$ plane for various values of $\tan\beta$.

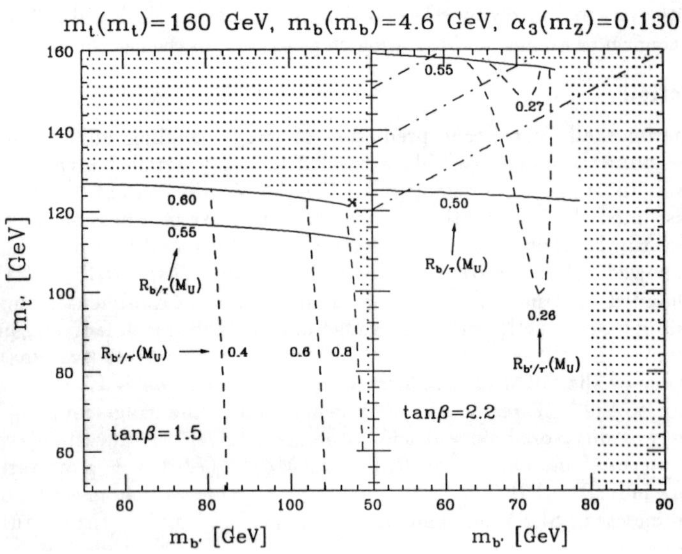

Figure 1: The allowed region in $m_{b'}$–$m_{t'}$ parameter space is illustrated for $m_{\tau'} = m_{\nu'} = 50\,\mathrm{GeV}$, $m_b = 4.6\,\mathrm{GeV}$, and $m_t = 160\,\mathrm{GeV}$. (All masses are $m(m)$ running masses.) The perturbatively disallowed region is dotted. Within the allowed undotted region, the horizontal solid (vertical dashed) lines correspond to fixed values of $R_{b/\tau}$ ($R_{b'/\tau'}$) as labelled. Results for $\tan\beta = 1.5$ and 2.2 are shown. The dash-dot lines in the $\tan\beta = 2.2$ window are contours of $\Delta\rho = 0.002$, 0.004 and 0.006 (in order of decreasing restrictiveness). The approximate location of the $\lambda_t, \lambda_{t'}, \lambda_{b'}$ Yukawa fixed point solution is marked by the $\times$ in the $\tan\beta = 1.5$ case.

Fig.1 illustrates the scope of the perturbatively-allowed solutions for $\tan\beta > 1$ and $R_{b/\tau} \neq 1$, adopting $m_b = 4.6\,\mathrm{GeV}$ and $m_t = 160\,\mathrm{GeV}$ (*i.e.* $m_t(pole) = 169\,\mathrm{GeV}$). Similar results apply for higher m_t. Clearly, there is still significant room for a fourth generation of quarks and leptons (with masses below m_t), and their superpartners (for a more detailed discussion of our results and procedure, see[12]). For $\tan\beta = 1.5$ we find a nearly square allowed region extending to $m_{b'}, m_{t'} \lesssim 120 - 125\,\mathrm{GeV}$, as shown in Fig.1a. For $\tan\beta = 2.2$ (overall, we establish an upper limit of $\tan\beta \lesssim 3.$), larger $m_{t'}$ values $\lesssim m_t$ are possible but only for $m_{b'} < m_Z$, as illustrated in Fig.1b. For both $\tan\beta$ cases, $R_{b/\tau}, R_{b'/\tau'}$ are simultaneously near-maximal (but < 1) in the large-$m_{b'}$, large-$m_{t'}$ corner of parameter space. In Fig.1a, $m_{t'}^{max}$ at fixed $m_{b'}$ ($m_{b'}^{max}$ at fixed $m_{t'}$) occurs when λ_t ($\lambda_{b'}$, with $\lambda_{\tau'}$ close behind) goes non-perturbative. In Fig.1b, $m_{t'}^{max}$ occurs when $\lambda_{t'}$ (followed very closely by $\lambda_{\nu'}$ and λ_t) goes non-perturbative; at $m_{b'}^{max}$ it is $\lambda_{\tau'}$ that goes non-perturbative. For higher $m_{\nu'} = m_{\tau'}$ (*e.g.* $\gtrsim 55\,\mathrm{GeV}$ at $\tan\beta = 1.5$) it is $\lambda_{\tau'}$ that goes non-perturbative at all boundary points.

We also find that approximate 'fixed-point'[13] behavior for the $\lambda_t, \lambda_{t'}, \lambda_{b'}$ Yukawas can be realized. For example, for $m_{\nu'} = m_{\tau'} = 50\,\mathrm{GeV}$ and $\tan\beta = 1.5$ the λ_t, $\lambda_{t'}$ and $\lambda_{b'}$ β-functions are simultaneously zero (and *defines* our use of the term 'fixed-point') for $m_t, m_{t'}, m_{b'} = 162.2,\ 126.3,\ 118\,\mathrm{GeV}$.

3. Experiments and a Fourth Generation

The requirement that all couplings remain perturbative places restrictive upper bounds on the t', b', ν' and τ' masses, and the question of an MSSM fourth generation may be settled in the next few years by searches at (i) the Tevatron for an appropriate set of new signatures (*i.e.* besides $t \to b + W$), and (ii) at LEP-II via detection of direct $\tau' + \bar{\tau}'$ and/or $\nu' + \bar{\nu}'$ production.

First, we find $\Delta\rho$ is not constraining (see Fig.1b). Secondly, LEP has set the firmest limit, $m_f \gtrsim M_Z/2$ for $f = b', t', \tau', \nu'$. The phenomenology at the Tevatron depends crucially on the mass heirarchy of the t', b' as well as the mixing pattern between the generations. If $m_{t'} > m_{b'}$ and $V_{b'c}$ or $V_{b'u}$ is significant, published data[14] already rule out $m_{b'} < 85\,\mathrm{GeV}$ at 95% C.L.. If unmixed with lower generations, the b' will decay via flavor-changing neutral current (FCNC) channels: $b' \to b\gamma, bg$ or bZ^* for $m_{b'} < m_Z$, with $b' \to bZ$ becoming dominant for $m_{b'} > m_Z$; present data may already rule this region out unless $b' \to bh$ is significant. All of these 'non-standard' decays should yield distinctly different signatures than the lepton-plus-jet signatures of charged current decays that are already mostly excluded. New search strategies are required. For example, when $m_{b'} < m_Z$ the process $p\bar{p} \to b'\bar{b}' \to b\gamma + \bar{b}\gamma$ yields a significant rate for 2jet$+2\gamma$ final states. Our rough Monte-Carlo study of the number of $t'\bar{t}'$ production and decay events in the dilepton-plus-jets and lepton-plus-jets channels is not inconsistent with the most recent CDF data (see[12] for details). On the other hand, if $m_{b'} > m_{t'}$, the situation could be quite different. Aside from some perhaps 'unnatural' mixing pattern among the quarks, $t' \to bW$ should dominate over $t' \to cX$ FCNC modes, and may already be excluded by t-quark searches.

To conclude, we find that a sequential fourth generation in the MSSM is perturbatively allowed, if one allows for $R_{b/\tau} \neq 1$, and there is no immediate conflict with CDF top signal analysis. Additional experimental analysis assuming non-standard decay modes seems warranted. For $\tan\beta < 1.5$, $R_{b/\tau}$ can be somewhat larger than shown in Fig.1 ($R_{b/\tau} \lesssim 0.6$) for our choice of $m_b(m_b) = 4.6\,\mathrm{GeV}$. Overall, we find that $m_{t'} \lesssim 185\,\mathrm{GeV}, m_{b'} \lesssim 155\,\mathrm{GeV}$ (for $m_t \gtrsim 150\,\mathrm{GeV}$), and that $m_{\tau',\nu'} \lesssim 85\,\mathrm{GeV}$. This suggests that a fourth generation in the gauge-unified MSSM context should be either confirmed our ruled out in the near future.

References

1. P. Langacker, in Proceedings of the PASCOS-90 Symposium, eds. P. Nath and S. Reucroft (World Scientific, 1990); P. Langacker and M. -X. Luo, *Phys. Rev.* **D44**, (1991) 817; U. Amaldi, W. de Boer, and H. Furstenau, *Phys. Lett.* **B260**, (1991) 447; J. Ellis, S. Kelley and D. V. Nanopoulos, *ibid.* **260**, (1991) 131.
2. S. Dimopoulos, L. Hall and S. Raby, *Phys. Rev. Lett.* **68**, (1992) 1984; *Phys. Rev.* **D45**, (1992) 4192; P. Ramond, R.G. Roberts and G. Ross, *Nucl. Phys.* **B406**, (1993) 19.
3. V. Barger, M. S. Berger, T. Han and M. Zralek, *Phys. Rev. Lett.* **68**, (1992) 3394; H. Arason, D. J. Castano, E. J. Piard and P. Ramond, *Phys. Rev.* **D47**, (1993) 232; V. Barger, M. S. Berger and P. Ohmann, *Phys. Rev.* **D47**, (1993) 1093; G. L. Kane, C. Kolda, L. Roszkowski and J. Wells, *Phys. Rev.* **D50**, (1994) 3498.
4. Mark II Collaboration, G. S. Abrams et al., *Phys. Rev. Lett.* **63**, 2173 (1989); L3 Collaboration, B. Advera et al., *Phys. Lett.* **B231**, 509 (1989); OPAL Collaboration, I. Decamp et al, *ibid.*, **231**, 519 (1989); DELPHI Collaboration, M. Z. Akrawy et al. *ibid.*, **231**, 539 (1989).
5. S. King, *Phys. Lett.* **B281**, (1992) 295.
6. M. Chanowitz, J. Ellis and M. Gaillard, *Nucl. Phys.* **B128**, (1977) 506; A. Buras, J. Ellis, M. Gaillard and D. V. Nanopoulos, *ibid.*, **135**, (1978) 66; M.B. Einhorn and D.R.T. Jones, *ibid.*, **196**, (1982) 475; J. Ellis, D. V. Nanopoulos and S. Rudaz, *ibid.*, **202**, (1992) 43; a recent analysis of both SM and the MSSM is given in H. Arason, D. J. Castano, E.J. Piard and P. Ramond, *Phys. Rev.* **D47**, (1990) 232.
7. D. V. Nanopoulos and D. Ross, *Phys. Lett.* **B118**, (1982) 99; J. Bjorkman and D. Jones, *Nucl. Phys.* **B259**, (1985) 533.
8. P. Langacker and N. Polonsky, *Phys. Rev.* **D47**, (1993) 4028; *ibid.*, **49**, (1994) 1454; V. Barger, M. Berger, P. Ohmann, and R. J. N. Phillips, *Phys. Lett.* **B314**, (1993) 351; G. Kane et al. Ref. [3]; G. Ross and R.G. Roberts, *Nucl. Phys.* **B377**, (1992) 571; B.D. Wright, MAD/PH/812.
9. J. L. Lopez, D. V. Nanopoulos and A. Zichichi, *Phys. Lett.* **B327**, (1994) 279.
10. P.H. Frampton, S. Nandi and J.G. Scanio, *Phys. Lett.* **B85**, (1979) 225.
11. See P. Langacker and N. Polonsky in.[8]
12. J.F. Gunion, D.W. McKay and H. Pois, *Phys. Lett.* **B334**, (1994) 339.
13. B. Pendelton and G. Ross, *Phys. Lett.* **B98**, (1981) 291; C. Hill, *Phys. Rev.* **D24**, (1981) 691; W. Bardeen, M. Carena, S. Pokorski and C. Wagner, *Phys. Lett.* **B320**, (1994) 110; J. Bagger, S. Dimopoulos and E. Masso, *Phys. Rev. Lett.* **55**, (1985) 920.
14. F. Abe *etal.*, *Phys. Rev. Lett.* **64**, (1990) 147, *Phys. Rev. Lett.* **68**, (1992) 447; T. Trippe, Particle Data Group, private communication.

TESTING THE $WW\gamma$ COUPLING AT e^+e^- COLLIDERS

Jan Kalinowski*
Institute of Theoretical Physics, Warsaw University
Hoża 69, 00681 Warsaw, Poland

ABSTRACT

The production of single photons plus missing energy at future e^+e^- colliders can provide a testing ground for non-standard $WW\gamma$ couplings. We show that even with conservative estimates of systematic errors there is still considerable sensitivity to anomalous couplings.

In spite of many experimental successes certain aspects of the standard model still await for direct tests. Among them is the non-abelian sector of the model and, in fact, a detailed investigation of gauge boson self-couplings will constitue one of the primary physics goals of the next linear colliders.[1] The processes with production of gauge bosons in the final state at future colliders should provide such crucial tests of electroweak theory. In particular, they should allow to improve LEP bounds on non-Yang-Mills like triple gauge boson vertices.

In this talk we suggest that the process $e^+e^- \to \bar{\nu}\nu\gamma$ containing a single isolated photon and missing energy may be used to study the precise structure of the $WW\gamma$ vertex.[2] Such final states have been studied at PETRA[3] and at LEP[4,5] as a means of determining the number of light neutrini. At these energies though, not much sensitivity to anomalous $WW\gamma$ form-factors should be expected, as s-channel processes mediated by virtual Z exchange also play an important role. However, at Next Linear Collider (NLC) energies ($\sqrt{s} = 500$ GeV), s channel contributions become less important and the bulk of the cross-section comes from t channel W exchange. The process $e^+e^- \to \bar{\nu}\nu\gamma$ from the point of view of $WW\gamma$ studies has been considered earlier.[6] The novel feature of our analysis, as we will show, is that it is possible to choose cuts which enhance the sensitivity of the observed cross-sections and differential distributions to the $WW\gamma$ form-factor.

The advantage of the process $e^+e^- \to \bar{\nu}\nu\gamma$ over the most studied[1] reaction $e^+e^- \to W^+W^-$ is that the latter, in spite of being a sensitive probe of non-standard $WW\gamma$ and WWZ couplings, suffers from the drawback that there is no obvious way to disentangle the effects of $WW\gamma$ and WWZ form-factors. Hence it is desirable to investigate other channels where such $\gamma - Z$ interference effects are not present.

The deviations of non-abelian vertices from the standard model couplings, as was shown by Hagiwara et al.,[7] can be parametrized by seven possible Lorentz and $U(1)_{em}$ invariant triple gauge boson form-factors. Requiring C and P invariance (in the absence of beam polarisation there is no way to detect CP violating asymmetries if the only particle detected is a photon of unknown polarisation) then only two, traditionally denoted by κ and λ, remain. In the standard model we have $\kappa = 1$ and

*Supported in part by the Polish Committee for Scientific Research Grant 2 P302 095 05

$\lambda = 0$. Deviations from the standard model are then parameterised by $\delta\kappa = \kappa - 1$ and λ. The modified Feynman rules for the $WW\gamma$ vertex may be obtained from Ref.[8] All other Feynman rules are standard ones.

Due to the complexity of the Feynman rules for the non-standard couplings it turns out to be convenient to calculate the matrix-element using the helicity amplitude formalism.[9,10] The results are presented in Ref.[2] As a further check we evaluated the helicity amplitudes using the formalisms of Ref.[11] and Ref.[10] and find excellent numerical agreement for various values of κ and λ.

Since we assume the electron to be massless we need to impose a minimum angle cut on the direction of the outgoing photon to avoid collinear singularities as well as a minimum energy cut to avoid IR problems. Setting $\theta_{min} = 20^o$ and $E_{min} = 10$ GeV we find for the standard model a cross-section of ~ 1.6 pb, which at projected NLC luminosities (~ 10 fb^{-1}) represents a sizable number of events. However, with these cuts alone the sensitivity to non-standard couplings is rather small because the bulk of the cross-section comes from initial state soft photon bremstrahlung which is independent of the non-abelian couplings. It is clear that only the more energetic photons will be sensitive to the anomalous form-factors which are associated with higher dimensional operators containing derivative interactions. Therefore we require that the minumum energy of the photon be 80 GeV. Further improvement can be obtained by eliminating the background from the Z exchange graphs, $e^+e^- \to Z\gamma \to \overline{\nu}\nu\gamma$. This can easily be achieved by means of a simple kinematical cut because in this case the photon is essentially monochromatic (in the limit that the Z width is negligible) with an energy close to half the centre of mass energy. Hence we require that the energy of the photon be less than 180 GeV in order not to reduce the signal from the W exchange graphs too much.

With these cuts on the photon energy, 80 – 180 GeV, the cross-section for the standard model is 0.21 pb, which still leads to an appreciable number of events at NLC luminosities. Cross-sections for non-standard values of $\delta\kappa$ and λ with the cuts mentioned above are presented in Fig.1, where we have varied $\delta\kappa$ and λ individually and not simultaneously in order to keep the analysis simple.

As we can see, the process $e^+e^- \to \overline{\nu}\nu$ is quite sensitive to the deviations from the standard model. The experimental limits that can be derived for $\delta\kappa$ and λ depend however on possible experimental and theoretical uncertainties. Statistical errors are probably quite small given the large number of events $\mathcal{O}(2000)$. Assuming that there are no experimental systematic errors, the main source of theoretical systematic errors lies in unknown higher order corrections. Note that the higher order corrections discussed in Ref.[4] are those which are dominant on the Z pole, and are therefore not adequate for our purposes. It is reasonable to assume that the bulk of the corrections are real and virtual QED corrections which integrated over the the phase space are probably quite small. However we are restricting ourselves to a limited region of the total phase space where radiative corrections may be sizable even though the total radiative corrections themselves are small. Making the conservative assumption that the overall systematic uncertainties are $\mathcal{O}(20\%)$ it is possible to put the folowing discovery bounds $-0.6 < \lambda < 0.6$ and $-0.6 < \delta\kappa < 2.2$ using the cross-section with the cuts mentioned above as only sensitive variable.

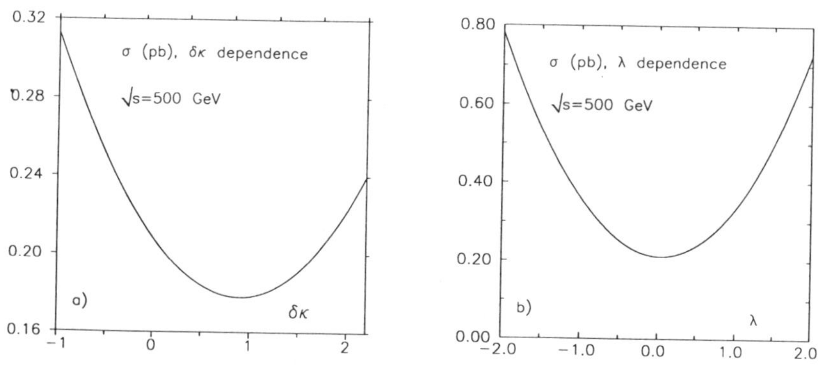

Fig. 1. Cross section for the process $e^+e^- \to \nu\bar{\nu}\gamma$ at $\sqrt{s} = 500$ GeV (a) as a function of $\delta\kappa$ for $\lambda = 0$ and (b) as a function of λ for $\delta\kappa = 0$

If higher luminosity can be achieved (~ 50 fb^{-1}) one can be more optimistic† about the systematic uncertainties and assuming errors to be $\mathcal{O}(5\%)$ better limits can be derived $-0.3 < \lambda < 0.3$ and $-0.2 < \delta\kappa < 0.2$ or $1.2 < \delta\kappa < 1.6$. The discovery limits derived by Couture and Godfrey,[12] where similar issues are discussed, are even more stringent than ours due to very optmistic assumptions about the size of theoretical systematic errors.

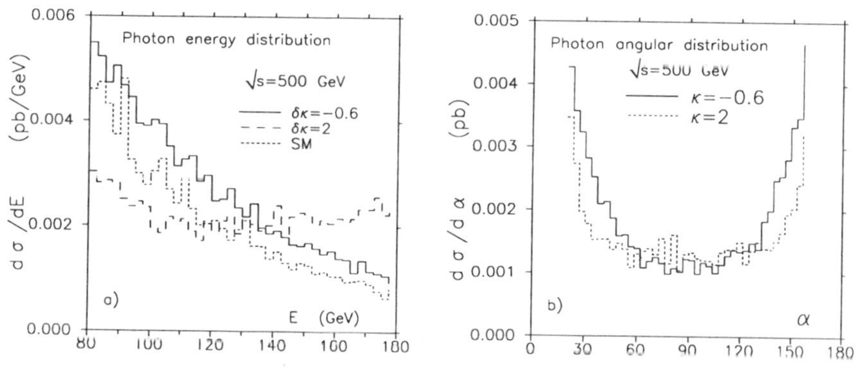

Fig. 2. (a) Energy spectrum and (b) angular distribution of the photon at $\sqrt{s} = 500$ GeV for the standard model and for $\delta\kappa = -0.6$ and 2.

†I thank R. Settles for discussion on this point

The total cross section measurement alone will not allow to determine the parameters $\delta\kappa$ and λ unambiguously. Further refinement is possible if one considers differential distributions. This is illustrated in Fig.2a where we have plotted the differential distribution with respect to photon energy for the standard model and for two values of $\delta\kappa$.

Although the cross-sections are almost the same the differential distributions are different. For the angular distribution, Fig.2b, the sensitivity to different values of $\delta\kappa$ is weaker. Similar effects are observed for λ keeping $\delta\kappa = 0$. However, to derive further discovery bounds on the basis of differential distributions, a detailed consideration of detector acceptances and higher order radiative corrections is necessary.

It is interesting to note that even our conservative results compare favourably with the bounds obtained by McKellar and He[13] on the basis of the recent CLEO measurement of $b \to s\gamma$.[14] Refering to $e^+e^- \to W^+W^-$ at LEP II at 190 GeV more stringent bounds of the order of ± 0.5 for λ and $\delta\kappa$ can be obtained based on theoretically favored three–parameter fits provided polarization properties of produced W bosons are fully measured.[15] Errors can be further reduced either by imposing additional constraints and performing fits with only one or two free parameters or employing initial beams polarization.[16] Considerably weaker bounds are deduced from unconstrained multi–parameter fits. However these bounds are set knowing that one-loop corrections are small leading to smaller theoretical systematic errors than what we have assumed. It is also worth pointing out that W pair production is sensitive to the time-like triple gauge boson form-factors, whereas the process we are studying probes the same form-factors in the space-like region. Therefore both measurements are in some sense complementary.

To conclude, we have demonstrated that $\sigma(e^+e^- \to \overline{\nu}\nu\gamma)$ at projected NLC energies and luminosities is sensitive to anomalous $WW\gamma$ couplings. Making conservative estimates for systematic errors involved it is possible to constrain λ and κ to lie within the regions $-0.6 < \lambda < 0.6$ and $-0.6 < \delta\kappa < 2.2$. These bounds can be improved through knowledge of currently unknown radiative corrections.

Acknowledgements

I would like to thank J. Abraham for many discussions and enjoyable collaboration.

References

1. For a review and references see "e^+e^- *Collisions at 500 GeV: The Physics Potential* ", Proc. Munich, Annecy, Hamburg Workshop, 1991-1993, DESY Reports 92-123A+B, 93-123C.
2. K.J. Abraham, J. Kalinowski, P. Ściepko, *Phys. Lett.* **B339** (1994) 136.
3. K.J.F. Gaemers, R.Gastmans, F.M. Renard, *Phys. Rev.* **D19** (1979) 1605.
4. F.A. Berends *et al.*, *Nucl. Phys.* **B301** (1988) 583.
5. OPAL Collab., R. Akers*et al.*, CERN-PPE/94-105.

6. G.V. Borisov, V.N. Larin, F.F. Tikhonin, *Z. Physik* **C41** (1988) 287.
7. K. Hagiwara *et al.*, *Nucl. Phys.* **B282** (1987) 253.
8. E. Yehudai, SLAC preprint, SLAC-383 (1991).
9. F.A. Berends, W. Giele, *Nucl. Phys.* **B294** (1987) 700.
10. J.F. Gunion, Z. Kunszt, *Phys. Lett.* **B161** (1985) 333.
11. K, Hagiwara, D. Zeppenfeld, *Nucl. Phys.* **B274** (1986) 1.
12. G. Couture, S. Godfrey, preprint OICP/C 94-4, UQAM-PHE-94-09.
13. X.G. He, B.H.J. McKellar, preprint UM-P-93/53, OZ-93/14.
14. E. Thorndike (CLEO Collab.), in Proc. of the American Physical Society Meeting, Washington DC, April 1993.
15. M. Bilenky, J.L. Kneur, F.M. Renard, D. Schildknecht, *Nucl. Phys.* **B409** (1993) 22.
16. A.A. Likhoded *et al.*, preprint IC/93/288, UTS-DFT-93-22, hep-ph 9309322.

UNIFICATION OF SPINS AND CHARGES IN GRASSMANN SPACE?

NORMA MANKOČ BORŠTNIK
Department of Physics, University of Ljubljana, Jadranska 19
and
J. Stefan Institute, Jamova 39
Ljubljana , 61 111, Slovenia

ABSTRACT

In a space of d $(d > 5)$ ordinary and d Grassmann coordinates, fields manifest in an ordinary four-dimensional subspace as fermionic (1/2, 3/2) or bosonic fields with the corresponding charges , according to two kinds of generators of the Lorentz transformations in the Grassmann space. Vielbeins and spin connections define gauge fields-gravitational and Yang-Mills. For $d = 15$ the theory offers the unification of all known charges - spins and Yang-Mills charges - of fermionic and bosonic fields and of all known interactions.

1. Introduction.

It is the hope that unification of all internal degrees of freedom (spins and Yang-Mills charges) and accordingly of all gauge interactions - Yang-Mills fields and gravity - may help to bring new insight into problems on the low energy region, which otherwise very successful phenomenological electroweak standard model cannot solve.

In this talk I present the theory in which *all internal degrees of freedom of particles* and *fields -spins* and *charges* -are defined by *two kinds* of *generators* of the *Lorentz transformations* in the *Grassmann* d-dimensional *space* (forming the algebra of the Lorentz group $SO(1, d-1)$). The generators of a spinorial character define spinors: generators with indices of the four dimensional subspace define spins of spinors, while the generators with indices of higher dimensions define charges of these spinors. All gauge fields - gravitational as well as Yang-Mills including electrodynamics - are defined by supervielbeins, whose spins and charges are determined by the generators of the Lorentz group of a vectorial charater.

The proposed theory, demonstrating many common features either with supersymmetric theories (since the supersymmetry is the unavoidable part of the theory), or string theories (since the supergeodesics is two dimensional) or the Kaluza-Klein theory (since all interactions are of gravitational origin), offers the additional view on the unification of Yang-Mills fields with gravity, since it unifies all internal degrees of freedom of fermionic and bosonic fields: spins and Yang-Mills charges. In this theory the mass and the electromagnetic charge of the particles are not connected when the generators of the Lorentz transformations in the Grassmann space define spins and charges (rather then the generators of the translations in the ordinary space, so that the old problem of the Kaluza-Klein theory of either too weak electromagnetic field or the Planck mass of the electron no longer appears).

In the proposed theory the space has d ordinary ($\{x^a\}$, *commuting*) and d Grassmann ($\{\theta^a\}$, *anticommuting*, $\theta^a\theta^b + \theta^b\theta^a = 0$) coordinates.

Both kinds of generators of translations in the Grassmann part of the space have an odd Grassmann character. While those of a vectorial character form the Grassmann odd Heisenberg algebra, form those of a spinorial character the Clifford algebra. Generators of translations and the Lorentz transformations of both kinds are after the canonical quantization of coordinates the differential operators in the Grassmann space, and the Grassmann space (and accordingly the ordinary space) has to have at least 5 dimensions in order that the Dirac γ^a operators are the Grassmann even differential operators, transforming spinors into spinors.

Vector space spanned over a d dimensional Grassmann space has the dimension 2^d . Half of vectors have an even, half of vectors have an odd Grassmann character, demonstrating the supersymmetry of the theory . The canonical quantization of fields quantizes the former to bosons, the later to fermions[1].

Generators of translations and the Lorentz transformations in the ordinary and the Grassmann space form the super-Poincaré algebra[1]. The super-Pauli-Ljubanski vector can be defined as a generalization of tho Pauli-Ljubanski vector with an odd Grassmann character[1]. It defines spinor charges.

Supervielbeins, depending on the ordinary and the Grassmann coordinates (and connecting supervectors of a freely falling coordinate system to supervectors of an external coordinate system[1]), define all gauge fields. A spin connection appearing as a superpartner of a vielbein has an odd Grassmann character and describes a fermionic part of a gravitational field.

Since the Lorentz algebra $SO(1,2n)$ has a subalgebra $SO(2m) \times SO(1,2n-2m)$, defines the choice $n = 7$ and $m = 5$, for example, the Lorentz $SO(1,4)$ subalgebra, generators of which determine scalars, spinors, vectors and tensors in the four dimensional part of the space, and $SO(10)$ subalgebra, generators of which define the electromagnetic, weak and colour charges (in the corresponding fundamental and adjoint representations). It is the dependence of fields on Grassmann coordinates, which determines spins and charges of all fields.

It is of course the open question whether or not there exists the topology allowing that four out of d ordinary dimensions would spread while the rest would stay compactified in the way which the above mentioned algebra suggests. Since the non compactified part of the ordinary space should decouple (almost) totally from the rest of the space, another open question arises, concerning the left-right symmetry and the CP violation. If the spontaneous break of these symmetries appear due to vielbeins or spin connections, which manifest in the ordinary four dimensional part of the space as (scalar) Higgs filds with the appropriate properties, then, for example, either mirror fermions exist (with W+A coupling to the ordinary weak bosons) or symmetry is broken on the level of constituents of quarks and leptons or even on a lower level, or asymmetry is due to some other mechanism, connected, for example, to the evolution of the Universe. Anyhow, this problem is the common problem of all theories, including strings, since there asymmetry is assumed on the level of the internal degrees of freedom of a string.

2. A particle in a freely falling coordinate system.

In the Grassmann space of the anticommuting coordinates $\{\theta^a\}$ the left derivatives have to be distinguished from the right derivatives. We shall make use of left derivatives only, defined as follows: $\frac{\overrightarrow{\partial}}{\partial\theta^a}\theta^b f = \delta^b_a f - \theta^b \frac{\overrightarrow{\partial}}{\partial\theta^a} f$.

It turns out that for a particle whose geodesics is defined in the space of ordinary and Grassmann coordinates [1], the Grassmann coordinates are proportional to their conjugate momenta $p^\theta_a := \frac{\overrightarrow{\partial} L}{\partial\dot{\theta}^a} = \epsilon^2\theta^a$. To see this, one has to start from the action for a super geodesics[1]. We shall present here results which follow from the actions for the supergeodesics if the action is integrated over the Grassmann odd parameter of the geodesics. Therefore the Lagrange function L depends on ordinary and on Grassmann coordinates and on the derivatives of these coordinates with respect to the ordinary time parameter , while we choose ϵ^2 equal either to $+i^2$ or to $-i^2$. It is appropriatelly to define generalized coordinates

$$\tilde{a}^a := i(p^{\theta a} - i\theta^a), \;\; \tilde{\tilde{a}}^a := -(p^{\theta a} + i\theta^a), \tag{2.1}$$

while the generators of the Lorentz transformations, which follow from the action $\int L d\tau$, are

$$M^{ab} = L^{ab} + S^{ab} \;,\; L^{ab} = x^a p^b - x^b p^a \;,\; S^{ab} = \theta^a p^{\theta b} - \theta^b p^{\theta a} \tag{2.2}$$

with the same parameters of the Lorentz transformations in both spaces. The generators may be written with respect to operators $\tilde{a}^a$ and $\tilde{\tilde{a}}^a$

$$S^{ab} = \tilde{S}^{ab} + \tilde{\tilde{S}}^{ab} \;,\; \tilde{S}^{ab} = -\frac{i}{4}(\tilde{a}^a\tilde{a}^b - \tilde{a}^b\tilde{a}^a) \;,\; \tilde{\tilde{S}}^{ab} = -\frac{i}{4}(\tilde{\tilde{a}}^a\tilde{\tilde{a}}^b - \tilde{\tilde{a}}^b\tilde{\tilde{a}}^a). \tag{2.2a}$$

The choice of ϵ makes either $\tilde{\tilde{a}}^a(\epsilon^2 = -i)$ or $\tilde{a}^a(\epsilon^2 = i)$ equal to zero and accordingly either $\tilde{\tilde{S}}^{ab} = 0$ or $\tilde{S}^{ab} = 0$.

In the canonical procedure the Poisson brackets follow, treating equivalently the ordinary and the Grassmann space [1]:

$$\{B, A\}_p = -\frac{\partial A}{\partial x^a}\frac{\partial B}{\partial p_a} + \frac{\partial A}{\partial p_a}\frac{\partial B}{\partial x^a} - (\frac{\overrightarrow{\partial} A}{\partial\theta^a}\frac{\overrightarrow{\partial} B}{\partial p^\theta_a} + \frac{\overrightarrow{\partial} A}{\partial p^\theta_a}\frac{\overrightarrow{\partial} B}{\partial\theta^a})(-1)^{n_A}, \tag{2.3}$$

where n_A is either one or two depending on whether A has an odd or an even Grassmann character, respectively.

In the quantization procedure$-i\{A, B\}_p$ goes to either a commutator or to an anticommutator, according to the Poisson brackets (2.3). The operators $\theta^a, p^{\theta a}$ (in the coordinate representation they become $\theta^a \longrightarrow \theta^a$, $p^\theta_a \longrightarrow -i\frac{\overrightarrow{\partial}}{\partial\theta^a}$) fulfil the Grassmann odd Heisenberg algebra, while the operators

$\tilde{a}^a$and $\tilde{\tilde{a}}^a$ fulfil the Clifford algebra:

$$\{\theta^a, p^{\theta b}\} = -i\eta^{ab}, \ \{\tilde{a}^a, \tilde{a}^b\} = 2\eta^{ab} = \{\tilde{\tilde{a}}^a, \tilde{\tilde{a}}^b\}, \tag{2.4}$$

with $\{\tilde{a}^a, \tilde{\tilde{a}}^b\} = 0 = \{\tilde{S}^{ab}, \tilde{\tilde{S}}^{cd}\}$ and $\tilde{S}^{ab} = -\frac{i}{4}[\tilde{a}^a, \tilde{a}^b]_-, \tilde{\tilde{S}}^{ab} = -\frac{i}{4}[\tilde{\tilde{a}}^a, \tilde{\tilde{a}}^b]_-$.

Either L^{ab} or S^{ab} or $\tilde{S}^{ab}$ or $\tilde{\tilde{S}}^{ab}$ form the Lie algebra of the Lorentz group.

It appears that S^{ab} define the adjoint representations while $\tilde{S}^{ab}$ and $\tilde{\tilde{S}}^{ab}$ define the fundamental representations of the Lorentz group.

The constraints which follow from the action lead to the Dirac and to the Klein-Gordon equation

$$p^a \tilde{a}_a |\tilde{\Psi}> = 0 \ , \ p^a p_a |\tilde{\Psi}> = 0, \ with \ p^a \tilde{a}_a p^b \tilde{a}_b = p^a p_a. \tag{2.5}$$

Since the operator $\tilde{a}^a$ (which is a differential operator in the Grassmann space) has an odd Grassmann character, it can not be recognized as the Dirac $\tilde{\gamma}^a$ operator. The dimension of the space d has therefore to be at least five ($d \geq 5$) (which means at least 5 odinary and 5 Grassmann coordinates) in order that the generators of the Lorentz transformations $\tilde{S}^{5m}, m = 0,1,2,3$ can be recognized as the Dirac γ^m operators of an even Grassmann charactor $\tilde{\gamma}^m = -\tilde{a}^5\tilde{a}^m = -2i\tilde{S}^{5m}$, $m = 0,1,2,3$.

It can be checked that in the four-dimensional subspace $\tilde{\gamma}^m$ fulfil the Clifford algebra $\{\tilde{\gamma}^m, \tilde{\gamma}^n\} = \eta^{mn}$, while $\tilde{S}^{mn} = -\frac{i}{4}[\tilde{\gamma}^m, \tilde{\gamma}^n]_-$. The operator $\tilde{\Gamma} = i\tilde{a}^0\tilde{a}^1\tilde{a}^2\tilde{a}^3 = i\tilde{\gamma}^0\tilde{\gamma}^1\tilde{\gamma}^2\tilde{\gamma}^3$ is one of the two Casimir operators of the Lorentz group $SO(1,3)$. We have seen, however, that not $SO(1,3)$ but rather $SO(1,4)$ is needed to properly define the Dirac algebra.

In the case that $d = 5$ and $< \tilde{\psi}|p^5|\tilde{\psi} >= m$ it follows

$$(\tilde{a}^m p_m - \tilde{a}^5 p^5)|\tilde{\psi} >= 0 = (\tilde{\gamma}^m p_m - m)|\tilde{\psi} > \ , \ m = 0,1,2,3. \tag{2.6}$$

For a large enough d not only the generators of the Lorentz transformations (of a spinorial character) in the Grassmann space define (after the canonical quantization of coordinates) spinorial degrees of freedom of a particle field in the four dimensional subspace, they define also quantum numbers of these fields which may be recognized as electromagnetic, weak and colour charges. For $d = 15$, for example, we can express the generators of the groups $U(1), SU(2)$ and $SU(3)$ in terms of the generators of the Lorentz group $\tilde{S}^{ab}$ with $a,b = 6,7,...,15$ while $SO(1,4)$ remains to define the spinorial degrees of freedom in the four dimensional subspace.

3. A particle in (Kaluza-Klein) gauge fields.

The dynamics of a point particle in a gravitational field can be found by transforming vectors from a freely falling to an external coordinate system[1] through vielbeins. To do this we introduce supervielbeins on the level of supergeodesics which manifest after integrating the action over the Grassmann odd parameter, as vielbeins $e^{ia}{}_\mu$ and spin connections$\theta^b e^{ia}{}_{\mu b}$ both depending on ordinary and on Grassmann coordinates, while index i appears due to two types of parameters on which the supergeodesics depends. The index a refers to a freely falling coordinate system, the index μ refers to an external coordinate system. Vielbeins with the Lorentz index smaller than five will determine ordinary gravitational fields, those with the Lorentz index higher than four will define Yang-Mills fields. Spin connections appear in our theory as (a part of) Grassmann odd fields.

Since $e^{ia}{}_{\mu}$ have an even Grassmann character it will describe the spin 2 part of a gravitational field. The fields $\epsilon\theta^{b}e^{ia}{}_{\mu b}$ with an odd Grassmann character (ϵ is again the complex constant) are candidates for spinorial part of a gravitational field.

It appears[1] that vielbeins and spin connections fulfil the relations

$$e^{ia}{}_{\mu}f^{i\mu}{}_{b}=\delta^{a}{}_{b}\ ,\ f^{i\mu}{}_{a}e^{ia}{}_{\nu}=\delta^{\mu}{}_{\nu}\ ,\ e^{ia}{}_{\mu b}f^{i\mu}{}_{c}=e^{ia}{}_{\mu}f^{i\mu}{}_{cb}\ ,\ i=1,2. \tag{3.1}$$

We find that metric tensor has the property $g^{i\mu\sigma}g^{i}{}_{\sigma\nu}=\delta^{\mu}{}_{\nu}$, with $g^{i}{}_{\mu\nu}=e^{ia}{}_{\mu}e^{i}{}_{a\sigma}$.

We find[1] vectors in a freely falling and in an external coordinate system to be connected as follows: $\dot{x}^{a}=e^{1a}{}_{\mu}\dot{x}^{\mu}\ ,\ \dot{x}^{\mu}=f^{1\mu}{}_{a}\dot{x}^{a}\ ,\ \theta^{a}=e^{2a}{}_{\mu}\theta^{\mu}\ ,\ \theta_{\mu}=f^{2\mu}{}_{a}\theta^{a}$, and $\dot{\theta}^{a}=e^{1a}{}_{\mu}\dot{\theta}^{\mu}+\theta^{b}e^{1a}{}_{\mu b}\dot{x}^{\mu}=(e^{2a}{}_{\mu}\theta^{\mu})^{\cdot}=e^{2a}{}_{\nu,\mu_{x}}\dot{x}^{\mu}\theta^{\nu}+e^{2a}{}_{\mu}\dot{\theta}^{\mu}+\dot{\theta}^{\mu}\overrightarrow{e^{2a}}{}_{\nu,\mu_{\theta}}\theta^{\nu}$.

The notation $e^{2a}{}_{\nu,\mu_{x}}=\frac{\partial}{\partial x^{\mu}}e^{2a}{}_{\nu},\overrightarrow{e^{2a}}{}_{\nu,\mu_{\theta}}=\frac{\overrightarrow{\partial}}{\partial\theta^{\mu}}e^{2a}{}_{\nu}$ was used.

The above equations define the following relations among fields: $e^{2a}{}_{\mu b}=0\ ,\ \overrightarrow{e^{2a}}{}_{\nu,\mu_{\theta}}\theta^{\nu}=e^{1a}{}_{\mu}-e^{2a}{}_{\mu}\ ,\ e^{1a}{}_{\mu b}=e^{2a}{}_{\nu,\mu_{x}}f^{2\nu}{}_{b}$, which means that a point particle with a spin sees a spin connection $\theta^{b}e^{ia}{}_{\mu b}$ related to a vielbein $e^{2a}{}_{\nu}$.

¿From the action for a superparticle in a freely falling coordinate system rewritten in terms of an external coordinate system, we get after integrating it over the Grassmann odd parameter of the supergeodesics the action

$$I=\int d\tau\{\frac{1}{N}g^{1}_{\mu\nu}\dot{x}^{\mu}\dot{x}^{\nu}-\epsilon^{2}\frac{2M}{N}\theta_{a}e^{1a}{}_{\mu}\dot{x}^{\mu}+\varepsilon^{2}\frac{1}{2}(\dot{\theta}^{\mu}\theta_{a}-\theta_{a}\dot{\theta}^{\mu})e^{1a}{}_{\mu}+\varepsilon^{2}\frac{1}{2}(\theta^{b}\theta_{a}-\theta_{a}\theta^{b})e^{1a}{}_{\mu b}\dot{x}^{\mu}\}. \tag{3.2}$$

The action (3.2) defines the two momenta of the system: $p_{\mu}=\frac{\partial L}{\partial\dot{x}^{\mu}}=p_{0\mu}+\varepsilon^{2}\theta^{a}\theta^{b}e^{1}_{a\mu b}$, $p^{\theta}_{\mu}=\varepsilon^{2}\theta_{a}e^{1a}{}_{\mu}=\varepsilon^{2}(\theta_{\mu}+\overrightarrow{e}^{2a}{}_{\nu,\mu_{\theta}}e^{2}{}_{a\alpha}\theta^{\nu}\theta^{\alpha})$. Here $p_{0\mu}$ are the covariant (canonical) momenta of a particle. For $p^{\theta}_{a}=p^{\theta}_{\mu}f^{1\mu}{}_{a}$ it follows that p^{θ}_{a} is proportional to θ_{a}. For a choice $\varepsilon^{2}=-i$, $\tilde{a}_{a}=i(p^{\theta}_{a}-i\theta_{a})$, while $\tilde{\tilde{a}}_{a}=0$. In this case we may write

$$p_{0\mu}=p_{\mu}-\frac{1}{2}\tilde{S}^{ab}e^{1}_{a\mu b}=p_{\mu}-\frac{1}{2}\tilde{S}^{ab}\omega_{ab\mu}\ ,\ \omega_{ab\mu}=\frac{1}{2}(e^{1}_{a\mu b}-e^{1}_{b\mu a}), \tag{3.3}$$

which is the ussual expression for the covariant momenta in gauge gravitational fields. One can find the two constraints

$$p^{\mu}_{0}p_{0\mu}=0=p_{0\mu}f^{1\mu}{}_{a}\tilde{a}^{a}. \tag{3.4}$$

To see how Yang-Mills fields enter into the theory, the Dirac equation Eq. (3.4) (after quantization) has to be rewritten in terms of fields which determine the gravitation in the four dimensional subspace and of those fields which determine the gravitation in higher dimensions. This should be done by requiring that coordinates with indices more than four should stay compacified. We shall suppose that a system manifests itself in a way that only some components of fields are different from zero and that they depend on all Grassmann coordinates (which determine spins and charges of fields) and on some of ordinary coordinatesx^α, $\alpha = 0,1,2,3$ while $m = -p_5(e^{15}{}_5)^{-1}$. We find

$$\tilde{\gamma}^a f^{1\mu}{}_a p_{0\mu} = \tilde{\gamma}^m f^{1\alpha}{}_m (p_\alpha - \frac{1}{2}\tilde{S}^{mn}\omega_{mn\alpha} + A_\alpha) + m, \tag{3.5}$$

where

$$A_\alpha = \sum_{A,i} \tilde{\tau}^{Ai} A^{Ai}_\alpha, \quad [\tilde{\tau}^{Ai}, \tilde{\tau}^{Bj}] = i\delta^{AB} f^{Aijk} \tilde{\tau}^{Ak}, \tag{3.5a}$$

with $\sum_{A,i} \tilde{\tau}^{Ai} A^{Ai}_\alpha = \frac{1}{2}\tilde{S}^{hk}\omega_{hk\alpha}$, $h,k = 6,7,8,..d$, according to the subalgebra of the Lie algebra of the Lorentz group $SO(d-5)$, provided that $\tilde{\tau}^{Ai}$ can be expressed as linear superpositions of operators $\tilde{S}^{ab}$

$$\tilde{\tau}^{Ai} = c^{Ai}{}_{ab}\tilde{S}^{ab}, \quad c^{Ai}{}_{ab} = -c^{Ai}{}_{ab}, \quad A = 1,n, \quad i = 1, n_A, \quad a,b = 6,d. \tag{3.5b}$$

Here f^{Aijk} are structure constants of n subgroups A , each with n_A operators, while constants $c^{Ai}{}_{ab}$ fulfil the following relations: $-4c^{Ai}{}_{ab}c^{Bjb}{}_c - \delta^{AB} f^{Aijk} c^{Ak}{}_{ac} = 0$.

This certainly can be done for $d = 15$, since $SO(1,14)$ has the subalgebra $SO(1,4)\times SO(10)$,while $SO(10)$ has the subalgebra $SU(3)\times SU(2)\times U(1)$. The commutator of Eq. (3.5a) is valide not only for operators of the spinorial character $\tilde{\tau}^{Ai}$ but also for operators of the vectorial character $\tau^{Ai} = c^{Ai}{}_{ab}S^{ab}$.

In Eq. (3.5) we require that $e^5{}_\mu = 0 = e^m{}_\sigma$, with $m = 0,1,2,3$ and $\sigma = 6,7,8,..d$. Fields $\omega_{hk\mu}$ therefore determine all Yang-Mills fields. A torsion and a curvature follow from the Poisson brackets $\{p_{0a}, p_{0b}\}_p$, with $p_{0a} = f^{1\mu}{}_a(p_\mu - \frac{1}{2}\tilde{S}^{cd}\omega_{cd\mu})$. We find

$$\{p_{0a}, p_{0b}\}_p = -\frac{1}{2}S^{cd}R_{cdab} + p_{0c}T^c{}_{ab}, \tag{3.6}$$

$$R_{cdab} = f^{1\mu}{}_{[a} f^{1\nu}{}_{b]}(\omega_{cd\nu,\mu^x} + \omega_c{}^e{}_\mu \omega_{ed\nu} + \overrightarrow{\omega}_{cd\mu,f\theta}\theta^e \omega_e{}^f{}_\nu),$$

$$T^c{}_{ab} = e^{1c}{}_\mu (f^{1\nu}{}_{[b} f^{1\mu}{}_{a],\nu} + \omega_{e\nu}{}^d \theta^e f^{1\nu}{}_{[b}\overrightarrow{f^{1\mu}}{}_{a],d^\theta}),$$

with $A_{[a}B_{b]} = A_a B_b - A_b B_a$. It has to be pointed out that the Poisson brackets $\{p_{0\mu}, p_{0\nu}\}_p$ can be written in terms of the odd Grassmann fields $\Psi^a{}_\mu = \theta^b\omega_{ab\mu}$ as well

$$\{p_{0\mu}, p_{0\nu}\}_p = \frac{i}{2}\tilde{a}^a \mathcal{D}_{[\mu}\Psi_{a\nu]}, \quad \mathcal{D}_\mu \Psi_{a\nu} = \Psi_{a\nu,\mu} + \frac{i}{2}\omega_{cd\mu}\tilde{S}^{cd}\Psi_{a\nu}, \tag{3.7}$$

where $\mathcal{D}_\mu$ appears as a *covariant derivative of a spinor- vector field* $\Psi^a{}_\mu$.

If the action for a free gravitational field is $I = \int d^dx d^d\theta \omega \mathcal{L}$, where ω is a scalar density in the Grassmann space[1], the Lagrange density $\mathcal{L}$ includes $det(e^{1a}{}_\mu)R$, $R =$

$R^{ab}{}_{ab}$, or (and) $det(e^{1a}{}_\mu)$ $T^a{}_{cd}T^{cd}{}_a$ as well as terms with $\bar{\Psi}^{a\nu}\tilde{\gamma}^\mu\mathcal{D}_{[\mu}\Psi_{a\nu]}$. If there exists the topology with four ordinary dimensions extended and with the rest $d-4$ dimensions compactified according to the suggested algebra, $\Psi^a{}_\mu$ and $f^{1\mu}$ describe all gauge fields of fermionic character and bosonic character.

4. Acknowledgements

The work was supported by Ministry of Science and Technology of Slovenia and the National Science Foundation through funds available to the US-Slovenia Joint Board for Scientific and Technological Cooperation (no.NSF 899). I would like to thank to Holger Bech Nielsen and Colin Froggatt for helpful discussions.

5. References

1. N. Mankoč -Borštnik,*Phys.Lett.* **B 292** (1992) 25, *Il nuovo Cimento* **A 105** (1992) 1461, *Journ. of Math. Phys.* **34** (1993) 8,*Int. Journal of Modern Phys.* **A 9** (1994) 1731,*IJS.TP.93/15* (Poincaré algebra in ordinary and Grassmann space and supersymmetry, submitted for publication, IJS.TP.94/8 (Grassmann space, gravity and electrodynamics, to appear in *Proceedings of the Minsk conference Quantum Systems*, New Trends and Methods, Minsk, May,1994

EXPERIMENTAL PROSPECTS for $\mathcal{SUSY}$ in e^+e^-

RONALD SETTLES
Max-Planck-Institut für Physik, Werner-Heisenberg-Institut, Föhringer Ring 6
D - 80805 Munich, Germany

ABSTRACT

Hopes for detecting supersymmetry in e^+e^- annihilations are reviewed using examples from several workshops on present and future colliders. $\mathcal{MSSM}$, the simplest model of minimal supersymmetry with grand unification is studied. Chances for discovering the higgs sector at LEP and at a future 500 GeV e^+e^- linear collider are cited along with ways for measuring the higgs' properties; similary for the supersymmetric partners of the presently-known elementary particles. Once discovered, examples are given as to how to attack the *spartner* spectroscopy and $\mathcal{MSSM}$ parameters at the e^+e^- linear collider. The complementarity of these measurements to those at pp colliders is emphasized.

1. Introduction

Supersymmetry is such a fascinating model that it would be a pity if it did not exist. It is just sitting there, a mathematically well-defined theory, computed in great detail, waiting to be discovered. So how could the discovery, that Mother Nature's name is $\mathcal{SUSY}$, be expedited in the scant duration of the human race on this globe? The answer is use the few colliders that are around, and to build a couple. The following table gives the status of these machines.

Table 1. The status of existing and future colliders.

Status→ **Collider↓**	Running	Under construction	Approved	Getting organized
LEP 1	√			
LEP 200		√		
Tevatron	√			
Tevatron upgrade			√	
LHC			√	
ILC[a]				√

[a]International Linear Collider, the 500 GeV e^+e^- linear collider, called NLC at SLAC, JLC at KEK, SBLC or TESLA at DESY, CLIC at CERN, and VLEPP at INP, which are the names of the R&D project at the different laboratories. The name ILC is used here to symbolize the international collaboration of leading laboratories, formed to coordinate the linear-collider R&D, which is in full swing throughout the world.

By now the physics at future colliders has been studied at many international workshops, SLAC,[1] La Thuile,[2] Snowmass,[3] Japan,[4] Europe,[5,6] Finland,[7] USA,[8] CERN,[9] and Hawaii,[10] among others.* There are yearly LCyy workshops devoted to progress on the machine technology, LC88 in SLAC, LC90 in KEK, LC91 in Protvino, LC92[11] in Garmisch-Partenkirchen, and LC93[12] in SLAC.

The next workshops are both in Japan in 1995: the machine workshop LC95 in March and the physics workshop, the International Linear Collider Workshop, in September.

The scope of this presentation is to hit some of the highlights which demonstrate the potential for the e^+e^- machines. The complementarity of the e^+e^- and *pp*-colliders for mapping out the $\mathcal{SUSY}$ world if it exists has been emphasized at all of the above-mentioned workshops, and recently demonstrated again in an interesting comparison by Michael Peskin,[13] who gives a nice overview of the model.

2. The Model

2.1. The particles

In what follows, the most predictive extension of the standard model ($\mathcal{SM}$) is assumed: minimal supersymmetry plus grand unification ($\mathcal{MSSM}$, $\mathcal{GUT}$). In the higgs sector, electroweak symmetry breaking in the minimal $\mathcal{SM}$ is mediated by a single higgs-doublet field, whereas in the $\mathcal{MSSM}$ there are two higgs-doublet fields ϕ_1 and ϕ_2, the first couples to down-type quarks and charged leptons, and the second couples to up-type quarks. Three of the eight degrees of freedom of these two fields are cannibalized to endow the gauge bosons with their masses, so that five physical states are left, a pair of CP-even neutral states h^0 and H^0, a CP-odd neutral state A^0, and and pair of charged states $H^\pm$.

In the particle sector, each of the well-known elementary particles acquires a superpartner (*sparticle*): bosons accompany fermions and vice-versa. Thus the *sleptons* ($\tilde{\ell}$) and *squarks* ($\tilde{q}$) are bosons, and the *photino* ($\tilde{\gamma}$) and *gluino* ($\tilde{g}$) fermions. The partners to the neutral gauge- and higgs-bosons mix to the *neutralino* ($\tilde{\chi}_i^0$,$i = 1,4$) states, the charged partners to the *chargino* ($\tilde{\chi}_j^\pm$,$j = 1,2$) states. The lightest supersymmetric particle, the *LSP*, is assumed to be the lightest neutralino ($\tilde{\chi}_1^0$) which is weakly interacting and stable.

2.2. Parameters and Relations

The parameters of supersymmetry and its grand unified version are the subject of many contributions to this workshop, so only the ones relevant to the ensuing discussion will be recapitulated here. In the higgs sector described above, the two higgs fields have the vacuum expectation values $v_1 = \langle\phi_1\rangle$ and $v_2 = \langle\phi_2\rangle$, their ratio is

*This year (1994) there were studies in the USA by "DPF Long Range Planning" working groups, to assess the particle-physics landscape in the aftermath of the SSC's demise; the proceedings will appear soon. (DPF is the Division of Particles and Fields of the American Physical Society.)

tanβ=v_1/v_2, a parameter α controls the mixing of the CP-even higgses, and finally there are the masses M_h, M_H, M_A, $M_{H^\pm}$. The $\mathcal{MSSM}$ imposes relations among these seven parameters, so that only two are left free, and the parameter space is frequently shown in the form of M_A -tanβ or M_A-M_h plots. Due to radiative loop effects, the predictions end up also depending on the top and *squark* masses.

In the simplest supersymmetric grand unification, the mass spectrum of the *sparticles* is determined by three parameters, m_0 – a universal scalar mass, $m_{1/2}$ = a universal gaugino mass, and μ = a supersymmetric higgs boson mass parameter, all about the size of M_W which is the scale of $\mathcal{SUSY}$ in "natural" models.[13]

At the $\mathcal{GUT}$ scale, masses and couplings are expected to unify, so one has m_0^2= $M_{\tilde{u}}^2$= $M_{\tilde{d}}^2$= .. $M_{\tilde{t}}^2$= .. $M_{\tilde{\ell}}^2$= .. . The gauge bosons of $SU(1) \times SU(2) \times SU(3)$ have fermionic partners, the masses of which unify to $m_{1/2} = M_1 = M_2 = M_3$. The coupling constants of $SU(1) \times SU(2) \times SU(3)$ are† $\alpha_1 = \frac{5}{3}\frac{\alpha}{\cos^2\theta_W}$, $\alpha_2 = \frac{\alpha}{\sin^2\theta_W}$, and $\alpha_3 = \alpha_s$, which unify to $\alpha_1 = \alpha_2 = \alpha_3 = \alpha_{GUT}$. Also $\frac{M_1}{\alpha_1} = \frac{M_2}{\alpha_2} = \frac{M_3}{\alpha_3} = \frac{m_{1/2}}{\alpha_{GUT}}$ at the level of the one-loop renormalization group equations.

We shall first look into the higgs sector, then at the supersymmetric partners.

3. The Higgs Sector

The searches for the neutral higgs at the e^+e^- colliders have been described many times, for example by Jean-François Grivaz[14] for LEP1, Daniel Treille[15,16] for LEP200, and Sachio Komamiya[17] and Patrick Janot[18,19] for the 500 GeV e^+e^- linear collider. The searches for the $\mathcal{SM}$ higgs also apply to the CP-even $\mathcal{MSSM}$ higgses, the h^0 or the H^0 being $\mathcal{SM}$-like depending on the mixing parameter α. The production mechanism is mainly bremsstrahlung from the Z^0 at lower energies and gauge-boson fusion at higher energies. If the energy is high enough, associated production of the h^0 or H^0 with the A^0 is possible. The Z^0 and the higgses decay into fermions (or gauge bosons for higgses with sufficient mass) with different rates depending on the couplings and the masses, so that searches for mass bumps in different channels (4 jets, 4 leptons, or 2 jets and 2 leptons) to cover all possibilities can easily be orchestrated.

3.1. LEP

At LEP1 all detectors have good efficiency (>30-50%) for higgs searches up to the kinematic limit.[14] The experiments have still found no signal up to $\sim$ 60 GeV and will saturate at $\sim$ 65 GeV at the end of the LEP1.[19] At LEP200 these limits can be extended up to approximately $\sqrt{s}$ – 100 GeV, so that if the lightest higgs is heavier than roughly 100 GeV, LEP will miss it.

3.2. The 500 GeV e^+e^- Linear Collider

The discovery of the $\mathcal{SM}$ higgs and the $\mathcal{MSSM}$ higgses has been simulated by

†α is the fine structure constant here. Unfortunately the same symbol is used for the mixing parameter of the CP-even higgses.

Patrick Janot[19] in quite some detail. Four-jet topologies can be used effectively and shown to be even more efficient with b-tagging since the higgs decays preferentially into the heaviest fermions. Similary, the $\tau^+\tau^- q\bar{q}$ final state is efficient since the h^0, H^0, and A^0decay into $\tau^+\tau^-$ 8% of the time, while the Z^0 does it only 3.4% of the time. The mass distributions for the $\tau^+\tau^-$ and $q\bar{q}$ systems for the events found in the topological search are shown in Fig.1 for two examples of $\mathcal{MSSM}$ parameters.

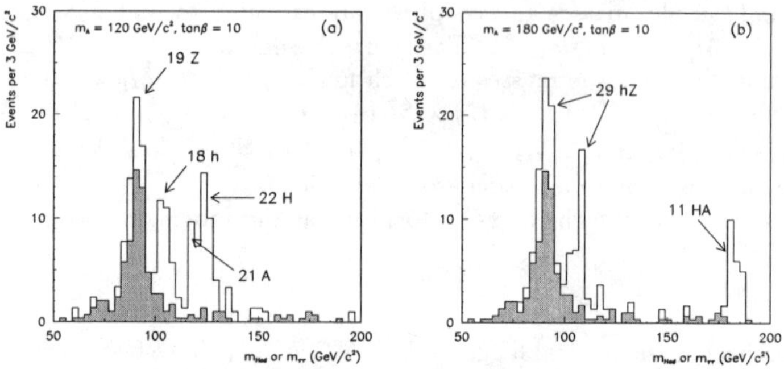

Fig. 1. Results for the search in the $\tau^+\tau^- q\bar{q}$ final state for (a) $M_A = 120$ GeV, tan$\beta = 10$ and (b) $M_A = 180$ GeV, tan$\beta = 10$.

Patrick Janot also demonstrates the discovery for the case that the LSP $\tilde{\chi}_1^0$ is light enough that the higgs can decay into it. This gives rise to an invisible higgs decay, which is highly suppressed in the $\mathcal{SM}$, so that this would consitute the simultaneous discovery of the higgs and $\mathcal{SUSY}$. The search is based on $e^+e^- \to Z^0h^0$ (or H^0) $\to f\bar{f}\tilde{\chi}_1^0\tilde{\chi}_1^0$, and the missing mass distribution in events with one-sided lepton pairs or jets reveals the existence of the $\tilde{\chi}_1^0$.

If the favorable situation just described does not exist, then it is still possible to disentangle whether the unfolding discovery higgs sector stems from the $\mathcal{SM}$ or $\mathcal{MSSM}$ for a substantial region of the parameter space. The situation for the neutral higgses is shown in Fig.2.[19] The technique here is to use the simultaneous production of more than one higgs, the observation of associated production of the h^0 or H^0 and the A^0, or the measurement of the braching ratio into b-quarks. New results of further studies of this type are presented by Janusz Rosiek and André Sopczak at this workshop.[20]

Michael Hildreth[21] addressed the measurement of the higgs' branching ratios as a handle to understanding the $\mathcal{SM}$-$\mathcal{MSSM}$ distinction. He used different combinations of lifetime tagging and jet masses to isolate the decay channels $b\bar{b}$, WW^*, $c\bar{c}$+gg, and $\tau^+\tau^-$. The results are shown in Fig.3.

Jack Gunion[22] concludes from this study that the accuracy should be sufficient to

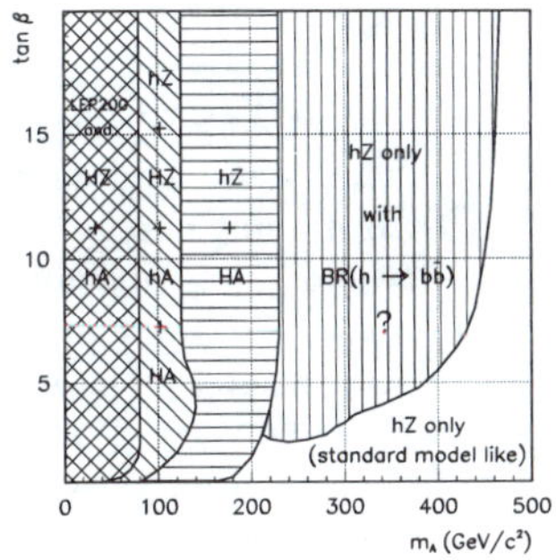

Fig. 2. Domains of the (M_A,tanβ) plane where the $\mathcal{SM}$ and the $\mathcal{MSSM}$ can be distinguished.

say whether there are one or two higgs-doublet fields, but that for the $\mathcal{MSSM}$ more accuracy may be needed. This can be seen in Fig.3, since the $\mathcal{SM}$ values are very close to those of the $\mathcal{MSSM}$ for tanβ<10.

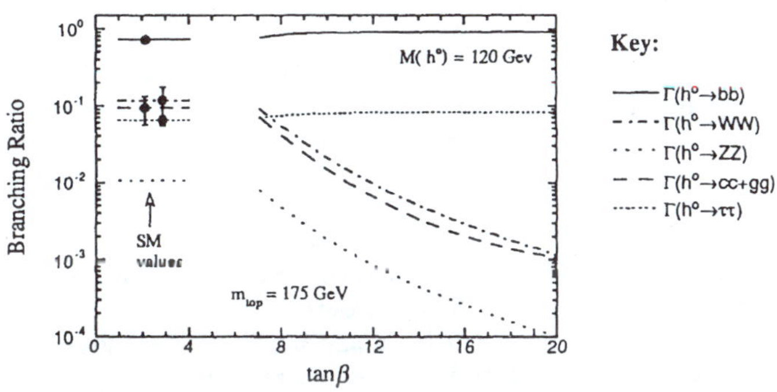

Fig. 3. Expected errors on the $\mathcal{SM}$ branching fractions along with the predicted branching fractions for a 120 GeV $\mathcal{MSSM}$ higgs.

Finally there is the charged higgs $H^{\pm}$, which cannot be discussed in this contribution due to time and space limitations. However, the discovery at the 500 GeV e^+e^- linear collider is equally straight-forward, as shown earlier by Paula Eerola and Jukka Sirkka[23] and more recently by André Sopczak,[24] to mention a few of the studies. Unfortunately it is impossible in a 30′ talk to give credit to all of the work that has gone into these higgs simulations.

4. The Supersymmetric Partners

The search philosophy for the simplest $\mathcal{MSSM}$ here is based on the fact that the *sparticles* decay down to the $\tilde{\chi}_1^0$, which is stable and does not interact in the detector. Thus the events are identified by their missing energy along with the jets and leptons produced: *stop* production $e^+e^-\to\tilde{t}\tilde{t}$ yields 6 jets plus missing energy or 2 or 4 jets plus leptons plus missing energy; *slepton* $e^+e^-\to\tilde{\ell}\tilde{\ell}$ and *chargino* $e^+e^-\to\tilde{\chi}_j^\pm\tilde{\chi}_j^\pm$production yield acoplanar pairs of leptons or jets, whereby beam polarization can help get rid of the W^+W^- background; *neutralino* $e^+e^-\to\tilde{\chi}_i^0\tilde{\chi}_i^0$ production gives rise to "ZEN"‡ events, which are monojets of hadrons or lepton pairs.

Again much work has gone into simulations for the 500 GeV e^+e^- linear collider, as reported by Jean-François Grivaz[25] at Finland and Shuji Orito[26] at Hawaii, or as found in the ee500 proceedings,[5] for example. But first we start with LEP.

4.1. LEP

Roughly the limits[27] found at LEP1 put the *sparticle* masses above the beam energy of 45 GeV for the *sleptons*, *squarks*, and *charginos*. For the *neutralinos*, the situation is more complicated as it depends on the coupling strengths and $\mathcal{SUSY}$ parameters, but the mass tends to be above 20 GeV in many cases. So there is nothing new here, and the limits will scale up with the beam energy as we ease up to LEP200.

However, the monojet search has recently been upgraded by ALEPH.[28] The search essentially involved selecting events with one hemisphere free of detector activity and with total momentum transverse to the beam axis of $5\%\sqrt{s}$. Three monojets events were found in the 2 million Z^0 events recorded up through 1993. One such event is seen in Fig.4.

At first glance, finding three monojets agrees with the number of 2.6 background events expected to fall within the cuts. However, two of the monojets have the kinematic properties of a mass around 5 GeV together with a transverse momentum of around 20 GeV/c, characteristics which are unusual for the background. A first estimate of the probability that events with these characteristics arise from background yielded 2.2%. After adding graphs involving W-exchange to the 4-fermion background generator, the probability increased to 4.8% that these events arise from background.

Nobody knows what these events are; there are too few of them. Their appearance is the same as the "ZEN" events mentioned above. This year's running (1994) has harvested another 2 million Z^0 events, so we should know soon whether the probability will rise or fall, that the $\mathcal{SM}$ produces a significant monojet background.

4.2. 500 GeV e^+e^- Linear Collider

So now let us be optimistic and assume that supersymmetry will be discovered. There has been much progress recently in understanding how to perform precision

‡What is the sound made by one hand clapping?

Made on 4-May-1994 19:14:51 by DUFLOT with DALI_D2.

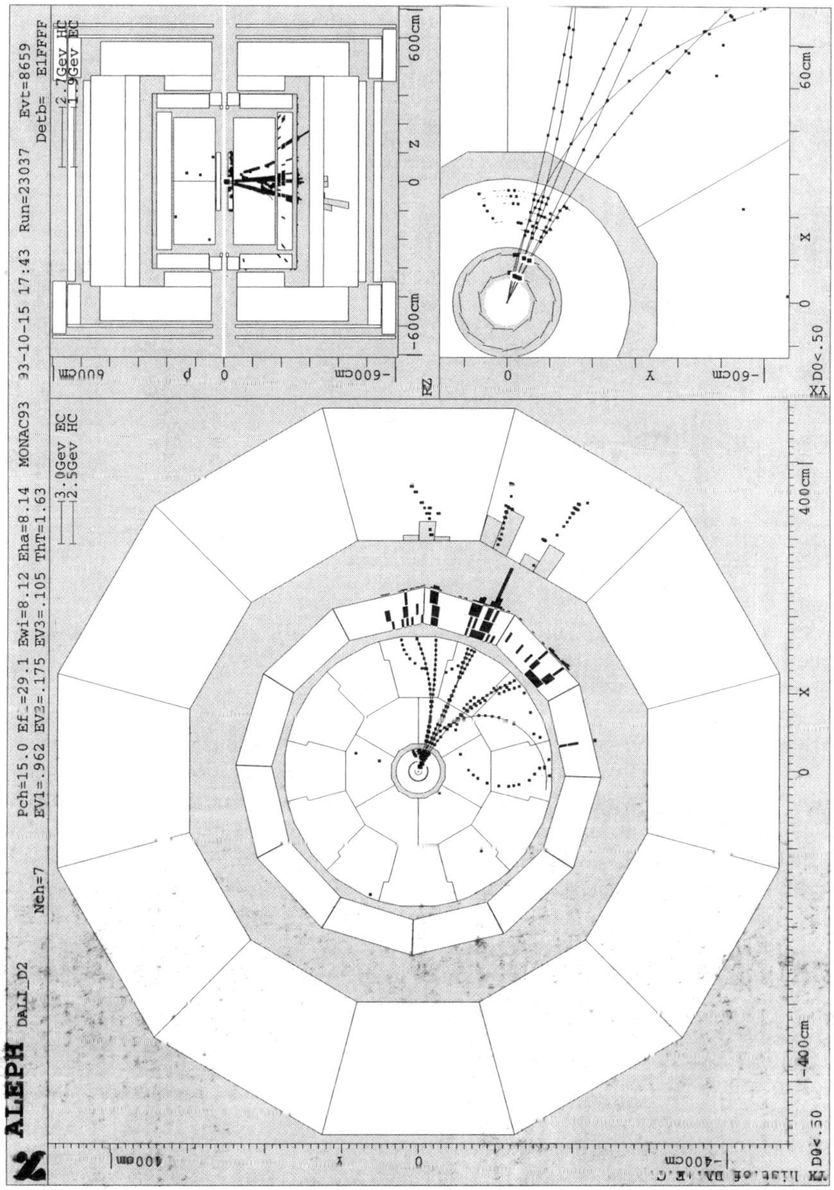

Fig. 4. One of the ALEPH monojet events.

measurements to map out the spectroscopy of the *sparticle* masses and use their properties to measure the $\mathcal{SUSY}$ parameters. This is summarized by Toshifumi Tsukamoto[29,30] and Shuji Orito[26] at Hawaii. For example, the results of one such study are seen in Fig.5. Here a search was simulated to find *slepton* production, specifically $e^+e^- \rightarrow \tilde{\mu}_R\tilde{\mu}_R \rightarrow \mu\tilde{\chi}_1^0\mu\tilde{\chi}_1^0$. The acoplanar-pair search is straight-forward, and beam polarization is used to eliminate the W background. The energy distribution of the muons is seen in Fig.5, and the end-points depend on the masses $M_{\tilde{\mu}_R}$ and $M_{\tilde{\chi}_1^0}$. The result of a fit for the two masses gives the *slepton* mass and the LSP mass to about ± 1.5 GeV and ± 1.0 GeV, respectively.

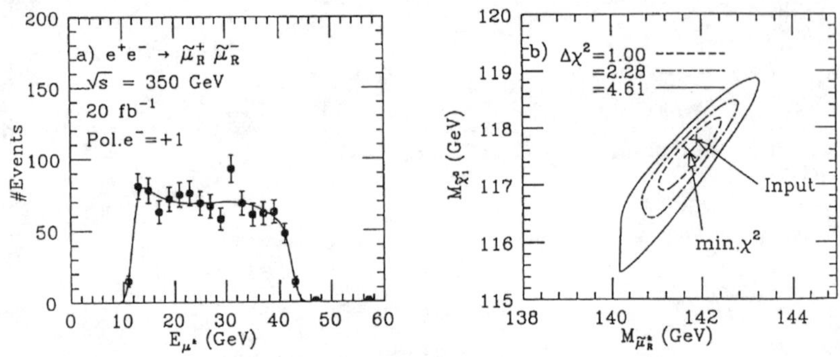

Fig. 5. (a) The energy distribution of the muons from $\tilde{\mu}_R\tilde{\mu}_R$ decays with the fit result shown by the solid curve. (b) χ^2-contours in the $M_{\tilde{\mu}_R}$ - $M_{\tilde{\chi}_1^0}$ plane obtained by fitting the muon energy spectrum.

Beam polarization is an important tool for these measurements. For example, *charginos* are a mixture of charged *wino* and *higgsino* states. By measuring the production of *charginos* for left- and right-handed e^- polarization, the angular distribution changes in that the rate of *higgsino* production changes, while the t-channel production of *winos* is turned off completely using right-handed electrons. Thus it is possible to decompose the states and deduce the mixing angles of *winos* and *higgsinos* in the *chargino*. A similar game can be played with the *neutralino*, which is a mixture of neutral *wino*, *bino*, and *higgsino* states.

In a further example, the determination of the chargino and neutralino mixing angles can then be used to test the unification assumptions listed above. With the mass of the *neutralino* decomposed into its *bino* (M_1) and *wino* (M_2) components, the one can test the relation $\frac{M_1}{\alpha_1} = \frac{M_2}{\alpha_2} = \frac{M_3}{\alpha_3} = \frac{m_{1/2}}{\alpha_{GUT}}$, as seen in Fig.6. Also shown is the test of a mass relation between *slepton* and *wino* masses predicted by a specific model for the masses of the superpartners.

Finally, the masses and production cross sections of the superpartners, depend in a redundant way on the parameters m_0, M_1, M_2, μ, tanβ, .. .[26] By combining the measurements of many channels, it is possible to deduce the parameters of $\mathcal{SUSY}$ and to have many cross-checks on their universality.

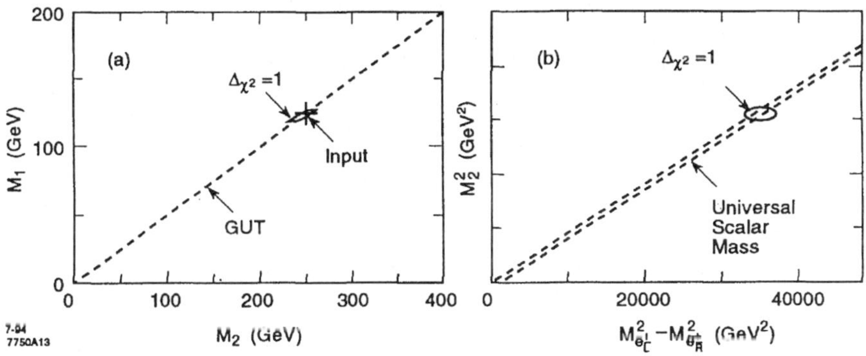

Fig. 6. Tests of $\mathcal{MSSM}$ grand unification at a 500 GeV e^+e^- linear collider. (a) Test of the mass relation between the $SU(2)$ and $U(1)$ gauge boson superpartners; (b) test of the mass relation involving the $SU(2)$ gauge boson superpartner and the electron superpartner. Both relations require the determination of mixing angles as well as physical particle masses. The χ^2 contours include reconstruction efficiencies and standard model backgrounds.

5. Conclusion

The general ideas in "natural" models[13] lead to a qualitative picture in which non-colored superpartners are light while the colored superpartners are heavy. These ideas underline the need for both the *pp* collider and the 500 GeV e^+e^- linear collider. The discussion in this paper hopefully demonstrates the vast potential of the latter for studying the light higgses and the $\tilde{\ell}$, $\tilde{\chi}_j^{\pm}$, and $\tilde{\chi}_i^0$ states, which will be light if these ideas of supersymmetry are close to reality. The heavy higgses and colored heavy states $\tilde{q}$ and $\tilde{g}$ will be measured by the *pp* collider LHC. And particle physics will be one step further.

6. Acknowledgements

I would like to thank my theorist colleagues for inventing supersymmetry, which is generating an exciting discussion in the world. This workshop was and is part of that discussion and was correspondingly stimulating, and thanks go to the organizers for inviting me.

7. References

1. C. Ahn et al.,*Opportunities and Requirements for Experimentation at a Very High Energy e^+e^- Collider*, SLAC-Report-329 (1988).
2. *Proceedings of the Workshop on Physics at Future Accelerators* (La Thuile 1987), ed. J.H. Mulvey, CERN 87-07 (1987).
3. *Proceedings of the Summer Study on High Energy Physics in the 1990s* (Snowmass, 1988), ed. S. Jensen (World Scientific, Singapore, 1989).
4. Three Workshops on the Japan Linear Collider (1989,1990,1992) resulting in: JLC Group, *JLC-I*, KEK Report 92-16 (1992).
5. *e^+e^--Collisions at 500 GeV: The Physics Potential, Parts A and B* (Workshops at Munich, Annecy, Hamburg, 1991), ed. P.M. Zerwas, DESY 92-123A&B (1992).
6. *e^+e^--Collisions at 500 GeV: The Physics Potential, Part C* (Workshops at Munich, Annecy, Hamburg, 1992-1993), ed. P.M. Zerwas, DESY 93-123C (1993).
7. *Physics and Experiments with Linear Colliders, Vols. I and II*, eds. R. Orava, P. Eerola and M. Nordberg (World Scientific, Singapore, 1992).
8. *Proceedings of the 1992 Workshops on High-Energy Physics with Colliding Beams* (Trilogy of Workshops at Yale, Wisconsin, Santa Cruz, 1992), ed. J. Rogers, SLAC-428 (1993).
9. *Ten Years of SUSY Confronting Experiment* (Workshop at CERN, September 1992), eds. J. Ellis, D.V. Nanopoulos, and A. Savoy-Navarro, CERN-TH.6707/92-PPE/92-180 (1992).
10. *Workshop on Physics and Experiments with Linear e^+e^- Colliders, Vols. I and II*, (Waikoloa, Hawaii, 1993), eds. F.A. Harris, S.L. Olsen, S. Pakvasa and X. Tata (World Scientific, Singapore, 1993).
11. *Proceedings of LC92, the ECFA Workshop on e^+e^- Linear Colliders, Vols. I and II*, ed. Ron Settles, MPI-PhE/93-14, ECFA 93-154 (1993).
12. *Proceedings of the Fifth International Workshop on Next-Generation Linear Colliders*, (LC93, SLAC, October 13-21, 1993), SLAC-436 (1993).
13. Michael E. Peskin, "Complementarity of e^+e^- and pp Colliders for the Exploration of Electroweak Symmetry Breaking", SLAC-PUB-6582 (August 1994).
14. J.-F. Grivaz, in Ref.[9], p. 172.
15. D. Treille, in Ref.[9], p. 203.
16. D. Treille, in Ref.[10], p. 1.
17. S. Komamiya, in Ref.[7], p. 277.
18. P. Janot, in Ref.[5], p. 107.
19. P. Janot, in Ref.[10], p. 192.
20. J. Rosiek and A. Sopczak, "Present and future searches with e^+e^- colliders for the neutral MSSM Higgs bosons – the complete 1-loop analysis", these proceedings.
21. M.D. Hildreth, in Ref.[10], p.635.

22. J.F. Gunion, in Ref.[10], p. 166.
23. P. Eerola and J. Sirkka, in Ref.[5], p. 133.
24. A. Sopczak, in Ref.[10], p. 666, and Ref.[6], p. 121.
25. J.-F. Grivaz, in Ref.[7], p. 353.
26. S. Orito, in Ref.[10], p. 230.
27. ALEPH Collaboration, *Physics Reports* **216** (1992) 253.
28. ALEPH Collaboration, *Phys. Lett.* **334B** (1994) 244.
29. T. Tsukamoto, in Ref.[10], p. 814.
30. T. Tsukamoto, K. Fujii, H. Murayama, M. Yamaguchi, and Y. Okada, KEK-Preprint-93-146 (1993).

PROBING ANOMALOUS GAUGE BOSON COUPLINGS AT LEP

S. DAWSON
Physics Department, Brookhaven National Laboratory, Upton, N. Y. 11973

and

G. VALENCIA
Department of Physics, Iowa State University, Ames, IA 50011

ABSTRACT

We bound anomalous gauge boson couplings using LEP data for the $Z \to \overline{f} f$ partial widths. We use an effective field theory formalism to compute the one-loop corrections resulting from non-standard model three and four gauge boson vertices. We find that measurements at LEP constrain the three gauge boson couplings at a level comparable to that obtainable at LEPII.

1 Introduction

High precision measurements at the Z pole at LEP have been used to place stringent limits on new physics beyond the standard model. Under the assumption that the dominant effects of the new physics would show up as corrections to the gauge boson self-energies, the LEP measurements have been used to parameterize the possible new physics in terms of three observables, S, T, U.[1, 2] A fourth observable corresponds to the partial width $Z \to b\bar{b}$.

In view of the extraordinary agreement between the standard model predictions and the observations, it seems reasonable to assume that the $SU(2)_L \times U(1)_Y$ gauge theory of electroweak interactions is essentially correct and that the only sector of the theory lacking experimental support is the symmetry breaking sector. There are many extensions of the minimal standard model that incorporate different symmetry breaking possibilities. One large class of models is that in which the interactions responsible for the symmetry breaking are strongly coupled and their effects would show up in experiments as deviations from the minimal standard model couplings. In this paper we use the measurements of the partial decay widths of the Z boson to place bounds on anomalous gauge boson couplings.

2 Formalism

We assume that the electroweak interactions are given by an $SU(2)_L \times U(1)_Y$ gauge theory with spontaneous symmetry breaking to $U(1)_{EM}$ and that we do not have any information on the symmetry breaking sector except that it is strongly interacting and that any new particles have masses higher than several hundred GeV. This scenario can be described with an effective Lagrangian with operators organized according to the number of derivatives or gauge fields they have. If we call Λ the scale at which the symmetry breaking physics comes in, this organization of operators corresponds to an expansion of amplitudes in powers of $(E^2 \text{ or } v^2)/\Lambda^2$. The lowest order effective Lagrangian for the symmetry breaking sector of the theory is: [3]

$$\mathcal{L}^{(2)} = \frac{v^2}{4} \mathrm{Tr}\Big[D^\mu \Sigma^\dagger D_\mu \Sigma\Big]. \tag{1}$$

In our notation W_μ and B_μ are the $SU(2)_L$ and $U(1)_Y$ gauge fields with $W_\mu \equiv W_\mu^i \tau_i$. The matrix $\Sigma \equiv \exp(i\vec{\omega}\cdot\vec{\tau}/v)$, contains the Goldstone bosons ω_i that give the W and Z their masses via the Higgs mechanism and the $SU(2)_L \times U(1)_Y$ covariant derivative is given by:

$$D_\mu \Sigma = \partial_\mu \Sigma + \frac{i}{2} g W_\mu^i \tau^i \Sigma - \frac{i}{2} g' B_\mu \Sigma \tau_3. \tag{2}$$

This non-renormalizable Lagrangian is interpreted as an effective field theory, valid below some scale $\Lambda \leq 3$ TeV. The lowest order interactions between the gauge bosons and fermions, as well as the kinetic energy terms for all fields, are the same as those in the minimal standard model.

For LEP observables, the operators that can appear at tree-level are those that modify the gauge boson self-energies. To order $\mathcal{O}(1/\Lambda^2)$ there are only three:[3, 4]

$$\mathcal{L}^{(2GB)} = \beta_1 \frac{v^2}{4}\Big(\mathrm{Tr}\Big[\tau_3 \Sigma^\dagger D_\mu \Sigma\Big]\Big)^2 + \alpha_8 g^2 \Big(\mathrm{Tr}\Big[\Sigma \tau_3 \Sigma^\dagger W_{\mu\nu}\Big]\Big)^2 + g g' \frac{v^2}{\Lambda^2} L_{10}\, \mathrm{Tr}\Big[\Sigma B^{\mu\nu} \Sigma^\dagger W_{\mu\nu}\Big], \tag{3}$$

which contribute respectively to T, U and S.

In this paper we will consider operators that affect the Z partial widths at the one-loop level. We will restrict our study to only those operators that appear at order $\mathcal{O}(1/\Lambda^2)$ in the gauge-boson sector and that respect the custodial symmetry in the limit $g' \to 0$. We will also consider an additional operator which is parity violating, but CP conserving. Because of the parity violating nature of this term it can lead to observable signals at LEPII via $e^+e^- \to W^+W^-$.[5] The Lagrangian we consider is thus:

$$\begin{aligned}\mathcal{L}^{(4)} &= \frac{v^2}{\Lambda^2}\Big\{ L_1 \Big(\mathrm{Tr}\Big[D^\mu \Sigma^\dagger D_\mu \Sigma\Big]\Big)^2 + L_2 \Big(\mathrm{Tr}\Big[D_\mu \Sigma^\dagger D_\nu \Sigma\Big]\Big)^2 \\ &\quad - i g L_{9L}\, \mathrm{Tr}\Big[W^{\mu\nu} D_\mu \Sigma D_\nu \Sigma^\dagger\Big] - i g' L_{9R}\, \mathrm{Tr}\Big[B^{\mu\nu} D_\mu \Sigma^\dagger D_\nu \Sigma\Big] \\ &\quad + g \hat{\alpha} \epsilon^{\alpha\beta\mu\nu} Tr\Big(\tau_3 \Sigma^\dagger D_\mu \Sigma\Big) Tr\Big(W_{\alpha\beta} D_\nu \Sigma \Sigma^\dagger\Big)\Big\},\end{aligned} \tag{4}$$

where the field strength tensors are given by:

$$\begin{aligned} W_{\mu\nu} &= \frac{1}{2}\Big(\partial_\mu W_\nu - \partial_\nu W_\mu + \frac{i}{2}g[W_\mu, W_\nu]\Big) \\ B_{\mu\nu} &= \frac{1}{2}\Big(\partial_\mu B_\nu - \partial_\nu B_\mu\Big)\tau_3. \end{aligned} \tag{5}$$

The operators in Eq. 3 and Eq. 4 would arise when considering the effects of those in Eq. 1 at the one-loop level or from the new physics responsible for symmetry breaking at a scale Λ at order $1/\Lambda^2$. We therefore explicitly introduce the factor v^2/Λ^2 in our definition of $\mathcal{L}^{(4)}$ so that the coefficients L_i are naturally of $\mathcal{O}(1)$. In the present paper we will compute the contribution from the operators of Eqs. 3 and 4 to the Z partial widths that are measured at LEP.

We will first perform a complete calculation to order $\mathcal{O}(1/\Lambda^2)$. That is, we will include the one-loop contributions from the operator in Eq. 1 (and gauge boson kinetic energies). The divergences generated in this calculation are absorbed by renormalization of the couplings in Eq. 3. This calculation will illustrate our method and as an example we use it to place bounds on L_{10}.

We will then place bounds on the couplings of Eq. 4 by considering their one-loop effects. The divergences generated in this one-loop calculation would be removed in general by renormalization of the couplings in the $\mathcal{O}(1/\Lambda^4)$ Lagrangian of those operators that modify the gauge boson self-energies at tree-level and perhaps by additional renormalization of the couplings in Eq. 3. Interestingly, we find that we can obtain a completely finite result for the $Z \to \bar{f}f$ partial widths using only the operators in Eq. 3 as counterterms when we consider the parity conserving operators. This is not the case, however, for the parity violating operator, $\hat{\alpha}$.

We first regularize the integrals in n space-time dimensions and remove all the poles in $n-4$ as well as the finite analytic terms by a suitable definition of the renormalized couplings. We base our analysis on the leading non-analytic terms proportional to $L_i \log\mu$. These terms determine the running of the $1/\Lambda^4$ couplings and cannot be generated by tree-level terms at that order. With a carefully chosen renormalization scale μ (such that the logarithm is of order one), these terms give us the correct order of magnitude for the size of the $1/\Lambda^4$ coefficients.[8] Quadratic divergences will be absorbed by the new physics arising at $\mathcal{O}(1/\Lambda^4)$ and cannot be used to limit the coefficients of the $\mathcal{O}(1/\Lambda^2)$ Lagrangian.[6] Clearly, the LEP observables do not measure the couplings in Eq. 4 and it is only from naturalness arguments like the one above that we can place bounds on the anomalous gauge-boson couplings. From this perspective, it is clear that these bounds are not a substitute for direct measurements in future high energy machines.

We will perform our calculations in unitary gauge, so we set $\Sigma = 1$ in Eqs. 1, 3 and 4. For the lowest order operators we use the conventional input parameters: G_F as measured in muon decay; the physical Z mass: M_Z; and $\alpha(M_Z) = 1/128.8$. Other lowest order parameters are derived quantities and we adopt one of the usual definitions for the mixing angle:

$$s_Z^2 c_Z^2 \equiv \frac{\pi \alpha(M_Z)}{\sqrt{2} G_F M_Z^2}. \tag{6}$$

We neglect the mass and momentum of the external fermions compared to the Z mass. The only fermion mass that is kept in our calculation is the mass of the top-quark when it appears as an intermediate state.

With this formalism we proceed to compute the $Z \to f\overline{f}$ partial width from the following ingredients.

- The $Z \to f\overline{f}$ vertex, which we write as:

$$i\Gamma_\mu = -i\frac{e}{4 s_Z c_Z}\gamma_\mu \Big[(r_f + \delta r_f)(1+\gamma_5) + (l_f + \delta l_f)(1-\gamma_5)\Big] \tag{7}$$

  where $r_f = -2Q_f s_Z^2$ and $l_f = r_f + T_{3f}$. The terms δl_f and δr_f occur at one-loop both at order $1/\Lambda^2$ and at order $1/\Lambda^4$ and can be found in Ref. [9].

- The renormalization of the lowest order input parameters. At order $1/\Lambda^2$ it is induced by tree-level anomalous couplings and one-loop diagrams with lowest order vertices. At order $1/\Lambda^4$ it is induced by one-loop diagrams with an anomalous coupling in one vertex.

- Tree-level and one-loop contributions to γZ mixing.

- Wave function renormalization.

With all these ingredients we can find our final expression for the physical partial width. We find:

$$\Gamma(Z \to f\overline{f}) \equiv \Gamma_{SM}(Z \to f\overline{f})\left(1 + \frac{\delta\Gamma_f^{L_i}}{\Gamma_0(Z \to f\overline{f})}\right). \tag{8}$$

where Γ_0 is the lowest order tree level result,

$$\Gamma_0(Z \to f\overline{f}) = N_{cf}(l_f^2 + r_f^2)\frac{G_F M_Z^3}{12\pi\sqrt{2}}. \tag{9}$$

The values for $\delta L_f^{L_i}$ are given in the following sections. Using Eq. 8 we introduce additional terms proportional to products of standard model one-loop corrections and corrections due to anomalous couplings. These are small effects that do not affect our results.

We will not attempt to obtain a global fit to the parameters in our formalism from all possible observables. Instead we use the partial Z widths:[12]

$$\begin{aligned} \Gamma_e &= 83.98 \pm 0.18\ MeV \\ \Gamma_\nu &= 499.8 \pm 3.5\ MeV \\ \Gamma_Z &= 2497.4 \pm 3.8\ MeV \\ R_h &= 20.795 \pm 0.040 \end{aligned} \tag{10}$$

The bounds on new physics are obtained by subtracting the standard model predictions at one-loop from the measured partial widths as in Eq. 8. We use the numbers of Langacker [13] which use the global best fit values for M_t and α_s with M_H in the range $60 - 1000$ GeV. The first error is from the uncertainty in M_Z and Δr, the second is from M_t and M_H, and the one in brackets is from the uncertainty in α_s.[13]

$$\begin{aligned}
\Gamma_e &= 83.87 \pm 0.02 \pm 0.10\ MeV \\
\Gamma_\nu &= 501.9 \pm 0.1 \pm 0.9\ MeV \\
\Gamma_Z &= 2496 \pm 1 \pm 3 \pm [3]\ MeV \\
R_h &= 20.782 \pm 0.006 \pm 0.004 \pm [0.03]
\end{aligned} \tag{11}$$

We add all errors in quadrature. [1]

3 Results

In this section we compute the corrections to the $Z \to f\overline{f}$ partial widths from the couplings of Eq. 4, and compare them to recent values measured at LEP. We treat each coupling constant independently, and compute only its *lowest* order contribution to the decay widths.

3.1 Bounds on L_{10} at order $1/\Lambda^2$

The operators in Eq. 3 are the only ones that induce a tree-level correction to the gauge boson self-energies to order $\mathcal{O}(1/\Lambda^2)$. These corrections are of course well known and correspond, *at leading order*, to the new physics contributions to S, T, U.[1]

We have regularized our one-loop integrals in n dimensions and isolated the ultraviolet poles $1/\epsilon = 2/(4-n)$. We find that we obtain a finite answer to order $1/\Lambda^2$ if we adopt the following renormalization scheme:

$$\begin{aligned}
\frac{v^2}{\Lambda^2} L_{10}^r(\mu) &= \frac{v^2}{\Lambda^2} L_{10} - \frac{1}{16\pi^2}\frac{1}{12}\left(\frac{1}{\epsilon} + \log\frac{\mu^2}{M_Z^2}\right) \\
\beta_1^r(\mu) &= \beta_1 - \frac{e^2}{16\pi^2}\frac{3}{2c_Z^2}\left(\frac{1}{\epsilon} + \log\frac{\mu^2}{M_Z^2}\right).
\end{aligned} \tag{12}$$

We thus replace the bare parameters L_{10} and β_1 with the scale dependent ones above. As a check of our answer, it is interesting to note that we would also obtain a finite answer by adding to our result the one-loop contributions to the self-energies obtained in unitary gauge in the minimal standard model with one Higgs boson in the loop.

Our result for L_{10} at order $1/\Lambda^2$ is then:

$$\frac{\delta\Gamma_f^{L_{10}}}{\Gamma_0(Z \to f\overline{f})} = \frac{e^2}{c_Z^2 s_Z^2} L_{10}^r(\mu) \frac{v^2}{\Lambda^2} \frac{2r_f(l_f + r_f)}{l_f^2 + r_f^2} \frac{c_Z^2}{s_Z^2 - c_Z^2} \tag{13}$$

[1] See also the contribution of A. Sopczak[14] to these proceedings for a discussion of extracting bounds on new physics from LEP measurements.

Once again we point out that, at this order, the contribution of L_{10} to the LEP observables occurs only through modifications to the self-energies that are proportional to q^2. At this order it is therefore possible to identify the effect of L_{10} with the oblique parameter S. If we were to compute the effects of L_{10} at one-loop (as we do for the $L_{9L,9R}$) comparison with S would not be appropriate.

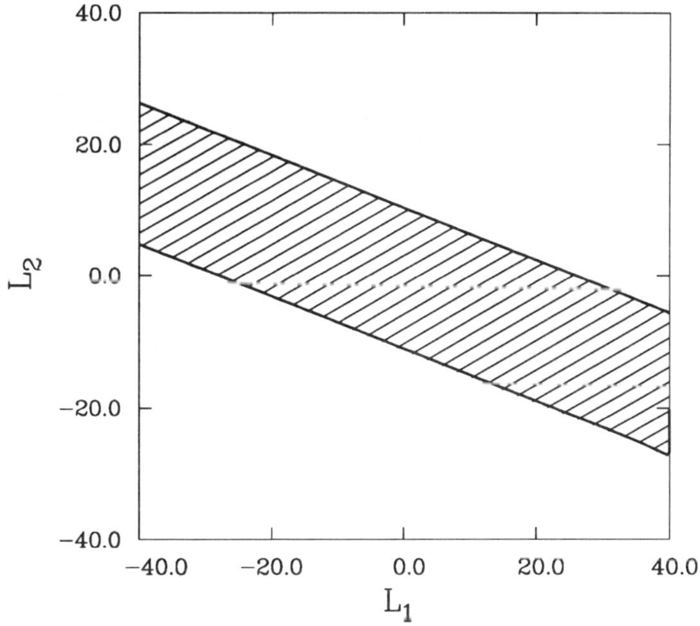

Figure 1: 90% confidence level bounds on $L_{1,2}$ from the $Z \to f\overline{f}$ partial widths, (Eq. 14). The allowed region is shaded.

Numerically we find the following 90% confidence level bounds on L_{10} when we take the scale $\Lambda = 2$ TeV:

$$\begin{aligned} \Gamma_e &\rightarrow -1.7 \leq L_{10}^r(M_Z)_{new} \leq 3.3 \\ R_h &\rightarrow -1.5 \leq L_{10}^r(M_Z)_{new} \leq 2.0 \\ \Gamma_Z &\rightarrow -1.1 \leq L_{10}^r(M_Z)_{new} \leq 1.5 \end{aligned} \tag{14}$$

3.2 Bounds on $L_{1,2}$ at order $1/\Lambda^4$

The couplings $L_{1,2}$ enter the one-loop calculation of the $Z \to f\overline{f}$ width through the four gauge boson couplings. Our prescription calls for using only the leading non-analytic contribution to the process $Z \to f\overline{f}$. This contribution can be extracted from the coefficient of the pole in $n-4$. Care must be taken to isolate the poles of ultraviolet origin (which are the only ones that interest us) from those of infrared origin that appear in intermediate steps of the calculation but that cancel as usual when one includes real emission processes as well.

Since in unitary gauge $L_{1,2}$ modify only the four-gauge boson couplings at the one-loop level, they enter the calculation of the Z partial widths only through the self-energy corrections and the renormalization of the lowest order couplings. These operators induce a non-zero value for $\Delta\rho \equiv \Pi_{WW}(0)/M_W^2 - \Pi_{ZZ}(0)/M_Z^2$. For the observables we are discussing, this is the *only* effect of $L_{1,2}$. We do not place bounds on them from global fits of the oblique parameter T, because we have not shown that this is the only effect of $L_{1,2}$ for the other observables that enter the global fits. The calculation to $\mathcal{O}(1/\Lambda^4)$ can be made finite with the following renormalization of β_1:

$$\beta_1^r(\mu) = \beta_1 + \frac{3}{4}\frac{\alpha^2(1+c_Z^2)}{s_Z^2 c_Z^4}\Big(L_1 + \frac{5}{2}L_2\Big)\frac{v^2}{\Lambda^2}\Big(\frac{1}{\epsilon} + \log\frac{\mu^2}{M_Z^2}\Big). \tag{15}$$

We obtain for the Z partial widths:

$$\frac{\delta\Gamma_f^{L_{1,2}}}{\Gamma_0(Z\to f\overline{f})} = -\frac{3}{2}\frac{\alpha^2(1+c_Z^2)}{s_Z^2 c_Z^4}\Big(L_1 + \frac{5}{2}L_2\Big)\frac{v^2}{\Lambda^2}\log\Big(\frac{\mu^2}{M_Z^2}\Big)\Big(1 + \frac{2r_f(l_f+r_f)}{l_f^2+r_f^2}\frac{c_Z^2}{s_Z^2-c_Z^2}\Big). \tag{16}$$

Using $\Lambda = 2$ TeV, and $\mu = 1$ TeV we find 90% confidence level bounds:

$$\begin{aligned}
\Gamma_e &\to -50 \le L_1 + \frac{5}{2}L_2 \le 26 \\
\Gamma_\nu &\to -28 \le L_1 + \frac{5}{2}L_2 \le 59 \\
R_h &\to -190 \le L_1 + \frac{5}{2}L_2 \le 130 \\
\Gamma_Z &\to -36 \le L_1 + \frac{5}{2}L_2 \le 27 \quad .
\end{aligned} \tag{17}$$

Combined, they yield the result:

$$-28 \le L_1 + \frac{5}{2}L_2 \le 26 \tag{18}$$

shown in Figure 1.

As mentioned before, the effect of $L_{1,2}$ in other observables is very different from that of β_1. It is only for the Z partial widths that we can make the $\mathcal{O}(1/\Lambda^4)$ calculation finite with Eq. 15.

3.3 Bounds on $L_{9L,9R}$ at order $1/\Lambda^4$

The calculation of the effects of the L_{9L} and L_{9R} operators can be carried out simultaneously with the one-loop effects of the lowest order effective Lagrangian, Eq. 1, because the form of the three and four gauge boson vertices induced by these two couplings is the same as that arising from Eq. 1. Performing the calculation in this way, we obtain a result that contains terms of order $1/\Lambda^2$ (those independent of $L_{9L,9R}$), terms of order $1/\Lambda^4$ proportional to $L_{9L,9R}$ and terms of order $1/\Lambda^4$ proportional to L_{10} and β_1.

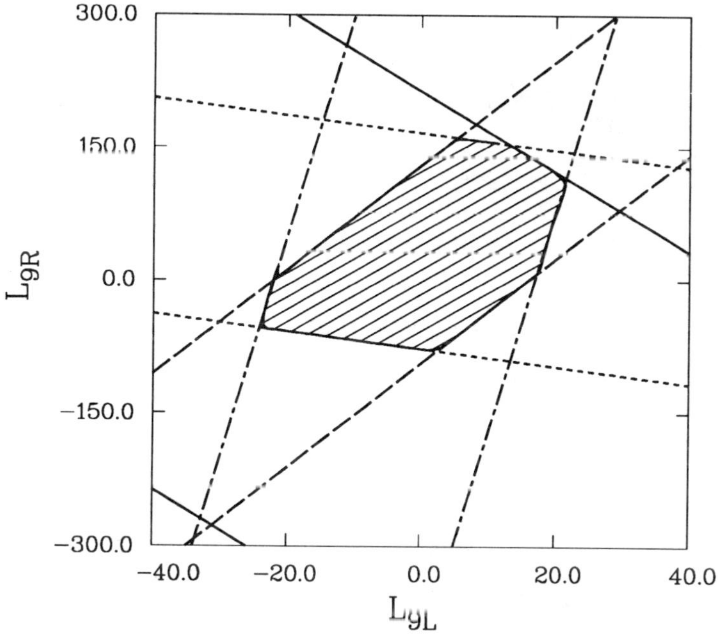

Figure 2: 90% confidence level bounds on $L_{9L,9R}$ from the $Z \to f\overline{f}$ partial widths, (Eq. 22). The allowed region is shaded. The solid, dotted, dashed, and dot-dashed lines are the bounds from Γ_e, Γ_ν, R_h, and Γ_Z, respectively.

It is amusing to note that the divergences generated by the operators $L_{9L,9R}$ in the one-loop (order $1/\Lambda^4$) calculation of the $Z \to \overline{f}f$ widths can all be removed by the following renormalization of the couplings in Eq. 3 (in the $M_t = 0$ limit):

$$\beta_1^r(\mu) = \beta_1 - \frac{\alpha}{\pi}\frac{e^2}{96 s_Z^4 c_Z^4}\frac{v^2}{\Lambda^2}\left[c_Z^2(1-20s_Z^2)L_{9L} + s_z^2(10-29c_Z^2)L_{9R}\right]\left(\frac{1}{\epsilon} + \log\frac{\mu^2}{M_Z^2}\right)$$

$$L_{10}^r(\mu) \;=\; L_{10} - \frac{\alpha}{\pi}\frac{1}{96 s_Z^2 c_Z^2}\Big[(1-24c_Z^2)L_{9L} + (32c_Z^2 - 1)L_{9R}\Big]\Big(\frac{1}{\epsilon} + \log\frac{\mu^2}{M_Z^2}\Big) \tag{19}$$

This proves our assertion that our calculation to order $\mathcal{O}(1/\Lambda^4)$ can be made finite by suitable renormalizations of the parameters in Eq. 3. However, we do not expect this result to be true in general. That is, we expect that a calculation of the one-loop contributions of the operators in Eq. 4 to other observables will require counterterms of order $1/\Lambda^4$. Thus, Eq. 19 does *not* mean that we can place bounds on $L_{9L,9R}$ from global fits to the parameters S and T. Without performing a complete analysis of the effective Lagrangian at order $1/\Lambda^4$ it is not possible to identify the renormalized parameters of Eq. 19 with the ones corresponding to S and T that are used for global fits.

Keeping only terms linear in $L_{9L,9R}$ we find,

$$\begin{aligned}\frac{\delta\Gamma^{L_9}}{\Gamma_0} \;=\; & \frac{\alpha^2}{24}\frac{1}{c_Z^4 s_Z^4}\frac{v^2}{\Lambda^2}\log\Big(\frac{\mu^2}{M_Z^2}\Big) \\ & \cdot\Big\{\Big[L_{9L}(1-24c_Z^2) + L_{9R}(-1+32c_Z^2)\Big]\frac{2r_f(l_f+r_f)}{l_f^2+r_f^2}\frac{c_Z^2}{s_Z^2-c_Z^2} \\ & +2\Big[L_{9L}c_Z^2(1-20s_Z^2) + L_{9R}s_Z^2(10-29c_Z^2)\Big]\Big(1+\frac{2r_f(l_f+r_f)}{l_f^2+r_f^2}\frac{c_Z^2}{s_Z^2-c_Z^2}\Big)\Big\} \\ & +\frac{\alpha^2}{12}\log\Big(\frac{\mu^2}{M_Z^2}\Big)\frac{1+2c_Z^2}{c_Z^4 s_Z^4(l_b^2+r_b^2)}\frac{v^2}{\Lambda^2}\frac{M_t^2}{M_Z^2}\Big[L_{9R}s_Z^2 - 7L_{9L}c_Z^2\Big]\delta_{fb}.\end{aligned} \tag{20}$$

The last term in Eq. 20 corresponds to the non-universal corrections proportional to M_t^2 that are relevant only for the decay $Z \to \bar{b}b$.

Using as before $\Lambda = 2$ TeV and $\mu = 1$ TeV we find 90% confidence level bounds:

$$\begin{aligned}\Gamma_e &\;\rightarrow\; -92 \le L_{9L} + 0.22L_{9R} \le 47 \\ \Gamma_\nu &\;\rightarrow\; -79 \le L_{9L} + 1.02L_{9R} \le 170 \\ R_h &\;\rightarrow\; -22 \le L_{9L} - 0.17L_{9R} \le 16 \\ \Gamma_Z &\;\rightarrow\; -22 \le L_{9L} - 0.04L_{9R} \le 17\end{aligned} \tag{21}$$

We show these inequalities in Figure 2.

If we bound one coupling at a time we can read from Figure 2 that:

$$\begin{aligned}-22 \le\; & L_{9L} \;\le 16 \\ -77 \le\; & L_{9R} \;\le 94 \;.\end{aligned} \tag{22}$$

In a vector like model with $L_{9L} = L_{9R}$ we have the 90% confidence level bound:

$$-22 < L_{9L} = L_{9R} < 18. \tag{23}$$

The limits from $\Gamma(Z \to b\bar{b})$ are not competitive with those of Eq. 21 despite the enhancement of M_t^2/M_Z^2.

3.4 Bounds on $\hat{\alpha}$ at order $1/\Lambda^4$

The one loop contributions to $Z \to f\overline{f}$ from the parity violating operator $\hat{\alpha}$ cannot be made finite by a renormalization of the operators of Eq. 3. The divergence in this case will be absorbed by renormalization of the couplings appearing at $\mathcal{O}(1/\Lambda^4)$. The leading non-analytic contribution can be extracted without explicitely performing that renormalization. It is, [5]

$$\frac{\delta\Gamma_f^5}{\Gamma_f^0} = \frac{3\alpha}{2\pi} g_5^Z \log\left(\frac{\mu}{M_W}\right)\left[\frac{2l_f}{l_f^2+r_f^2}\frac{c_Z^2}{s_Z^2} + \left(1 + \frac{2r_f(l_f+r_f)}{l_f^2+r_f^2}\frac{c_Z^2}{s_Z^2-c_Z^2}\right)\right] \quad , \tag{24}$$

where

$$g_5^Z \equiv \frac{4M_Z^2\hat{\alpha}}{\Lambda^2} \quad . \tag{25}$$

We find the results for $\Lambda = 2\ TeV$ and $\mu = 1\ TeV$,

$$\begin{aligned} \Gamma_e &\rightarrow -.1 \leq g_5^Z \leq .05 \\ \Gamma_\nu &\rightarrow -.08 \leq g_5^Z \leq .04 \\ R_h &\rightarrow -.07 \leq g_5^Z \leq .1 \\ \Gamma_Z &\rightarrow -.9 \leq g_5^Z \leq 1.2 \quad . \end{aligned} \tag{26}$$

4 Conclusion

We can compare our results[2] with bounds that future colliders are expected to place on the anomalous couplings.

In Fig. 3, we compare our 95% confidence level bounds on L_{9L} and L_{9R} with those which can be obtained at LEPII with $\sqrt{s} = 190$ GeV and an integrated luminosity of 500 pb^{-1}. [7, 11] We find that LEP and LEPII are sensitive to slightly different regions of the L_{9L} and L_{9R} parameter space, with the bounds from the two machines being of the same order of magnitude. We again emphasize our caveat that the bounds from LEP rely on naturalness arguments and are no substitute for measurements in future colliders.

The limits presented here on the four point couplings L_1 and L_2 are the first available for these couplings. They will be measured directly at the LHC.

Computing the leading contribution of each operator, and allowing only one non-zero coefficient at a time, our 90 % confidence level bounds are:

$$\begin{array}{rcl} -1.1 < & L_{10}^r(M_Z)_{new} & < 1.5 \\ -28 < & L_1 & < 26 \\ -11 < & L_2 & < 11 \end{array}$$

[2]Our normalization of the L_i is different from that of Ref. [7]. We have translated their results into our notation.

$$
\begin{array}{rcl}
-22 < & L_{9L} & < 16 \\
-77 < & L_{9R} & < 94 \\
-.07 < & g_5^Z & < .04.
\end{array}
\tag{27}
$$

Two parameter bounds on (L_1, L_2) and (L_{9L}, L_{9R}) are given in the text.

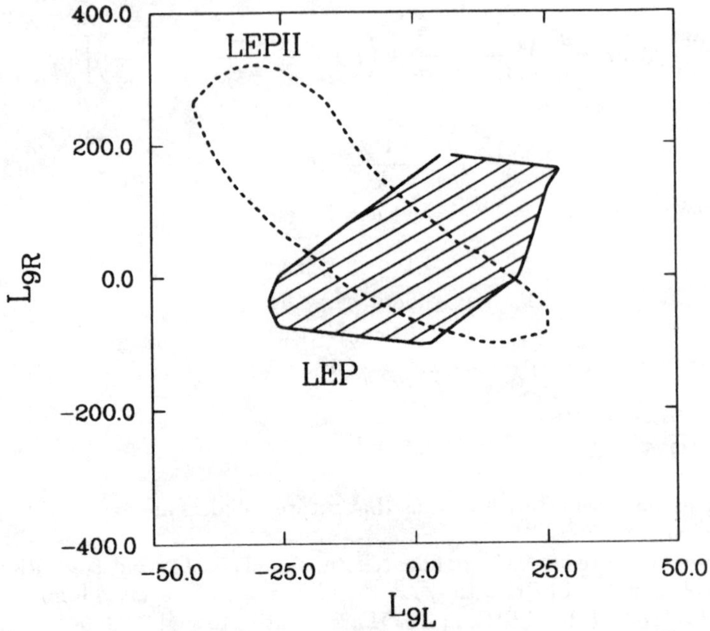

Figure 3: Comparison of the 95% confidence level bounds from the Z partial widths (shaded region) with that obtainable at LEPII with $\sqrt{s} = 190\ GeV$ and $\int \mathcal{L} = 500\ pb^{-1}$, (dotted contour) [7].

5 Acknowledgements

The work of G. V. was supported in part by a DOE OJI award. G.V. thanks the theory group at BNL for their hospitality while part of this work was performed. We are grateful to W. Bardeen, J. F. Donoghue, E. Laenen, W. Marciano, A. Sirlin, and A. Sopczak for useful discussions. We thank P. Langacker for providing us with his latest numbers and F. Boudjema for providing us with the data file for the LEPII bounds in Figure 3.

References

[1] D. C. Kennedy and B. W. Lynn, *Nucl. Phys.* **B322** 1 (1989); M. Peskin and T. Takeuchi, *Phys. Rev. Lett.* **65** 964 (1990).

[2] G. Altarelli, and R. Barbieri, and F. Caravaglios, *Nucl. Phys.* **B405** 3 (1993).

[3] T. Appelquist and C. Bernard, *Phys. Rev.* **D22** 200 (1980); A. Longhitano, *Nucl. Phys.* **B188** 118 (1981).

[4] T. Appelquist and G.-H. Wu, *Phys. Rev.* **D48** 3235 (1993).

[5] S. Dawson and G. Valencia,*Phys. Lett.* **333B** 207 (1994); *Phys. Rev.* **D49** 94 (192)188.

[6] M. Einhorn and J. Wudka, UM-TH-92,25; C. Burgess and D. London, *Phys. Rev. Lett.* **69** 3428 (1992); *Phys. Rev.* **D48** 4337 (1993).

[7] F. Boudjema, *Proceedings of Physics and Experiments with Linear c^+c^- Col liders*, ed. by F. A. Harris *et al.*, (1993), p. 713, and references therein.

[8] A. Manohar and H. Georgi, *Nucl. Phys.* **B234** 189 (1984).

[9] S.Dawson and G. Valencia, BNL-60949, hep-ph 9410364.

[10] K. Hagiwara, *et. al.*, *Nucl. Phys.* **B282** 253 (1987), C. Burge *et. al.*, McGill-93/14 (1993).; K. Hagiwara, *et. al.*, *Phys. Lett.* **283B** 353 (1992); P. Hernandez and F. Vegas, *Phys. Lett.* **307B** 116 (1993); E.. Eboli, *et. al.*, MAD/PH/836.

[11] G. Gounaris *et.al.* in *Proc. of the Workshop on e^+e^- Collisions at 500 GeV: The Physics Potential*, DESY-92-123B. p. 735, ed. P. Zerwas; M. Bilenky *et.al.* BI-TP 92/44 (1993).

[12] G. Altarelli, CERN-TH-7319/94; D. Schaile, Plenary talk presented at the ICHEP-94 meeting, Glasgow (1994).

[13] P. Langacker, UPR-0624T; and private communication. The central value for Γ_ν is from J. Bernabeu, A. Pich and A. Santamaria, *Nucl. Phys.* **B363** 326 (1991) for $M_t = 170$ GeV. The error is from P. Langacker, Precision Tests of the Standard Model, Lectures given at TASI-92, Boulder Co. (1992).

[14] A. Sopczk, *Limits on Physics Beyond the Minimal Standard Model From Combined LEP Lineshape Data*, CERN preprint, 1994.

SUPERGRAVITY DARK MATTER

Pran Nath†
Department of Physics, Northeastern University, Boston, MA 02115

and

R. Arnowitt
Center for Theoretical Physics, Department of Physics
Texas A & M University
College Station, TX 77843-4242

ABSTRACT

A review is given of the analyses of neutralino dark matter in supergravity unified theories with radiative breaking of the electro-weak symmetry. It is shown that the event rates in neutralino-nucleus scattering exhibit a sharp dip when the neutralino mass is $\sim M_Z/2$ or $\sim M_h/2$ when the accurate method for the relic density computation is used and the constraints of CHDM as implied by the COBE data are imposed. The effect of the recent measurement of $b \to s\gamma$ by the CLEO collaboration on dark matter analyses is also discussed. It is found that the event rates for $\mu\rangle 0$ are affected significantly while the rates for $\mu\langle 0$ are not appreciably affected. The effect of recent Δq determination from new data on polarized $\mu - p$ and $e - p$ scattering, on event rates in neutralino-nucleus scattering is also discussed.

I. Introduction

The recent resurgence of interest in supersymmetry is at least in part due to the encouraging results from the the high precision LEP data which point in the direction of supersymmetric unification[1]. Thus one may now pose the concrete question, as to what the correct model of particle physics is up to the unification scale M_G. The first possibility obviously is SUSY GUTs. However, in ordinary SUSY grand unification there is currently no phenomenologically viable way of breaking SUSY. Thus to achieve a viable theory one must add soft SUSY breaking terms to the theory by hand. However, the number of possibilities is enormous. In SU(5) theory the number of parameters is 57 if one does not impose any constraints. This number reduces to 25 when one imposes reality and matter parity constraints on the coupling constants. For an $SU(3)_C \times SU(2)_L \times U(1)_Y$ invariant theory, the number of arbitrary parameters one can add to the theory is 137. This number reduces to 48 if one imposes reality and matter parity constraints on the couplings.

† Conference Speaker

Thus SUSY GUTS is unpredictive in sectors which involve the SUSY spectrum. We consider next supergravity grand unification[2]. This model contains a viable spontaneous breaking of supersymmetry and the parameter space of this theory is four dimensional. It should be noted that the so called "no-scale" models and the "dilaton dominated" string scenarios[3] are subcases of this parameter space.

II. Supergravity GUT Basics

Ideas of supergravity grand unification have a sound theoretical foundation. We discuss them here briefly. First the effective lagrangian of this theory below the Planck scale is assumed to arise from coupling of N=1 supergravity with N=1 gauge and matter fields. In the limit when Planck mass $\to \infty$, the theory correctly reduces to a globally supersymmetric theory. Sugra GUT manufactures a spontaneous breaking of supersymmetry and after the breaking there is growth of soft SUSY breaking terms in the physical sector of the theory Terms which involve product of single derivatives of the Kähler potential lead to a universal mass for the scalar fields under the assumption of a generation independent Kähler potential. Similarly product of single derivative of Kähler potential and the superpotential lead to trilinear and bilinear soft SUSY breaking terms. Finally a non-trivial gauge dependence of the gauge kinetic energy term leads to a non-vanishing gaugino mass.

We now formalize the sugravity GUT construction. One assumes that the theory has a grand unification group G which breaks to $SU(3)_C \times SU(2)_L \times U(1)_Y$ at the GUT scale and that the spectrum below the GUT scale is the spectrum of MSSM with two doublets of Higgs and their superpartners. For models of the above type, after spontaneous breaking of supersymmetry and of GUT symmetry, and integrating over the hidden and heavy sector fields, the effective potential of the theory below the GUT scale can be described in terms of the following parameters: α_G, M_G, m_o, $m_{1/2}$, A_o, B_o, μ_o where m_o is the universal scalar mass, $m_{1/2}$ is the universal gaugino mass, A_o and B_o are co-efficients of the trilinear and bilinear SUSY breaking terms and μ_o in the co-efficient of the H_1H_2 term in the superpotential. The μ_o that appears in the low energy theory consists of two parts. The first part $\bar{\mu}_o$ is a remnant of a dimension two $\bar{\mu}_o 5_H \bar{5}_H$ terms before spontaneous breaking of SUSY. The analysis shows, however, that there is a shift in the μ-parameter that depends on GUT parameters and on the soft breaking parameters. If there was a constraint which set $\bar{\mu}_o = 0$, then μ_o will be completely determined. This is what happens in a class of models with a higher GUT symmetry[4].

One of the very non-trivial aspects of supergravity GUT analysis is that all terms of $O(M_G/M_{Planck})^n M_G$ cancel in the low energy domain after spontaneous breaking of supersymmetry and GUT symmetry. If this were not the case, the soft SUSY breaking terms could be of size $O(M_G)$ and low energy physics would be destroyed. However, M_G does enter low energy but only via boundary conditions in the renormalization group equations.

The phenomenon of radiative breaking of the electro-weak symmetry in supergravity GUT framework is now all very familiar and we do not discuss this here in any detail. We wish to point out, however, that there are many subtle points in carrying out the analysis. These have to do with color and charge conservation of the global minima. The basic purpose of this exercise is that the spontaneous breaking of $SU(2)\times U(1)$ can be deduced as a consequence of soft SUSY breaking terms. The radiative breaking of the electroweak symmetry also leads to a reduction by one in the number of parameters. Additionally using the LEP data information on α_1, α_2, α_3 and on M_Z, one can reduce the 7 dimensional parameter space of the model to a 4 dimensional space. We choose this set to consist of m_o, $m_{1/2}$, A_o and $\tan\beta$ where $\tan\beta = v_2/v_1$, where v_2 gives mass to the up quark and v_1 gives mass to the down quarks and leptons. Finally, one needs to limit the parameter space in m_o and $m_{\frac{1}{2}}$ using a naturalness criterion. In this analysis we assume this criterion to imply that m_o, $m_{\tilde{g}} \langle 1 TeV$. Next we discuss the nature of the superparticle spectrum that results from the above analysis since the properties of this spectrum enter in the dark matter analysis to be discussed later. We begin by noting that there are 32 SUSY particles in the theory and there are only 4 parameters. Thus one ends up with 28 sum rules. These sum rules[5-11] are valid for a large class of models including those where special values are assigned to the soft SUSY breaking parameters. An interesting phenomenon is the appearance of scaling laws over much of the parameter space. These arise dynamically as a consequence of radiative electro-weak breaking and can be deduced by taking the limit $|\mu| \rangle\rangle M_Z$[5]:

$$
\begin{aligned}
&m_{\tilde{Z}_1} \approx \tfrac{1}{2} m_{\tilde{Z}_2},\ m_{\tilde{Z}_2} \approx m_{\tilde{W}_1},\ m_{\tilde{Z}_3} \approx m_{\tilde{Z}_4} \approx m_{\tilde{W}_2} \\
&m_{\tilde{W}_1} \approx \tfrac{1}{3} m_{\tilde{g}} (\mu \langle 0);\ m_{\tilde{W}_1} \approx \tfrac{1}{4} m_{\tilde{g}} (\mu \rangle 0) \\
&m_H \approx m_A \approx m_{H^+}
\end{aligned}
\tag{2.1}
$$

In supergravity there are typically three low lying states $\tilde{e}_R$, $\tilde{\nu}$, $\tilde{Z}_1$. Occassionally the lighter stop can also become a low lying state. In the scaling limit masses of $\tilde{e}_R$, $\tilde{\nu}$ and $\tilde{Z}_1$ are given by:

$$
\begin{aligned}
&m_{\tilde{Z}_1} \approx \left(\frac{\alpha_1}{\alpha_G}\right) m_{1/2} \\
&m_{\tilde{e}_R}^2 = m_o^2 + 0.15\, m_{\frac{1}{2}}^2 + 0.5\, M_Z^2 \cos 2\beta \\
&m_{\tilde{\nu}^2} = m_o^2 + 0.52\, m_{\frac{1}{2}}^2 + 0.5\, \mathrm{M}_Z^2 \cos 2\beta\ .
\end{aligned}
\tag{2.2}
$$

Conditions under which $\tilde{\nu}$ is LSP are very stringent. One requires that at least $m_{\tilde{Z}_1} \langle 58$ GeV, $m_o \langle 60$. A similar situation holds for $\tilde{e}_R$. Thus over most of the parameter space $\tilde{Z}_1$ is the LSP.

III. Dark Matter of Supergravity

As is well known there is currently very strong astronomical evidence for the existence of dark matter in the universe. Several dark matter candidates (brown dwarfs, neutron stars, blackholes etc.) have been discussed in astronomy, and there are similarly several candidates (axions, neutrinos, neutralino, sneutrino) in particle physics.

As discussed in Sec. II in supergravity it is the neutralino which is most likely the LSP according to renormalization group analysis and for the rest of the analysis we shall assume that this indeed is the case. We should thus require that at least $\Omega_{\tilde{Z}_1} h^2 \langle 1$ so as not to overclose the universe. We may also wish to impose a stronger constraint as indicated by the COBE data that the mix of cold dark matter and hot dark matter is about ≈ 2 if one uses the full power spectrum of density fluctuations. Further if one uses inflationary scenario requiring $\sum \Omega_i = 1$ and for baryonic matter $\Omega_{B \simeq 0.1}$, and the assumption that h lies in the interval $0.5 \langle h \langle 0.75$ as implied by the current astrophysical data, one is then led to the constraint $0.1 \langle \Omega_{\tilde{Z}_1} h^2 \langle 0.35$ we shall call this the CHDM contraint.

To take account of the constraints discussed above, one must compute the neutralino relic density at the current temperature. Here one follows the standard methods for the computation of the neutralino mass density and uses the relation.

$$\rho_{\tilde{Z}_1} = 4.75 \times 10^{-40} \left(\frac{T_{\tilde{Z}_1}}{T_\gamma} \right)^3 \left(\frac{T\gamma}{2.75} \right)^3 \frac{N_f^{\frac{1}{2}}}{J(x_f)} \frac{g}{cm^3}$$
$$J(x_f) = \int_o^{x_f} dx \langle \sigma v \rangle Gev^{-2}, \; x = \frac{kT}{m_{\tilde{Z}_1}} \tag{3.1}$$

There is however a subtle issue regarding computation of $J(x_f)$ which involves the integral over the thermally averaged quantity $\langle \sigma v \rangle$. Normally one carries out this integration by first expanding $\langle \sigma v \rangle$ in a power series expansion of the type $(a + bx)$. However, it is well known that this expansion breaks down near thresholds and poles[13]. Now for neutralino annihilation this precisely is the situation often as one may be close to the Z-pole or the higgs-pole. In fact it turns out that the effect of the pole is felt a significant distance from the pole in thermal averaging. Thus one needs to carefully integrate over the poles. What one finds is a rather large effect sometimes by orders of magnitude when one takes into account the pole effects[14].

IV. Detection of Dark Matter

Current estimates of dark matter in the solar neighborhood give a mass density of 0.3 GeV cm^{-3}, and the average velocity with which they impact targets are estimated to be $\langle v \rangle \approx 320\, km/s$. The basic unit for the detection of the neutralino is quark via neutralino-quark scattering. Realistically, since quarks are bound in nucleons and nucleons in nuclei, one has to consider neutralino-nucleus scattering. The recoil energies associated with elastic scattering with nuclei fall in the range (1-100 KeV). Currently there are various detector possibilities which are sensitive to recoil energies in this range.

There are two types of interactions which are central to the computation of the elastic scattering of neutralinos from quarks. The first are the spin-dependent interactions that arise from exchange of the Z-boson and from the exchange of chiral squarks. The second are the spin independent interactions which arise from the Higgs exchange and from squark exchange with mass insertions. Thus the effective low energy Lagrangian which governs the neutralino-quark scattering is given by[15-17]

$$L_{eff} = \bar{\tilde{Z}}_1 \gamma_\mu \gamma_5 \tilde{Z}_1 \bar{q} \gamma^\mu \left(A_q P_L + B_q P_R \right) q + \bar{\tilde{Z}}_1 \tilde{Z}_1 m_q \bar{q} C_q q \tag{4.1}$$

Here the scalar coupling is the spin independent part and the axial coupling is the spin dependent part.

In the analysis carried out we have used the standard procedures on the computation of quark operators between nucleus states. However, for the SUSY part of the calculation we have made certain improvements[18-19]. These include first of all using the accurate density as discussed in Sec. III. Second we have included the heavy Higgs exchange which is often times neglected. Third we have included loop corrections to the Higgs mixing angle and finally all the analysis is done in the radiative electro-weak breaking scheme. Next we discuss some details of SUSY physics in DM analysis. The lightest neutralino is a superposition of four neutral states in the theory: SU(2) and U(1) gauginos and two neutral Higgsinos.

$$\tilde{Z}_1 = n_1 \tilde{W}_3 + n_2 \tilde{B} + n_3 \tilde{H}_1^0 + n_4 \tilde{H}_2^0 \tag{4.2}$$

The indices n_1, n_2, n_3, n_4 enter importantly into the event rate analysis. In the scaling region $\left(|\mu| \gg M_Z \right)$ one can compute n_i using the second order perturbation theory in M_Z/μ. One finds

$$\begin{pmatrix} n_1 \\ n_2 \\ n_3 \\ n_4 \end{pmatrix} \underset{|\mu|\rangle\rangle M_Z}{\rightarrow} \begin{pmatrix} -\dfrac{M_Z^2}{2m_{\tilde{Z}_1}\mu}\sin 2\theta_W \sin\beta \\ 1-\dfrac{1}{2}\dfrac{M_Z^2}{\mu^2}\sin^2\theta_W \\ -\dfrac{M_Z}{\mu}\sin\theta_W\sin\beta \\ \dfrac{M_Z}{\mu}\sin\theta\cos\beta \end{pmatrix} \tag{4.3}$$

where θ_W is the weak angle. From the above one can deduce the following ordering of the indices in the scaling region[18,19]

$$\begin{aligned} &|n_3|/|n_4| = \tan\beta\rangle 1 \\ &n_2\rangle|n_1|,\ n_2\rangle|n_3|\rangle|n_4| \end{aligned} \tag{4.4}$$

The above relations can be verified by numerical renormalization group analyses in the scaling region.

Next we discuss relative contributions of the light and heavy Higgs. The Higgs contribute to C_q and their contribution is given by

$$C_q^{Higgs} = \frac{g_2^2}{4M_W}\left[\begin{Bmatrix} \dfrac{\cos\alpha}{\sin\beta}\dfrac{F_h}{m_h^2} \\ -\dfrac{\sin\alpha}{\cos\beta}\dfrac{F_h}{m_h^2} \end{Bmatrix} + \begin{Bmatrix} \dfrac{\sin\alpha}{\sin\beta}\dfrac{F_H}{m_H^2} \\ \dfrac{\cos\alpha}{\cos\beta}\dfrac{F_H}{m_H^2} \end{Bmatrix}\right] \begin{matrix} u-quark \\ \\ d-quark \end{matrix} \tag{4.5a}$$

where α is the Higgs mixing angle. In Eq. (4.5a) F_h and F_H are given by

$$\begin{aligned} F_H &= (n_1 - n_2\tan\theta_w)(-n_3\sin\alpha + n_4\cos\alpha) \\ F_h &= (n_1 - n_2\tan\theta_w)(-n_3\cos\alpha + n_4\sin\alpha) \end{aligned} \tag{4.5b}$$

where θ_W is the Weak angle. Naively one might say that the H Higgs contribution is negligible since it is so heavy. However, this statement is correct for the up quark but not for the down quark. The reasons for this is that the Higgs mixing angle α is typically small so there is a suppression of the light Higgs contribution vs. the heavy Higgs in the d-quark sector.

Regarding the computation of the Higgs mixing angle itself one must be careful how one computes it. If one uses formulae which hold at the tree level, and uses loop corrected values of masses there, one can get results wrong up to an order of 10. The formula for the mixing angle which extends correctly to the loop level is given by[20]

$$\cos 2\alpha = \frac{M_{11} - M_{22}}{\sqrt{(M_{11} - M_{22})^2 + 4M_{12}^2}} \tag{4.6}$$

where (M)ij is the Higgs (mass)^2 matrix in the CP-even sector. Actually, loop corrections to α turn out to be typically small if one computes α consistently to a given loop order. The total event rate in neutralino-nucleus scattering is sum of the incoherent (spin dependent) and coherent (spin dependent) parts and is given by

$$R = [R_{coh} + R_{inc}]\left[\frac{\rho_{\tilde{Z}_1}}{0.3GeV\, cm^{-3}}\right]\left[\frac{\langle v_{\tilde{Z}_1}\rangle}{320\, km/s}\right]\frac{\text{events}}{\text{kg da}} \tag{4.7a}$$

where

$$R_{coh} = \frac{16 m_{\tilde{Z}_1} M_N^3 M_Z^4}{\left[M_N + m_{\tilde{Z}_1}\right]^2}|A_{coh}|^2 \tag{4.7b}$$

$$R_{inc} = \frac{16_{m_{\tilde{Z}_1}} M_N}{\left[M_N + m_{\tilde{Z}_1}\right]^2}\lambda^2 J(J+1)|A_{inc}|^2 \tag{4.7c}$$

where A_{inc} is due to the spin dependent part and A_{coh} arises from the spin independent part (i.e.,. $(A_{inc} \sim Bq - Aq$ and $A_{coh} \sim\ Cq)$, J is spin of the nucleus, M_N is the nucleus mass and λ is determined using the magnetic moment of the nucleus. From Eq.(4.7) it is seen that for large M_N, R_{coh} part behaves as $O(M_N)$ and R_{inc} part behaves as $O(1/M_N)$. Since R is a sum of two terms, one can conceive of two types of limits: one where R_{inc} is large and the other where R_{coh} is large. In actuality most often one is always in the situation where R_{coh} dominates. Next we discuss details of the event rate analysis. First of all, here one observes that rapid annihilation near Z and Higgs poles leads to values of Ωh^2 which fall below the CHDM limit and have to be eliminated. The eliminated part of the parameter space contains the large events rates. Thus one sees a dip in the event rates in the vicinity of $m_{\tilde{Z}_1} \sim M_Z/2$ and $m_h/2$. We have analyzed several target materials He^3, CaF_2, 50% mixture of Ge^{73} and Ge^{76}, NaI, GaAs and Pb. We find that except for the lightest target materials i.e. ^3He and CaF_2, the incoherent part of R i.e. R_{inc} is very small ~o(a few %). We have also analyzed the effect of the recent new measurements of quark-spin matrix elements Δq on R_{inc}. These quark-spin matrix elements enter in the computation of R_{inc}. The old values of Δq were extracted from the EMC data which give for Δq the values, $\Delta u = 0.77 \pm 0.08$, $\Delta d = 0.49 \pm 0.08$ and $\Delta s = -0.15 \pm 0.08$. Values of Δq given by the recent data from the spin-Muon collaboration (SMU) are $\Delta u = 0.83 \pm 0.03$, $\Delta d = -0.43 \pm 0.03$, and $\Delta s = -0.10 \pm 0.03$ and lie within 1 sigma of the values from EMC. Using the new values of Δq one finds that R_{inc} does change significantly. However, since R_{inc} is a small component of the total, we find that the total event rates are not significantly affected except for the lightest target materials.

The analysis shows that there are significant regions of the parameter space where R>0.1 and thus this region is accessible to observation with current technology. However, to probe the full parameter space one needs to have an increase in the detector sensitivity of $O(10^2\text{-}10^3)$ or more.

These results are exhibited in Figs. 1-2 for $m_t = 168 GeV$. Here the maximum and minimum of event rates is plotted for $\mu \rangle 0$ (Fig. 1) and $\mu \langle 0$ (Fig. 2) as a function of gluino mass for three different target materials: CaF_2 (dash-dot), Ge (dashed) and Pb (solid) when all other parameters $(m_o,\ A_t,\ \tan\beta \leq 20)$ are integrated out over their allowed region of the parameter space.

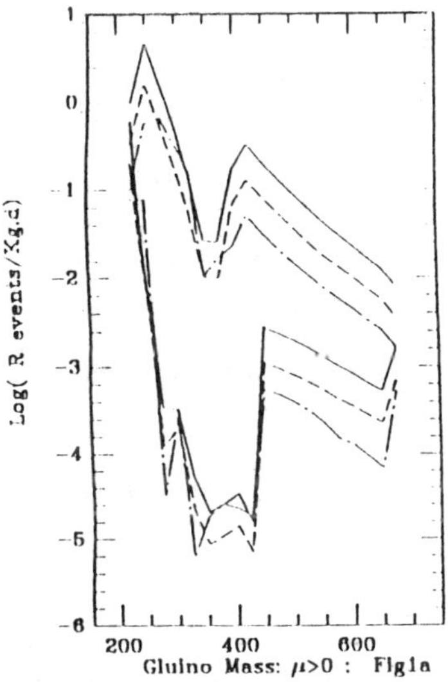

Fig. 1: $\mu \rangle 0$ case

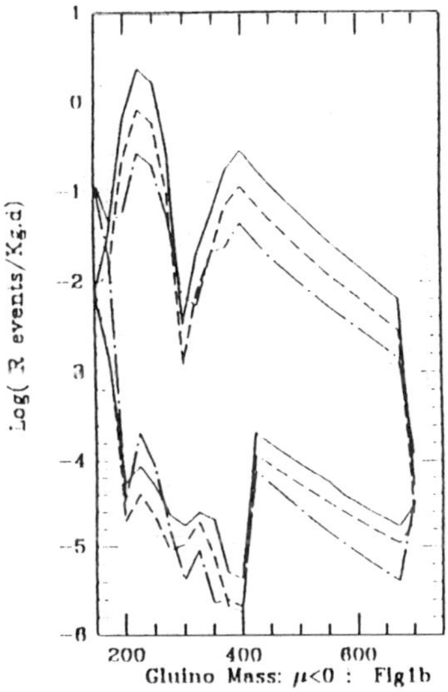

Fig. 2: $\mu \langle 0$ case

V. Effect of $b \to s\gamma$ Measurement on Event Rates

At the HEP94 Conference at Glasgow, CLEO announced[23] the first measurement of the photonic penguin decay process $b \to s\gamma$ with a branching ratio

$$BR(b \to s\gamma) = (2.32 \pm 0.51 \pm 0.29 \pm 0.32) \times 10^{-4} \tag{5.1}$$

Theoretically to leading order QCD calculation, one finds[24]

$$\frac{BR(b \to s\gamma)}{BR(b \to ce\overline{\nu})} = \frac{6\alpha}{\pi\rho\lambda} \frac{\left|V_{ts}^* V_{tb}\right|^2}{\left|V_{cb}\right|^2} \left|\overline{C}_7(m_b)\right| \tag{5.2}$$

where V_{ts} etc. are KM matrix elements, ρ is a phase-space factor, λ is a QCD correction to the semileptonic decay and $\overline{C}_7(m_b)$ is the effective Wilson co-efficient at scale m_b, and is given by

$$\overline{C}_7(m_b) = \eta^{16/23} C_7(M_W) + \frac{8}{3}\left(\eta^{14/23} - \eta^{16/23}\right) C_8(M_W) + C_2 \tag{5.3}$$

Here $C_7(C_8)$ are the Wilson co-efficients at scale M_W for the photonic (gluonic) penguins and $\eta = \alpha_s(M_W)/\alpha_s(m_b)$. In the SM, $\overline{C}_7$ receives contributions from W-exchange. In SUSY theory, there are additional contributions from H^+, $\tilde{W}$, $\tilde{Z}$ and $\tilde{g}$. However, there are many uncertainties in the theoretical evaluations. These include uncertainties in quark masses, in α_s, in the ratio of $\left|V_{ts}^* V_{tb}/V_{cb}\right|$, and in the next to leading order (NLO) QCD corrections. These corrections are of 0(30%). There is some debate in the literature[25,26] regarding what the branching ratio is in the SM. An accurate analysis would require knowing the NLO corrections completely. In the meantime one can use a bench mark value of $(2.75 \pm 0.8) \times 10^{-4}$ quoted in the CLEO paper.

To discuss the SUSY effects on the dark matter analyses it is useful to introduce the parameter r_{SUSY} defined by

$$r_{SUSY} = BR(b \to s\gamma)_{SUSY} / BR(b \to s\gamma)_{SM} \tag{5.4}$$

Theoretical analyses within radiative electro-weak symmetry breaking show that r_{SUSY} has a very large range, i.e., $r_{SUSY} \approx (0, 10)$ due to the possibility of both constructive and destructive interference of the SUSY contributions with the standard model contributions. Experimentally using the data of Eq. (5.1) one finds that r_{exp} lies in the rang (0.4, 2.2). Thus we see that the CLEO experiment puts a significant constraint on the theory. In dark matter analyses one finds[27] that 0(65%) of the parameter space is eliminated by $\mu \rangle o$, and 0(20%) for $\mu \langle o$. Similarly, one finds a significant effect on the event rates for $\mu \rangle o$. For

$\mu\langle o$, the effect is much less. These results are exhibited in Figs. 3 and 4 which are similar to Figs. 1 and 2 in all other respects except that we have imposed the additional constraint $r_{SUSY} \leq 1.33$. One can see by comparison of Figs. 1 and 3 that the effect of this constraint is rather dramatic for the $\mu\rangle o$ case. However, by comparison of Figs. 2 and 4 we find that the effect of the $r_{SUSY} \leq 1.33$ constraint is not very strong for $\mu\langle o$.

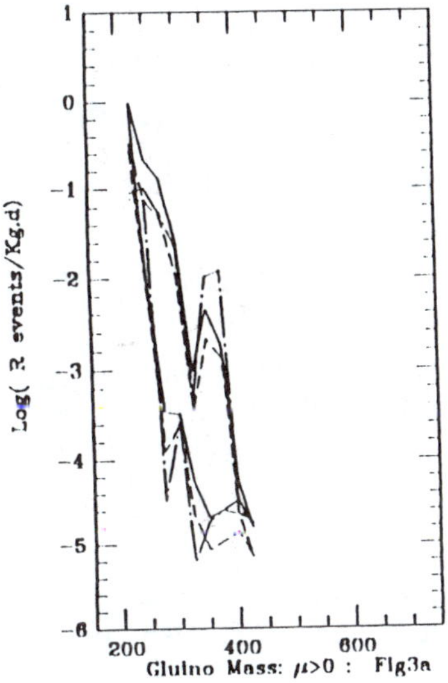

Fig. 3: $\mu\rangle o$

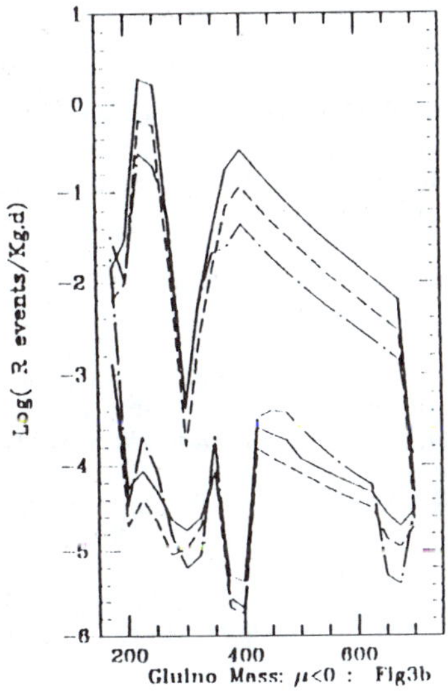

Fig. 4: $\mu\langle o$

VI. Conclusion

We have given here an analysis of neutralino relic density and of event rates in neutralino-nucleus scattering within the framework of supergravity grand unification. We have demonstrated the existence of a new phenomena of a dip in the events rates when the neutralino mass is $\sim M_Z/2$ or $m_h/2$ and the constraints of CHDM are impoed. Except for the lightest detector materials $\left({}^3H_e, CaF_2\right)$, R_{coh} is seen to dominate R_{inc}. Thus the overall effect of using new vs. old values of quark spin matrix elements Δq, which affect only R_{inc}, is found to be not very significant for the total event rates. The analysis shows that there exists a significant part of the parameter space where the event rate $R\rangle 0.1$ and thus this region of the parameter space is accessible with current technology[28]. However, it is argued that more sensitive detectors with sensitivities greater than $O(10^2\text{-}10^3)$ over what the current detectors have are needed to map the full parameter space. The effect of the recent CLEO measurement of $b \to s\gamma$ on event rate is also analyzed. It is found that event rates for $\mu\rangle o$ are significantly affected, while the effects on the event rates for $\mu\langle o$ is small.

Acknowledgments

This research was supported in part by NSF grant numbers PHY-19306906 and PHY-9411543 .

References

1. For a recent review see W. de Boer, Karlsruhe preprint IEKP-KA/94-01(94)

2. A.H. Chamseddine, R. Arnowitt and P. Nath, Phys. Rev. Lett. 49 (1982) 970.For a review see P. Nath, R. Arnowitt an A. H. Chamseddine, "Applied N=1 Supergravity", (World Scientific, Singapore, 1984);H.P. Nilles, *Physics Reports* **110**, 1 (1984)..

3. For a review see J. Lopez, D.V. Nanopoulos and A. Zichichi, CERN-TH.7/36/94.

4. K. Inoue, A. Kakuto and T. Tankano, *Prog. Theor. Phys.***75** (1986) 664; A. Anselm and A. Johansen, *Phys. Lett.* **B200**, (988) 331; R. Barbieri, G. Dvali and A. Strumia, *Nucl. Phys.* **B391** (1993) 487.

5. R. Arnowitt and P. Nath, *Phys. Rev. Lett.* **69** (1992) 725; P. Nath and R. Arnowitt, *Phys. Lett.* **B289** (1992) 368.

6. G.G. Ross and Roberts, *Nucl. Phys.* **B377** (1992) 971; M. Drees and M. M. Nojiri, Nucl. Phys. B369 (1992) 54; K. Inoue, et. al., *Phys. Rev.* **D45** (1992) 387

7. J. Lopez, D.V. Nanopoulos and A. Zichichi, CERN-PPE/94-01.

8. M. Carena, S. Pokorski and C.E.M. Wagner, Report MPI-PH/93-10(1993) and references therein.

9. S.P. Martin and P. Ramond, *Phys. Rev.* **D48** (1993) 289.

10. G. Kane, C. Kolda, L. Rozskowski, J.D. Wells, *Phys. Rev.* **D49** (1994) 6173.

11. V. Barger, M.S. Berger and P. Ohman, Univ. of Wisconsin Preprint MAD/PH/801 (1993).

12. B.W. Lee and S. Weinberg, *Phys. Rev. Lett.* **39**, 165 (1977); D.A. Dicus, E. Kolb and V. Teplitz, *Phys. Rev. Lett.* **39**, 168 91977); H. Goldberg, *Phys. Rev. Lett.* **50** (1983) 1419; J. Ellis, J.S. Hagelin, D.V. Nanopoulos, K. Olive and M. Srednicki, *Nucl. Phys.* **B238**, 453 (1984); J. Ellis, L. Roszkowski and Z. Lalak, *Phys. Lett.* **B245** (1990) 545; M. Drees and M. M. Nojiri, *Phys. Rev.* **D47** (1993) 376..

13. P. Gondolo and G. Gelmini, *Nucl. Phys.* **B360** 145 (1991); K. Griest and D. Seckel, *Phys. Rev.* **D43** (1991) 3191.

14. R. Arnowitt and P. Nath, *Phys. Lett.* **B299**, 58(1993); E. B303, 403 (1993); P. Nath and R. Arnowitt, *Phys. Rev. Lett.* **70**, 3696 (1993); S. Kelley, J. Lopez, D.V. Nanopoulos, H. Pois and K. Yuan, *Phys. Rev.* **D47**, 2461 (1993)

15. W. Goodman and E. Witten, *Phys. Rev.* **D31**, 3059 (1985); K. Griest, Phys. Rev. Lett. 62, 666 (1988); *Phys. Rev.* **D38**, 2357 (1988); R. Barbieri, M. Frigeni and G. F. Giudice, *Nucl.. Phys.* **B313**, 725 (1989); M. Srednicki and R. Watkins, Phys. Rev. Lett., B225, 140 (1989); G. F. Giudice and E. Roulet, *Nucl. Phys.* **B316**, 429 (1989); R. Flores, K. Olive and M. Srednicki, *Phys. Lett.* **237**, 72 (1990); M. Kamionkowski, *Phys. Rev.* **D44**, 3021 919910; T. Falk, R. Madden, K. Olive and M. Srednicki, *Phys. Lett.*, **B318**, 354 (1993); M.T. Ressel, M. B. Aufderheide, S.D. Bloom, K. Griest, G.T. Mathews, and D. A. Resler, *Phys. Rev.* **D48** (1993) 5519..

16. J. Ellis and R. Flores, *Phys. Lett.* **B263**, 259 (1991); B300, 175 (1993); *Nucl. Phys.* **B400**, 25 91993); A. Bottino, V.de Alfero, N. Fornengo, G. Mignola and S. Scopel and the Beijing, Roma and Saclay collaboration, *Phys. Lett.* **B295**, 330 (1992); A. Bottino, V. de Alfaro, N. Fornengo, G. Mignola and M. Pignone, *Astroparticle Phys.* **2** (1992) 67.

17. M. Drees and M. Nojiri, *Phys. Rev.* **D48**, 3483 (1993).

18. R. Arnowitt and P. Nath, CERN TH.7362/94 (1994)/hep-ph/9408226.

19. P. Nath and R. Arnowitt, CERN .7363/94 (1994)

20. H. Haber and R. Hempfling, *Phys. Rev. Lett.* **66** (1991) 1815; *Phys. Rev.* **D48**, 4280 (1993).

21. J. Ellis and M. Karliner, CERN TH.7324/94.

22. M. Kamionkowski, L. Krauss, and M. Ressel, IASSNS-HEP-94/14.

23. E.H. Thorndike, Talk at ICHEP94, Glasgow, 1994 (ICHEP94 REF. GLS0392)

24. S. Bertolini, F. Borzumati, A. Masiero and G. Ridolfi, *Nucl. Phys.* **B353**, 591 (1991)

25. M. Ciuchini, Talk at ICHEP94 Conference, Glasgow, 1994; M. Ciuchini, E. Franco, G. Martinelli L. Reina and L. Silvestrini, CERN TH.7283/94 (1994).

26. A.J. Buras, M. Misiak, M. Munz and S. Pokorski, MPI-Ph/93.

27. P. Nath and R. Arnowitt, CERN TH.7214/94 (1994) (to appear in Phys. Lett. B); F. Borzumati, M. Drees and M. Nojiri, Preprint MAD/PH/835

28. D.O. Caldwell, in "Supersymmetry and Unification of Fundamental Interactions: SUSY-93", edited by P. Nath, World Scientific, Singapore 1993.

SEARCHING FOR RELIC NEUTRALINOS

A. BOTTINO, N. FORNENGO, G. MIGNOLA
Dipartimento di Fisica Teorica, Università di Torino and
INFN, Sezione di Torino, via P.Giuria 1, 10125 Torino, Italy

and

S. SCOPEL
Dipartimento di Fisica, Università di Genova and
INFN, Sezione di Genova, via Dodecaneso 33, 16146 Genova, Italy

presented by A. Bottino

ABSTRACT

Theoretical expectations for direct and indirect searches for relic neutralinos are presented. Complementarity among various investigation means is discussed in connection with the values of the neutralino relic abundance.

1. Introduction

In the present report we discuss the theoretical expectations for detection of relic neutralinos either by the direct method (*e.g.* by measuring the nuclear recoil energy due to neutralino–nucleus elastic scattering) or by detecting indirect signals due to neutralino–neutralino annihilation occurring in the halo or inside celestial bodies (Sun and Earth).

Evidence for the presence of these relic neutralinos would obviously be of primary importance as a test of the standard Big Bang theory, independently of the fact that these Susy relic particles would or would not provide a large contribution to the total mass of the Universe. Should the relic neutralinos contribute substantially to the total density parameter Ω, their detection would also provide an extraordinary hint for a clarification of the long–standing dark matter issue. In fact, in this case, relic neutralinos would play a significant role as Cold Dark Matter (CDM) constituents.

In the following we adopt the usual assumptions that: i) R–parity is conserved, ii) the neutralino is the lightest supersymmetric particle (LSP). Very convenient theoretical frameworks where dark matter neutralino phenomenology may be easily studied are provided by the Minimal Supersymmetric Standard Model (MSSM) and by its implementation in a Supergravity theory (SUGRA) [1,2,3]. Here MSSM is meant to denote the minimal supersymmetric extension of the standard model where sleptons and squarks are taken as degenerate, with the exceptions for the stop particles [3].

The neutralino (χ) is defined as the lowest–mass linear combination of photino, zino and higgsinos

$$\chi = a_1\tilde{\gamma} + a_2\tilde{Z} + a_3\tilde{H}_1^0 + a_4\tilde{H}_2^0 \tag{1}$$

Here $\tilde{\gamma}$ and $\tilde{Z}$ are the fields obtained from the original U(1) and SU(2) neutral gauginos, $\tilde{B}$ and $\tilde{W}_3$, by a rotation in terms of the Weinberg angle.

The neutralino mass m_χ and the coefficients a_i depend on the parameters: μ (Higgs mixing parameter), M_1, M_2 (masses of $\tilde{B}$ and of $\tilde{W}_3$, respectively) and $\tan\beta = v_u/v_d$ (v_u and v_d are the v.e.v.'s which give masses to up–type and down–type quarks). It is customary to employ the standard GUT relationship between M_1 and M_2: $M_1 = (5/3)\tan^2\theta_W M_2 \simeq 0.5\ M_2$. We use this assumption here.

In the following for the parameters M_2 and μ we will consider the ranges: 20 GeV$\leq M_2 \leq 6$ TeV, 20 GeV$\leq |\mu| \leq 3$ TeV. $\tan\beta$ will be taken at the representative value $\tan\beta = 8$.

For the evaluation of the neutralino relic abundance and of the event rates for direct and indirect neutralino detections one has also to assign values to the masses of a large number of particles, namely to the Higgs bosons and to the Susy scalar partners of leptons and quarks: sleptons ($\tilde{l}$) and squarks ($\tilde{q}$). In the MSSM scheme we consider here these values are assigned arbitrarily: a standard procedure consists in assuming mass degeneracy both for sleptons and for squarks except for the stop particles [3]. As for the neutral Higgs bosons we recall that in the MSSM there are three neutral Higgs particles: two CP–even bosons h and H (of masses m_h, m_H with $m_H > m_h$) and a CP–odd one A (of mass m_A). Once a value for one of these masses (say, m_h) is assigned, the other two masses (m_A, m_H) are derived through mass relationships depending on radiative effects.

Implementation of MSSM with supergravity sets a much more constrained phenomenological framework, since SUGRA establishes strict relations between all the masses in play and the few fundamental theoretical parameters: A and B (appearing in the soft symmetry–breaking interaction terms), m_0 (common scalar mass at the GUT scale), $m_{1/2}$ (common gaugino mass at the GUT scale) and μ. Furthermore, other specific theoretical requirements (features of the symmetry breaking, condition that the neutralino be the LSP, ...) strongly restrict the whole parameter space. This has important consequences; for instance, it constrains the neutralino to compositions with dominance of the gaugino components. It has to be noted that important constraints on the neutralino parameter space may be inferred from the recent data on $b \to s + \gamma$ process [4]. The precise nature of these constraints still requires further investigation, especially in view of the large uncertainties due to QCD–effects [5].

In the present note the emphasis is on the investigation power of various detection methods and on their complementarity. On purpose, the model adopted here for the calculations is as simple as possible (not too much constrained by theoretical requirements) with choices for the free masses that are only restricted by experimental bounds.

Our main concern is to discuss the minimal sensitivity required in experimental devices in order to undertake a significant investigation of neutralino dark matter. For this reason we present here evaluations where the smallest values compatible with experimental lower bounds are assigned to the unknown masses; this usually provides maximal values for the signals. To be definite, in the following we will set the sfermion masses at the value $m_{\tilde{f}} = 1.2\, m_\chi$, when $m_\chi > 45$ GeV, $m_{\tilde{f}} = 45$ GeV otherwise. Only the mass of the top squarks are assigned a larger value of 1 TeV. The Higgs mass m_h is set at the value of 50 GeV. The top mass is fixed at $m_t = 170$ GeV.

2. Neutralino Relic Abundance

For the computation of the direct and indirect event rates for neutralino one has to use a specific value for the neutralino density ρ_χ. Obviously, it would be inappropriate to assign to the neutralino local density ρ_χ the standard value for the total dark matter density $\rho_l = 0.3$ GeV cm^{-3}, unless one specifically verifies that the neutralino relic abundance $\Omega_\chi h^2$ turns out to be at the level of an $(\Omega h^2)_{min}$ consistent with ρ_l. This is why a correct evaluation of the event rates for χ detection also requires a calculation of its relic abundance.

Thus we evaluate $\Omega_\chi h^2$ and we determine ρ_χ by adopting a standard procedure [6]: when $\Omega_\chi h^2 \geq (\Omega h^2)_{min}$, we put $\rho_\chi = \rho_l$; when $\Omega_\chi h^2$ turns out to be less than $(\Omega h^2)_{min}$, we take

$$\rho_\chi = \rho_l \frac{\Omega_\chi h^2}{(\Omega h^2)_{min}} \,. \tag{2}$$

Here $(\Omega h^2)_{min}$ is set equal to 0.03.

For the neutralino relic abundance $\Omega_\chi h^2$ we employ the results of our previous work [7]. In Fig.1 we display regions of the M_2, μ plane which are characterized by different values of $\Omega_\chi h^2$. Also shown in this figure are the iso–mass curves (dashed lines) and the iso–compositions curves (solid lines). Along an iso–composition line the composition parameter P defined as the gaugino fractional weight, *i.e.* $P = a_1^2 + a_2^2$, is kept fixed. In this figure, as well as in the following ones, only results for positive μ are displayed.

The results of Fig.1 can conveniently be reported in a $\Omega_\chi h^2$ vs. m_χ plot, for fixed values of P. This is done in Fig.2 where $\Omega_\chi h^2$ is plotted as a function of m_χ for three representative neutralino compositions: i) a gaugino–dominated composition ($P = 0.9$), ii) a composition of maximal gaugino–higgsino mixing ($P = 0.5$), iii) a higgsino–dominated composition ($P = 0.1$). As expected, out of the three compositions displayed in Fig.2, the gaugino–dominated state provides the largest values of $\Omega_\chi h^2$. In order to have more substantial values of $\Omega_\chi h^2$, one has to consider purer gaugino compositions ($P \gtrsim 0.99$). The very pronounced dips in the plots of Fig.2 are due to the s–poles in the χ–χ annihilation cross section (exchange of the Z and of the neutral Higgses). The sharp decrease at 80–90 GeV is due to the opening of the W^+W^- and ZZ final states in the χ–χ annihilation. We remind that $\Omega_\chi h^2 \propto (<\sigma_{\rm ann} v>_{\rm int})^{-1}$, where $<\sigma_{\rm ann} v>_{\rm int}$ is the annihilation cross section times relative velocity, averaged over the neutralino thermal distribution, integrated from the freezing temperature down to the present temperature.

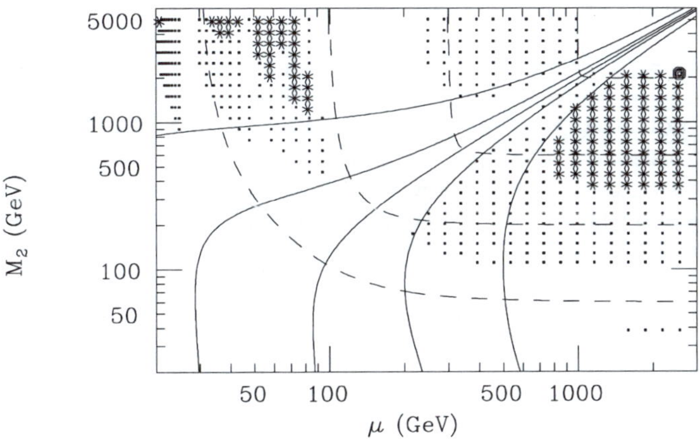

Figure 1. In the M_2–μ plane are shown the regions where the neutralino relic abudance $\Omega_\chi h^2$ belongs to the intervals $0.1 < \Omega_\chi h^2 \leq 1$ (stars) and $0.01 < \Omega_\chi h^2 \leq 0.1$ (dots). The full-circle denotes the region where $\Omega_\chi h^2 > 1$. The analysis is performed for the neutralino mass in the range 20 GeV $\leq m_\chi \leq$ 1 TeV. Dashed lines denote iso-mass curves with $m_\chi = 30, 100, 300, 1000$ GeV. Solid lines denote curves of iso-composition (here $P = 0.01, 0.1, 0.5, 0.9, 0.99$ are shown).

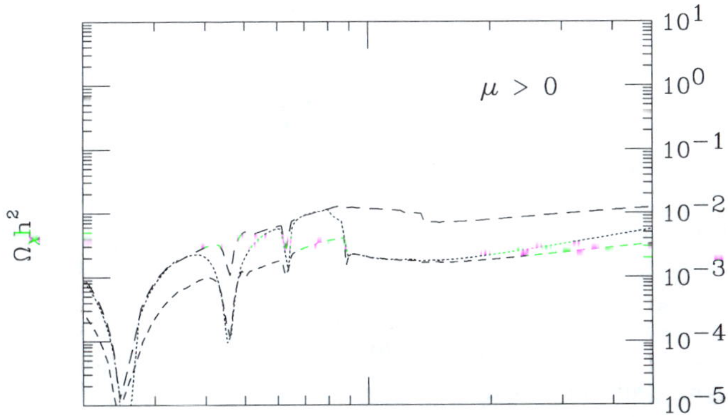

Figure 2. Neutralino relic abundance $\Omega_\chi h^2$ as a function of m_χ for the three representative neutralino compositions $P = 0.1$ (dotted line), 0.5 (short-dashed line), 0.9 (long-dashed line).

3. Indirect Detection at Neutrino Telescopes

Let us turn now to the indirect search for neutralino dark matter which can be performed by means of neutrino telescopes [10,11,12]. Neutralinos, if present in our galactic halo as dark matter components, would be slowed down by elastic scattering off the nuclei of the celestial bodies (Sun and Earth) and then gravitationally trapped inside them. Due to the process of neutralino capture these macroscopic bodies could accumulate neutralinos which would subsequently annihilate in pairs. An important outcome of this χ–χ annihilation would be a steady flux of neutrinos from these celestial bodies.

The differential neutrino flux at a distance d from the annihilation region is given by

$$\frac{dN_\nu}{dE_\nu} = \frac{\Gamma_A}{4\pi d^2} \sum_{F,f} B^{(F)}_{\chi f} \frac{dN_{f\nu}}{dE_\nu} \tag{3}$$

where Γ_A is the annihilation rate and F denotes the χ–χ annihilation final states which are: 1) fermion–antifermion pairs , 2) pairs of neutral and charged Higgs bosons, 3) one gauge boson–one Higgs boson pairs, 4) pairs of gauge bosons; $B^{(F)}_{\chi f}$ denotes the branching ratio into the fermion f (heavy quark or τ lepton), in the channel F; $dN_{f\nu}/dE_\nu$ denotes the differential distribution of the neutrinos generated by the semileptonic decays of the fermion f. The ν_μ's, crossing the Earth, would convert into muons and generate a signal of up–going muons inside a neutrino telescope. Calculations of this muon flux from the original neutrino flux may be performed using standard procedures [10,11,12].

Particular care has to be taken in the evaluation of the annihilation rate Γ_A. This quantity is given by [13]

$$\Gamma_A = \frac{C}{2} \tanh^2\left(\frac{t}{\tau_A}\right) \tag{4}$$

where t is the age of the macroscopic body ($t = 4.5\ Gyr$ for Sun and Earth), $\tau_A = (CC_A)^{-1/2}$, C is the capture rate of neutralinos in the macroscopic body and C_A is the annihilation rate per effective volume of the body. The capture rate C is provided by the formula [14]

$$C = \frac{\rho_\chi}{v_\chi} \sum_i \frac{\sigma_{el,i}}{m_\chi m_i} (M_B f_i) \langle v^2_{esc} \rangle_i X_i, \tag{5}$$

where v_χ is the neutralino mean velocity, $\sigma_{el,i}$ is the cross section of the neutralino elastic scattering off the nucleus i of mass m_i (for some properties of the elastic χ–nucleus cross section see next Sect.4 and Ref. 15), $M_B f_i$ is the total mass of the element i in the body of mass M_B, $\langle v^2_{esc} \rangle_i$ is the square escape velocity averaged over the distribution of the element i, X_i is a factor which takes account of kinematical properties occurring in the neutralino–nucleus interactions. C_A is given by [13]

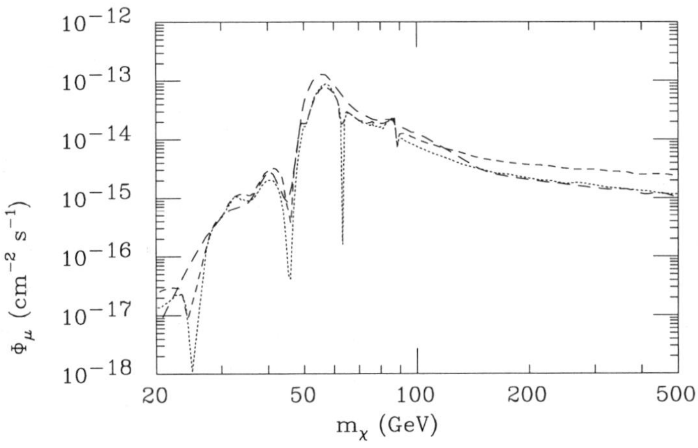

Figure 3. Fluxes of the up–going muons as functions of m_χ for χ–χ annihilation in the Earth, for the three representative neutralino compositions $P = 0.1$ (dotted line), 0.5 (short dashed line), 0.9 (long–dashed line). The threshold for the muon energy is $E_\mu^{\rm th} = 2$ GeV.

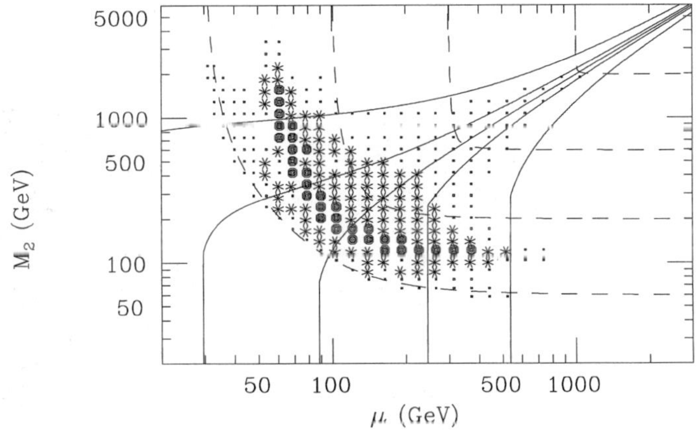

Figure 4. In the M_2–μ plane are shown the regions explored by Kamiokande (full circles) and the domains explorable by an improvement factor of 10 (asterisks) and of 100 (dots) in sensitivity, for χ–χ annihilation in the Earth. Dashed lines denote iso–mass curves with m_χ = $30, 100, 300, 1000$ GeV. Solid lines denote curves of iso–composition (here $P = 0.01, 0.1, 0.5, 0.9, 0.99$ are shown).

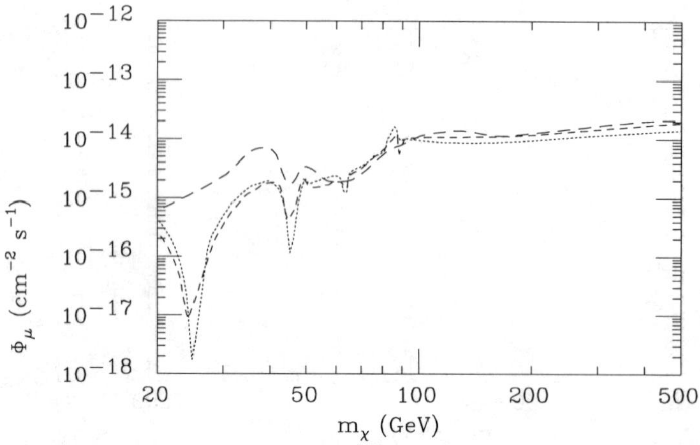

Figure 5. Fluxes of the up-going muons as functions of m_χ for χ–χ annihilation in the Sun, for the three representative neutralino compositions $P = 0.1$ (dotted line), 0.5 (short-dashed line), 0.9 (long-dashed line). The threshold for the muon energy is $E_\mu^{\rm th} = 2$ GeV.

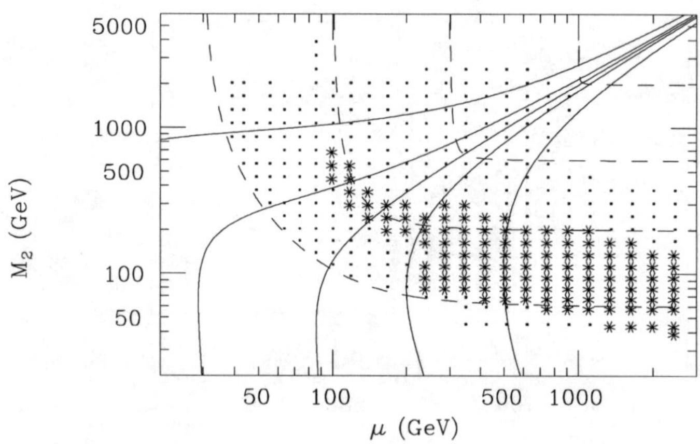

Figure 6. In the M_2–μ plane are shown the regions which may be explored by a neutrino telescope whose sensitivity is improved by a factor of 10 (asterisks) and of 100 (dots) as compared to the one of Kamiokande, for χ–χ annihilation in the Sun. Dashed lines denote iso-mass curves with $m_\chi = 30, 100, 300, 1000$ GeV. Solid lines denote curves of iso-composition (here $P = 0.01, 0.1, 0.5, 0.9, 0.99$ are shown).

$$C_A = \frac{<\sigma_{\rm ann} v>_0}{V_0} \left(\frac{m_\chi}{20 \text{ GeV}}\right)^{3/2} \tag{6}$$

where $<\sigma_{\rm ann} v>_0$ is the annihilation cross section times relative velocity, averaged over the neutralino thermal distribution, at present temperature. V_0 is defined as $V_0 = (3m_{Pl}^2 T/(2\rho \times 10 \text{ GeV}))^{3/2}$ where T and ρ are the central temperature and the central density of the celestial body. For the Earth ($T = 6000$ K, $\rho = 13 \text{ g} \cdot \text{cm}^{-3}$) $V_0 = 2.3 \times 10^{25} \text{cm}^3$, for the Sun ($T = 1.4 \times 10^7$ K, $\rho = 150 \text{ g} \cdot \text{cm}^{-3}$) $V_0 = 6.6 \times 10^{28} \text{ cm}^3$.

For the computation of the capture rate (and then also of τ_A) one has to use a specific value for the neutralino density ρ_χ according to the procedure explained in Sect.2.

The results of our evaluations for the quantities C, C_A and Γ_A are reported in Ref. 11,12. From these results it turns out that: i) for the Sun, the equilibrium between capture and annihilation is reached over the whole m_χ range; ii) in the case of the Earth, equilibrium is not reached for $m_\chi \gtrsim m_W$, because of the substantial suppression introduced in Γ_A by the factor $\tanh^2(t/\tau_A)$.

Now we report some of our results about the flux of the up–going muons in the case of χ–χ annihilation in the Earth. In Fig.3 we show the fluxes of the up going muons as functions of m_χ for a number of values of the neutralino composition P, for χ–χ annihilation in the Earth. The threshold for the muon energy is $E_\mu^{th} = 2$ GeV. We recall here that the present experimental upper bound for signals coming from the Earth is $4.0 \cdot 10^{-14} \text{cm}^{-2}\text{s}^{-1}$ (90 % C.L.) [16]. By comparing this upper limit with our fluxes we see that the regions explored by Kamiokande (at our representative point: $\tan\beta = 8, m_h = 50$ GeV) concern the mass range 50 GeV $\lesssim m_\chi \lesssim$ 65 GeV. These regions are illustrated in Fig.4 in a M_2–μ plot. In this figure we also display the regions which could be explored by a neutrino telescope with an improvement factor of 10 (and of 100) in sensitivity.

The location and the shape of the most easily explorable regions in the M_2–μ plane depend on the Earth chemical composition and on the neutralino composition in terms of the gaugino, higgsino components. In fact the capture rate of neutralinos is more effective when neutralino mass matches the mass of some of the main components of the Earth and when the neutralino is a large gaugino–higgsino mixture. Because of these two properties the signal is maximal along iso–mass lines in the range 50–65 GeV, with elongations along iso–composition lines of sizeable mixing.

In Fig.5 we report the fluxes for up–going muons as functions of m_χ due to χ–χ annihilation in the Sun. As before the threshold for the muon energy is $E_\mu^{th} =$ 2 GeV. The evaluated fluxes are below the present experimental upper limit of Kamiokande: $6.6 \cdot 10^{-14} \text{cm}^{-2}\text{s}^{-1}$ (90 % C.L.) [16]. The regions explorable by a neutrino telescope with an improvement factor of 10 (and of 100) in sensitivity respect to the present Kamiokande sensitivity are displayed in Fig.6. Here the regions that are more easely explorable, expand toward gaugino–dominated compositions since spin–dependent cross sections (with exchange of light squarks) are important in the capture of neutralinos by the Sun.

From these results it can be concluded that neutrino telescopes with an area above 10^5 m^2 are very powerful tools for investigating neutralino dark matter in large regions of the parameter space. It also emerges from the previous results that the signals from the Earth and from the Sun somewhat complement each other to allow an exploration about DM neutralino over a wide range of m_χ. In Ref. 12 has been derived the relation between the exposure At of a neutrino telescope (A being the telescope area, t the live time) and the explorable range in the neutralino parameter space, when the signal–to–background ratio is optimized by appropriate angular selections.

4. Direct Detection

Another way to search for dark matter neutralinos is the direct detection which relies on the measurement of the recoil energy of nuclei of a detector, due to elastic scattering of χ's. The relevant quantities to calculate are the differential rate (in the nuclear recoil energy E_r):

$$\frac{dR}{dE_r} = N_T \frac{\rho_\chi}{m_\chi} \int_{v_{min}(E_r)}^{v_{max}} dv f(v) v \frac{d\sigma_{el}}{dE_r}(v, E_r) \tag{7}$$

and the integrated rate R_{int}, which is the integral of Eq. (7) from the threshold energy E_r^{th}, which is a characteristic feature of the detector, up to a maximal energy E_r^{max}. In Eq.(7) N_T denotes the number of target nuclei, $d\sigma_{el}/dE_r$ is the differential elastic cross section and $f(v)$ is the distribution of χ velocities in the Galaxy. It is important to note again that the local density ρ_χ is evaluated here according to the procedure discussed in Sect.2. In general, the χ–nucleus cross section has two contributions: a coherent contribution, depending on A^2 (A is the mass number of the nucleus) which is due to Higgs and $\tilde{q}$ exchange diagrams; a spin–dependent contribution, arising from Z and $\tilde{q}$ exchange, proportional to $\lambda^2 J(J+1)$.

By way of example, let us remind the expression of the coherent cross section due to the Higgs–exchange [18]:

$$\sigma_{el,H} = \frac{8G_F^2}{\pi} \alpha_H^2 A_i^2 \frac{m_Z^2}{m_h^4} \frac{m_i^2 m_\chi^2}{(m_i + m_\chi)^2} \tag{8}$$

where α_H is a quantity depending on the neutralino–Higgs and the Higgs–quarks couplings. It is worth mentioning that α_H depends rather sensitively on the χ–composition and on a number of parameters, such as $\tan\beta$ and the Higgs masses.

Except for very special points in the parameter space, the coherent contribution to elastic cross section strongly dominates over the spin–dependent one. For a detailed analysis on the calculation of the direct event rates see Ref. 19 and references quoted therein. For an experimental overview about dark matter detectors see Ref. 20.

Here, as an example, we simply report in Fig.7 the event rates R_{int} for a Germanium detector as a function of m_χ for neutralino compositions $P = 0.1, 0.5, 0.9$. Rates are calculated by integrating the differential rate of Eq.(7) over the electron–equivalent

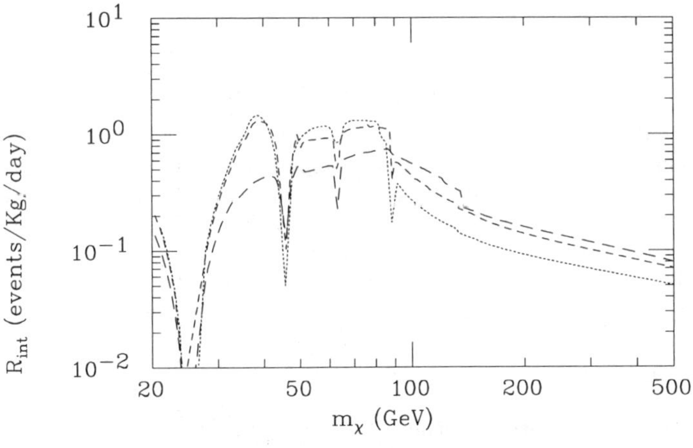

Figure 7. Event rates for detection of neutralino dark matter for a (natural composition) Germanium detector for the three representative neutralino compositions $P = 0.1$ (dotted line), 0.5 (short–dashed line), 0.9 (long–dashed line). The rate refers to the events with an electron–equivalent energy in the range (2–4) KeV.

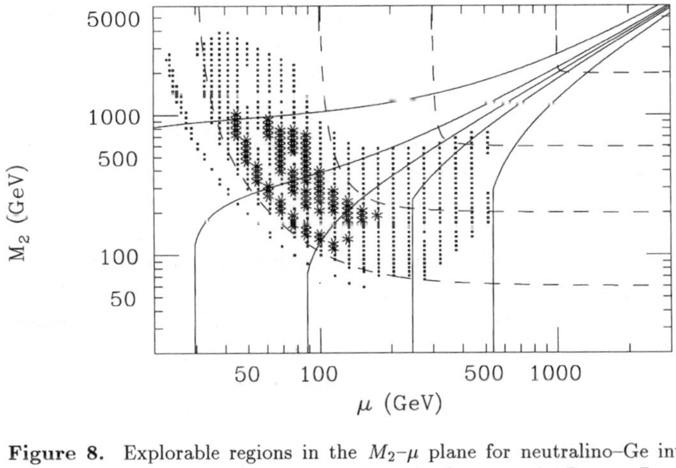

Figure 8. Explorable regions in the M_2–μ plane for neutralino–Ge interaction. Asterisks denote regions where the integrated rate $R_{\rm int}$ is in the range $0.1R_{\rm exp} < R_{\rm int} \le R_{\rm exp}$, $R_{\rm exp}$ being the present experimental bound. Dots denote regions where the integrated rate $R_{\rm int}$ is in the range $0.01R_{\rm exp} < R_{\rm int} \le 0.1R_{\rm exp}$,

energy range (2–4) KeV. In Fig.8 we show the regions of the M_2–μ parameter space which can be explored with an improvement of one and two orders of magnitude in the sensitivity of the detector.

As for the shape of these regions we refer to the comments presented above, in Sect.3, in connection with Fig.4. Again, the signals are higher along the iso–mass line with an m_χ close to the mass of the nuclei composing the detector. Thus, using detectors of different compositions allows explorations of the M_2–μ plane over a wide range in m_χ. For instance, investigation of regions with small m_χ values (of order of 10 GeV) with very low threshold detectors [21] would be very interesting. In fact this m_χ range (which is excluded by accelerator data only under a number of assumptions) is out of reach for the indirect detection discussed in the previous Section.

5. Comparison between Direct and Indirect Signals

Fig.s 3,5,7 provide a comparison between the capabilities of the direct detection versus the detection by neutrino telescopes. It has to be noted that these figures refer to an arbitrary representative point of the parameter space. However, since both (direct and indirect) signals depend mainly on a common quantity, i.e. the χ–nucleus elastic scattering, the relative size is approximately independent of the variation of the model parameters. The main departures from this feature occur in the case of the signal from the Earth in regions where the factor $\tanh^2(t/\tau_A)$ is small and for the signal from the Sun in regions where spin–dependent effects in the capture rate are dominant. Fig.s 3,5,7 also display how the the theoretical predictions compare with the present sensitivity for each individual method. A comparison of the relative power of the two detection methods for exploring the M_2–μ plane is provided by Fig.s 4,6,8.

6. Conclusions

In the previous sections we have examined two types of detection methods for relic neutralinos, one based on direct measurement and one on the detection of indirect signals due to χ–χ annihilation in celestial bodies. As we have seen by the explicit expressions given in Sect.s 3–4, in both cases the detection rates are proportional to the neutralino local density and to the elastic neutralino–nucleus cross section: $R \propto \rho_\chi \sigma_{el}$ (for simplicity, here and in the following, we consider the neutralino mass fixed at some arbitrary value m_χ; the other free parameters may vary in the parameter space along iso–mass curves).

Because of the properties discussed in Sect.2, once the neutralino relic abundance is evaluated, one has to distinguish between two cases:

a) $\Omega_\chi \geq \Omega_{min}$ which entails $R \propto \sigma_{el}$

b) $\Omega_\chi < \Omega_{min}$ which entails $R \propto \sigma_{el} / < \sigma_{\rm ann} v >_{\rm int}$

Now, if the strength of the couplings increases, and hence σ_{el} and $\sigma_{\rm ann}$ increase, it turns out that also in case b) the rates R usually increase even if at a much less

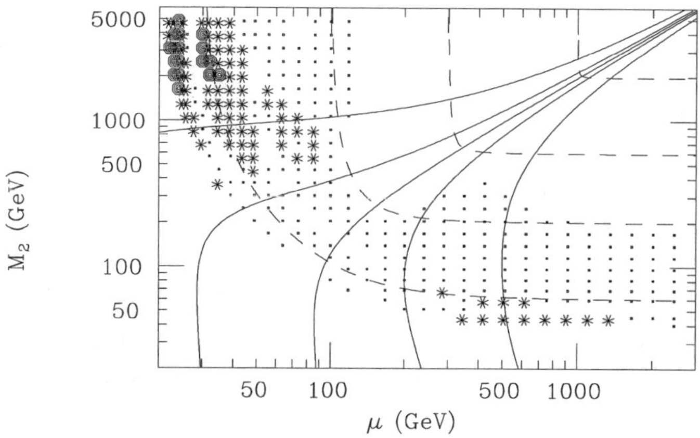

Figure 9. In the M_2–μ plane are shown the regions explored by $\bar{p}/p$ measurements [23] (full circles) and the domains explorable by an improvement factor of 10 (asterisks) and of 100 (dots) in sensitivity. Dashed lines denote iso–mass curves with $m_\chi = 30, 100, 300, 1000$ GeV. Solid lines denote curves of iso–composition (here $P = 0.01, 0.1, 0.5, 0.9, 0.99$ are shown).

extent than in case a). It then happens that for the two detection methods discussed so far, the largest values of the detection rates for relic neutralinos occur for models where $\tan\beta$ is large and neutralino is largely mixed (this last condition has to be somewhat relaxed for the signal from the Sun, as discussed above). For these models, because of the large value of the annihilation cross section, the neutralino relic density is small. This is a scenario where the detection methods under discussion have good chances to detect relic neutralinos, even if these particles cannot provide a substantial contribution to the total Ω of the Universe. Regions of the parameter space where this situation occurs are defined by some authors as cosmologically uninteresting and consequently disregarded. On the contrary, we believe that they deserve a careful investigation, since much interesting physics could occur there. The relevance of this point is further confirmed by some recent results derived from SUGRA theories where some of the universality properties, usually required at the GUT scale, are relaxed [22]. In fact it turns out that, in this case, very interesting schemes with large neutralino mixing, large $\tan\beta$ and small relic abundance naturally emerge.

As we have seen, when Ω_χ is large the two detection techniques previously discussed have less chances to succeed (except for the signal from the Sun, which however requires neutrino telescopes of very large area). In order to cover this case adequately, one may resort to different detection methods. These are based on the measurements of indirect signals due to neutralino–neutralino annihilation in the Galactic halo. These have been widely discussed in the literature; we refer to Ref. 23 for previous references and for a new analysis aimed at the evaluation of the antiproton to proton ratio ($\bar{p}/p$) in cosmic rays. A common feature of this class of measurements is that the event rates are typically proportional to the square of the local neutralino density times the annihilation cross section: $R \propto \rho_\chi^2 < \sigma_{\rm ann} v >_0$. This implies that, in this case, detection rates are high in regions of the parameter space where the neutralino relic abundance is large. By way of example, we give in Fig.9 the regions of M_2–μ plane that can be explored by measuring the ($\bar{p}/p$) ratio [23].

In conclusion we may state that complementarity among various detection methods may potentially offer a good coverage of the neutralino parameter space.

7. References

1. H.P. Nilles, *Phys. Rep.* **110** (1984) 1;
   H.E. Haber and G.L. Kane, *Phys. Rep.* **117** (1985) 75.
2. G.L. Kane, C. Kolda, L. Roszkowski and J.D. Wells, UM–TH–93–24 preprint;
   V. Barger, M.S. Berger and P. Ohmann, MAD/PH/801 preprint.
3. R. Arnowitt and P. Nath, CTP–TAMU–52/93 preprint.
4. J. Wu, R. Arnowitt and P. Nath, CTP–TAMU–03/94 preprint;
   F.M. Borzumati, M. Drees and M.M. Nojiri, KEK–TH–400 preprint;
   V. Barger, M.S. Berger, P. Ohmann and R.J.N. Phillips, MAD–PH–842 preprint.
5. A. Ali and C. Greub, *Z. Phys.* **C60** (1993) 433;
   A.J. Buras, M. Misiak, M.Münz and S. Pokorski, MPI–PH/93–77 preprint.
6. T.K. Gaisser, G. Steigman and S. Tilav, *Phys. Rev.* **D34** (1986) 2206.
7. A. Bottino, V.de Alfaro, N. Fornengo, G. Mignola and M. Pignone, *Astroparticle Physics* **2** (1994) 67. References to other recent evaluations of the neutralino relic abundance are given in [8,9].
8. K. Griest, M. Kamionkowski and M.S. Turner, *Phys. Rev.* **D41** (1990) 3565;
   J. McDonald, K.A. Olive and M. Srednicki, *Phys. Lett.* **B283** (1992) 80.
9. P. Gondolo, M. Olechowski and S. Pokorski, MPI–PH/92–81 preprint;
   M. Drees and M.M. Nojiri, *Phys. Rev.* **D47** (1993) 376;
   P. Nath and R. Arnowitt, CTP–TAMU–66/92 preprint;
   J.L. Lopez, D.V. Nanopoulos, and K. Yuan, CTP–TAMU–14/93 preprint;
   R.G. Roberts and L. Roszkowski, *Phys. Lett.* **B309** (1993) 329.
10. G.F. Giudice and E. Roulet, *Nucl. Phys.* **B316** (1989) 429;
    G.B. Gelmini, P. Gondolo and E. Roulet, *Nucl. Phys.* **B351** (1991) 623;
    H. Kamionkowski, *Phys. Rev.* **D44** (1991) 3021;
    A. Bottino, V. de Alfaro, N. Fornengo, G. Mignola and M. Pignone, *Phys. Lett.* **B265** (1991) 57;

M. Mori et al., *Phys. Rev.* **D48** (1993) 5505;
M. Drees, G. Jungman, M. Kamionkowski and M.M. Nojiri, *Phys. Rev.* **D49** (1994) 636;
R. Gandhi, J.L. Lopez, D.V. Nanopoulos, K. Yuan and A. Zichichi, CERN–TH 6999/93.

11. A. Bottino, N. Fornengo, G. Mignola and S. Scopel, Proc. of the International Workshop on "Neutrino Telescopes", Venice 1994.
12. A. Bottino, N. Fornengo, G. Mignola and L. Moscoso, DFTT 34/94 preprint (to appear in *Astroparticle Physics*).
13. K. Griest and D. Seckel, *Nucl. Phys.* **B283** (1987) 681.
14. A. Gould, *Ap. J.* **321** (1987) 571; *Ap. J.* **328** (1988) 919; *Ap. J.* **368** (1991) 610;
15. A. Bottino, V. de Alfaro, N. Fornengo, A. Morales, J. Puimedon and S. Scopel, *Mod. Phys. Lett.* **A7** (1992) 733.
Notice that in the present note, for the Higgs boson–nucleon coupling, we use the evaluation of J. Gasser, H. Leutwyler and M. E. Sainio, *Phys. Lett.* **B253** (1991) 252.
16. M. Mori et al. (Kamiokande coll.), *Phys. Lett.* **B289** (1992) 463.
17. J.G. Learned, *Nucl. Phys.* (Proc. Suppl.) **B31** (1993) 456 (Proc. Neutrino 92, Ed. A. Morales);
F. Halzen, Proc. *Fourth International Workshop on Neutrino Telescopes* (1992), Ed. M. Baldo Ceolin;
L. Resvanis, Proc. *Fifth International Workshop on Neutrino Telescopes* (1993), Ed. M. Baldo Ceolin.
18. R. Barbieri, M. Frigeni and G.F. Giudice, *Nucl. Phys.* **B313** (1989) 725.
19. A. Bottino, V.de Alfaro, N. Fornengo, G. Mignola and S. Scopel, *Astroparticle Physics* **2** (1994) 77.
20. L. Mosca, Proc. of the XIVth Moriond Workshop *Particle Astrophysics, Atomic Physics and Gravitation*, to appear;
See also the Proceedings of the Workshop on *The Dark Side of the Universe*, Rome, June 1993, (Ed.s R. Bernabei and C. Tao) for extensive literature about experiments of direct detection.
21. S. Cooper et al., proposal of Max–Planck–Institut für Physik and Technische Universität München, MPI–PhE/93–29.
22. M. Olechowski and S. Pokorski, MPI–PHT/94–40 preprint and these Proceedings.
23. A. Bottino, C. Favero, N. Fornengo and G. Mignola, DFTT 35/94 preprint (to appear in *Astroparticle Physic*).

SUPERSYMMETRIC "SOLUTIONS" TO COSMOLOGICAL PROBLEMS: BARYOGENESIS AND DARK MATTER

KEITH A. OLIVE
School of Physics and Astronomy, University of Minnesota
Minneapolis, MN 55455, USA

ABSTRACT

The possible role of supersymmetry in our understanding of big bang baryogenesis and cosmological dark matter is explored. The discussion will be limited to the out-of equilibrium decay scenario in SUSY GUTs, the decay of scalar condensates, and lepto-baryogenesis as a means for generating the observed baryon asymmetry. Attention will also be focused on neutralino dark matter.

1. Introduction

There are several outstanding problems in cosmology which rely on particle physics solutions. If supersymmetry (broken as it may be) is realized in nature, then it is not unreasonable to expect that supersymmetry plays a non-trivial role in the solutions to these problems. The two specific problems that I will concentrate upon here are: the origin of the baryon asymmetry and the nature of dark matter. The former problem has historically been associated with Grand Unified Theories (GUTs) and among the original ideas to generate the asymmetry was the out-of-equilibrium decay scenario[1]. I will begin, therefore, with a look back at the supersymmetric versions of this scenario. There are also purely supersymmetric solutions to baryogenesis, most notably is the decay of scalar condensates known as the the Affleck- Dine (AD) scenario[2] which will also be briefly discussed. I will comment on the role of cosmological inflation on both the out-of-equilibrium decay and the AD scenarios. Finally, it is no longer sufficient to generate a baryon asymmetry, but one must preserve it in the face of baryon number violating interactions associated with the standard electroweak model[3]. These interactions, however, open up new possibilities for generating an asymmetry such as the out-of-equilibrium decay of superheavy leptons[4]. These possibilities (in the context of supersymmetry) will also be discussed.

There are many possible solutions to the dark matter problem, many of which do not involve supersymmetry (nor any new particle physics candidate). However, the minimal supersymmetric standard model (MSSM) with unbroken R-parity does offer (in much of the parameter space) a cosmologically interesting dark matter candidate, the lightest supersymmetric particle or LSP[5,6]. The most likely choice being the supersymmetric partner of the U(1)-hypercharge gauge boson, the bino. Though the "entire" supersymmetric parameter space will be surveyed, I will focus on the bino as the LSP. A curious possibility that the LSP is a light photino which is nearly degenerate with the lighter stop quark[7,8] will also be discussed.

2. Baryogenesis

Our best information on the cosmological baryon density comes from big bang nucleosynthesis. In order to achieve consistency with the observational determinations of the light element abundances of deuterium through ^{7}Li, the baryon-to photon ratio is restricted to lie in the range[9]

$$2.8 \times 10^{-10} < \eta < 4 \times 10^{-10} \tag{1}$$

where $\eta = n_B/n_\gamma$. Combined with the lack of any observed antimatter (in primary form), our understanding of this small number is the problem which big bang baryogenesis attempts to solve.

2.1. The out-of-equilibrium decay scenario

The production of a net baryon asymmetry requires baryon number violating interactions, C and CP violation and a departure from thermal equilibrium[10]. The first two of these ingredients are contained in GUTs, the third can be realized in an expanding universe where it is not uncommon that interactions come in and out of equilibrium. In SU(5), the fact that quarks and leptons are in the same multiplets allows for baryon non-conserving interactions such as $e^- + d \leftrightarrow \bar{u} + \bar{u}$, etc., or decays of the supermassive gauge bosons X and Y such as $X \to e^- + d, \bar{u} + \bar{u}$. Although today these interactions are very ineffective because of the very large masses of the X and Y bosons, in the early Universe when $T \sim M_X \sim 10^{15}$ GeV these types of interactions should have been very important. C and CP violation is very model dependent. In the minimal SU(5) model the magnitude of C and CP violation is too small to yield a useful value of η and in general the C and CP violation comes from the interference between tree level and one loop corrections.

The departure from equilibrium is very common in the early Universe when interaction rates cannot keep up with the expansion rate. In fact, the simplest (and most useful) scenario for baryon production makes use of the fact that a single decay rate goes out of equilibrium. It is commonly referred to as the out of equilibrium decay scenario[1]. The basic idea is that the gauge bosons X and Y (or Higgs bosons) may have a lifetime long enough to insure that the inverse decays have already ceased so that the baryon number is produced by their free decays.

More specifically, let us call X, either the gauge boson or Higgs boson, which produces the baryon asymmetry through decays. Let α be its coupling to fermions. For X a gauge boson, α will be the GUT fine structure constant, while for X a Higgs boson, $(4\pi\alpha)^{1/2}$ will be the Yukawa coupling to fermions. If the decay rate for X, $\Gamma_D \simeq \alpha M_X$ is less than the expansion rate of the Universe, $H \simeq \sqrt{N}T^2/M_P$ (where N is the number of relativistic particles at temperature T and M_P is the Planck mass) at a temperature $T \sim M_X$ the decays will occur the out-of-equilibrium. Thus the condition on the superheavy mass is determined from, $\Gamma_D < H$ at $T = M_X$, or

$$M_X \gtrsim \alpha M_P (N(M_X))^{-1/2} \sim 10^{18} \alpha \mathrm{GeV} \tag{2}$$

In this case, we would expect a maximal net baryon asymmetry to be produced and is given by

$$\frac{n_B}{s} \sim \frac{n_B}{N n_\gamma} \sim 10^{-2}\epsilon \tag{3}$$

where s is the entropy density (a better quantity to compare to in an adiabatically expanding universe) and ϵ is the baryon asymmetry produced by an $X, \bar{X}$ decay and represents the degree of CP violation in the decay.

At least two Higgs five-plets are required to generate sufficient C and CP violation[11]. (It is possible within minimal SU(5) to generate a non- vanishing ϵ at 3 loops, however its magnitude would be too small for the purpose of generating a baryon asymmetry.) With two five-plets, H and H', the interference of diagrams of the type in figure 1, will yield a non-vanishing ϵ,

$$\epsilon \propto \text{Im}(a'^{\dagger} a b' b^{\dagger}) \neq 0 \tag{4}$$

if the couplings $a \neq a'$ and $b \neq b'$.

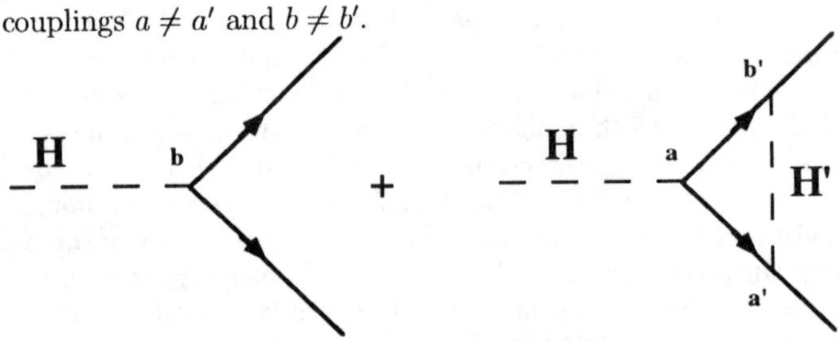

Figure 1: One loop contribution to the C and CP violation with two Higgs five-plets.

The out-of-equilibrium decay scenario discussed above did not include the effects of an inflationary epoch. In the context of inflation[12], one must in addition ensure baryogenesis after inflation as any asymmetry produced before inflation would be inflated away along with magnetic monopoles and any other unwanted relic. Reheating after inflation, may require a Higgs sector with a relatively light $O(10^{10} - 10^{11})GeV$ Higgs boson. To see this, consider a simple model in which the inflaton potential depends on only a single dimensionful parameter μ. In this case the energy density perturbations produced by inflation can be roughly estimated to be[13]

$$\frac{\delta\rho}{\rho} \sim O(100)\frac{\mu^2}{{M_P}^2} \tag{5}$$

which when matched to the observed quadrupole moment observed in the microwave background anisotropy[14]

$$\frac{\delta\rho}{\rho} = (5.4 \pm 1.6) \times 10^{-6} \tag{6}$$

fixes the coefficient μ of the inflaton potential[13]:

$$\frac{\mu^2}{M_P^2} = \text{few} \times 10^{-8} \tag{7}$$

Fixing (μ^2/M_P^2) has immediate general consequences for inflation[15]. For example, the Hubble parameter during inflation, $H^2 \simeq (8\pi/3)(\mu^4/M_P^2)$ so that $H \sim 10^{-7}M_P$. The duration of inflation is $\tau \simeq M_P^3/\mu^4$, and the number of e-foldings of expansion is $H\tau \sim 8\pi(M_P^2/\mu^2) \sim 10^9$. If the inflaton decay rate goes as $\Gamma \sim m_\eta^3/M_P^2 \sim \mu^6/M_P^5$, the universe recovers at a temperature $T_R \sim (\Gamma M_P)^{1/2} \sim \mu^3/M_P^2 \sim 10^{-11}M_P \sim 10^8 GeV$. Thus, the light Higgs is necessary since the inflaton, η, is typically light ($m_\eta \sim \mu^2/M_P \sim O(10^{11})$ GeV), and the baryon number violating Higgs would have to be produced during inflaton decay. Note that a "light" Higgs is acceptable from the point of view of proton decay due to its reduced couplings to fermions. The out-of-equilibrium decay scenario would now be realized by Higgs boson decay rather than gauge boson decay and a different sequence of events. First the inflaton would be required to decay to Higgs bosons (triplets?) and subsequently the triplets would decay rapidly by the processes shown in figure 1. These decays would be well out of equilibrium as at reheating $T \ll m_H$ and $n_H \sim n_\gamma$[16]. In this case, the baryon asymmetry is given simply by

$$\frac{n_B}{s} \sim \epsilon\frac{n_H}{T_R{}^3} \sim \epsilon\frac{n_\eta}{T_R{}^3} \sim \epsilon\frac{T_R}{m_\eta} \sim \epsilon\left(\frac{m_\eta}{M_P}\right)^{1/2} \sim \epsilon\frac{\mu}{M_P} \sim 10^{-4}\epsilon \tag{8}$$

where I have substituted for $n_\eta = \rho_\eta/m_\eta \sim \Gamma^2 M_P{}^2/m_\eta$.

In a supersymmetric grand unified SU(5) theory, the superpotential F_Y can be expressed in terms of SU(5) multiplets

$$F_Y = h_d\mathbf{H_2}\ \bar{\mathbf{5}}\ \mathbf{10} + h_u\mathbf{H_1}\ \mathbf{10}\ \mathbf{10} \tag{9}$$

where $10, \bar{5}, H_1$ and H_2 are chiral supermultiplets for the 10, and $\bar{5}$ plets of SU(5) matter fields and the Higgs 5 and $\bar{5}$ multiplets respectively. There are now new dimension 5 operators which violate baryon number and lead to proton decay as shown in figure 2. The first of these diagrams leads to effective dimension 5 Lagrangian terms such as

$$\mathcal{L}_{\text{eff}}^{(5)} = \frac{h_u h_d}{M_H}(\tilde{q}\tilde{q}ql) \tag{10}$$

and the resulting dimension 6 operator for proton decay[17]

$$\mathcal{L}_{\text{eff}} = \frac{h_u h_d}{M_H}\left(\frac{g_2^2}{M_{\tilde{W}}}\text{or}\frac{g_1^2}{M_{\tilde{B}}}\right)(qqql) \tag{11}$$

As a result of these diagrams the proton decay rate scales as $\Gamma \sim h^4 g^4/M_H^2 M_{\tilde{G}}^2$ where M_H is the triplet mass, and $M_{\tilde{G}}$ is a typical gaugino mass of order $\lesssim$ 1 TeV. This rate however is much too large if $M_H \sim 10^{10}$ GeV.

It is however possible to have a lighter ($O(10^{10} - 10^{11})$ GeV) Higgs triplet needed for baryogenesis in the out-of-equilibrium decay scenario with inflation. One needs two pairs of Higgs five-plets (H_1, H_2 and H_1', H_2') which is anyway necessary to have sufficient C and CP violation in the decays. By coupling one pair (H_2 and H_1') only to the third generation of fermions via[18]

$$a\mathbf{H_1}\mathbf{10}\mathbf{10} + b\mathbf{H_1'}\mathbf{10_3}\mathbf{10_3} + c\mathbf{H_2}\mathbf{10_3}\bar{\mathbf{5}}_\mathbf{3} + d\mathbf{H_2'}\mathbf{10}\bar{\mathbf{5}} \tag{12}$$

proton decay can not be induced by the dimension five operators. Triplet decay will however generate a baryon asymmetry proportional to $\epsilon \sim \mathrm{Im} dc^\dagger ba^\dagger$.

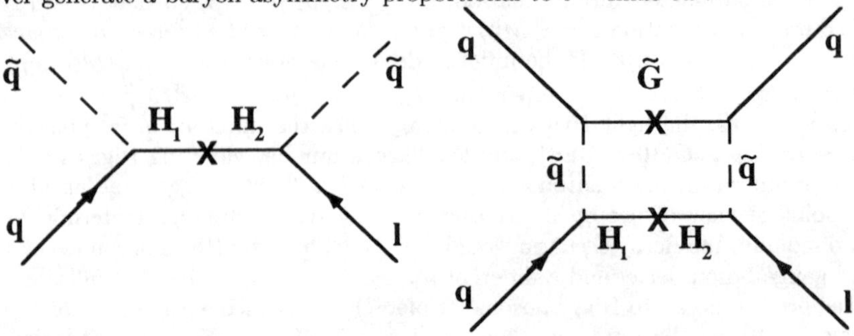

Figure 2: Dimension 5 and induced dimension 6 graphs violating baryon number.

2.2. The Affleck-Dine Mechanism

Another mechanism for generating the cosmological baryon asymmetry is the decay of scalar condensates as first proposed by Affleck and Dine[2]. This mechanism is truly a product of supersymmetry. It is straightforward though tedious to show that there are many directions in field space such that the scalar potential vanishes identically when SUSY is unbroken. SUSY breaking lifts this degeneracy so that

$$V \simeq \tilde{m}^2 \phi^2 \tag{13}$$

where $\tilde{m}$ is the SUSY breaking scale and ϕ is the direction in field space corresponding to the flat direction. For large initial values of ϕ, $\phi_o \sim M_{GUT}$, a large baryon asymmetry can be generated[2,19]. This requires the presence of baryon number violating operators such as $O = qqql$ which are naturally provided for in superymmetric GUTs and such that $\langle O \rangle \neq 0$. The decay of these condensates through such an operator with an effective quartic coupling of order $\tilde{m}^2/(\phi_o^2 + M_X^2)$ can lead to a net baryon asymmetry.

The baryon asymmetry produced, is computed by tracking the evolution of the sfermion condensate, which is determined by

$$\ddot{\phi} + 3H\dot{\phi} = -\tilde{m}^2 \phi \tag{14}$$

If it is assumed that the energy density of the Universe is dominated by ϕ, then the oscillations will cease, when

$$\Gamma_\phi \simeq \frac{\tilde{m}^3}{\phi^2} \simeq H \simeq \frac{\rho_\phi^{1/2}}{M_P} \simeq \frac{\tilde{m}\phi}{M_P} \tag{15}$$

or when the amplitude of oscillations has dropped to $\phi_D \simeq (M_P \tilde{m}^2)^{1/3}$. Note that the decay rate is suppressed as fields coupled directly to ϕ gain masses $\propto \phi$. It is now straightforward to compute the baryon to entropy ratio,

$$\frac{n_B}{s} = \frac{n_B}{\rho_\phi^{3/4}} \simeq \frac{\lambda \phi_o^2 \phi_D^2}{\tilde{m}^{5/2} \phi_D^{3/2}} = \frac{\lambda \phi_o^2}{\tilde{m}^2} \left(\frac{M_P}{\tilde{m}} \right)^{1/6} \tag{16}$$

and after inserting the quartic coupling,λ,

$$\frac{n_B}{s} \simeq \epsilon \frac{\phi_o^2}{(M_X^2 + \phi_o^2)} \left(\frac{M_P}{\tilde{m}} \right)^{1/6} \tag{17}$$

which could be quite large.

In the context of inflation, a couple of significant changes to the scenario take place. First, it is more likely that the energy density is dominated by the inflaton rather than the sfermion condensate. Second, the the initial value (after inflation) of the condensate ϕ can be determined by the inflaton mass m_η, ${\phi_o}^2 \simeq H^3 \tau \simeq m_\eta M_P$. The baryon asymmetry in the Affleck-Dine scenario with inflation becomes[15]

$$\frac{n_B}{s} \sim \frac{\epsilon {\phi_o}^4 {m_\eta}^{3/2}}{{M_X}^2 {M_P}^{5/2} \tilde{m}} \sim \frac{\epsilon m_\eta^{7/2}}{{M_X}^2 {M_P}^{1/2} \tilde{m}} \sim (10^{-6} - 1)\epsilon \tag{18}$$

for $m \sim (10^{-17} - 10^{-16}) M_P$, and $M_X \sim (10^{-4} - 10^{-3}) M_P$ and $m_\eta \sim (10^{-8} - 10^{-7}) M_P$

2.3. Lepto-baryogenesis

2.3.1. Preservation of the asymmetry

The realization[3] of significant baryon number violation at high temperature within the standard model, has opened the door for many new possibilities for the generation of a net baryon asymmetry. Electroweak baryon number violation occurs through non-perturbative interactions mediated by "sphalerons", which violate $B + L$ and conserve $B - L$. For this reason, any GUT produced asymmetry with $B - L = 0$ may be subsequently erased by sphaleron interactions[20].

With $B - L = 0$, it is relatively straightforward to see that the equilibrium conditions including sphaleron interactions gives zero net baryon number[21]. By assigning each particle species a chemical potential, and using gauge and Higgs interactions as conditions on these potentials (with generation indices suppressed),

$$\begin{array}{lll} \mu_- + \mu_0 = \mu_W & \mu_{u_R} - \mu_{u_L} = \mu_0 & \mu_{d_R} - \mu_{d_L} = -\mu_0 \\ \mu_{l_R} - \mu_{l_L} = -\mu_0 & \mu_{d_L} - \mu_{u_L} = \mu_W & \mu_{l_L} - \mu_\nu = \mu_W \end{array} \tag{19}$$

one can write down a simple set of equations for the baryon and lepton numbers and electric charge which reduce to:

$$\begin{aligned} B &= 12\mu_{u_L} \\ L &= 3\mu - 3\mu_0 \\ Q &= 6\mu_{u_L} - 2\mu + 14\mu_0 \end{aligned} \tag{20}$$

where $\mu = \sum \mu_{\nu_i}$. In (20), the constraint on the weak isospin charge, $Q_3 \propto \mu_W = 0$ has been employed. Though the charges B, L, and Q have been written as chemical potentials, since for small asymmetries, an asymmetry $(n_f - n_{\bar{f}})/s \propto \mu_f/T$, we can regard these quantities as net number densities.

The sphaleron process yields the additional condition,

$$9\mu_{u_L} + \mu = 0 \tag{21}$$

which allows one to solve for L and $B-L$ in terms of μ_{u_L}, ultimately giving

$$B = \frac{28}{79}(B-L) \tag{22}$$

Thus, in the absence of a primordial $B-L$ asymmetry, the baryon number is erased by equilibrium processes. Note that barring new interactions (in an extended model) the quantities $\frac{1}{3}B - L_e$, $\frac{1}{3}B - L_\mu$, and $\frac{1}{3}B - L_\tau$ remain conserved.

With the possible erasure of the baryon asymmetry when $B-L=0$ in mind, since minimal SU(5) preserves $B-L$, electroweak effects require GUTs beyond SU(5) for the asymmetry generated by the out-of-equilibrium decay scenario to survive. GUTs such as SO(10) where a primordial $B-L$ asymmetry can be generated becomes a promising choice. The same holds true in the Affleck-Dine mechanism for generating a baryon asymmetry. In larger GUTs there are baryon number violating operators and associated flat directions[22]. A specific example in SO(10) was worked out in detail by Morgan[23].

Another possibility for preserving a primordial baryon asymmetry when $B-L=0$ arises if the asymmetry produced by scalar condensates in the Affleck-Dine mechanism is large[24] ($n_B/s \gtrsim 10^{-2}$). After the decay of the A-D condensate, the baryon number is shared among fermion and boson superpartners. However, in equilibrium, there is a maximum chemical potential $\mu_f = \mu_B = \tilde{m}$ and for a large asymmetry, the baryon number density stored in fermions, $n_{B_f} = \frac{g_f}{6}\mu_f T^2$ is much less than the total baryon density. The bulk of the baryon asymmetry is driven into the $p=0$ bosonic modes and a Bose-Einstein condensate is formed[25]. The critical temperature for the formation of this condensate is given by $n_B \simeq n_{B_b} + n_{B_c} = \frac{g_b}{3}\tilde{m}T_c^2$ so that,

$$n_{B_c} = \frac{g_b}{3}\left(1 - \left(\frac{T}{T_c}\right)^2\right)T_c^2 \tag{23}$$

At $T < T_c$, most of the baryon number remains in a condensate and for large n_B, the condensate persists down to temperatures of order 100 GeV. Thus sphaleron interactions are shut off and a primordial baryon asymmetry is maintained even with $B-L=0$. One should note however that additional sources of entropy are required to bring η down to acceptable levels.

2.3.2. Generating a baryon asymmetry from a primordial lepton asymmetry

Sphaleron interactions also allow for new mechanisms to produce a baryon asymmetry. The simplest of such mechanisms is based on the decay of a right handed neutrino-like state[4]. This mechanism is certainly novel in that does not require grand unification at all. By simply adding to the Lagrangian a Dirac and Majorana mass term for a new right handed neutrino state,

$$\mathcal{L} \ni M\nu^c\nu^c + \lambda H L \nu^c \tag{24}$$

the out-of-equilibrium decays $\nu^c \to L + H^*$ and $\nu^c \to L^* + H$ will generate a non-zero lepton number $L \neq 0$. The out-out-equilibrium condition for these decays translates to $10^{-3}\lambda^2 M_P < M$ and M could be as low as $O(10)$ TeV. (Note that once again in order to have a non-vanishing contribution to the C and CP violation in this process at 1-loop, at least 2 flavors of ν^c are required. For the generation of masses of all three neutrino flavors, 3 flavors of ν^c are required.) Sphaleron effects can transfer this lepton asymmetry into a baryon asymmetry since now $B - L \neq 0$. A supersymmetric version of this scenario has also been described[13,26].

The survival of the asymmetry, of course depends on whether or not electroweak sphalerons can wash away the asymmetry. The persistence of lepton number violating interactions in conjunction with electroweak sphaleron effects could wipe out[27] both the baryon and lepton asymmetry in the mechanism described above through effective operators of the form $\lambda^2 LLHH/M$. In terms of chemical potentials, this interaction adds the condition $\mu_\nu + \mu_0 = 0$. The constraint comes about by requiring that this interaction be out of equilibrium at the time when sphalerons are in equilibrium. Otherwise, the additional condition on the chemical potentials would force the solution $B = L = 0$. To prevent the erasure of the baryon asymmetry, the constraint on $M/\lambda^2 \gtrsim 3 \times 10^9$ GeV obtained by requiring the $B + L$ violating operators to remain out of equilibrium at least until right-handed electrons come into equilibrium[28] leads to a bound on neutrino masses, $m_\nu \sim \lambda^2 v^2/M \lesssim 20$ keV, where $v = 247$ GeV is the Higgs vev. Similar constraints can be derived on R-parity violating operators[29].

In addition to the mechanism described earlier utilizing a right-handed neutrino decay, several others are now also available. In a supersymmetric extension of the standard model including a right-handed neutrino, there are numerous possibilities. Along the lines of the right-handed neutrino decay, the scalar partner[13] or a condensate[26] of $\tilde{\nu}^c$'s will easily generate a lepton asymmetry. Furthermore if the superpotential contains terms such as $\nu^{c3} + \nu^c H_1 H_2$, there will be a flat direction violating lepton number[30,13] à la Affleck and Dine. While none of these scenarios require GUTs, those that involve the out-of equilibrium decay of either fermions, scalars or condensates must have the mass scale of the right-handed neutrino between 10^9 and about 10^{11} GeV, to avoid washing out the baryon asymmetry later and to be produced after inflation respectively. In contrast, the decay of the flat direction condensate (which involves other fields in addition to $\tilde{\nu}^c$) only works for $10^{11} < M < 10^{15}$ GeV.

3. Dark Matter

There are several reasons for postulating the existence of dark matter. On the theoretical side, if the cosmological density parameter is one, then the upper bound on the fraction of Ω in baryons is restricted by nucleosynthesis to take values[9] $\Omega_B < 0.08$ leaving the remainder as non-baryonic dark matter. Also on the theoretical side, is the effect of dark matter on the growth of density perturbations. The problem of making galaxies and clusters is exasperated without dark matter. There are also several observational pieces of evidence which include: galactic rotation curves, X-ray emitting hot gas from elliptical galaxies and clusters, as well as gravitational lensing by dark halos. What portion of the dark matter is truly non-baryonic is still unknown, but if in fact $\Omega = 1$, most of the dark matter would be in the form a new particle candidate. I will here concentrate only the supersymmetric candidates. For a more general recent review see: ref. (31).

Supersymmetric theories introduce several possible candidates. If R-parity (which distinguishes between "normal" matter and the supersymmetric partners) is unbroken there is at least one supersymmetric particle which must be stable. I will assume R-parity conservation. The stable particle (usually called the LSP) is most probably some linear combination of the only $R = -1$ neutral fermions, the neutralinos[5]: the wino $\tilde{W}^3$, the partner of the 3rd component of the $SU(2)_L$ gauge boson; the bino, $\tilde{B}$, the partner of the $U(1)_Y$ gauge boson; and the two neutral Higgsinos $\tilde{H}_1$, and $\tilde{H}_2$. Gluinos are expected to be heavier, $m_{\tilde{g}} = (\frac{\alpha_3}{\alpha}) \sin^2\theta_W M_2$ and do not mix with the other states (M_2 is the soft SUSY breaking $SU(2)$ gaugino mass). The sneutrino[32] is also a possibility but has been excluded as a dark matter candidate by direct[33] searches, indirect[34] and accelerator[35] searches. For a recent examination of very heavy sneutrino candidates see ref.(36).

The the only parameters which determine the mass and composition of the LSP are; M_2, μ and $\tan\beta$ (assuming the GUT relations among the soft SUSY breaking gaugino masses). The latter two are the supersymmetric Higgsino mixing mass and the ratio of the Higgs scalar vevs respectively. However, for the relic abundance of LSP's, it is necessary to specify the Higgs (scalar) masses, and the sfermions masses. The LSP can be expressed as a linear combination

$$\chi = \alpha\tilde{W}^3 + \beta\tilde{B} + \gamma\tilde{H}_1 + \delta\tilde{H}_2 \tag{25}$$

Pure state LSP possibilities are: The photino[37], when $M_2 \to 0$

$$\tilde{\gamma} = \tilde{W}^3 \sin\theta_W + \tilde{B}\cos\theta_W \tag{26}$$

and

$$m_{\tilde{\gamma}} \to \frac{8}{3}\frac{{g_1}^2}{({g_1}^2 + {g_2}^2)} M_2 \tag{27}$$

the Higgsino, $\tilde{S}^0$ [5], when $\mu \to 0$

$$\tilde{S}^0 = \tilde{H}_1 \cos\beta + \tilde{H}_2 \sin\beta \tag{28}$$

and

$$m_{\tilde{S}} \to \frac{2v_1 v_2}{v^2}\mu \tag{29}$$

When M_2 is large and $M_2 \ll \mu$ then the bino[6], $\tilde{B}$, is the LSP and

$$m_{\tilde{B}} \simeq M_1 = \frac{5}{3}\frac{\alpha_1}{\alpha_2} M_2 \tag{30}$$

Finally when μ is large and $\mu \ll M_2$ either the Higgsino state[6]

$$\tilde{H}_{(12)} = \frac{1}{\sqrt{2}}(\tilde{H}_1^0 + \tilde{H}_2^0) \qquad \mu < 0 \tag{31}$$

or the state

$$\tilde{H}_{[12]} = \frac{1}{\sqrt{2}}(\tilde{H}_1^0 - \tilde{H}_2^0) \qquad \mu > 0 \tag{32}$$

is the LSP depending on the sign of ϵ and

$$m_{\tilde{H}} \simeq |\mu| \tag{33}$$

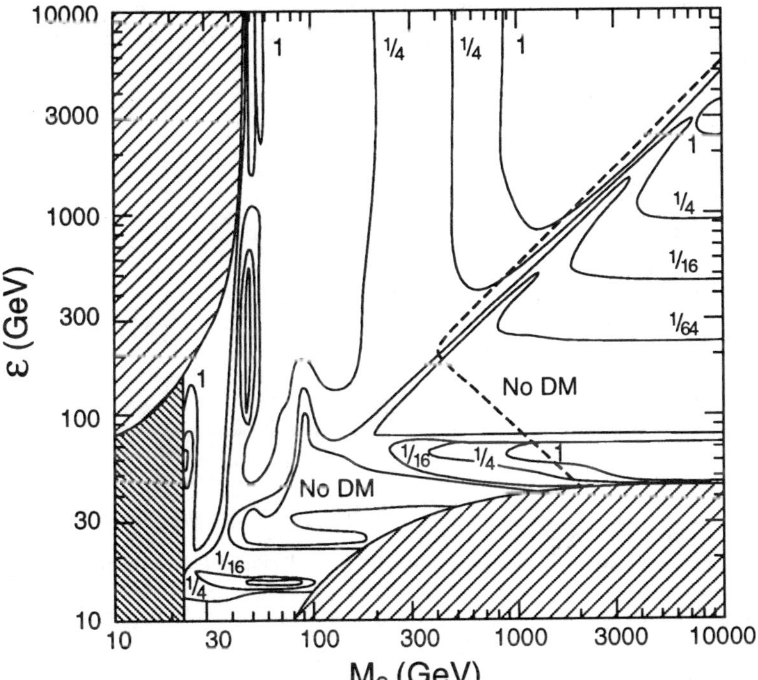

Figure 3: The relic neutralino density, Ωh^2, in the $M_2 - \mu(= -\epsilon)$ plane.

The relic abundance of LSP's is controlled by annihilations until freeze out. The value of Ωh^2 is roughly proportional to $1/\langle \sigma v \rangle_{\rm ann}$ and is determined by solving the

Boltzmann equation for the LSP number density in an expanding Universe. The technique[38] used is similar to that for computing the relic abundance of massive neutrinos[39]. For binos, as was the case for photinos, it is possible to adjust the sfermion masses $m_{\tilde{f}}$ to obtain closure density. Adjusting the sfermion mixing parameter allows even greater freedom[40]. In figure 3[41], the relic abundance (Ωh^2) is shown in the $M_2 - \mu$ plane with $\tan\beta = 2, \mu < 0$, the Higgs pseudoscalar mass $m_0 = 50GeV$, $m_t = 100GeV$ and $m_{\tilde{f}} = 200GeV$. Binos (which occupy the upper triangular quarter of figure 3 as the LSP), are cosmologically significant in the mass range 25– $\sim$ 300 GeV. The lower bound coming from the requirement that for large μ, $M_2 \gtrsim 45$ GeV to aviod a light chargino (the shaded regions at either large μ or M_2) and the upper bound coming from the bound on Ωh^2 (heavier binos would require sfermions with masses $m_{\tilde{f}} < m_{\tilde{B}}$). As annihilations as well as scatterings proceed through sfermion exachange, detection rates for binos are expected to be somewhat low[46], $\lesssim 0.1$/kg/day. Clearly the minimal model offers sufficient room to solve the dark matter problem. Similar results have been found by other groups[42,43,44]. In figure 3, in the higgsino sector $\tilde{H}_{12}$ marked off by the dashed line, co-annihilations[45,43] between $\tilde{H}_{(12)}$ and $\tilde{H}_{[12]}$ were not included. These tend to lower significantly the relic abundance in much of this sector.

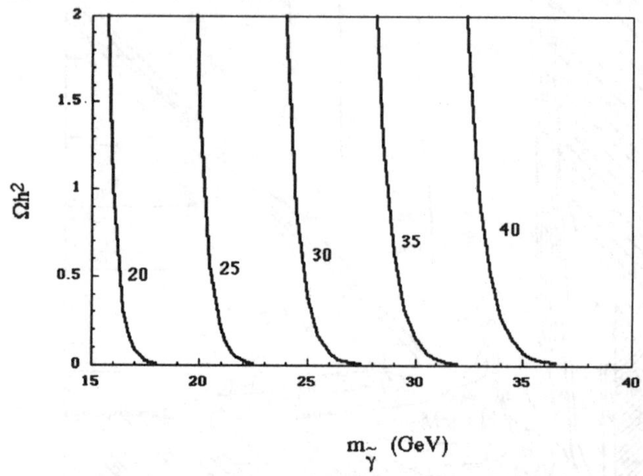

Figure 4: The relic photino density, for several values of the stop mass.

There is also a curious possibility which has been recently suggested[7] in which the photino is the LSP and is light and nearly degenerate with a light stop. For example, it is still experimentally possible that the lighter stop quark has a mass in the range 20-40 GeV if the stop mixing angle $\theta_t \simeq 0.98$. At or near this value, the stop does not couple to the Z^o. For a photino with a mass in the range 16-33 GeV (i.e. nearly degenerate with the stop), the stop is nearly invisible. The relic density of the light photinos is acceptable even though all other sfermion masses may be very high, because the co-annihilation process[45] $\tilde{\gamma} + c \to \tilde{t}$ and $\tilde{t}\tilde{t}^* \to X$ is efficient if $m_{\tilde{t}} - m_{\tilde{\gamma}} \sim 3GeV$. The relic density of photinos in this case[8] is shown in figure 4. However, if all other SUSY mass scales are high, this photino is virtually undectable[8] although this sector may have consequences for the top quark branching ratio.

4. Acknowledgements

This work was supported in part by DOE grant DE-FG02-94ER40823.

5. References

1. S. Weinberg, *Phys. Rev. Lett.* **42**, (1979) 850; D. Toussaint, S. B. Treiman, F. Wilczek, and A. Zee, *Phys. Rev.* **D19** (1979) 1036.
2. I. Affleck and M. Dine, *Nucl. Phys.* **B249** (1985) 361.
3. V. Kuzmin, V. Rubakov and M. Shaposhnikov, *Phys. Lett.* **B155** (1985) 36; for a review see; A.G. Cohen, D.B. Kaplan, and A.E. Nelson, *Ann. Rev. Nucl. Part. Sci.* **43** (1993) 27.
4. M. Fukugita and T. Yanagida, *Phys. Lett.* **B174** (1986) 45.
5. J.Ellis, J. Hagelin, D.V. Nanopoulos, K.A.Olive and M. Srednicki, *Nucl. Phys.* **B238** (1984) 453.
6. K.A. Olive and M. Srednicki, *Phys. Lett.* **B230** (1989) 78; *Nucl. Phys.* **B355** (1991) 208.
7. M. Fukugita, H. Murayama, M. Yamaguchi, and T. Yanagida, Phys. Rev. Lett. **72** (1994) 3009.
8. K.A. Olive and S. Rudaz, Phys. Lett. **B** (1994) in press.
9. T.P. Walker, G. Steigman, D.N. Schramm, K.A. Olive, and H.-S. Kang, *Ap.J.* **376** (1991) 51.
10. A.D. Sakharov, *JETP Lett.* **5** (1967) 24.
11. D.V. Nanopoulos and S. Weinberg, *Phys. Rev.* **D20** (1979) 2484; S. Barr, G. Segré and H.A. Weldon, *Phys. Rev.* **D20** (1979) 2494; A. Yildiz and P.H. Cox, *Phys. Rev.* **D21** (1980) 906.
12. for reviews see: A.D. Linde, Particle Physics And Inflationary Cosmology Harwood (1990); K.A. Olive, *Phys.Rep.* **C190**(1990) 307.
13. B. Campbell, S. Davidson, and K.A. Olive, *Nucl.Phys.* **B399** (1993) 111.
14. G.F. Smoot et al. *Ap.J* **396** (1992) L1; E.L. Wright et al. *Ap.J.* **396** (1992) L13.
15. J. Ellis, K. Enqvist, D.V. Nanopoulos, and K.A. Olive, *Phys. Lett.* **B191**(1987) 343.
16. A.D. Dolgov, and A.D. Linde, *Phys. Lett.* **B116** (1982) 329; D.V. Nanopoulos, K.A. Olive, and M. Srednicki, *Phys. Lett.* **B127** (1983) 30.
17. J. Ellis, D.V. Nanopoulos, and S. Rudaz, *Nucl. Phys.* **B202** (1982) 43.
18. D.V. Nanopoulos and K. Tamvakis, *Phys. Lett.* **B114** (1982) 235.
19. A.D. Linde, *Phys. Lett.* **B160** (1985) 243.

20. P. Arnold and L. McLerran, *Phys. Rev.* **D36** (1987) 581; **D37** (1988) 1020.
21. J.A. Harvey and M.S. Turner, *Phys. Rev.* **D42** (1990) 3344.
22. B. Campbell, J. Ellis, D.V. Nanopoulos, and K. A. Olive, *Mod. Phys. Lett.* **A1** (1986) 389.
23. D. Morgan, *Nucl. Phys.* **B364** (1991) 401.
24. S. Davidson, H. Murayama, and K.A. Olive, *Phys. Lett.* **B328** (1994) 354.
25. A.D. Dolgov and D.P. Kirilova, *Sov. J. Nucl. Phys.* **50** (1989) 1006.
26. H. Murayama, H. Suzuki, T. Yanagida, and J. Yokoyama, *Phys. Rev. Lett.* **70** (1993) 1912.
27. M. Fukugita and T. Yanagida, *Phys. Rev.* **D42** (1990) 1285.
28. J. Cline, K. Kainulainen, and K.A. Olive, *Phys. Rev. Lett.* **71** (1993) 2372 ; and *Phys. Rev.* **D49** (1994) 6394.
29. B. Campbell, S. Davidson, J. Ellis, and K. A. Olive, *Phys. Lett.* **B256** (1991) 457; Astropart. Phys. 1 (1992) 77; W. Fischler, G.F. Giudice, R.G. Leigh and S. Paban, *Phys. Lett.* **B258** (1991) 45; L.E. Ibanez and F. Quevedo, *Phys. Lett.* **B283** (1992) 261; H. Dreiner and G.G. Ross, *Nucl. Phys.* **B407** (1993) 188.
30. B. Campbell, S. Davidson, and K.A. Olive, *Phys. Lett.* **B303** (1993) 63.
31. K.A. Olive, in the *Proceedings of the TASI-93 Summer School in Elemtary Particle Physics*, eds. S. Raby and T. Walker, UMN-TH-1221/93.
32. L.E. Ibanez, *Phys. Lett.* **137B** (1984) 160; J. Hagelin, G.L. Kane, and S. Raby, *Nucl., Phys.* **B241** (1984) 638.
33. S. Ahlen, et. al., *Phys. Lett.* **B195** (1987) 603; D.D. Caldwell, et. al., *Phys. Rev. Lett.* **61** (1988) 510.
34. see e.g. K.A. Olive and M. Srednicki, *Phys. Lett.* **205B** (1988) 553.
35. see e.g. B. Adeva, et. al., *Phys. Lett.* **B231** (1989) 509; D. Decamp, et. al., *Phys. Lett.* **B231** (1989) 519; M.Z. Akrawy, et. al., *Phys. Lett.* **B231** (1989) 530; P. Aarnio, *Phys. Lett.* **B231** (1989) 539.
36. T. Falk, K.A. Olive, and M. Srednicki, *Phys. Lett.* **B** (1994) in press.
37. H. Golberg, *Phys. Rev. Lett.* **50** (1983) 1419; L.M. Krauss, *Nucl. Phys.* bf B227 (1983) 556.
38. R. Watkins, M. Srednicki and K.A. Olive, *Nucl. Phys.* **B310** (1988) 693.
39. P. Hut, *Phys. Lett.* **69B** (1977) 85; B.W. Lee and S. Weinberg, *Phys. Rev. Lett.* **39** (1977) 165.
40. T. Falk, R. Madden, K.A. Olive, and M. Srednicki, *Phys. Lett.* (in press) (1993).
41. J. McDonald, K. A. Olive and M. Srednicki, *Phys. Lett.* **B283** (1992) 80.
42. K. Greist, M. Kamionkowski, and M.S. Turner, *Phys. Rev.* **D41** (1990) 3565.
43. M. Drees and M.M. Nojiri, *Phys. Rev.* **D47** (1993) 376.
44. J. Lopez, D.V. Nanopoulos, and K. Yuan, *Nucl. Phys.* **B370** (1992) 445 and preprint 1993.
45. K. Greist and D. Seckel, *Phys. Rev.* **D43** (1991) 3191; S. Mizuta and M. Yamaguchi, *Phys. Lett.* **B298** (1993) 120.
46. R. Flores, K.A. Olive and M. Srednicki, *Phys. Lett.* **B237** (1990) 72.

CLASSICAL AD QUANTUM IN COSMOLOGY

J. KOWALSKI-GLIKMAN
Institute for Theoretical Physics, University of Wrocław
Pl. Maxa Born 9, 50204 Wrocław, Poland

ABSTRACT

There are several regimes in chaotic inflationary cosmology where some part of the system is classical and some other quantum. I describe how to deal with such systems and how to disentangltheir dynamics into classical behaviour and quantum corrections. I also discuss the conditions for quantum corrections to be small.

Cosmology is one of the most exciting fields of modern high energy physics. On the one hand, it provides the only possible experimental ground for testing theories of high energy particle physics beyond the energy scale of accelerators (see other contributions to this proceedings). On the other hand, because of the simple geometry of realistic cosmological models, it makes it possible to test consistency of physical theories at the energy scales close to the Planck scale of quantum gravity.

The remarkable progress in our understanding of early cosmology (by which I understand the models describing the universe at the energies between 10^{15} and 10^{19}GeV) resulted in wide recognition of the chaotic inflationary model[1] as the (without doubts) the most simple and (possibly) the only realistic model of the early evolution of the universe.

The basic features of the chaotic inflation are the following (see the figure)

1. The universe emerges from the "quantum gravity abyss" with the scale factor and the energy density of order 1.*

2. The initial conditions for the inflaton field are such that $\ddot{\phi} \ll V'(\phi)$, $(\partial_i\phi)^2 \ll 1$, $V(\phi) \sim 1$. For the simple massive scalar model with $m^2 \sim 10^{-12}$ (as required to secure the correct magnitude of the spectrum for large scale structures formation), this means that the initial value of the field ϕ, $\phi_0 \sim 10^6$, and ϕ doesn't vary too much in space and time.

3. Subsequently, the field ϕ rolls down the potential well, and the universe is undergoing inflation ($a(t) \sim e^{Ht}$, with H slowly varying function of time.)

4. The inflation ends at the point $\phi \sim 1$ and is followed by very rapid oscillations around the minimum $\phi = 0$, which are interpreted as giving rise to reheating leading to creation of thermal hot matter in the universe (hot big bang.)

*In what follows I use the Planck units i.e., the energy unit is 10^{19}GeV, the length unit is 10^{-33}cm etc.

Within the inflationary regime, quantum fluctuations emerge superimposed on the classical motion down the potential well. In the interval $1 \leq \phi \leq 10^3$ these fluctuations are small

$$\Delta_Q \phi < \delta_{class} \phi, \tag{1}$$

where $\Delta_Q \phi$ is the effective magnitude of the quantum fluctuations which cummulate during some natural time interval, and $\delta_{class}\phi$ is the classical distance covered by the field ϕ during the same time. These fluctuations give rise to the seeds for large scale formation in the standard way.

However, for $10^3 \leq \phi \leq 10^6$

$$\Delta_Q \phi > \delta_{class} \phi, \tag{2}$$

and the quantum fluctuations dominate the evolution of the inflaton field, giving rise to the extremely interesting global structure of the universe (see the paper[2] and the references therein.)

There are therefore at least three periods during the early evolution of the inflaton field when there is a significant interplay between classical and quantum behaviour of the constituents of the system:

1. Emergence of the universe from the quantum gravity era. It is believed that at the Planck energy scale gravity becomes classical, however the inflaton field may be still very much quantum. How this process looks in details? More importantly, are the initial conditions for the chaotic inflation listed above likely or not?

2. In the subsequent inflationary phase some "part" of the inflaton field is "classical" and some part is "quantum." How the detailed evolution of ϕ looks like?

3. At the end of inflation the universe enters the reheating phase with the classical oscillating inflaton and quantum radiation field. What are details of this process?†

This kind of problems has been analyzed in the recent papers[4] and[5]. Following these papers, let me discuss the quantum system consisting of the scale factor a and the inflaton field ϕ. The quantum hamiltonian of this system is the one-dimensional Wheeler–De Witt operator which, at the same time, is a constraint of the theory. For the wave function of the universe we have the equation

$$\left[\frac{\partial}{\partial a}\frac{1}{a}\frac{\partial}{\partial a} - \frac{1}{a^3}\frac{\partial^2}{\partial \phi^2} + V(\phi)a^3\right]\Psi(\phi, a) = 0. \tag{3}$$

This equation comrises the whole information obout quantum universe. Observe that there is no time in this equation, and this is a general feature of all theories with general coordinate invariance. (For comprehensive discussion of the time problem in classical and quantum gravity see the article[6].) Thus the wave function of the universe provides us with correlations between the scale factor and the inflaton field. This representation is not convenient for two reasons. First, we would like to be able to trace the time evolution of the universe, and second we would like to see how the classical inflationary universe emerges from the quantum gravity state. Actually, we

†Recently there is a renewed intererst in analyzing the reheating process. See[3] and references therein.

have a very good idea of what the classical behaviour is. This is just the classical dynamics of the system. Therefore, should we be able to decompose the whole dynamics of the system into

$$(classical\ \ motion) = (corrections),$$

we would know that the (*corrections*) are of the quantum origin.

To solve these problems we proceed as follows. First we write the wave function in the form

$$\Psi(a,\phi) = R(a)e^{iS(a)}\chi(a,\phi), \tag{4}$$

with real R and S. This representation can be made unambiguous by imposing certain conditions on χ. Substituting this into Eq. 3 we find two equations, one of whose has the form

$$-i\frac{1}{a}\frac{\partial S}{\partial a}\frac{\partial \chi}{\partial a} - H_M = \ldots, \tag{5}$$

where $H_M = \frac{1}{a^3}\frac{\partial^2}{\partial\phi^2} + V(\phi)a^3$ is the matter part of the hamiltonian, where "..." denote other correction terms. It is clear that one can recover the standard Schrödinger equation (with correction terms) for χ, if one introduces the new variable $t(a)$ such that

$$\frac{1}{a}\frac{\partial S}{\partial a}\frac{\partial \chi}{\partial a} \stackrel{def}{=} \frac{\partial \chi}{\partial t}, \tag{6}$$

that is, if

$$\frac{\partial S}{\partial a} = -a\dot{a}, \qquad \dot{a} = \frac{da}{dt}. \tag{7}$$

Then it can be shown that Eq. 3 is *strictly equivalent* to the following set of equations

$$a\dot{a}^2 = < H_M > + (Q_G + C_E), \tag{8}$$

$$i\dot{\chi} - H_M\chi = C_S, \tag{9}$$

where $< H_M >$ is the quantum average of H_M with respect to χ and Q_G, C_E, C_S are various correction terms corresponding to gravitational quantum correction to Einstein equations, matter quantum corrections to Einstein equations, and matter quantum corrections to the Schrödinger equation, respectively. The exact form of these correction terms can be found in[4].

Let me pause for a moment to discuss the logic which led us to Eqs. 8, 9. Starting from Eq. 3, we represented the wave function in the form Eq. 6, recovered time with the help of Eq. 7, and ended up with the set of equations Eq. 8, 9. Let me stress once again that no approximations have been made during this procedure, Eqs. 8, 9 are completely equivalent to the initial Eq. 3. What we have now is the system of two equations: the Einstein equation and the Schrödinger equation supplemented by quantum correction terms. This system is clearly convenient for our purposes. In the regime where gravity is almost classical and the inflaton field is still in the quantum regime, we can easily see what the corrections are and under which conditions are they small.

So what is the form and relative magnitude of the correction terms? Typically, the quantum gravity corrections Q_G and the terms C_S are proportional to a^{-3}, so they are getting rapidly diluted in the inflationary phase (and in the Friedmann phase as well.) Further, the C_E term is proportional to the energy dispersion, so it is small if quantum fluctuations of the energy are small as compared to the energy itself. It should be noted however that if these fluctuations are not small, there is a huge contribution from these fluctuations to Einstein equations. One expects therefore that if these fluctuations are not small initially, they would prevent the emergence of the classical universe. The next point which should be emphasized is that in the derivation I implicitely assumed that the wave function Ψ is known. This wave function is hard to find, of course, though some approximate solutions are being currently under investigation[7]. On the other hand, one can, for example, neglect all the correction terms, solve the resulting equations, and then check explicitly if for the particular solution the correction terms are small indeed, and then treat them as perturbations.

The procedure leading to Eqs. 8, 9 is very general, of course, and can be applied to many systems. For example, we might be interested in the question concerning how the evolution of the inflaton field during inflation splits into classical and quantum part. To do that, we first neglect the correction term C_S, which is justified by the fact that during inflation this term is rapidly approaching zero. Then we make the polar decomposition of χ

$$\chi(\phi,t)=\rho(\phi,t)e^{i\sigma(\phi,t)}$$

and substitute the result into Eq. 8 with $C_S=0$. The procedure is described in[4], the result is

$$\ddot{\phi}+3\frac{\dot{a}}{a}\dot{\phi}+\frac{\partial V_M}{\partial\phi}=-\frac{\partial Q_M}{\partial\phi}, \tag{10}$$

where the quantity Q_M encodes all quantum correction and is called the quantum potential. We ended up therefore with the purely classical motion with the only difference that the classical potential $V(\phi)$ is now supplemented by the quantum counterpart Q_M. Thus the evolution of the system is described by purely classical trajectories (in the modified potential, though.) One may wonder what happened to the probabilistic picture of quantum mechanics. This part happens to be hidden in the initial conditions to our evolutionary equations. It turns out that there is a well defined probability measure on the set of these conditions (given by the initial wave function), but once we pick up a single one, the subsequent evolution is completely probabilistic.[‡]

One may wonder if it is necessary to go through the intermediate step, Eq. 9, to obtain Eq. 10. In fact, as it was shown in[5], one can obtain the set of equations

$$\frac{1}{2}a\dot{a}^2=a^3\left(\frac{\dot{\phi}}{2}+V_M(\phi)\right)+Q \tag{11}$$

$$\ddot{\phi}+3\frac{\dot{a}}{a}\dot{\phi}+\frac{\partial V_M}{\partial\phi}=-\frac{\partial Q}{\partial\phi}, \tag{12}$$

directly from Eq. 3, without making any approximations. This means that we are able to rewrite the equation governing the quantum universe in the equivalent, but much more convenient form, in which the analysis can be made in terms of classical objects and interpretation is much easier.

It is worth mentioning that the procedure described above can be readily extended to cover any bosonic quantum field theory. For example, in the forthcoming paper[8], this procedure is applied to the case of the quantum field theory of massive scalar

‡For the careful discussion of this point, see[5].

field on the De Sitter background. It could be also applied to the case of Yang-Mills theory and full quantum theory of gravity.

There is one more interesting area where the above procedure can be easily applied. In the reheating phase, we have to do with the classical oscillating inflaton field ϕ and the thermal field ρ. We again encounter the system with the classical part (inflaton) and the quantum subsystem (radiation). Here again one starts with the hamiltonian

$$H = H_\phi + H_\rho + H_{int}$$

and proceeds through the steps similar to those indicated above. Then one can proceed with the analysis of various posible outputs of reheating.

It is interesting to note in this context that assuming that the interaction hamiltonian is of the form $\phi^2 H_I(\rho, t)$, the effective equation for the inflaton field ϕ the leading term looks as follows

$$\frac{1}{2}\dot{\phi}^2 + \frac{1}{2}m^2\phi^2 =< H_I > \phi^2. \tag{13}$$

Above $< H_I >$ is an average computed with respect to the radiation field, and m is the effective mass of the inflaton field. The form of this equation suggests the exciting possibility of creation of the universe in laboratory. By appropriately preparing the radiation field, one may make $< H_I >$ into the source for parametric resonance for ϕ in order to efficiently pump energy into this field. It is worth observing that once the field ϕ reaches the magnitude of order one, and provided that it is sufficiently homogeneous in space, the inflation starts and the region starts expanding rapidly. It happens also the this region covers itself by the horizon so it, in fact, becomes the new universe effectively disconected from our own.

Even thogh the creation of the universe is clearly impossible using modern technology, the future experimental high energy physicist preparing this experiment should not forget to place the announcement "We appologise for all the inconveniences"[9] somewhere in the newly created universe.

References

1. A.D. Linde, *Phys. Lett.* **129B** (1983) 177; see also A.D. Linde, *Particle Physics and Inflationary Cosmology* (Harwood, Chur, 1990) and J. Kowalski-Glikman *Eight Lectures on Cosmology* (Wrocław University Preprint, 1994).
2. A.D. Linde, D.A. Linde, and A. Mezhlumian, *Phys. Rev.* **D49** (1994) 1783.
3. L. Kofman, A.D. Linde, and A.A. Starobinsky, *Reheating After Inflation* hep-th 9405187.
4. J. Kowalski-Glikman and J. Vink, *Class. Quant. Grav.* **7** (1990) 701.
5. J. Vink, *Nucl. Phys* **B369** (1992) 707.
6. C.J. Isham, *Canonical Gravity and the Problem of Time* in *Integrable Systems, Quantum Groups, Field Theories*, L.A. Ibort and M.A. Rodriguez Eds., (Kluwer Academic Publishers, Dordrecht/Boston/London, 1993).
7. A. Błaut and J. Kowalski-Glikman, in preparation.
8. J. Kowalski-Glikman, in preparation.
9. D. Adams, *So Long and Thanks for All the Fish.*

THE $b \to s + \gamma$ DECAY IN SUPERGRAVITY

R. ARNOWITT[†]
Center for Theoretical Physics, Department of Physics,
Texas A&M University, College Station, Texas 77843-4242

PRAN NATH
Department of Physics, Northeastern University
Boston, MA 02115

Abstract

The $b \to s + \gamma$ decay is a powerful tool for testing models of new physics because the new physics diagrams enter in the same loop order as the Standard Model ones. The current experimental and theoretical status of this decay is reviewed. Predictions based on the minimal supergravity model (MSGM) in the leading order (LO) are discussed. It is shown that results are sensitive to the value of m_t and α_G. The current experimental value for the $b \to s + \gamma$ rate already very likely eliminates part of the SUSY parameter space when both m_o and $m_{\tilde{g}}$ are small and when A_t and μ have the same sign. Dark matter detection rates for $\tilde{Z}_1$ cold dark matter for $\mu < 0$ are only minimally affected by the current data, as are proton decay predictions for models consistent with current proton lifetime and $\tilde{Z}_1$ relic density bounds.

1. Introduction

It is generally expected that the TeV energy domain will bring forth an array of new physics. There are many speculations as to what form this new physics will take: supersymmetry, technicolor, compositeness, additional W and Z bosons, etc. Unfortunately, most of the precision LEP measurements are not very sensitive to new physics as the Standard Model contributions enter at the tree level, while possible new physics contributions begin at the loop level. Thus the most one might hope for in these measurements is a few percent correction from new physics.

The recently discovered decay by CLEO of $b \to s + \gamma$ is an exception to this as this process is sensitive to new physics and is observable for several reasons:

- Being a FCNC process, it begins at the loop level so that Standard Model loops and new physics loops enter at the same level. Thus there can be large new physics corrections.
- It is of size $G_F^2\alpha$, where G_F is the Fermi constant (rather then $G_F^2\alpha^2$ as is usual for FCNC processes)
- QCD corrections are large and enhance the rate by a factor ≈ 3.

[†]Conference Speaker

In spite of this, it is difficult at this time to make detailed statements concerning how well theory and experiment agree. First the experimental errors are still quite large. Further, the large QCD corrections mean that next to leading order (NLO) QCD effects are important and these are difficult to do. Thus theoretical errors are currently also quite large. One can, however, still learn interesting things at this level of knowledge, and as experiment and theory get refined, more precise statements will become available.

2. $b \to s + \gamma$ Decay

The $b \to s + \gamma$ branching ratio measured by CLEO is[1]

$$BR(B \to X_S\gamma) = (2.32 \pm 0.5 \pm 0.29 \pm 0.32) \times 10^{-4} \tag{1}$$

where the first error is statistical, and the last two are systematic. If one combined errors in quadrature one obtains $BR(B \to X_s\gamma) \cong (2.32 \pm 0.66) \times 10^{-4}$ which has an error of about 30%. In the spectator approximation one may relate the B meson decays to the b quark decays:

$$\frac{BR(B \to X_s\gamma)}{BR(B \to X_c e\bar{\nu}_e)} \cong \frac{\Gamma(b \to s + \gamma)}{\Gamma(b \to c + e + \bar{\nu}_e)} \equiv R \tag{2}$$

where $BR(B \to X_c e\bar{\nu}_e) = (10.7 \pm 0.5)\%$. [The spectator model has corrections that begin at $0(1/m_b^2)$.]

At the electroweak scale $\mu = O(M_W)$, the elementary diagrams involve the W-t-quark loop for the Standard Model plus additional loops (H^--t-quark, $\tilde{W}^- - \tilde{t}$-squark) for the supersymmetric generalization (Fig. 1). Thus at $\mu \approx M_W$ the interaction can

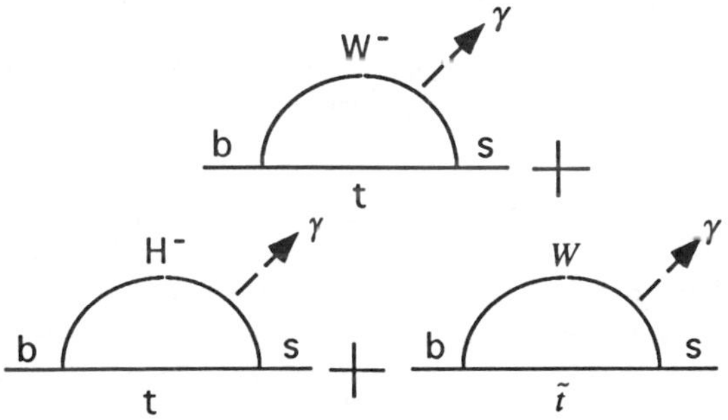

Fig. 1 Elementary diagrams for $b \to s + \gamma$ decay at scale $\mu \approx M_W$. Only the third generation quark and squark contribute significantly.

be described by an effective Hamiltonian[2]

$$H_{eff} = V_{tb}V_{ts}^* \frac{G_F}{\sqrt{2}} C_7(M_W) Q_7 \tag{3}$$

where $Q_7 = (e/24\pi^2)m_b\bar{s}_L\sigma^{\mu\nu}b_R F_{\mu\nu}$. Here $F_{\mu\nu}$ is the electromagnetic field strength and m_b is the b quark mass. However, the decay occurs at $\mu \approx m_b$, and one must use the renormalization group equations (RGE) to go from M_W to m_b. This produces operator mixing with the color transition magnetic moment operator $Q_8 = (g_3/16\pi^2)m_b\bar{s}_R\sigma^{\mu\nu}T^A b_L G^A_{\mu\nu}$ where T^A and $G^A_{\mu\nu}$ are the gluon generator and field strength) and with six 4-quark operators $Q_1 .... Q_6$. The ratio of Eq. (2) at $\mu = m_b$ becomes then

$$R = |\frac{V_{ts}^* V_{tb}}{V_{cb}}|^2 \frac{6\alpha}{\pi I(z)} \frac{|C_7^{eff}(m_b)|^2}{[1 - \frac{2}{3\pi}\frac{\alpha_3(m_b)}{\zeta} f(z)]} \tag{4}$$

where $z = m_c/m_b = 0.313 \pm 0.013$, $\eta = \alpha_3(M_Z)/\alpha_3(m_b) = 0.548$, I(z)=1-8z$^2 + 8z^6 - z^8 - 24z^4 lnz$ is a phase space factor for the $b \to ce\bar{\nu}_e$ decay, and the denominator bracket (f(z)≅2.41) is a QCD correction to $b \to ce\bar{\nu}_e$. In LO, $C_7^{eff}(m_b)$ is given by[2,3]

$$C_7^{eff}(m_b) = \eta^{\frac{16}{23}} C_7(M_W) + \frac{8}{3}(\eta^{\frac{14}{23}} - \eta^{\frac{16}{23}})C_8(M_W) + C_2(M_W) \tag{5}$$

where C_2 represents the operator mixing with the 4-quark operators.

The advantage of using the ratio R is that poorly known CKM matrix elements and a $(m_b)^5$ factor cancel out and one is left with the relatively well known ratios $|V_{ts}^* V_{tb}/V_{cb}|^2 = 0.95 \pm 0.04$ and $z = m_c/m_b$. There remain, however, a number of errors and uncertainties in the above results which we now list:

(i) Existing errors in the input parameters $\alpha_3(M_Z) = 0.12 \pm 0.01; z; BR(B \to X_c e\bar{\nu}_e)$; and the CKM matrix element ratio.

(ii) Use of spectator model. [Corrections are of O($1/m_b^2$)].

(iii) Neglect of the next to leading order (NLO) corrections. This is the largest error since QCD corrections to this process are large.

An estimate of the size of the NLO corrections can be obtained by finding the change in the LO result between running the RGE to $\mu = m_b/2$ and $\mu = 2m_b$. Thus when higher order corrections are included, the μ dependence should disappear, and so the μ dependence of the LO gives a measure of the size of the NLO corrections. The neglect of the NLO corrections is then estimated to cause an error of about[4] $\pm$ 25%.

Combining the above errors in quadrature, the LO calculation yield for the Standard Model for m_t=174 GeV the result[4,5]

$$BR[B \to X_s\gamma] \cong (2.9 \pm 0.8) \times 10^{-4} \tag{6}$$

which is about a 30% error. Comparing Eqs. (1) and (6), we see at this point that it is not possible to distinguish between the Standard Model being in agreement with experiment or differing from it by a factor of 2. Some of the NLO corrections have now been calculated[5,6], but a clear answer requires a full calculation of the NLO corrections which is not easy (as they involve calculating the finite parts of two loop and divergent parts of three loop diagrams).

There is an additional theoretical correction due to the existence of heavy thresholds. (This is really part of the NLO corrections but requires special treatment.) The LO analysis considers the effective theory at $\mu = M_W$ and integrates the RGE down to $\mu \approx m_b$. However, other particles in the loop are not degenerate with W boson, and one really should start at a higher mass scale and integrate out each particle as one crosses its mass threshold. For the Standard Model W-top graph, this leads to about a 15% enhancement[7,8]. For the SUSY case, it is estimated[8] that the effect is small for the H-top loop, and perhaps $\pm$ 15% for the $\tilde{W} - \tilde{t}$ loop. These effects depend on the mass spectrum of the SUSY particles, and thus may be important as SUSY diagrams in $b \to s + \gamma$ decay can occur with opposite sign to the Standard Model one.

3. Minimal Supergravity Model (MSGM)

In order to obtain the SUSY prediction for the $b \to s + \gamma$ branching ratio, one needs to specify the SUSY particle masses. The MSSM is not very useful in this respect as it depends on too many (about 20) arbitrary parameters. We will here make use of the minimal supergravity model[9] (MSGM) which has much greater predictive power.

The MSGM has already exhibited a number of accomplishments, and has made a number of predictions which can be tested in forthcoming experiments. We list some of these here:

- The MSGM accounts for the unification of couplings at the GUT scale $M_G \approx 10^{16}$ GeV implied by the LEP data[10]
- It allows for spontaneous breaking supersymmetry in the "hidden" sector (which cannot be achieved in a phenomenologically satisfactory way in the MSSM).
- Flavor changing neutral interactions are naturally suppressed.
- The masses of the 32 new SUSY particles and all their interactions are predicted in terms of only four parameters [m_o, the universal scalar mass; $m_{1/2}$, the universal gaugino mass; A_o, the universal cubic soft breaking parameter; tan$\beta \equiv < H_2 > / < H_1 >$] and the sign of μ, the Higgs mixing parameter. [Here (H_1, H_2) are the Higgs that gives rise to (down, up) quark masses.] Thus the theory makes many predictions that can be tested at[9] LEP2 and the LHC and at an upgraded Tevatron[11].
- Models with R parity yield a natural candidate for cold dark matter, the lightest neutralino $\tilde{Z}_1$, with relic abundances consistent with COBE and other astrophysical data.

- If representations breaking the GUT group are not large, the Gut threshold corrections are small, and hence low energy predictions are mostly independent of the choice of Gut group.
- The spontaneous breaking of supersymmetry at M_G naturally triggers the breaking of SU(2)$\times$U(1) at the electroweak scale by radiative breaking. Thus the MSGM offers a natural explanation of electroweak breaking.
- For models with proton decay, the decay rate is suppressed by a factor of $\approx 10^4$ relative to Standard Model Guts, and hence is consistent with current data. (However, such models generally predict rates that should be detectable in the next round of p-decay experiments.)

It is convenient, in the following discussion to trade $m_{1/2}$ and A_o for the two low energy parameters, $m_{\tilde{g}}$ (the gluino mass) and A_t (the t-quark A parameter at the electroweak scale). We then explore the entire parameter space over the range

$$100GeV \leq m_o, m_{\tilde{q}} \leq 1\ TeV;\ -6 \leq A_t/m_o \leq 6;\ tan\beta \leq 20 \tag{7}$$

subject to the constraints that there be no violation of the LEP and Tevatron bounds on SUSY masses, and that radiative breaking of SU(2)$\times$U(1) occurs. We chose a mesh with Δm_o = 100GeV, $\Delta m_{\tilde{g}}$= 25 GeV, ΔA_t=0.5, $\Delta(tan\beta)$=2, 4. One then determines the masses and couplings of the SUSY particles for each of the parameter points, which allows calculation of the $b \to s + \gamma$ rate in the LO approximation.

4. Parameter Dependence

The dominant contributions to the $b \to s + \gamma$ loops come from the third generation quarks and squarks. The $\tilde{W} - \tilde{t}$ loops play an important role as the two stop states, $\tilde{t}_1$ and $\tilde{t}_2$ ($m_{\tilde{t}_1} < m_{\tilde{t}_2}$) are very split. The stop mass matrix reads

$$\begin{pmatrix} m^2_{\tilde{t}_L} & m_t(A_t m_o + \mu ctn\beta) \\ m_t(A_t m_o + \mu ctn\beta) & m^2_{\tilde{t}_R} \end{pmatrix} \tag{8}$$

Because m_t is large, the $\tilde{t}_1$ can become light, and in fact large regions of the parameter space get excluded when $\tilde{t}_1$ is driven tachyonic.

As is well known[12], the $H^- - t$ loop adds constructively to the Standard Model to increase the $b \to s + \gamma$ branching ratio. However, the $\tilde{W} - \tilde{t}$ loop can enter with either sign increasing or decreasing the total amplitude, as is seen in Fig. 2. Note that the effect of the $\tilde{W} - \tilde{t}_1$ graph is larger for light $\tilde{t}_1$, as one might expect, and in the MSGM there exist parameter points where the total branching ratio can be below or above the Standard Model result.

Solving for the eiginvalues of Eq. (7) shows that $m^2_{\tilde{t}_1}$ is smaller if A_t and μ have the same sign or when $m^2_{\tilde{t}_R}$ is negative. The latter can happen if $A_t < 0$. These effects can be seen in Figs. (3-5)[13,14]. Fig. 3 shows that the branching ratio is largest when

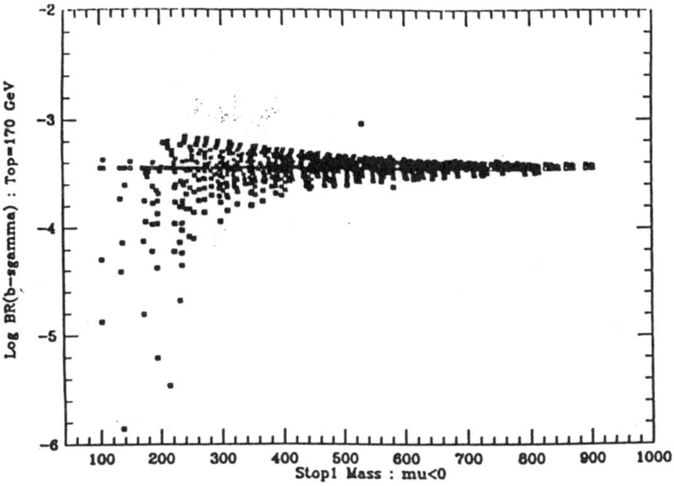

Fig. 2 Scatter plot for LO BR($b \to s + \gamma$) as a function of $m_{\tilde{t}_1}$.

when m_o and $m_{\tilde{g}}$ are small ($m_{\tilde{W}_1} \simeq (\frac{1}{3} - \frac{1}{4})m_{\tilde{g}}$) and that it is larger when A_t and μ

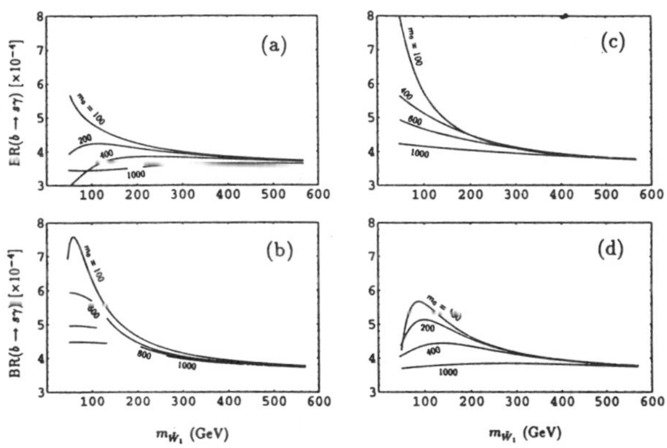

Fig. 3 $BR(b \to s + \gamma)$ vs. $m_{\tilde{W}_1}$, $m_t = 165$ GeV, $\tan\beta$=5, $|A_t/m_o|$=0.5, α_G^{-1}=24.11. Graphs (a) and (b) are for $A_t < 0$, (c) and (d) for $A_t > 0$. Graphs (a) and (c) are for $\mu > 0$ and (b) and (d) are for $\mu < 0$.

have the same sign then when they have opposite signs (as one would expect since

$m_{\tilde{t}_1}$ is then smaller). The gaps in the graphs (a) and (b) occur due to the fact that the $m_{\tilde{t}_1}$ has turned tachyonic (which can occur when A_t is negative). Fig. 4 shows that the branching ratio increases as m_t increases (since $m_{\tilde{t}_1}$ shrinks when the off

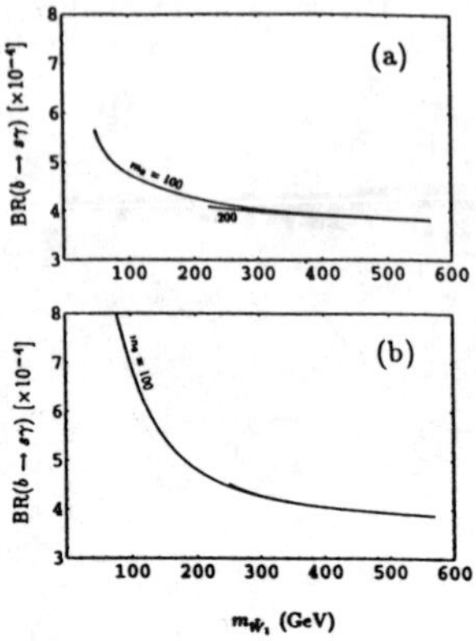

Fig. 4 Same as Fig. 3 for $m_t = 170$ GeV

diagonal elements Eq. (7) grow) and in fact for $A_t < 0$ the region for tachyonic $m_{\tilde{t}_1}$ has grown. Note that the $A_t > 0$ curves are only slightly dependent on m_t.

Fig. 5 shows the sensitivity of the branching ratio to α_G. For $A_t < 0$ a larger portion of the parameter space is excluded because $\tilde{t}_1$ is driven tachyonic when α_G

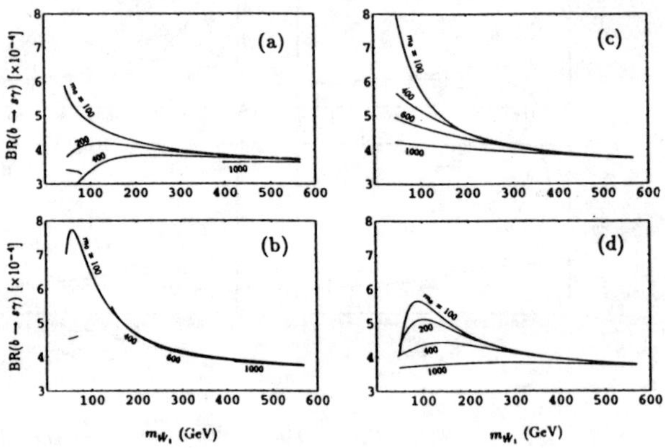

Fig. 5 Same as Fig. 3 with $\alpha_G^{-1} = 24.5$

decreases (because h_t, the t Yukawa coupling constant, is driven closer to its Landau pole. (Again, for $A_t > 0$, there is little effect due to this small change in α_G.) If one increases both m_t and α_G^{-1}, the effects combine to eliminate the $A_t < 0$ part of the parameter space for $m_o > 100$ GeV for this case.

5. $b \to s + \gamma$ Constraints on Dark Matter Detection

The experimental results on $b \to s + \gamma$ can impose constraints on dark matter analyses[15]. We saw in Sec. 4, that when μ and A_t have the same sign, the $b \to s + \gamma$ decay rate is larger than when μ and A_t have opposite signs[13]. This increase is particularly marked when m_o and $m_{\tilde{g}}$ are small. This is also the region of largest dark matter detection rates. Thus one expects that for such type situations, the experimental $b \to s + \gamma$ rate of Eq. (1) might already be able to rule out some of the SUSY parameter space where the dark matter detection rates are expected to be highest and hence most accessible experimentally.

To examine this in more detail, we consider the experimental branching ratio $BR(b \to s + \gamma) = 2.32 \pm 0.66 \times 10^{-4}$ and ask, what part of the dark matter detection rates correspond to parameter points where the $b \to s + \gamma$ theoretical rate is within the 95% CL of the experimental value. We use here the LO calculation of the $b \to s + \gamma$ rate as a figure of merit[16]. Fig. 6 shows the maximum and minimum event rates for a Pb dark matter detector as a function of A_t for $\mu < 0$. The cosmological con-

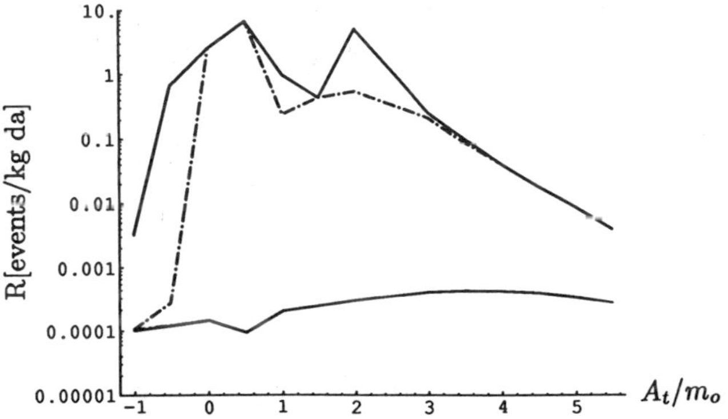

Fig. 6 The maximum and minimum detection rates for a Pb dark matter detector as a function of A_t for $\mu < 0$. The solid curve is without the $b \to s + \gamma$ constraint while the dot-dash curve imposes the constraint that the theoretical LO branching ratio be within the 95% CL of the experimental value.

straints on the relic density require that the allowed parameter space be mainly for $A_t > 0$. Thus for $\mu < 0$, the predicted $b \to s + \gamma$ rate is large only in the small region where $A_t < 0$, and it is only in this region where the experimental value of the

$b \to s + \gamma$ branching ratio eliminates a significant part of the parameter space. Note that the minimum event rates are unaffected.

Fig. 7 shows the corresponding graphs for $\mu > 0$. Here the majority of parameter

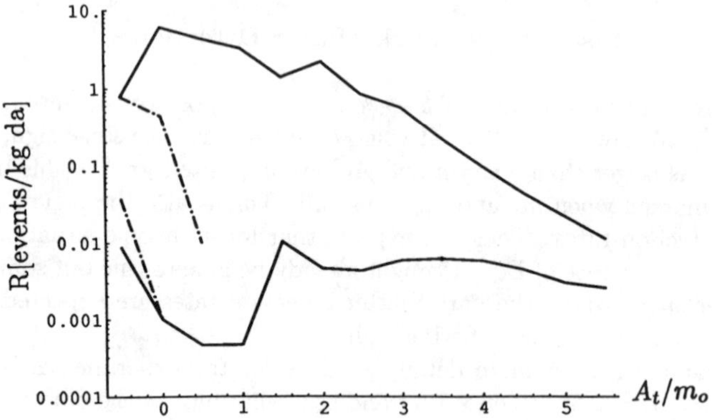

Fig. 7 Same as Fig. 6 for $\mu > 0$.

space occurs for A_t and μ having the same sign, and hence the theoretical LO $b \to s+\gamma$ decay rate is larger then the experimental rate over most of the parameter space. In fact, the 95% CL bound eliminates all of the parameter space except in the band $-0.5 \leq A_t/m_o \leq 0.5$.

6. $b \to s + \gamma$ Decay Constraints on Proton Decay

We consider next supergravity Gut models that allow for proton decay, and restrict the discussion to SU(5)-type models where proton decay is mediated by a superheavy

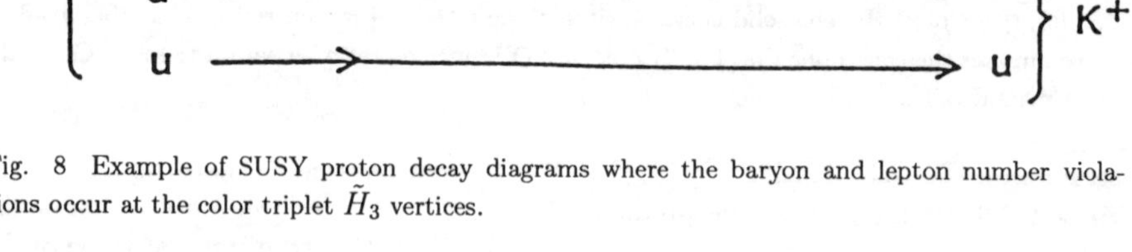

Fig. 8 Example of SUSY proton decay diagrams where the baryon and lepton number violations occur at the color triplet $\tilde{H}_3$ vertices.

Higgsino color triplet $\tilde{H}_3$ with mass M_{H_3}=O(M_G). The basic diagram is shown in Fig. 8 where the chargino $\tilde{W}$ exchange "clothes" the Higgsino interactions. The decay rate can be written in the form $\Gamma(p \to \bar{\nu}K^+) = const|B|^2/M_{H_3}^2$ where B represents the clothing loop amplitude. The current experimental bound on the proton lifetime is[17] $\tau(p \to \bar{\nu}K^+) > 1 \times 10^{32}$ yr which places a bound on B of[18]

$$|B| \lesssim 100(\frac{M_{H_3}}{M_G})GeV^{-1}; \quad M_G \equiv 2 \times 10^{16} GeV \tag{9}$$

In the following we limit M_{H_3} to obey $M_{H_3} \leq 10M_G$, so that there not be unknown Planck scale physics entering into the analysis.

We examine here those parts of the parameter space that simultaneously satisfies the proton decay lifetime bound and the cosmological bound that the $\tilde{Z}_1$ be the cold component of dark matter, i.e.[19] $0.10 \lesssim \Omega_{\tilde{Z}_1} h^2 \lesssim 0.35$ where $\Omega_{\tilde{Z}_1} = \rho_{\tilde{Z}_1}/\rho_c$ and h=H/(100 km/sec Mpc). Here $\rho_{\tilde{Z}_1}$ is the relic mass density of the $\tilde{Z}_1$, ρ_c is the critical mass density that closes the Universe, and H is the Hubble constant. We then ask what fraction of the parameter points satisfying both the proton decay bound and the cosmological constraints are consistent with the CLEO measurement of the $b \to s + \gamma$ decay rate. Using the leading order (LO) predictions for the $b \to s + \gamma$ decay rate, we find that 95% of the parameter points that satisfy both the proton lifetime and cosmological bounds lie within the 95% CL bounds of the CLEO data. (13% lie within 68% CL bounds of the CLEO data). The parameter points satisfying the proton decay bound generally give a value of BR($b \to s + \gamma$) which lies above the central CLEO value. However, a number of the NLO corrections[5] do decrease the theoretical value. Thus the $b \to s + \gamma$ decay does not as yet seriously constrain proton decay models. We note that the parameter points satisfying the proton decay, cosmological and $b \to s + \gamma$ decay bounds require

$$m_{\tilde{g}} < 375GeV; \; m_o > 400GeV; \; 0 \leq A_t/m_o \leq 0.5; \; tan\beta \leq 10 \tag{10}$$

Because the proton decay bound requires tanβ to be small the dark matter event rates will remain small, i.e.

$$R < 0.01 event/kg\ da \tag{11}$$

Thus if proton decay were seen at the next round of proton decay experiments (e.g. at Super Kamiokande) these models would predict that $\tilde{Z}_1$ dark matter would not be observable with current dark matter detectors.

7. Conclusions and Summary

The $b \to s + \gamma$ decay is a process which is sensitive to new physics and thus is an excellent place to test models of new physics. However, both theory and experiment need improvement if quantitative tests are to be made. Using the leading order

(LO) approximation to the theory, a number of semi-quantitative results do exist:

The rate for $b \to s + \gamma$ is largest when both m_o and $m_{\tilde{g}}$ are small and when μ and A_t have the same sign. Thus these are the parts of the parameter space which will get eliminated if the experimental value of the branching ratio stays below the Standard Model number.

Large parts of the parameter space for $A_t < 0$ gets deleted when m_t and α_G^{-1} increases, as the $\tilde{t}_1$ squark turns tachyonic. This is a consequence of m_t being large.

The current $b \to s + \gamma$ experimental bound appears to eliminate most of the parameter space which is consistent with the $\tilde{Z}_1$ as cold dark matter for $\mu > 0$, but does not effect the $\mu < 0$ part very much.

Most of the parameter points which satisfy both current proton decay bounds and $\tilde{Z}_1$ relic density bounds, are also consistent with present $b \to s+\gamma$ experimental rates. However, as the data improves, the $b \to s + \gamma$ decay will make significant impact on models that predict proton decay.

One may expect significant constraints on new physics models once the data and theoretical predictions on the $b \to s + \gamma$ decay become more accurate.

Acknowledgements

This work was supported in part by NSF grant numbers PHY-9411543 and PHY-19306906.

References

1. E.H. Thorndike, Talk at ICHEP94, Glasgow, 1994.

2. S. Bertolini, F. Borzumati and A. Masiero, *Phys. Rev. Lett.* **59**, 180 (1987); B. Grinstein, R. Springer and M.B. Wise, *Nucl. Phys.* **B339**, 269 (1990); S. Bertolini, F. Borzumati, A. Masiero and G. Ridolfi, *Nucl. Phys.* **B353**, 591 (1991).

3. R. Barbieri and G. Giudice, *Phys. Lett.* **B309**, 86 (1993); M. Misiak, *Phys. Lett.* **B269**, 161 (1991); *Nucl. Phys. Phys.* **B393**, 23 (1993).

4. A.J. Buras, M.Misiak, M. Münz, and S. Pokorski, *Nucl. Phys.* **B424**, 374 (1994).

5. M. Ciuchini, E. Franco, G. Martinelli, L. Reina and L. Silverstrini, *Phys. Lett.* **B316**, 127 (1993).

6. M. Misiak and M. Münz, TUM-T31-79/94.

7. P. Cho and B. Grinstein, *Nucl. Phys.* **B365**, 279 (1991); Err. (in print).

8. H. Anlauf, SLAC-PUB-6525 (1994).

9. A.H. Chamseddine, R. Arnowitt and P. Nath, *Phys. Rev. Lett.* **49**, 970 (1982). For reviews see P. Nath, R. Arnowitt and A.H. Chamseddine, *"Applied N=1 Supergravity"* (World Scientific, Singapore 1984); H.P. Nilles, Phys. Rep. **110**, 1 (1984); R. Arnowitt and P. Nath, *Proc. of VII J.A. Swieca Summer School* (World Scientific, Singapore, 1994).

10. P. Langacker, Proc., PASCOS-90, Eds. P. Nath and S. Reucroft (World Scientific, Singapore 1990); J. Ellis, S. Kelley and D.V. Nanopoulos, *Phys. Lett.* **B249**, 441 (1990); U. Amaldi, W. de Boer and H. Furstenau, *Phys. Lett.* **B260**, 447 (1991); F. Anselmo, L. Cifarelli, A. Peterman and A. Zichichi, *Nuov. Cim.* **104A**, 1817 (1991); **115A**, 581 (1992).

11. T. Kamon, J.L. Lopez, P. McIntyre and J. White, *Phys. Rev.* **D50**, 5676 (1994).

12. J.L. Hewett, *Phys. Rev. Lett.* **70**, 1045 (1993); V. Barger, M.S. Berger and R.J.N. Phillips, *Phys. Rev. Lett.* **70**, 1368 (1993).

13. J. Wu, R. Arnowitt and P. Nath, CERN-TH.7316/94-CTP-TAMU-03/94-NUB-TH-3092/94.

14. Some authors[4,5] define the LO without the denominator bracket in Eq. (4). To be in accord with this definition one should reduce the curves of Figs. (3-5) by about 15%.

15. P. Nath and R. Arnowitt, *Phys. Lett.* **B336**, 395 (1994); CERN-TH.7363/94-NUB-TH-3099/94-CTP-TAMU-38/94.

16. As discussed in Sec. 2, the LO theoretical error is about the same as the experimental error.

17. Particle Data Group, *Phys. Rev.* **D50**, 1173 (1994).

18. P. Nath, R. Arnowitt and A.H. Chamseddine, *Phys. Rev.* **D32**, 2348 (1985); R. Arnowitt and P. Nath, *Phys. Rev.* **D49**, 1479 (1994).

19. R. Arnowitt and P. Nath, CERN-TH.7362/94-CTP-TAMU-37/94 NUB-TH-3098/94.

COMPLETE WEAK-SCALE THRESHOLD CORRECTIONS IN THE MINIMAL SUPERSYMMETRIC STANDARD MODEL

J. BAGGER, K. MATCHEV AND D. PIERCE
Department of Physics and Astronomy
The Johns Hopkins University
Baltimore, MD 21218, USA

ABSTRACT

We compute the full set of weak-scale gauge and Yukawa threshold corrections in the minimal supersymmetric standard model, and use them to study the effects of the supersymmetric particle spectrum on gauge and Yukawa coupling unification.

1. Introduction

The idea of supersymmetric unification has received a significant boost from the recent observation that the gauge couplings unify in the minimal supersymmetric standard model.[1] Indeed, this observation has given great impetus to the search for supersymmetric particles at present and future accelerators. In this talk we will assume that supersymmetry has been found, and ask what the supersymmetric particle spectrum can tell us about unification-scale physics.

To lowest order, the answer is not much, because the one-loop renormalization group equations (RGE's) do not depend on the physics of the unification scale. To second order, the story is different. This is because the two-loop RGE's must be used in conjunction with the one-loop threshold corrections. The threshold conditions depend on the weak-scale supersymmetric spectrum *and* on the unknown unification-scale physics. Therefore to second order, the superparticle spectrum is correlated with the physics of the unification scale.

In this talk we will report on results of a complete next-to-leading-order analysis of supersymmetric unification. Our results are based on the full set of two-loop supersymmetric renormalization group equations, together with the complete one-loop threshold conditions, including finite parts. We use a two-sided differential equation solver which allows us to impose boundary conditions at the weak scale, M_Z, and at the unification scale, $M_{\rm GUT}$.

At the weak scale, we specify the $\overline{DR}$ gauge couplings $\hat{g}_1(M_Z)$ and $\hat{g}_2(M_Z)$, and the $\overline{DR}$ vev $\hat{v}$, which we extract from the Fermi constant, G_F, the Z-boson mass, M_Z, and the electromagnetic coupling $\alpha_{\rm EM}$. We also specify the top and tau $\overline{DR}$ Yukawa couplings $\hat{\lambda}_t(M_Z)$ and $\hat{\lambda}_\tau(M_Z)$, which we compute from $\tan\beta \equiv \tan\hat{\beta} = \hat{v}_2/\hat{v}_1$ and the top and tau pole masses, m_t and m_τ.

We define $M_{\rm GUT}$ to be the scale where $\hat{g}_1(M_{\rm GUT}) = \hat{g}_2(M_{\rm GUT})$. At $M_{\rm GUT}$ we specify the $\overline{DR}$ supersymmetric parameters $M_{1/2} \equiv \hat{M}_{1/2}(M_{\rm GUT})$, $M_0 \equiv \hat{M}_0(M_{\rm GUT})$,

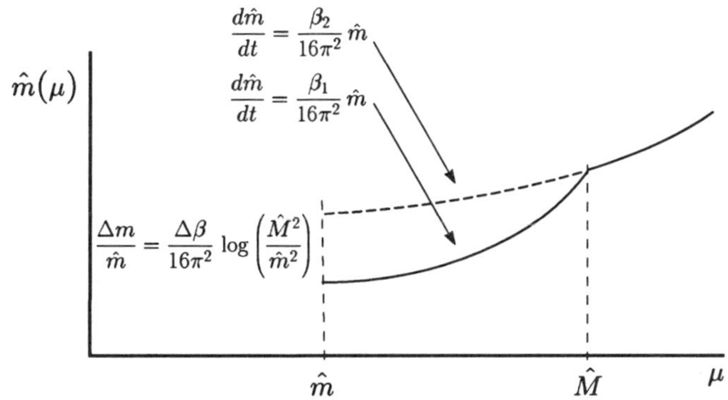

Fig. 1. The $\overline{DR}$ mass $\hat{m}(\mu)$ vs. the $\overline{DR}$ scale $\mu = e^t$.

and $A_0 \equiv \hat{A}_0(M_{\rm GUT})$. We also define the $\overline{DR}$ strong coupling $\hat{g}_3(M_{\rm GUT})$ by the unification condition

$$\hat{g}_3(M_{\rm GUT}) \;=\; \hat{g}_1(M_{\rm GUT})(1+\epsilon_g)\;, \tag{1}$$

where ϵ_g parametrizes the gauge-coupling threshold correction at the unification scale. In a similar way, we impose bottom-tau unification and define the bottom-quark $\overline{DR}$ Yukawa coupling to be

$$\hat{\lambda}_b(M_{\rm GUT}) \;=\; \hat{\lambda}_\tau(M_{\rm GUT})(1+\epsilon_b)\;, \tag{2}$$

where ϵ_b parametrizes the unification-scale Yukawa coupling threshold correction.

We solve the supersymmetric renormalization group equations self-consistently, subject to the boundary conditions at M_Z and $M_{\rm GUT}$. In this way we find a prediction for the $\overline{DR}$ couplings $\hat{g}_3(M_Z)$ and $\hat{\lambda}_b(M_Z)$. We then apply the weak-scale threshold corrections to determine the $\overline{MS}$ strong coupling, $\alpha_s(M_Z)$, the bottom-quark pole mass, m_b, and the full supersymmetric particle spectrum. Our results allow us to correlate the unification-scale thresholds ϵ_g and ϵ_b with m_b, $\alpha_s(M_Z)$ and the superparticle spectrum.

2. Threshold Corrections

2.1. The Match-and-Run Procedure

To study unification-scale physics, it is important to use the correct weak-scale threshold corrections. Typically, these corrections are computed using the so-called match-and-run technique, which is based on the successive decoupling of particles at the scale of their masses. As an example of this technique, let us consider a running $\overline{DR}$ mass $\hat{m}$, whose full β-function depends on all the particles in the supersymmetric standard model. For large renormalization scales μ, $\hat{m}$ evolves according to its complete supersymmetric RGE

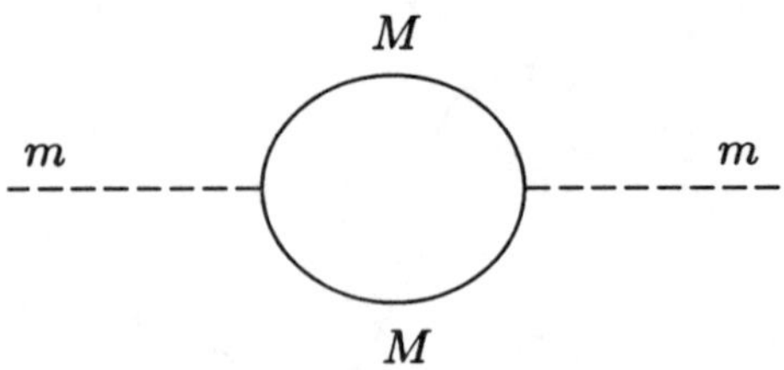

Fig. 2. The one-loop diagram that yields the threshold correction for a particle of mass m from a particle of mass M.

Let us now run $\hat{m}$ towards the weak scale, M_Z. Along the way, we eventually encounter the scale of the squark masses. According to the match-and-run procedure, we must stop the evolution and construct a new effective theory in which the squarks are integrated out. We must then continue the evolution, using the new β-function, without the squark contribution, subject to the matching condition $\hat{m}(m_{sq}^-) = \hat{m}(m_{sq}^+)$. We repeat this procedure at each new threshold, finally stopping at the scale $\mu = \hat{m}(\mu)$. The quantity $\hat{m}(\hat{m})$ is the match-and-run approximation to the physical pole mass.

To test this procedure, let us compute the threshold correction to a particle of mass $\hat{m}$ from a particle of mass M, with $M > m$. According to the match-and-run procedure, the decoupling of the heavy particle gives the correction

$$\frac{\Delta m}{\hat{m}} = \frac{\Delta\beta}{16\pi^2} \log\left(\frac{\hat{M}^2}{\hat{m}^2}\right) , \tag{3}$$

as shown in Figure 1. In this expression, $\Delta\beta$ is the difference of the β-functions, before and after decoupling.

The exact one-loop result can be found by computing the diagram of Figure 2. It is given by

$$\frac{\Delta m}{\hat{m}} = \frac{\Delta\beta}{16\pi^2} \int_0^1 dx \, \log\left(\frac{|\,M^2 - x(1-x)\,m^2\,|}{\mu^2}\right) . \tag{4}$$

For $\mu = m$, this reduces to

$$\frac{\Delta m}{\hat{m}} = \frac{\Delta\beta}{16\pi^2} \left[\log\left(\frac{M^2}{m^2}\right) + \text{finite}\right] , \tag{5}$$

where the finite contribution does not contain any logarithms.

These results indicate that the match-and-run procedure gives a good approximation to the pole mass when $\hat{M} \gg \hat{m}$. In this case the large logarithm, proportional to $\log(\hat{M}^2/\hat{m}^2)$, dominates the threshold correction.

When $\hat{m} \simeq \hat{M}$, however, the finite term is typically as large as the logarithm. The finite correction is completely missed by the match-and-run procedure. In supersymmetric unified models with universal boundary conditions, it is quite possible for all the supersymmetric masses to be near M_Z, in which case the match-and-run procedure gives an error of $\mathcal{O}(1)$.

2.2. Gauge Coupling Thresholds

It is important to use the correct weak-scale threshold corrections because supersymmetric unification depends sensitively on the weak-scale boundary conditions. This can be easily understood from the one-loop renormalization group equations for the gauge couplings. Assuming unification, the RGE's imply

$$\frac{\beta_2-\beta_3}{\hat{g}_1^2(\mu)}+\frac{\beta_3-\beta_1}{\hat{g}_2^2(\mu)}+\frac{\beta_1-\beta_2}{\hat{g}_3^2(\mu)} = 0 \,, \tag{6}$$

at any scale μ, where the β_i are the one-loop β-functions for the running gauge couplings. Solving for $\alpha_s(M_Z)$ and varying the inputs $\hat{g}_1(M_Z)$ and $\hat{g}_2(M_Z)$, one finds

$$\frac{\delta\alpha_s}{\alpha_s(M_Z)} \simeq -\,10.6\,\frac{\delta\hat{g}_1}{\hat{g}_1(M_Z)}+12.2\,\frac{\delta\hat{g}_2}{\hat{g}_2(M_Z)}\,. \tag{7}$$

This implies that an error in the determination of $\hat{g}_1(M_Z)$ or $\hat{g}_2(M_Z)$ of 1% percent can lead to an error in the evaluation of $\alpha_s(M_Z)$ of more than 10%.

The fact the RGE's can naturally amplify small errors means that one must determine the weak-scale threshold corrections to complete one-loop accuracy. Therefore, in our analysis, we compute $\hat{g}_1(M_Z)$ and $\hat{g}_2(M_Z)$ directly from the Fermi constant, $G_F = 1.16639\times10^{-5}$ GeV^{-2}, the Z-boson mass $M_Z = 91.187$ GeV, the electromagnetic coupling $\alpha_{\rm EM} = 1/137.036$, the top-quark mass m_t, and the parameters that describe the supersymmetric model. We include logarithmic and finite contributions, so our results are more accurate than those based on the match-and-run procedure. Note that we do not use the best-fit value of the standard-model weak mixing angle, $\sin^2\hat{\theta}_{\rm SM}$. This is because the standard-model value of $\sin^2\hat{\theta}_{\rm SM}$, extracted from a combined fit to the data, cannot be used in a supersymmetric analysis when finite corrections are important. (See also [2]. We did not include the complete gauge coupling thresholds in the preliminary results presented at the conference.)

We start our analysis by computing the electromagnetic coupling $\hat{\alpha}$ and the supersymmetric weak mixing angle $\hat{s}^2$ in the $\overline{DR}$ renormalization scheme,

$$\hat{\alpha} = \frac{\alpha_{\rm EM}}{1-\Delta\hat{\alpha}}\,, \qquad\qquad \hat{s}^2\hat{c}^2 = \frac{\pi\alpha_{\rm EM}}{\sqrt{2}G_F M_Z^2(1-\Delta\hat{r})}\,, \tag{8}$$

where $\hat{c}^2 = 1-\hat{s}^2$,

$$\Delta\hat{\alpha} = 0.0658\pm0.0007 - \frac{\alpha_{\rm EM}}{2\pi}\left\{-\frac{7}{4}\log\left(\frac{M_W}{M_Z}\right)+\frac{16}{9}\log\left(\frac{m_t}{M_Z}\right)+\frac{1}{3}\log\left(\frac{m_{H^+}}{M_Z}\right)\right. \tag{9}$$

$$\left.+\sum_{i=1}^{6}\frac{4}{9}\log\left(\frac{m_{\tilde{u}_i}}{M_Z}\right)+\sum_{i=1}^{6}\frac{1}{9}\log\left(\frac{m_{\tilde{d}_i}}{M_Z}\right)+\sum_{i=1}^{3}\frac{1}{3}\log\left(\frac{m_{\tilde{e}_i}}{M_Z}\right)+\sum_{i=1}^{2}\frac{4}{3}\log\left(\frac{m_{\chi_i^+}}{M_Z}\right)\right\},$$

and[3]

$$\Delta\hat{r} \;=\; \Delta\hat{\alpha} \;+\; \frac{\hat{\Pi}_W(0)}{M_W^2} \;-\; \frac{\hat{\Pi}_Z(M_Z)}{M_Z^2} \;+\; \text{vertex} \;+\; \text{box}\,. \tag{10}$$

Equation (9) includes the light quark contribution extracted from experimental data,[4] together with the leptonic contribution. It also contains the logarithms of the W-boson, top-quark, charged-Higgs, squark, slepton, and chargino masses. In eq. (10), the $\hat{\Pi}$ denote the real and transverse parts of the gauge boson self-energies, evaluated in the $\overline{DR}$ scheme. Equation (10) also includes the vertex and box contributions that renormalize the Fermi constant, as well as the leading higher-order m_t^4 and QCD standard-model corrections given in ref. 5. (For more details about our calculation see [6].)

From these results we find the weak gauge couplings $\hat{g}_1(M_Z)$ and $\hat{g}_2(M_Z)$ using the $\overline{DR}$ relations

$$\hat{g}_1(M_Z) \;=\; \sqrt{\frac{5}{3}}\frac{\hat{e}}{\hat{c}}\,, \qquad \hat{g}_2(M_Z) \;=\; \frac{\hat{e}}{\hat{s}}\,, \tag{11}$$

where $\hat{\alpha} = \hat{e}^2/4\pi$. These couplings serve as the weak-scale boundary conditions for the two-loop supersymmetric RGE's.[7] Note that they depend on the masses of the supersymmetric particles, which are determined self-consistently through the solution to the renormalization group equations. The $\overline{DR}$ values for $\hat{g}_1(M_Z)$ and $\hat{g}_2(M_Z)$ contain the full one-loop threshold corrections, including the logarithmic and finite contributions.

2.3. Mass Thresholds

Since the exact one-loop self-energies have been computed for all particles in the minimal supersymmetric standard model,[8] it is possible to include all the logarithmic and finite corrections in the threshold corrections. In this section we will discuss the leading corrections for the top- and bottom-quark masses.

The relation between the top-quark pole mass and the running $\overline{DR}$ mass is given by

$$m_t \;=\; \hat{m}_t(\mu) \;-\; \hat{\Sigma}_t(m_t)\,, \tag{12}$$

where $\hat{m}_t$ is the $\overline{DR}$ mass, and $\hat{\Sigma}_t$ is the top-quark self-energy, defined by $\hat{\Sigma}_t = \hat{\Sigma}_1 + m_t\hat{\Sigma}_\gamma$, and the quark self-energy is written $\hat{\Sigma}_1 + \hat{\Sigma}_\gamma \not{p} + \hat{\Sigma}_{\gamma 5}\not{p}\gamma_5 + \hat{\Sigma}_5\gamma_5$.

Equation (12) relates the running mass to the pole mass at *any* scale μ near m_t. In a unification analysis, this leads to a computational simplification because the RGE's can be run down to a single scale μ, and then all the threshold corrections applied simultaneously. For the case at hand, we take the scale $\mu = M_Z$.

The top quark threshold correction receives its most important contributions from gluon and gluino/squark loops. The correction from the gluon loop is well-known,*

$$m_t \;=\; \hat{m}_t(\mu)\,\left\{1 \;+\; \frac{\alpha_s}{3\pi}\left[5 + 6\log\left(\frac{\mu}{m_t}\right)\right]\right\}\,, \tag{13}$$

*Actually this correction is more commonly seen as $4\alpha_s/(3\pi)$, which is the result for $\mu = m_t$ in the $\overline{MS}$ scheme.

where $\hat{m}_t(\mu)$ is the running $\overline{DR}$ mass evaluated at the scale μ. For $\mu = m_t$, this correction is about 6%. The gluino/squark contribution is given by

$$\begin{aligned}\Delta m_t^{\tilde{g}\tilde{q}} &= -\frac{\alpha_s}{3\pi}\left\{\mathrm{Re}\left[B_1(m_t, m_{\tilde{g}}, m_{\tilde{t}_1}) + B_1(m_t, m_{\tilde{g}}, m_{\tilde{t}_2})\right]\right. \qquad (14)\\ &\left. - \frac{2m_{\tilde{g}}\,(A_t + \bar{\mu}\cot\beta)}{m_{\tilde{t}_1}^2 - m_{\tilde{t}_2}^2}\,\mathrm{Re}\left[B_0(m_t, m_{\tilde{g}}, m_{\tilde{t}_1}) - B_0(m_t, m_{\tilde{g}}, m_{\tilde{t}_2})\right]\right\}\hat{m}_t(\mu)\ ,\end{aligned}$$

where B_0 and B_1 are the two point functions

$$B_n(p, m_1, m_2) \;=\; -\int_0^1 dx\, x^n \log\left(\frac{(1-x)\,m_1^2 + x\,m_2^2 - x(1-x)\,p^2}{\mu^2}\right)\ , \qquad (15)$$

and $\bar{\mu}$ is the Higgs mass parameter. This expression contains a finite plus a logarithmic piece for $m_{\tilde{g}} > m_t$ and/or $m_{\tilde{t}} > m_t$. The logarithmic contribution, which can be larger than the gluon contribution (13), is given correctly by the match-and-run procedure. The finite piece is of order 1%.

The bottom-quark mass also receives a significant threshold correction, similar to (13), from the diagram with a gluon loop. For $\mu = M_Z$, the large logarithm $\log(m_b/M_Z)$ must be resummed to determine the pole mass. For the bottom quark, the correction from the gluino/squark diagram is also important; the second line in (14) becomes

$$\frac{\Delta m_b}{m_b} \;\simeq\; -\frac{2\alpha_s}{3\pi}\frac{\bar{\mu} m_{\tilde{g}}}{m_{\tilde{b}}^2}\,\tan\beta \qquad (16)$$

for large $\tan\beta$. This correction,[9] which is completely missed in the match-and-run procedure, can be as large as 50%. In the following, we choose $\bar{\mu} > 0$ to minimize the bottom-quark mass.

2.4. Yukawa Coupling Thresholds

To discuss supersymmetric unification, and in particular, bottom-tau Yukawa coupling unification, one needs to find the running $\overline{DR}$ Yukawa couplings. In this section we will show how to use the mass threshold corrections to find the running $\overline{DR}$ Yukawa couplings. The $\overline{DR}$ Yukawa couplings contain all the weak-scale threshold corrections.

Let us illustrate our procedure for the case of the top quark. As above, we find the running $\overline{DR}$ mass from the physical pole mass using (12). We then relate the $\overline{DR}$ mass to the $\overline{DR}$ vev and Yukawa coupling by the simple relation

$$\hat{m}_t(\mu) \;=\; \frac{1}{\sqrt{2}}\hat{\lambda}_t(\mu)\hat{v}_2(\mu)\ . \qquad (17)$$

The $\overline{DR}$ vev is determined from the gauge couplings and the Z-boson mass,

$$M_Z^2 \;=\; \frac{1}{4}\left(\hat{g}(\mu)^2 + \hat{g}'(\mu)^2\right)\hat{v}^2 \;-\; \hat{\Pi}_Z(M_Z)\ . \qquad (18)$$

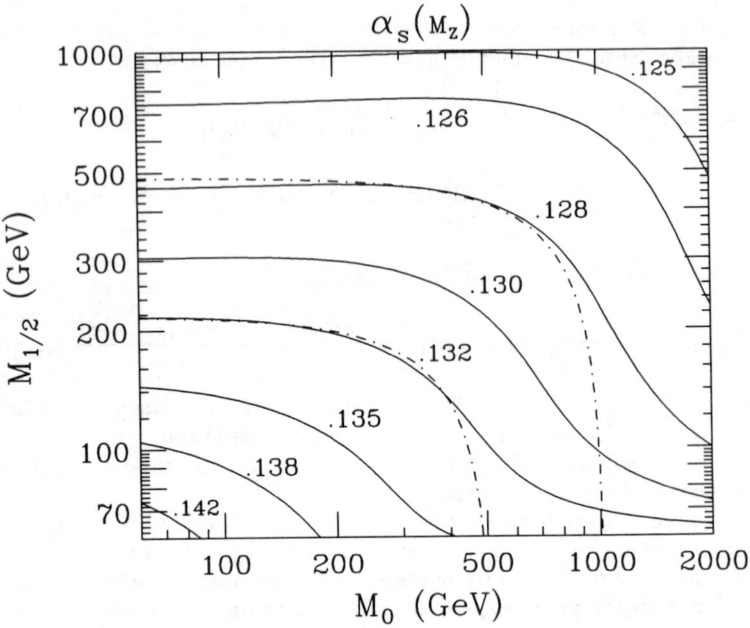

Fig. 3. Contours of $\alpha_s(M_Z)$ and one-loop light-squark masses in the M_0, $M_{1/2}$ plane, with $m_t = 170$ GeV, $\tan\beta = 2$ and $A_0 = 0$. The dot-dashed contours indicate squark masses of 500 and 1000 GeV. The squark masses obey the heuristic relation $m_{\tilde{q}}^2 \simeq M_0^2 + 4.5\, M_{1/2}^2$.

Combining these expressions, we find the threshold-corrected Yukawa coupling,

$$\hat{\lambda}_t(\mu) \;=\; \frac{\hat{g}_2(M_Z)}{\sqrt{2}\,\hat{c}}\,\frac{m_t}{M_Z}\left[1 + \frac{\hat{\Sigma}_t(m_t)}{m_t} - \frac{1}{2}\,\frac{\hat{\Pi}_Z(M_Z)}{M_Z^2}\right], \tag{19}$$

where m_t is the top-quark pole mass. Equation (19) converts the measured top-quark mass into a threshold-corrected Yukawa coupling at the scale μ.

Following this procedure, we can find all the threshold-corrected Yukawa couplings in terms of two-point functions. The great advantage of this approach is that the final formulae are much simpler than those obtained using three-point diagrams.

3. Numerical Analysis

3.1. Gauge Coupling Unification

Now that we have the full set of one-loop threshold corrections, we can carry out a consistent numerical evaluation of the two-loop supersymmetric RGE's. In what follows we will study correlations between the low-energy supersymmetric spectrum and the physics of the unification scale.

As a point of reference, we show in Figure 3 our results for $\alpha_s(M_Z)$ in the M_0, $M_{1/2}$ plane, with no unification-scale threshold corrections, for $m_t = 170$ GeV, $\tan\beta = 2$,

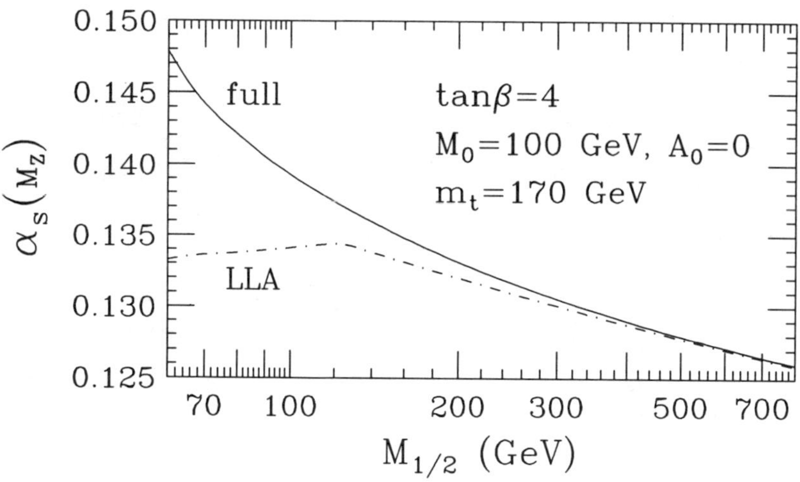

Fig. 4. The $\overline{MS}$ coupling $\alpha_s(M_Z)$ vs. $M_{1/2}$. The curve labeled LLA shows the result if we include only the logarithms of the supersymmetric masses (the leading logarithm approximation), while the solid line corresponds to the full result including all finite corrections.

and $A_0 = 0$. Comparing with the value from the particle data group,[10] $\alpha_s(M_Z) = 0.117 \pm 0.005$, we see that $\alpha_s(M_Z)$ is rather large.

The value of $\alpha_s(M_Z)$ is correlated with the low-energy supersymmetric spectrum through its dependence on M_0 and $M_{1/2}$. In Figure 3 we also show representative values of the light-squark masses in the M_0, $M_{1/2}$ plane. If, for naturalness, we require the squark masses to be below 1 TeV, we find the lower bound $\alpha_s(M_Z) > 0.127$ from Figure 3.

The values of $\alpha_s(M_Z)$ quoted here are larger than in many previous analyses for two reasons. First, during the past few years, the best-fit value for the top-quark mass has been increasing. This implies a decreasing central value for the standard-model weak mixing angle, which then leads to an increase in $\alpha_s(M_Z)$. Second, the finite corrections decrease s^2, which leads to a further increase in $\alpha_s(M_Z)$. The finite corrections are important in the region $M_{1/2} \lesssim 200$ GeV where $\alpha_s(M_Z)$ is appreciably larger than in the leading logarithmic approximation, as shown in Figure 4.

Of course, the value of $\alpha_s(M_Z)$ can be reduced by a unification-scale threshold correction with $\epsilon_g < 0$. This is illustrated in Figure 5, where we show the upper and lower bounds on ϵ_g necessary to obtain $\alpha_s(M_Z) = 0.117 \pm 0.01$. We find that ϵ_g of just -2% is sufficient for a wide range of supersymmetric masses. For squark masses of order 1 TeV, no unification thresholds are required.

3.2. Yukawa Coupling Unification

In typical SU(5) models, the bottom and tau Yukawa couplings unify at the scale $M_{\rm GUT}$. This leads to a prediction for the bottom-quark pole mass, m_b, as a function of

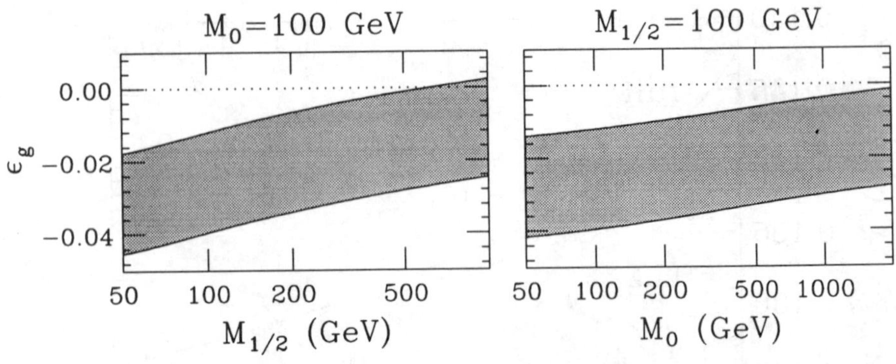

Fig. 5. The shaded regions indicate the allowed values of the gauge coupling threshold correction, ϵ_g, necessary to obtain $\alpha_s(M_Z) = 0.117 \pm 0.01$, with $m_t = 170$ GeV, $\tan\beta = 2$ and $A = 0$.

m_t and $\tan\beta$. For $m_t \simeq 170$ GeV and arbitrary $\tan\beta$, one typically finds a bottom-quark mass above the range experiment,[10] which we take to be $4.7 < m_b < 5.2$ GeV. There are two regions of $\tan\beta$ where m_b is smaller and successful bottom-tau unification is easier to achieve: $\tan\beta \lesssim 2$ and $\tan\beta \gtrsim 40$.

In the region $\tan\beta \lesssim 2$, the top-quark Yukawa coupling is large, approaching the perturbative limit. The large coupling drives electroweak symmetry breaking and significantly affects the RGE for the bottom-quark Yukawa. It pushes $\hat{\lambda}_b(M_Z)$ down and decreases the bottom-quark mass. In the region of large $\tan\beta$, both the top- and bottom-quark Yukawa couplings are large. They both make significant contributions to the RGE's. Indeed, for large $\tan\beta$, the large bottom-quark Yukawa actually drives *down* the value of the bottom-quark mass. Therefore we have the possibility of successful bottom-tau unification for small and large $\tan\beta$.

In Figure 6 we show m_b and $\alpha_s(M_Z)$ versus m_t, for small $\tan\beta$ and various values of M_0, and $M_{1/2}$, with no unification-scale thresholds. The figure corresponds to $\hat{\lambda}_t(M_{\rm GUT}) = 3$. Smaller values of $\hat{\lambda}_t(M_{\rm GUT})$ give rise to larger values of m_b, so the curves can be interpreted as lower limits on the bottom-quark mass (in the small $\tan\beta$ region). We see that m_b is large unless the squark masses are of order 1 TeV.

In Figure 7 we show m_b and $\alpha_s(M_Z)$ versus $\tan\beta$, in the large $\tan\beta$ region, for $m_t = 170$, and various values of M_0, and $M_{1/2}$, with no unification-scale thresholds. For each point, we choose the maximum value of $\tan\beta$ subject to the physical requirements $m_A > 22$ GeV, $m_{\tilde{\tau}} > 45$ GeV and $m_{\chi_1^+} > 47$ GeV, where A is the CP-odd Higgs boson, $\tilde{\tau}$ is the tau slepton, and χ_1^+ is the lightest chargino. From the figure we see that the smallest values of m_b occur for squark masses near 1 TeV.

As with $\alpha_s(M_Z)$, the picture is changed by unification-scale threshold corrections. To understand their effects, note the striking similarity between the m_b and $\alpha_s(M_Z)$ curves in Figures 6 and 7. This tells us that the value of m_b is tightly correlated with the value of $\alpha_s(M_Z)$. It leads us to expect that the gauge threshold correction will have an important effect on m_b.

This expectation is confirmed in Figure 8, where we show the band of unification-

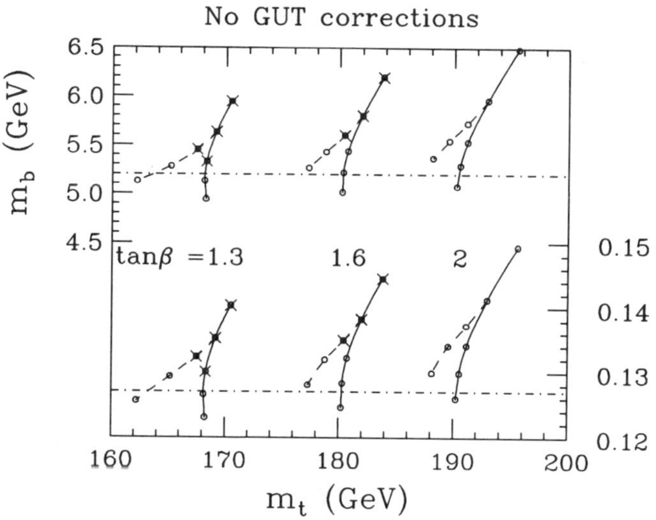

Fig. 6. The bottom-quark pole mass and $\alpha_s(M_Z)$ vs. m_t, with no unification-scale thresholds, for various values of $\tan\beta$, for $A_0 = 0$ and $\hat{\lambda}_t(M_{\rm GUT}) = 3$. The right (solid) leg in each pair of lines corresponds to $M_{1/2}$ varying from 60 to 1000 GeV, with M_0 fixed at 60 GeV. The left (dashed) leg corresponds to M_0 varying from 60 to 1000 GeV, with $M_{1/2} = 100$ GeV. On the solid lines the circles mark, from top to bottom, $M_{1/2} = 60$, 100, 200, 400, and 1000 GeV, and on the dashed lines the circles mark $M_0 = 60$, 200, 400, and 1000 GeV. The lowest point on each left leg and the second-to-lowest point on each right leg corresponds to $m_{\tilde{q}} \simeq 1$ TeV. The ×'s mark points with one-loop Higgs mass $m_h < 60$ GeV.

scale thresholds ϵ_b that are necessary to bring m_b into the preferred range. For the figure, we first choose ϵ_g to fix $\alpha_s(M_Z)$ at its central value, $\alpha_s(M_Z) = 0.117$. We then vary ϵ_b to obtain $4.7 < m_b < 5.2$. The two bands of ϵ_b correspond to the regions of small and large $\tan\beta$. In the first region, we set $\tan\beta = 2.5$, and in the second, $\tan\beta = 30$. The figure indicates that Yukawa threshold corrections of less than 10% are sufficient to ensure successful unification, provided the gauge thresholds are such that $\alpha_s(M_Z)$ agrees with experiment.

4. Conclusions

In this talk we discussed the complete one-loop weak-scale threshold corrections in the minimal supersymmetric standard model. We illustrated how they affect gauge and Yukawa unification, with and without unification-scale threshold corrections. In the absence of such corrections, we find that $\alpha_s(M_Z)$ and m_b are large unless the squark masses are of order one TeV. With small threshold corrections, we find acceptable values for $\alpha_s(M_Z)$ and m_b for heavy or light supersymmetric spectra.

We would like to thank R. Zhang for collaboration during the early stages of this work, and S. Pokorski for helpful discussions. J.B. would like to thank Z. Płuciennik for extensive discussions during the conference. This work was supported by the U.S.

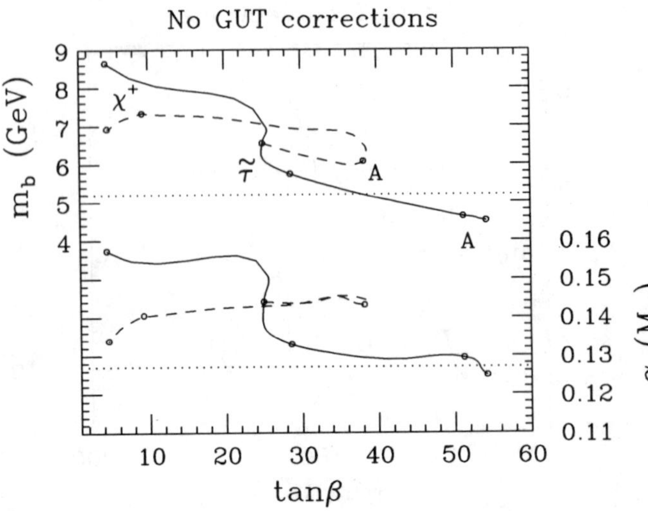

Fig. 7. The bottom-quark pole mass and $\alpha_s(M_Z)$ vs. the maximum $\tan\beta$, with no unification-scale thresholds, for $m_t = 170$ GeV and $A_0 = 0$. The solid line corresponds to $M_{1/2}$ varying from 60 to 1000 GeV, with M_0 fixed at 60 GeV. The dashed line corresponds to M_0 varying from 60 to 1000 GeV, with $M_{1/2} = 100$ GeV. On the solid lines the circles mark, from top to bottom, $M_{1/2} = 60$, 100, 200, 400, and 1000 GeV, and on the dashed lines the circles mark $M_0 = 60$, 200, 400, and 1000 GeV. The maximum value of $\tan\beta$ is determined by the experimental limits on the masses of the A, $\tilde{\tau}$ and the χ_1^+, as indicated.

National Science Foundation under grant NSF-PHY-9404057.

5. References

1. H. Georgi, H.R. Quinn and S. Weinberg, *Phys. Rev. Lett.* **33** (1974) 451; S. Dimopoulos, S. Raby and F. Wilczek, *Phys. Rev.* **D24** (1981) 1681; L. Ibanez and G. Ross, *Phys. Lett.* **B105** (1981) 439; U. Amaldi, A. Bohm, L. Durkin, P. Langacker, A. Mann, W. Marciano, A. Sirlin and H. Williams, *Phys. Rev.* **D36** (1987) 1385; J. Ellis, S. Kelley and D. Nanopoulos, *Phys. Lett.* **B249** (1991) 441; **B260** (1991) 131; P. Langacker and M. Luo, *Phys. Rev.* **D44** (1991) 817; U. Amaldi, W. de Boer and H. Furstenau, *Phys. Lett.* **B260** (1991) 447; F. Anslemo, L. Cifarelli, A. Peterman and A. Zichichi, *Nuovo Cimento* **104A** (1991) 1817.
2. P. Chankowski, Z. Płuciennik and S. Pokorski, Warsaw preprint IFT–94/19. See also the contributions of these authors to these proceedings.
3. G. Degrassi, S. Fanchiotti and A. Sirlin, *Nucl. Phys.* **B351** (1991) 49, and references therein.

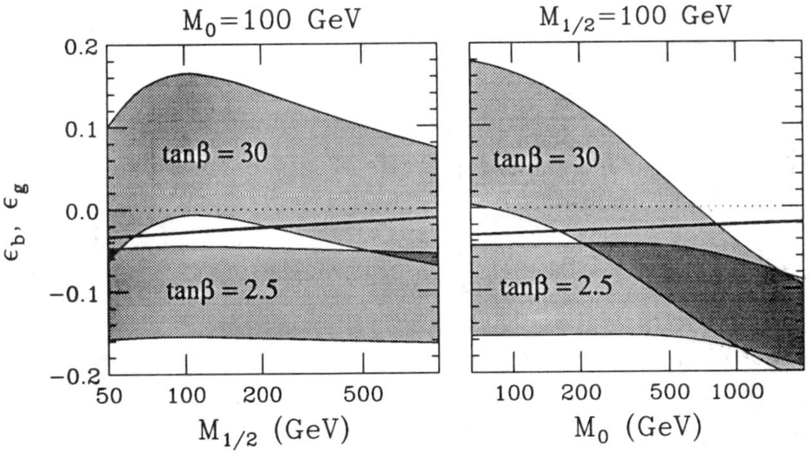

Fig. 8. Values of the bottom-quark Yukawa coupling threshold correction, ϵ_b, necessary to obtain $4.7 < m_b < 5.2$, assuming ϵ_g is adjusted to give $\alpha_s(M_Z) = 0.117$. The solid line marks ϵ_g, while the two bands indicate ϵ_b for different values of $\tan\beta$, for $A_0 = 0$ and $m_t = 180$ GeV.

4. We use the result of the most recent analysis of Jegerlehner as quoted in B.A. Kniehl, preprint KEK-TH-412 (1994).
5. S. Fanchiotti, B. Kniehl and A. Sirlin, *Phys. Rev.* **D48** (1993) 307.
6. J. Bagger, K. Matchev, D. Pierce and R. Zhang, to appear.
7. M.E. Machacek and M.T. Vaughn, *Nucl. Phys.* **B222** (1983) 83; *ibid.* **B236** (1984) 221; *ibid.* **B249** (1985) 70;
I. Jack, *Phys. Lett.* **B147** (1984) 405;
S. Martin, M. Vaughn, Northeastern preprint NUB-3081-93-TH (1993);
Y. Yamada, *Phys. Rev.* **D50** (1994) 3537.
8. The MSSM gauge and Higgs sector self energies can be found in P.H. Chankowski, S. Pokorski and J. Rosiek, *Nucl. Phys.* **B423** (1994) 437.
The gaugino/Higgsino mass corrections are in D. Pierce and A. Papadopoulos, *Phys. Rev.* **D50** (1994) 565; *Nucl. Phys.* **B430** (1994) 278.
The threshold functions for the entire MSSM spectrum are in ref. 6.
9. L.J. Hall, R. Rattazzi and U. Sarid, *Phys. Rev.* **D50** (1994) 7048;
M. Carena, M. Olechowski, S. Pokorski and C.E.M. Wagner, *Nucl. Phys.* **B426** (1994) 269;
R. Hempfling, *Z. Phys.* **C63** (1994) 309.
10. Particle Data Group Review of Particle Properties, *Phys. Rev.* **D50** (1994) 1.

Fermion Masses
and
SO(10) SUSY GUTs

Stuart Raby
*Department of Physics, The Ohio State University, 174 W. 18th Ave.
Columbus, OH 43210*

ABSTRACT

In this talk I summarize published work on a systematic operator analysis for fermion masses in a class of effective supersymmetric SO(10) GUTs[1] [a] Given a minimal set of four operators at M_G, we have just 6 parameters in the fermion mass matrices. We thus make 8 predictions for the 14 low energy observables (9 quark and charged lepton masses, 4 quark mixing angles and $\tan\beta$). Several models, i.e. particular sets of dominant operators, are in quantitative agreement with the low energy data. In the second half of the talk I discuss the necessary ingredients for an SO(10) GUT valid below the Planck (or string) scale which reproduces one of our models. [b] This complete GUT should still be interpreted as an effective field theory, i.e. perhaps the low energy limit of a string theory.

[a]This work is in collaboration with G. Anderson, S. Dimopoulos, L.J. Hall, and G. Starkman.

[b]These are preliminary results of work in progress with Lawrence Hall.

1. Introduction

The Standard Model[SM] provides an excellent description of Nature. Myriads of experimental tests have to date found no inconsistency.

Eighteen phenomenological parameters in the SM are necessary to fit all the low energy data[LED]. These parameters are not equally well known. $\alpha, \sin^2(\theta_W)$, m_e, m_μ, m_τ and M_Z are all known to better than 1% accuracy. On the other-hand, m_c, m_b, V_{us} are known to between 1% and 5% accuracy, and $\alpha_s(M_Z), m_u, m_d$, $m_s, m_t, V_{cb}, V_{ub}/V_{cb}, m_{Higgs}$ and the Jarlskog invariant measure of CP violation J are not known to better than 10% accuracy. One of the main goals of the experimental high energy physics program in the next 5 to 10 years will be to reduce these uncertainties. In addition, theoretical advances in heavy quark physics and lattice gauge calculations will reduce the theoretical uncertainties inherent in these parameters. Already the theoretical uncertainties in the determination of V_{cb} from inclusive B decays are thought to be as low as 5%[2]. Moreover, lattice calculations are providing additional determinations of $\alpha_s(M_Z)$ and heavy quark masses[3].

Accurate knowledge of these 18 parameters is important. They are clearly not a random set of numbers. There are distinct patterns which can, if we are fortunate, guide us towards a fundamental theory which predicts some (if not all) of these parameters. Conversely these 18 parameters are the LED which will test any such theory. Note, that 13 of these parameters are in the fermion sector. So, if we are to make progress, we must necessarily attack the problem of fermion masses.

In the program discussed in this talk, we define a procedure for finding the dominant operator set reproducing the low energy data. In the minimal operator sets we have just six parameters in the fermion mass matrices. We use the six best known low energy parameters as input to fix these six unknowns and then predict the rest. These theories are supersymmetric[SUSY] SO(10) grand unified theories[GUTs][4]. In the next two sections I want to briefly motivate these choices.

2. Why SUSY GUTs?

Looking back at the history of particle physics, it is clear that much of our understanding comes from using symmetries. This is because, even without a complete understanding of the dynamics, symmetries can be used to relate different observables. Here too we want to correlate the known low energy data, the three gauge couplings and the fermion masses and mixing angles. We want to describe these 16 parameters in terms of fewer fundamental numbers. GUTs allow us to do just that. In fact using this symmetry we can express the low energy data as follows –

$$Observable = Input\ parameters \times Boundary\ condition\ at\ M_G \times RG\ factor$$

where the *observable* is the particular low energy data we want to calculate, the *input parameters* is the set of fundamental parameters defined at the GUT scale and the last factor takes into account the renormalization group running of the experimental observable from M_G to the low energy scale. The grand unified symmetry SU(5) (or SO(10), E(6) etc.) determines the *boundary conditions at* M_G[5]. These are given in terms of Clebsch-Gordan coefficients relating different observables. Of course, these relations are only valid at the GUT scale and the RG equations are necessary to relate them to experiment. It is through the RG equations that supersymmetry enters. We will assume that only the states present in the minimal supersymmetric standard model[MSSM] are in the theory below M_G. We assume this because it works. Consider the GUT expression for the gauge couplings –

$$\alpha_i(M_Z) = \alpha_G\ R_i(\alpha_G, \frac{M_G}{M_Z})$$

where the boundary condition is $R_i(\alpha_G, 1) \equiv 1 + \cdots$. The input parameters are α_G and M_G and the Clebschs in this case are all one. Thus we obtain the well known result that given α and $\sin^2\theta_W$ measured at M_Z we predict the value for $\alpha_s(M_Z)$[6](For recent analysis of the data, see [7]). Note that SUSY without GUTs makes no prediction, since there is no symmetry to specify the boundary conditions and GUTs without SUSY makes the wrong prediction.

I should also point out that the SO(10) operator analysis for fermion masses that I am about to describe is not new. This analysis was carried out 10 years ago with the result that the favored value of the top quark mass was about 35 GeV[8].

3. Why SO(10)?

There are two reasons for using SO(10).

1. It is the smallest group in which all the fermions in one family fit into one irreducible representation, i.e. the **16**. Only one additional state needs to be added to complete the multiplet and that is a right-handed neutrino. In larger gauge groups, more as yet unobserved states must be introduced to obtain complete multiplets. Thus we take $\mathbf{16}_i \supset \{U_i, D_i, E_i, \nu_i\}, i = 1, 2, 3$ for the 3 families with the third family taken to be the heaviest. *Since SO(10) Clebschs can now relate U, D, E and ν mass matrices, we can in principle reduce the number of fundamental parameters in the fermion sector of the theory.* We return to this point below.

2. In any SUSY theory there are necessarily two higgs doublets – H_u and H_d. Both these states fit into the **10** of SO(10) and thus their couplings to up and down type fermions are also given by a Clebsch. There are however six additional states in the **10** which transform as a $\mathbf{3} + \overline{\mathbf{3}}$ under color. These states contribute to proton decay and must thus be heavy. The problem of giving these color triplet states large mass of order M_G while keeping the doublets light is sometimes called the second gauge hierarchy problem. This problem has a natural solution in SO(10) which we discuss later[9].

Note that the gauge group SO(10) has to be spontaneously broken to the gauge group of the SM – $SU(3) \times SU(2) \times U(1)$. This GUT scale breaking can be accomplished by a set of states including $\{\ \mathbf{45}, \mathbf{16}, \overline{\mathbf{16}}, \cdots\}$. The **45**(the adjoint representation) enters into our construction of effective fermion mass operators, thus I will discuss it in more detail in the next section.

I promised to return to the possibility of reducing the number of fundamental parameters in the fermion sector of the theory. Recall that there are 13 such parameters. Using symmetry arguments we can now express the matrices $\mathbf{D}, \mathbf{E}$, and ν in terms of one complex 3x3 matrix, U. Unfortunately, this is not sufficient to solve our problem. There are 18 arbitrary parameters in this one matrix. In order to reduce the number of fundamental parameters we must have zeros in this matrix. We thus need new *family symmetries* to enforce these zeros.

4. The Big Picture

Let us consider the big picture(see Fig. 1). Our low energy observer measures the physics at the electroweak scale and perhaps an order of magnitude above. Once the SUSY threshold is crossed we have direct access to the effective theory at M_G, the scale where the 3 gauge couplings meet. Of course the GUT scale $M_G \sim 10^{16}$ GeV is still one or two orders of magnitude below some more fundamental scale such as the Planck or string scales (which we shall refer to as M). Between M and M_G there may be some substructure. In fact, we may be able to infer this substructure by studying fermion masses.

In our analysis we assume that the theory below the scale M is described by a SUSY SO(10) GUT. Between M_G and M, at a scale v_{10}, we assume that the gauge group SO(10) is broken spontaneously to SU(5). This can occur due to the vacuum expectation value of an adjoint scalar in the X direction (i.e. corresponding to the U(1) which commutes with SU(5)) and the expectation values of a 16 and a $\overline{16}$(denoted by Ψ and $\overline{\Psi}$ respectively). Then SU(5) is broken at the scale $v_5 = M_G$ to the SM gauge group. This latter breaking can be done by different adjoints (45) in the Y, B-L or T_{3R} directions.

Why consider 4 particular breaking directions for the 45 and no others? The X and Y directions are orthogonal and span the two dimensional space of U(1) subgroups of SO(10) which commute with the SM. B-L and T_{3R} are also orthogonal and they span the same subspace. Nevertheless we consider these four possible breaking directions and these are the *only directions* which will enter the effective operators for fermion masses. Why not allow the X and Y directions or any continous rotation of them in this 2d subspace of U(1) directions . The answer is that there are good dynamical arguments for assuming that these and only these directions are important. The X direction breaks SO(10) to an intermediate SU(5) subgroup and it is reasonable to assume that this occurs at a scale $v_{10} \geq v_5$. Whether v_{10} is greater than v_5 or equal will be determined by the LED. The B-L direction is required for other reasons. Recall the color triplet higgs in the 10 which must necessarily receive large mass. As shown by Dimopoulos and Wilczek[9], this doublet-triplet splitting can naturally occur by introducing a 10 45 10 type coupling in the superspace potential. Note that the higgs triplets carry non-vanishing B-L charge while the doublets carry zero charge. Thus when the 45 gets a vacuum expectation value[vev] in the B-L direction it will give mass to the color triplet higgs at v_5 and leave the doublets massless. Thus in any SO(10) model which solves this second hierarchy problem, there must be a 45 pointing in the B-L direction. We thus allow for all 4 possible breaking vevs X, Y, B-L and T_{3R}. Furthermore we believe this choice is "natural" since we know how to construct theories which have these directions as vacua without having to tune any parameters.

Our fermion mass operators have dimension ≥ 4. *From where would these higher dimension operators come?* Note that by measuring the LED we directly probe the physics in some effective theory at M_G. This effective theory can, and likely will, include operators with dimension greater than 4. Consider, for example, our big picture looking down from above. String theories are very fundamental. They can in principle describe physics at all scales. Given a particular string vacuum, one can obtain an effective field theory valid below the string scale M. The massless sector can include the gauge bosons of SO(10) with scalars in the 10, 45 or even 54 dimensional representations. In addition, we require 3 families of fermions in the 16. Of course, in a string context when one says that there are 3 families of fermions what is typically meant is that there are 3 more 16s than $\overline{16}$s. The extra 16 + 16 pairs are assumed to get mass at a scale $\geq M_G$, since there is no symmetry which prevents this. When these states are integrated out in order to define the effective field theory valid below M_G they will typically generate higher dimension operators.

Consider the tree diagram in Fig. 2. The state $\mathbf{16_2}$ contains the second generation of fermions. It has off-diagonal couplings to heavy fermions $\mathbf{16_i}, \overline{\mathbf{16}}_\mathbf{i}, \mathbf{i} = \mathbf{1}, \mathbf{2}$. If, for example, the vev $\mathbf{45_X} > M_G$ then it will be responsible for the dominant contribution to the mass of the states labelled 1 and 2. These two states however mix by smaller off-diagonal mass terms given by the vev $\mathbf{45_{B-L}}$ or the singlet vev (or mass term) denoted by M_G. When these heavy states are integrated out one generates to leading order the operator $\mathbf{O_{22}} = 16_2 10 \frac{M_G 45_{B-L}}{(45_X)^2} 16_2$ plus calculable corrections of order $(v_5/v_{10})^2$. *It is operators of this type (which can be obtained by implicitly integrating out heavy 16's) which we use to define our operator basis for fermion masses in the effective theory at* M_G.

5. Operator Basis for Fermion Masses at M_G

Let us now consider the general **operator basis for fermion masses.** We include operators of the form

$$\mathbf{O_{ij}} = \mathbf{16_i} \; (\cdots)_n \; \mathbf{10} \; (\cdots)_m \; \mathbf{16_j}$$

where

$$(\cdots)_n = \frac{M_G^k \; 45_{k+1} \cdots 45_n}{M_P^l \; 45_X^{n-l}}$$

and the 45 vevs in the numerator can be in any of the 4 directions, $\mathbf{X}, \mathbf{Y}, \mathbf{B-L}, \mathbf{T_{3R}}$ discussed earlier.

It is trivial to evaluate the Clebsch-Gordon coefficients associated with any particular operator since the matrices $X, Y, B-L, T_{3R}$ are diagonal. Their eigenvalues on the fermion states are given in Table 1.

6. Dynamic Principles

Now consider the dynamical principles which guide us towards a theory of fermion masses.

0. At zeroth order, we work in the context of a SUSY GUT with the MSSM below M_G.

1. We use SO(10) as the GUT symmetry with three families of fermions $\{16_i \;\; i = 1, 2, 3\}$ and the minimal electroweak Higgs content in one 10. SO(10) symmetry relations allow us to reduce the number of fundamental parameters.

2. We assume that there are also family symmetries which enforce zeros of the mass matrix, although we will not specify these symmetries at this time. As we will make clear in section 12, these symmetries will be realized at the level of the fundamental theory defined below M.

Table 1. Quantum numbers of the four 45 vevs on fermion states. Note, if u denotes a left-handed up quark, then $\bar{u}$ denotes a left-handed charge conjugate up quark.

	X	Y	B − L	T_{3R}
u	1	1/3	1	0
$\bar{u}$	1	-4/3	-1	-1/2
d	1	1/3	1	0
$\bar{d}$	-3	2/3	-1	1/2
e	-3	-1	-3	0
$\bar{e}$	1	2	3	1/2
ν	-3	-1	-3	0
$\bar{\nu}$	5	0	3	-1/2

3. Only the third generation obtains mass via a dimension 4 operator. The fermionic sector of the Lagrangian thus contains the term $A\ O_{33} \equiv A\ \ 16_3\ 10\ 16_3$. This term gives mass to t, b and τ. It results in the symmetry relation — $\lambda_t = \lambda_b = \lambda_\tau \equiv A$ at M_G. This relation has been studied before by Ananthanarayan, Lazarides and Shafi[10] and using m_b and m_τ as input it leads to reasonable results for m_t and $\tan\beta$.

4. All other masses come from operators with dimension > 4. As a consequence, the family hierarchy is related to the ratio of scales above M_G.

5. [**Predictivity requirement**] We demand the minimal set of effective fermion mass operators at M_G consistent with the **LED**.

7. Systematic Search

Our goal is to find the *minimal* set of fermion mass operators consistent with the LED. With any given operator set one can evaluate the fermion mass matrices for up and down quarks and charged leptons. One obtains relations between mixing angles and ratios of fermion masses which can be compared with the data. It is easy to show, however, without any detailed calculations that the minimal operator set consistent with the LED is given by

$$O_{33} + O_{23} + O_{22} + O_{12} \quad - - - \text{“22” texture}$$

or

$$O_{33} + O_{23} + O'_{23} + O_{12} \quad - - - \text{“23′” texture}$$

It is clear that at least 3 operators are needed to give non-vanishing and unequal masses to all charged fermions, i.e. $det(m_a) \neq 0$ for $a = u, d, e$. That the operators must be in the [33, 23 and 12] slots is not as obvious but is not difficult to show. It is then easy to show that 4 operators are required in order to have CP violation. This is because, with only 3 SO(10) invariant operators, we can redefine the phases of the three 16s of fermions to remove the three arbitrary phases. With one more operator, there is one additional phase which cannot be removed. A corollary of this observation is that this minimal operator set results in just 5 arbitrary parameters in the Yukawa matrices of all fermions, 4 magnitudes and one phase.* This is the minimal parameter set which can be obtained without solving the remaining problems of the fermion mass hierarchy, one overall real mixing angle and a CP violating phase. We should point out however that the problem of understanding the fermion mass hierarchy and mixing has been rephrased as the problem of understanding the hierarchy of scales above M_G.

From now on I will just consider models with "22" texture. This is because they can reproduce the observed hierarchy of fermion masses without fine-tuning.† Models with "22" texture give the following Yukawa matrices at M_G (with electroweak doublet fields on the right) –

$$\lambda_{\mathbf{a}} = \begin{pmatrix} 0 & z'_a\, C & 0 \\ z_a\, C & y_a\, E\, e^{i\phi} & x'_a\, B \\ 0 & x_a\, B & A \end{pmatrix}$$

with the subscript $a = \{u, d, e\}$. The constants $x_a, x'_a, y_a, z_a, z'_a$ are Clebschs which can be determined once the 3 operators (O_{23}, O_{22}, O_{12}) are specified. Recall, we have taken $O_{33} = A\, 16_3\, 10\, 16_3$, which is why the Clebsch in the 33 term is independent of a. Finally, combining the Yukawa matrices with the Higgs vevs to find the fermion mass matrices we have 6 arbitrary parameters given by A, B, C, E, ϕ and $\tan\beta$ describing 14 observables. We thus obtain 8 predictions. We shall use the best known parameters, $m_e, m_\mu, m_\tau, m_c, m_b, |V_{cd}|$, as input to fix the 6 unknowns. We then predict the values of $m_u, m_d, m_s, m_t, \tan\beta, |V_{cb}|, |V_{ub}|$ and J.

Note: since the predictions are correlated, our analysis would be much improved if we minimized some χ^2 distribution and obtained a best fit to the data. Unfortunately this has not yet been done. In the paper however we do include some tables (see for example Table 4 in this talk) which give all the predictions for a particular set of input parameters.

8. Results

The results for the 3rd generation are given in Fig. 3. Note that since the parameter A is much bigger than the others we can essentially treat the 3rd generation independently. The small corrections, of order $(B/A)^2$, are however included in the

*This is two fewer parameters than was necessary in our previous analysis (see [11])

†For more details on this point, see section 9 below or refer to [1].

complete analysis. We find the pole mass for the top quark $M_t = 180 \pm 15$ GeV and $\tan\beta = 56 \pm 6$ where the uncertainties result from variations of our input values of the $\overline{MS}$ running mass $m_b(m_b) = 4.25 \pm 0.15$ and $\alpha_s(M_Z)$ taking values .110 – .126. We used two loop RG equations for the MSSM from M_G to M_{SUSY}; introduced a universal SUSY threshold at $M_{SUSY} = 180$ GeV with 3 loop QCD and 2 loop QED RG equations below M_{SUSY}. The variation in the value of α_s was included to indicate the sensitivity of our results to threshold corrections which are necessarily present at the weak and GUT scales. In particular, we chose to vary $\alpha_s(M_Z)$ by letting $\alpha_3(M_G)$ take on slightly different values than $\alpha_1(M_G) = \alpha_2(M_G) = \alpha_G$.

The following set of operators passed a straightforward but coarse grained search discussed in detail in the paper[1]. They include the diagonal dimension four coupling of the third generation –

$$O_{33} = \quad 16_3\ 10\ 16_3 \tag{1}$$

The six possible O_{22} operators –

$$O_{22} = \tag{2}$$
$$\begin{array}{lr} 16_2\ \frac{45_X}{M}\ 10\ \frac{45_{B-L}}{45_X}\ 16_2 & (a) \\ 16_2\ \frac{M_G}{45_X}\ 10\ \frac{45_{B-L}}{M}\ 16_2 & (b) \\ 16_2\ \frac{45_X}{M}\ 10\ \frac{45_{B-L}}{M}\ 16_2 & (c) \\ 16_2\ 10\ \frac{45_{B-L}}{45_X}\ 16_2 & (d) \\ 16_2\ 10\ \frac{45_X\ 45_{B-L}}{M^2}\ 16_2 & (e) \\ 16_2\ 10\ \frac{45_{B-L}\ M_G}{45_X^2}\ 16_2 & (f) \end{array}$$

Note: in all cases the Clebschs y_i (defined by O_{22} above) satisfy

$$y_u \cdot y_d : y_e = 0 : 1 : 3.$$

This is the form familiar from the Georgi-Jarlskog texture[12]. Thus all six of these operators lead to *identical* low energy predictions.

Finally there is a unique operator O_{12} consistent with the LED –

$$O_{12} - \quad 16_1 \left(\frac{45_X}{M}\right)^3\ 10 \left(\frac{45_X}{M}\right)^3 16_2 \tag{3}$$

The operator O_{23} determines the KM element V_{cb} by the relation –

$$V_{cb} = \chi \sqrt{\frac{m_c}{m_t}} \times (RG factors)$$

where the Clebsch combination χ is given by

$$\chi \equiv \frac{|x_u - x_d|}{\sqrt{|x_u x'_u|}}$$

m_c is input, m_t has already been determined and the renormalization group[RG] factors are calculable. Demanding the experimental constraint $V_{cb} < .054$ we find the constraint $\chi < 1$. A search of all operators of dimension 5 and 6 results in the 9 operators given below. Note that there only three different values of $\chi = 2/3,\ 5/6,\ 8/9$

$$O_{23} = \tag{4}$$

$\chi = 2/3$

$$(1)\quad 16_2 \frac{45_Y}{M} 10 \frac{M_G}{45_X} 16_3$$
$$(2)\quad 16_2 \frac{45_Y}{M} 10 \frac{45_{B-L}}{45_X} 16_3$$
$$(3)\quad 16_2 \frac{45_Y}{45_X} 10 \frac{M_G}{45_X} 16_3$$
$$(4)\quad 16_2 \frac{45_Y}{45_X} 10 \frac{45_{B-L}}{45_X} 16_3$$

$\chi = 5/6$

$$(5)\quad 16_2 \frac{45_Y}{M} 10 \frac{45_Y}{45_X} 16_3$$
$$(6)\quad 16_2 \frac{45_Y}{45_X} 10 \frac{45_Y}{45_X} 16_3$$

$\chi = 8/9$

$$(7)\quad 16_2\ 10 \frac{M_G^2}{45_X^2} 16_3$$
$$(8)\quad 16_2\ 10 \frac{45_{B-L} M_G}{45_X^2} 16_3$$
$$(9)\quad 16_2\ 10 \frac{45_{B-L}^2}{45_X^2} 16_3$$

We label the operators (1) - (9), and we use these numbers also to denote the corresponding models. *Note, all the operators have the vev* 45_X *in the denominator. This can only occur if* $v_{10} > M_G$.

At this point, there are no more simple criteria to reduce the number of models further. We have thus performed a numerical RG analysis on each of the 9 models (represented by the 9 distinct operators O_{23} with their calculable Clebschs $x_a, x'_a, a = u, d, e$ along with the unique set of Clebschs determined by the operators O_{33}, O_{22} and O_{12}). We then iteratively fit the 6 arbitrary parameters to the six low energy inputs and evaluate the predictions for each model as a function of the input parameters. The results of this analysis are given in Figs. (4 - 10).

Let me make a few comments. Light quark masses (u,d,s) are $\overline{MS}$ masses evaluated at 1 GeV while heavy quark masses (c,b) are evaluated at (m_c, m_b) respectively. Finally, the top quark mass in Fig. 3 is the pole mass. Figs. 4 and 5 are self evident. In Fig. 6, we show the correlations for two of our predictions. The ellipse in the m_s/m_d vs. m_u/m_d plane is the allowed region from chiral Lagrangian analysis[13]. One sees that we favor lower values of $\alpha_s(M_Z)$. For each fixed value of $\alpha_s(M_Z)$, there are 5 vertical line segments in the V_{cb} vs. m_u/m_d plane. Each vertical line segment represents a range of values for m_c (with m_c increasing moving up) and the different line segments represent different values of m_b (with m_b increasing moving to the left). In Figure 9 we test our agreement with the observed CP violation in the K system. The experimentally determined value of $\epsilon_K = 2.26 \times 10^{-3}$. Theoretically it is given by an expression of the form $B_K \times \{m_t, V_{ts}, \cdots\}$. B_K is the so-called Bag constant which has been determined by lattice calculations to be in the range $B_K = .7 \pm .2$[14]. In Fig. 9 we have used our predictions for fermion masses and mixing angles as input, along with the experimental value for ϵ_K, and fixed B_K for the 9 different models. One sees that model 4 is inconsistent with the lattice data. In Fig. 10 we present the predictions for each model, for the CP violating angles which can be measured in B decays. The interior of the "whale" is the range of parameters consistent with the SM found by Nir and Sarid[15] and the error bars represent the accuracy expected from a B factory.

Note that model 4 appears to give too little CP violation and model 9 has uncomfortably large values of V_{cb}. Thus these models are presently disfavored by the data. I will thus focus on model 6 in the second part of this talk.

9. Summary

We have performed a systematic operator analysis of fermion masses in an effective SUSY SO(10) GUT. We use the LED to lead us to the theory. Presently there are 3 models (models 4, 6 & 9) with "22" texture which agree best with the LED, although as mentioned above model 6 is favored. In all cases we used the values of α and $\sin^2\theta_W$ (modulo threshold corrections) to fix $\alpha_s(M_Z)$.

Table 2 shows the virtue of the "22" texture. In the first column are the four operators. In the 2nd and 3rd columns are the parameters in the mass matrix relevant for that particular operator and the input parameters which are used to fix these parameters. Finally the 4th column contains the predictions obtained at each level. One sees that each family is most sensitive to a different operator‡

Consider the theoretical uncertainties inherent in our analysis.

1. The experimentally determined values of m_b, m_c, and $\alpha_s(M_Z)$ are all subject to strong interaction uncertainties of QCD. In addition, the predicted value of $\alpha_s(M_Z)$ from GUTs is subject to threshold corrections at M_W which can only be calculated once the SUSY spectrum is known and at M_G which requires

‡This property is not true of "23" textures.

Table 2. Virtue of "22" texture.

Operator	Parameters	Input	Predictions
O_{33}	$\tan\beta$ A	b τ	t $\tan\beta$
O_{23}	B	c	V_{cb}
O_{22}	E	μ	s
O_{12}	C ϕ	e V_{us}	u d $\frac{V_{ub}}{V_{cb}}$ J

knowledge of the theory above M_G. We have included these uncertainties (albeit crudely) explicitly in our analysis.

2. In the large $\tan\beta$ regime in which we work there may be large SUSY loop corrections which will affect our results. The finite corrections to the b and τ Yukawa couplings have been evaluated[16,17]. They depend on ratios of soft SUSY breaking parameters and are significant in certain regions of parameter space.[§] In particular it has been shown that the top quark mass can be reduced by as much as 30%. Note that although the prediction of Fig. 3 may no longer be valid, there is still necessarily a prediction for the top quark mass. It is now however sensitive to the details of the sparticle spectrum and to the process of radiative electroweak symmetry breaking[18]. This means that the observed top quark mass can now be used to set limits on the sparticle spectrum. This analysis has not been done. Moreover, there are also similar corrections to the Yukawa couplings for the s and d quarks and for e and μ. These corrections are expected to affect the predictions for V_{cb}, m_s, m_u, m_d. It will be interesting to see the results of this analysis.

3. The top, bottom and τ Yukawa couplings can receive threshold corrections at M_G. We have not studied the sensitivity to these corrections.

4. Other operators could in principle be added to our effective theory at M_G. They might have a dynamical origin. We have assumed that, if there, they are subdominant. Two different origins for these operators can be imagined. The first is field theoretic. The operators we use would only be the leading terms in a power series expansion when defining an effective theory at M_G by integrating out heavier states. The corrections to these operators are expected to be about 10%. We may also be sensitive to what has commonly been referred to as Planck slop[19], operators suppressed by some power of the Planck (or string) scale M. In fact the operator O_{12} may be thought of as such. The question is why aren't our results for the first and perhaps the second generation, hopelessly sensitive to this unknown physics? This question will be addressed in the next section.

[§]There is a small range of parameter space in which our results are unchanged[16]. This requires threshold corrections at M_G which distinguish the two Higgs scalars.

10. Where are we going?

In the first half of Table 3 I give a brief summary of the good and bad features of the effective SUSY GUT discussed earlier. Several models were found with just four operators at M_G which successfully fit the low energy data. If we add up all the necessary parameters needed in these models we find just 12. This should be compared to the SM with 18 or the MSSM with 21. Thus these theories, minimal effective SUSY GUTs[MESG], are doing quite well. Of course the bad features of the MESG is that it is not a fundamental theory. In particular there are no symmetries which prevent additional higher dimension operators to spoil our results. Neither are we able to calculate threshold corrections, even in principle, at M_G.

It is for these reasons that we need to be able to take the MESG which best describes the LED and use it to define an effective field theory valid at scales $\leq M$. The good and bad features of the resulting theory are listed in the second half of Table 3.

- In the effective field theory below M we must incorporate the *symmetries* which guarantee that we reproduce the MESG with no additional operators¶

  Moreover, the necessary combination of discrete, U(1) or R symmetries may be powerful enough to restrict the appearance of Planck slop.

- Finally, the *GUT symmetry breaking* sector must resolve the problems of natural doublet-triplet splitting (the second hierarchy problem), the μ problem, and give predictions for proton decay, neutrino masses and calculable threshold corrections at M_G.

- On the bad side, it is still not a fundamental theory and there may not be a unique extension of the MESG to higher energies.

11. String Threshold at M_S

Upon constructing the effective field theory $\leq M_S$, we will have determined the necessary SO(10) states, symmetries and couplings which reproduce our fermion mass relations. This theory can be the starting point for constructing a realistic string model. String model builders could try to obtain a string vacuum with a massless spectra which agrees with ours. Of course, once the states are found the string will determine the symmetries and couplings of the theory. It is hoped that in this way a *fundamental* theory of Nature can be found. Work in this direction by several groups is in progress[20]. String theories with SO(10), three families plus additional $16 + \overline{16}$ pairs, 45's, 10's and even some 54 dimensional representations appear possible. One

¶This statement excludes the unavoidable higher order field theoretic corrections to the MESG which are, in principle, calculable.

Table 3.

	Good	Bad
Eff. F.T. $\leq M_G$	4 op's. at $M_G \Rightarrow$ **LED** 5 para's. $\Rightarrow$ 13 observables + 2 gauge para's. $\Rightarrow$ 3 observables + 5 soft SUSY breaking para's. $\Rightarrow \cdots$ Total **12** parameters	Not fundamental No symmetry $\Rightarrow$ Why these operators? (F.T. + Planck slop) Threshold corrections?
Eff. F.T. $\leq M$ $M = M_{string}$ or M_{Planck}	Symmetry i) gives Eff. F.T. $\leq M_G$ + corrections ii) constrains other operators GUT symmetry breaking i) d - t splitting ii) μ problem iii) proton decay iv) neutrino masses v) threshold corrections at M_G	Not fundamental Not unique?

of the first results from this approach is the fact that only one of the three families has diagonal couplings to the 10, just as we have assumed.

12. Constructing the Effective Field Theory below M_S

In this section I will discuss some preliminary results obtained in collaboration with Lawrence Hall[21]. I will describe the necessary ingredients for constructing model 6. Some very general results from this exercise are already apparent.

- *States* — We have constructed a SUSY GUT which includes all the states necessary for GUT symmetry breaking and also for generating the 45 vevs in the desired directions. A minimal representation content below M_S includes 54s + 45s + 3 16s + n($\overline{16}$ + 16) pairs + 2 10s.

- *Symmetry* — In order to retain sufficient symmetry the superspace potential in the visible sector W necessarily has a number of flat directions. In particular the scales v_5 and v_{10} can only be determined when soft SUSY breaking and quantum corrections are included. An auxiliary consequence is that the vev of $W_{visible}$ vanishes in the supersymmetric limit.

- *Couplings* — As an example of the new physics which results from this analysis I will show how a solution to the μ problem, the ratio λ_b/λ_t and proton decay may be inter-related.

In Table 4 are presented the predictions for Model 6 for particular values of the input parameters.

Table 4: Particular Predictions for Model 6 with $\alpha_s(M_Z) = 0.115$

Input Quantity	Input Value	Predicted Quantity	Predicted Value
$m_b(m_b)$	4.35 GeV	M_t	176 GeV
$m_\tau(m_\tau)$	1.777 GeV	$\tan\beta$	55
$m_c(m_c)$	1.22 GeV	V_{cb}	.048
m_μ	105.6 MeV	V_{ub}/V_{cb}	.059
m_e	0.511 MeV	$m_s(1GeV)$	172 MeV
V_{us}	0.221	$\hat{B}_K$	0.64
		m_u/m_d	0.64
		m_s/m_d	24.

In addition to these predictions, the set of inputs in Table 4 predicts: $\sin 2\alpha = -.46$, $\sin 2\beta = .49$, $\sin 2\gamma = .84$, and $J = 2.6 \times 10^{-5}$.

Model 6

The superspace potential for Model 6 has several pieces - $W = W_{fermion} + W_{symmetry\ breaking} + W_{Higgs} + W_{neutrino}$.

12.1. Fermion sector

The first term must reproduce the four fermion mass operators of Model 6. They are given by

$$\begin{aligned} O_{33} &= 16_3\ 10_1\ 16_3 \\ O_{23} &= 16_2\ \tfrac{A_2}{\tilde{A}}\ 10_1\ \tfrac{A_2}{\tilde{A}}\ 16_3 \\ O_{22} &= 16_2\ \tfrac{\tilde{A}}{M}\ 10_1\ \tfrac{A_1}{\tilde{A}}\ 16_2 \\ O_{12} &= 16_1\left(\tfrac{\tilde{A}}{M}\right)^3 10_1 \left(\tfrac{\tilde{A}}{M}\right)^3 16_2 \end{aligned} \tag{5}$$

There are two 10s in this model, denoted by $10_i, i = 1,2$ but only 10_1 couples to the ordinary fermions. The A fields are different 45s which are assumed to have vevs in the following directions – $\langle A_2\rangle = 45_Y$, $\langle A_1\rangle = 45_{B-L}$, and $\langle\tilde{A}\rangle = 45_X$. As noted earlier, there are 6 choices for the 22 operator and we have just chosen one of them, labelled a, arbitrarily here. In Figure 11, we give the tree diagrams which reproduce the effective operators for Model 6 to leading order in an expansion in the ratio of small to large scales. The states $\overline{\Psi}_a, \Psi_a, a = 1, \cdots, 9$ are massive $\overline{16}, 16$ states respectively with mass given by $\langle S_M\rangle \sim M$. Each vertex represents a separate Yukawa interaction in $W_{fermion}$ (see below). Field theoretic corrections to the effective GUT operators may be obtained by diagonalizing the mass matrices for the heavy states and integrating them out of the theory.

$$W_{fermion} = \tag{6}$$

$$16_3 16_3 10_1 + \bar{\Psi}_1 A_2 16_3 + \bar{\Psi}_1 \tilde{A} \Psi_1 + \Psi_1 \Psi_2 10_1$$

$$+\bar{\Psi}_2 \tilde{A} \Psi_2 + \bar{\Psi}_2 A_2 16_2 + \bar{\Psi}_3 A_1 16_2$$

$$+\bar{\Psi}_3 \tilde{A} \Psi_3 + \Psi_3 \Psi_4 10_1 + \mathcal{S}_M \sum_{a=4}^{9} (\bar{\Psi}_a \Psi_a)$$

$$+\bar{\Psi}_4 \tilde{A} 16_2 + \bar{\Psi}_5 \tilde{A} \Psi_4 + \bar{\Psi}_6 \tilde{A} \Psi_5$$

$$+\Psi_6 \Psi_7 10_1 + \bar{\Psi}_7 \tilde{A} \Psi_8 + \bar{\Psi}_8 \tilde{A} \Psi_9 + \bar{\Psi}_9 \tilde{A} 16_1$$

- *Note that the vacuum insertions in the effective operators above cannot be rearranged, otherwise an inequivalent low energy theory would result. In order to preserve this order naturally we demand that each field carries a different value of a U(1) family charge (see Fig. 11).* Note also that the particular choice of a 22 operator will affect the allowed U(1) charges of the states. Some choices may be acceptable and others not.

- Consider $W_{fermion}$. It has many terms, each of which can have different, in principle, complex Yukawa couplings. *Nevertheless the theory is predictive because only a very special linear combination of these parameters enters into the effective theory at M_G. Thus the observable low energy world is simple, not because the full theory is particularly simple, but because the symmetries are such that the effective low energy theory contains only a few dominant terms.*

12.2. Symmetry breaking sector

The symmetry breaking sector of the theory is not particularly illuminating. Two 54 dimensional representations, S, S' are needed plus several singlets denoted by $\mathcal{S}_i, i = 1, \cdots, 7$. They appear in the first two terms and are responsible for driving the vev of A_1 into the B-L direction, the third term drives the vev of the $\overline{16}, 16$ fields $\bar{\Psi}, \Psi$ into the right-handed neutrino like direction breaking SO(10) to SU(5) and forcing $\tilde{A}$ into the X direction. The fourth, fifth and sixth terms drive A_2 into the Y direction. Finally the last two terms are necessary in order to assure that all non singlet states under the SM gauge interactions obtain mass of order the GUT scale. All primed fields are assumed to have vanishing vevs.

Note if $\langle \mathcal{S}_3 \rangle \approx M_S$ then two of these adjoints state may be heavy. Considerations such as this will affect how couplings run above M_G.

$$W_{symmetry\ breaking} = \tag{7}$$

$$A'_1(SA_1 + \mathcal{S}_1 A_1) + S'(\mathcal{S}_2 S + A_1^2)$$

$$+\tilde{A}'(\bar{\Psi}\Psi + \mathcal{S}_3 \tilde{A})$$

$$+A'_2(\mathcal{S}_4 A_2 + S\tilde{A} + (\mathcal{S}_1 + \mathcal{S}_5)\tilde{A})$$

$$+\bar{\Psi}' A_2 \Psi + \bar{\Psi} A_2 \Psi'$$

$$+A_1 A_2 \tilde{A}' + \mathcal{S}_6 (A'_1)^2$$

12.3. Higgs sector

The Higgs sector is introduced below. It does not at the moment appear to be unique, but it is crucial for understanding the solution to several important problems – doublet-triplet splitting, μ problem and proton decay – and these constraints may only have one solution. The $10_1 A_1 10_2$ coupling is the term required by the Dimopoulos-Wilczek mechanism for doublet-triplet splitting. Since A_1 is an anti-symmetric tensor, we need at least two 10s.

The couplings of 10_1 to the 16s are introduced to solve the μ problem. After naturally solving the doublet-triplet splitting problem one has massless doublets. One needs however a small supersymmetric mass μ for the Higgs doublets of order the weak scale. This may be induced once SUSY is broken in several ways.

- The vev of the field A_1 may shift by an amount of order the weak scale due to the introduction of the soft SUSY breaking terms into the potential. In this theory the shift of A_1 appears to be too small.

- There may be higher dimension D terms in the theory of the form, eg.

$$\frac{1}{M_{Pl}} \int d^4\theta 10_1^2 (A_2^*).$$

  Then supergravity effects might induce a non-vanishing vev for the F term of A_2 of order the $m_W M_G$. This will induce a value of μ of order $m_W M_G / M_{Pl}$. The shift in the F-terms also appear to be negligible.

- Higher dimension D-terms with hidden sector fields may however work. Consider $\frac{1}{M_{Pl}} \int d^4\theta 10_1^2 z^*$ where z is a hidden sector field which is connected with soft SUSY breaking. It would then be natural to have $F_z \approx \mu M_{Pl}$.

- One loop effects may induce a μ term once soft SUSY breaking terms are introduced[22]. In this case we find $\mu \sim \frac{A\lambda^4}{16\pi^2}$ where λ^4 represents the product of Yukawa couplings entering into the graph of Figure 12.

We use the last mechanism above for generating μ in the example which follows.

$$W_{Higgs} = \tag{8}$$

$$+\bar{\Psi}' A_2 \Psi + \bar{\Psi} A_2 \Psi'$$

$$+10_1 A_1 10_2 + \mathcal{S}_7 10_2^2$$

$$+\bar{\Psi}\bar{\Psi}' 10_1 + \Psi\Psi' 10_1$$

Note that the first two terms already appeared in the discussion of the symmetry breaking sector. They are included again here since as you will see they are important for the discussion of the Higgs sector as well. The last two terms are needed to incorporate the solution to the μ problem. As a result of these couplings to $\overline{\Psi}, \Psi$ the Higgs doublets in 10_1 mix with other states. The mass matrix for the SU(5) $\mathbf{\bar{5}}, \mathbf{5}$ states in $\mathbf{10_1}, \mathbf{10_2}, \mathbf{\Psi}, \mathbf{\Psi'}, \overline{\mathbf{\Psi}}, \overline{\mathbf{\Psi}}'$ is given below.

$$\begin{array}{c} \\ 5_1 \\ 5_2 \\ 5_{\overline{\Psi}} \\ 5_{\overline{\Psi}'} \end{array} \begin{array}{c} \bar{5}_1 \ \bar{5}_2 \ \bar{5}_{\Psi} \ \bar{5}_{\Psi'} \\ \begin{pmatrix} 0 & A_1 & 0 & \Psi \\ A_1 & \mathcal{S}_7 & 0 & 0 \\ 0 & 0 & 0 & A_2 \\ \overline{\Psi} & 0 & A_2 & 0 \end{pmatrix} \end{array}$$

Higgs doublets In the doublet sector the vev A_1 vanishes. Since the Higgs doublets in 10_1 now mix with other states, the boundary condition $\lambda_b/\lambda_t = 1$ is corrected at tree level. The ratio is now given in terms of a ratio of mixing angles.

Proton decay The rate for proton decay in this model is set by the quantity $(M^t)^{-1}_{11}$ where M^t is the color triplet Higgsino mass matrix[23]. We find $(M^t)^{-1}_{11} = \frac{\mathcal{S}_7}{A_1^2}$. This may be much smaller than $\frac{1}{M_G}$ for $\mathcal{S}_7$ sufficiently smaller than M_G. Note there are no light color triplet states in this limit. Proton decay is suppressed since in this limit the color triplet Higgsinos in 10_1 become Dirac fermions (with mass of order M_G), *but they do not mix with each other.*

12.4. Symmetries

The theory has been constructed in order to have enough symmetry to restrict the allowed operators. This is necessary in order to reproduce the mass operators in the effective theory, as well as to preserve the vacuum directions assumed for the 45s and have natural doublet-triplet splitting. Indeed the construction of the symmetry breaking sector with the primed fields allows the 45s to carry nontrivial U(1) charges. This model has several unbroken U(1) symmetries which do not seem to allow any new mass operators. It has a discrete Z_4 R parity in which all the primed fields, $\mathcal{S}_{6,7}$ and 10_2 are odd and $16_i, i = 1, 2, 3$ and $\overline{\Psi}_a, \Psi_a, a = 1, \cdots, 9$ go into i times themselves. This guarantees that the odd states (and in particular, 10_2) do not couple into the fermion mass sector. There is in addition a Family Reflection Symmetry (see Dimopoulos- Georgi [6]) which guarantees that the lightest supersymmetric particle is stable. Finally, there is a continuous R symmetry which is useful for two reasons, (1) as a consequence, only dimension 4 operators appear in the superpotential and (2) this R symmetry is an anomalous Peccei-Quinn U(1) which naturally solves the strong CP problem.

Neutrino sector The neutrino sector seems to be very model dependent. It will constrain the symmetries of the theory, but I will not discuss it further here.

13. Work in Progress

The results for fermion masses and mixing angles should be considered as zeroth order results. These results receive calculable corrections from physics at the weak and GUT scales. Before we add new operators to try and improve the agreement with experiment we must evaluate the leading order corrections. In the following I will discuss preliminary results of work in progress.

1. New parameter! — We predict $\tan\beta$ to be large. It is interesting to ask whether we can lower the value of $\tan\beta$ with the addition of one new parameter. This parameter can appear naturally in any SO(10) model if the light higgs doublets are mixtures of the doublets in 10_1 and doublets contained in other states (consider for example the higgs sector discussed in this talk). Let H_u, H_d be the higgs doublets in the MSSM coupling to up, down quarks respectively and $d_{10}, \overline{d_{10}}$ be the higgs doublets in 10_1. Then we introduce a parameter ω by the expression
$$d_{10} = H_u, \; \overline{d_{10}} = \omega \; H_d.$$
In this case we expect
$$\tan\beta \sim \omega(\frac{m_t}{m_b})$$
with $\omega \leq 1$. We find[24](in a systematic operator search with 4 operators) that solutions only exist for $\omega > 0.5$ and $\tan\beta > 30$. Thus large $\tan\beta$ is unavoidable.

2. Finite SUSY corrections of order $\tan\beta$ Large radiative corrections to the bottom quark mass have been known for some time[16,17]. They can be expressed by
$$\delta m_b = (\epsilon_1 + \epsilon_2 |V_{tb}|^2) m_b$$

where the first term comes from gluino loops and the second from higgsino loops. ϵ_1 and ϵ_2 are proportional to $\tan\beta$. In the regime of large $\tan\beta$ these corrections can be significant. They depend on the details of the SUSY spectrum. We have considered the radiative corrections to the down quark mass matrix. These corrections can be expressed as corrections to quark masses and CKM matrix elements. We find[25]

$$\delta m_a = \epsilon_1 m_a, \text{ for } a = s, d$$

and

$$\delta V_{cb} = -\epsilon_2 V_{cb} \qquad \delta \left|\frac{V_{ub}}{V_{cb}}\right| \approx 0$$
$$\delta J = -2\,\epsilon_2 J \quad \delta\alpha \approx \delta\beta \approx \delta\gamma \approx 0$$

For non-universal scalar masses $-\epsilon_1 \sim \epsilon_2$ can be $\sim 5-10\%$ or even larger.

3. GUT threshold corrections — The boundary conditions at M_G for the two loop gauge running equations depend on α_G, M_G ; and $\Delta_i(GUT spectrum), i = 1, 2, 3$. The threshold corrections in Δ_i depend logarthmically on the spectrum of massive states at M_G[5]. This affects the prediction for $\alpha_s(M_Z)$ given α and $\sin^2\theta_W$. In fact, there may be an interesting constraint on the proton decay rate using the observed value of $\alpha_s(M_Z)$. This is because in order to suppress the proton decay rate we need $(M^t)^{-1}_{11} = \frac{\mathcal{S}_7}{A_1^2} \leq \frac{1}{\tan\beta\, M_G}$. However the two doublets in 10_2 get mass of order $\mathcal{S}_7$, while the triplets all have mass of order M_G. Thus there may be a lower limit on the vev of $\mathcal{S}_7$ from gauge unification. This work is now in progress[26]. Note there may be compensating effects from allowing $\mathcal{S}_6 << M_G$ and thus I can not make any definite statements at this time.

4. New χ^2 analysis — Since the predictions for fermion masses are strongly correlated, the analysis done in [1] using the 6 best known low energy parameters to fix the 6 input parameters is not satisfactory. Blažek is now redoing this analysis. He is using a χ^2 analysis in an attempt to find the best fit to the low energy data.

14. Conclusion

In this talk, I have presented a class of supersymmetric SO(10) GUTs which are in *quantitative* agreement with the low energy data. With improved data these particular models may eventually be ruled out. Nevertheless the approach of using low energy data to ascertain the dominant operator contributions at M_G seems robust. Taking it seriously, with *quantitative* fits to the data and including the leading order corrections to the zeroth order results, may eventually lead us to the correct theory.

What is the proverbial *smoking gun* for the theories presented here ? There are three observations which combined would confirm SUSY GUTs.

1. Gauge coupling unification consistent with the observed values of $\alpha, sin^2\theta_W, \alpha_s$.

2. Observation of SUSY particles.

3. Observation of proton decay into the modes $p \to K^+\bar{\nu}$ and $p \to K^0\mu^+$ [23]. Although SUSY GUTs may not predict the rate for this process, nevertheless the observation of this process would confirm SUSY GUTs.

In addition, the minimal SO(10) models presented here all demand large $\tan\beta$. Thus observation of large $\tan\beta$ would certainly strengthen these ideas. Finally, if the calculable corrections to the predictions of one of these models improve the agreement with the data, it would be difficult not to accept this theory as a true description of nature.

1. G. Anderson, S. Dimopoulos, L.J. Hall, S. Raby and G. Starkman, *Phys. Rev.* **D49** 3660 (1994).
2. M. Shifman, N.G. Uraltsev and A. Vainshtein, "$|V_{cb}|$ from OPE Sum Rules for Heavy Flavor Transitions," University of Minnesota preprint, TPI-MINN-94/13-T, (LANL hep-ph/9405207) (1994).
3. C.T.H. Davies, K. Hornbostel, A. Langnau, G.P. Lepage, A. Lidsey, C.J. Morningstar, J. Shigemitsu, and J. Sloan, "A New Determination of M_b using LATTICE QCD," Ohio State preprint OHSTPY-HEP-T-94-004, (LANL hep-lat/9404012)(1994); for a review on α_s from the lattice, see Aida X. El-khadra, "The Strong Coupling from Quarkonia," Ohio State preprint OHSTPY-HEP-T-93-020, (LANL hep-lat/9312080)(1993). For recent results, see for example, C.R. Allton, M. Ciuchini, M. Crisafulli, E. Franco, V. Lubicz, and G. Martinelli, "Quark masses from lattice QCD at the next-to-leading order," Rome preprint 94/1018 (LANL hep-ph/9406263) (1994).
4. H. Georgi and S. Glashow, *Phys. Rev. Lett.* **32** 438 (1974); J. Pati and A. Salam, *Phys. Rev.* **D8** 1240 (1973).
5. H. Georgi, H. Quinn and S. Weinberg, *Phys. Rev. Lett.* **33** 451 (1974); S. Weinberg, *Phys. Lett.* **91B** 51 (1980); C.H. Llewellyn-Smith, G.G. Ross and J.F. Wheater, *Nucl. Phys.* **B177** 263 (1981); L.J. Hall, *Nucl. Phys.* **B178** 75 (1981); P. Binetruy and T. Schucker, *Nucl. Phys.* **B178** 293 (1981).
6. S. Dimopoulos, S. Raby and F. Wilczek, *Phys. Rev.* **D24** 1681 (1981); S. Dimopoulos and H. Georgi, *Nucl. Phys.* **B 193** 150 (1981); L. Ibanez and G.G. Ross, *Phys. Lett.* **105B** 439 (1981); M. B. Einhorn and D. R. T. Jones, *Nucl. Phys.* **B196** 475 (1982); W. J. Marciano and G. Senjanovic, *Phys. Rev.* **D 25** 3092 (1982).
7. For recent analyses, see P. Langacker and N. Polonsky, *Phys. Rev.* **D47** 4028 (1993); ibid., **D49** 1454 (1994); M. Carena, S. Pokorski, C.E.M. Wagner, *Nucl. Phys.* **B406** 59 (1993).
8. S. Dimopoulos, *Phys. Lett.* **129 B** 417 (1983); J. Bagger,S. Dimopoulos, H. Georgi and S. Raby, Proceedings of the Fifth Workshop on Grand Unification, Providence, Rhode Island, April 12 - 14, 1984, World Scientific Publishing Co., Singapore (1984). For early work in this direction, see C.D. Froggatt and H.B.

Nielsen, *Nucl. Phys.* **B147** 277 (1979); Z. G. Berezhiani, *Phys. Lett.* **129 B** 99 (1983); ibid., **150 B** 177 (1985).

9. S. Dimopoulos and F. Wilczek, Proceedings Erice Summer School, Ed. A. Zichichi (1981). For a recent discussion, see K.S. Babu and S.M. Barr, Delaware preprint BA-94-04 (1994).
10. B. Ananthanarayan, G. Lazarides and Q. Shafi, *Phys. Rev.* **D44** 1613 (1991); H. Arason et al., *Phys. Rev. Lett.* **67** 2933 (1991); S. Kelley, J.Lopez and D. Nanopoulos, *Phys. Lett.* **B274** 387 (1992). See also the earlier work of M. Olechowski and S. Pokorski, *Phys. Lett.* **B214** 393 (1988).
11. S. Dimopoulos, L.J. Hall and S. Raby, *Phys. Rev. Lett.* **68** 1984 (1992);ibid., *Phys. Rev.* **D45** 4192 (1992);ibid., **D46** R4793 (1992); P. Ramond, UFIFT-92-4 (1992); H. Arason, D. Castano, E.J. Piard and P. Ramond, *Phys. Rev.* **D47** 232 (1993); G. Anderson, S. Raby, S. Dimopoulos, and L. J. Hall, Phys. Rev. **D47** 3702 (1993); V. Barger, M. S. Berger, T. Han and M. Zralek, Phys. Rev. Lett. **68** 3394 (1992); V. Barger, M. S. Berger and P. Ohmann, Phys. Rev. **D47** 1093 (1993);ibid., 2038 (1993).
12. H. Georgi and C. Jarlskog, *Phys. Lett.* **86B** 297 (1979); H. Georgi and D.V. Nanopoulos, *Nucl. Phys.* **B159** 16 (1979); J. Harvey, P. Ramond and D.B. Reiss, *Phys. Lett.* **92B** 309 (1980); *Nucl. Phys.* **B199** 223 (1982).
13. D. Kaplan and A. Manohar, *Phys. Rev. Lett.* **56** 2004 (1986); H. Leutwyler, *Nucl. Phys.* **B337** 108 (1990).
14. R. Gupta, D. Daniel, G. W. Kilcup, A. Patel, and S. R. Sharpe, *Phys. Rev.* **D47** 5113 (1993); S. Sharpe, in Lattice '91, Proceedings of the International Symposium, Tsukuba, Japan, 1991, edited by M. Fukugita et. al. [Nucl. Phys. B (Proc. Suppl.) **26** 197 (1992).]
15. Y. Nir and U. Sarid, *Phys. Rev.* **D47** 2818 (1993).
16. R. Rattazzi, U. Sarid and L.J. Hall, "Yukawa Unification: the Good, The Bad and The Ugly," Stanford preprint SU-ITP-94-15 (LANL hep-ph/9405313)(1994).
17. M. Carena, S. Pokorski, M. Olechowski, and C.E.M. Wagner, "Electroweak Symmetry Breaking and Bottom-Top Yukawa Unification," CERN Theoretical Report No. CERN-Th.7163/94 (LANL hep-ph/9402253)(1994).
18. L.E. Ibanez and G.G. Ross, *Phys. Lett.* **110B** 215 (1982) ; L. Alvarez-Gaume, J. Polchinsky and M. Wise, *Nucl. Phys.* **B221** 495 (1983) ; L.E. Ibanez and C. Lopez, *Phys. Lett.* **126B** 54 (1983),ibid., *Nucl. Phys.* **B233** 511 (1984) ; C. Kounnas, A.B. Lahanas, D.V. Nanopoulos and M. Quiros, *Nucl. Phys.* **B236** 438 (1984); L.E. Ibanez, C. Lopez and C. Munoz, *Nucl. Phys.* **B256** 218 (1985).
19. J. Ellis and M.K. Gaillard, *Phys. Lett.* **88B** 315 (1979).
20. S. Chaudhuri, S. Chung and J.D. Lykken, FERMILAB PUB-94/137-T (LANL hep-ph/9405374)(1994); G. Cleaver, Talk given at Paschos 94; L. Ibañez, Talk given at DESY Workshop 94.
21. L.J. Hall and S. Raby, in preparation.
22. L.J. Hall, in Supersymmetry and Supergravity/Nonperturbative QCD, Proceedings of the Winter School in Theoretical Physics, Mahabaleshwar, India,

1984, edited by P. Roy and V. Singh, Lecture Notes in Physics Vol. 208 (Springer, Berlin, 1984).
23. S. Dimopoulos, S. Raby and F. Wilczek, *Phys. Lett.***112B** 133 (1982); J. Ellis, D.V. Nanopoulos and S. Rudaz, *Nucl. Phys.***B202** 43 (1982). For recent analyses, see R. Arnowitt and P. Nath, *Phys. Rev.***D49** 1479 (1994); J. Hisano, H. Murayama, and T. Yanagida, *Nucl. Phys.***B402** 46 (1993); P. Langacker, "Proton Decay," (LANL hep-ph/9210238) talk given at The Benjamin Franklin Symposium in Celebration of the Discovery of the Neutrino, Philadelphia, PA, 29 Apr - 1 May (1992).
24. M. Carena, S. Dimopoulos, S. Raby and C. Wagner, in preparation.
25. T. Blažek, S. Raby and S. Pokorski, in preparation.
26. V. Lucas and S. Raby, work in progress.
27. T. Blažek, work in progress.

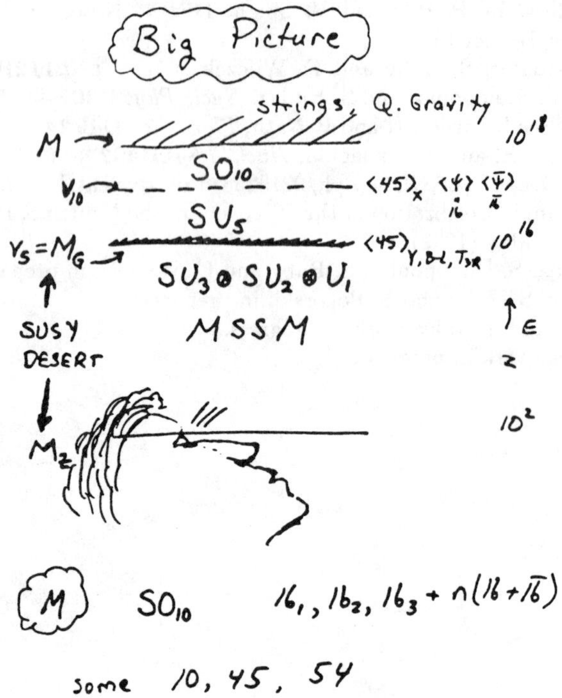

Figure 1

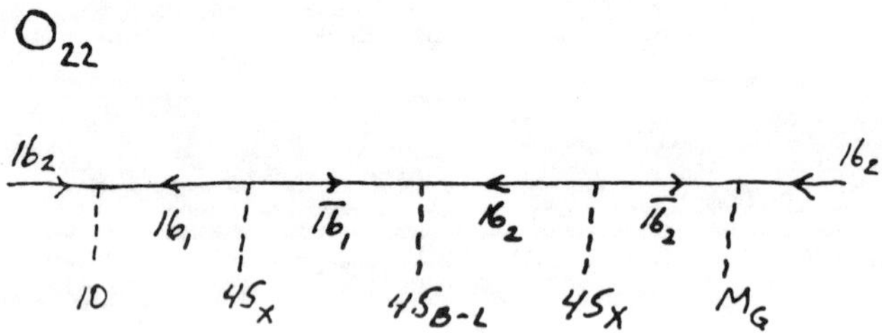

Figure 2

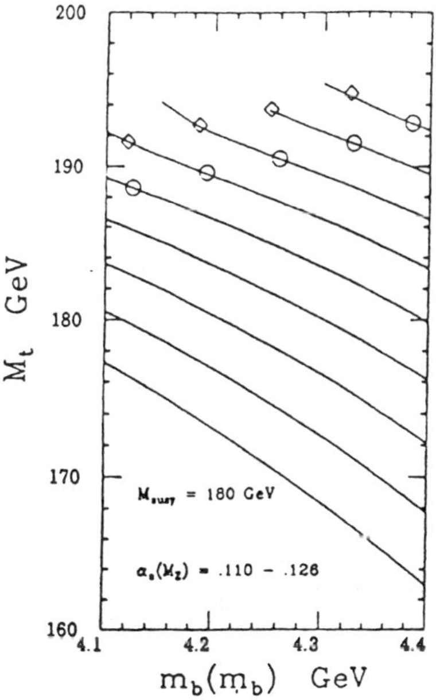

Figure 3a Top Mass Prediction

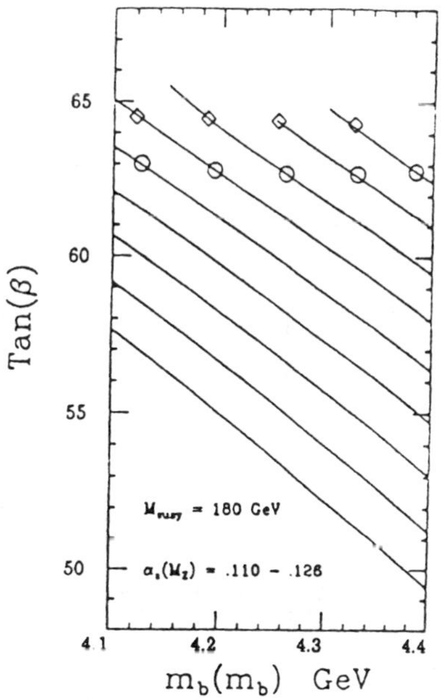

Figure 3b Prediction for $\tan\beta$

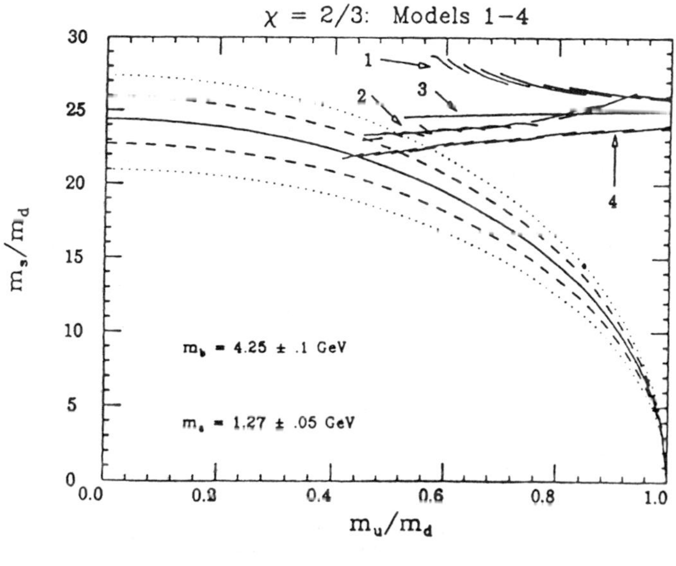

Figure 4a Prediction for $\frac{m_s}{m_d}$ and $\frac{m_u}{m_d}$.

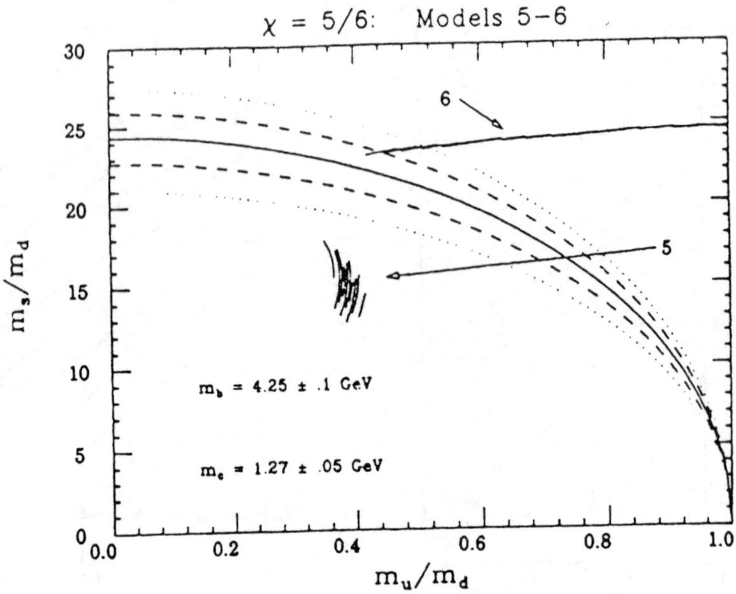

Figure 4b Prediction for $\frac{m_s}{m_d}$ and $\frac{m_u}{m_d}$.

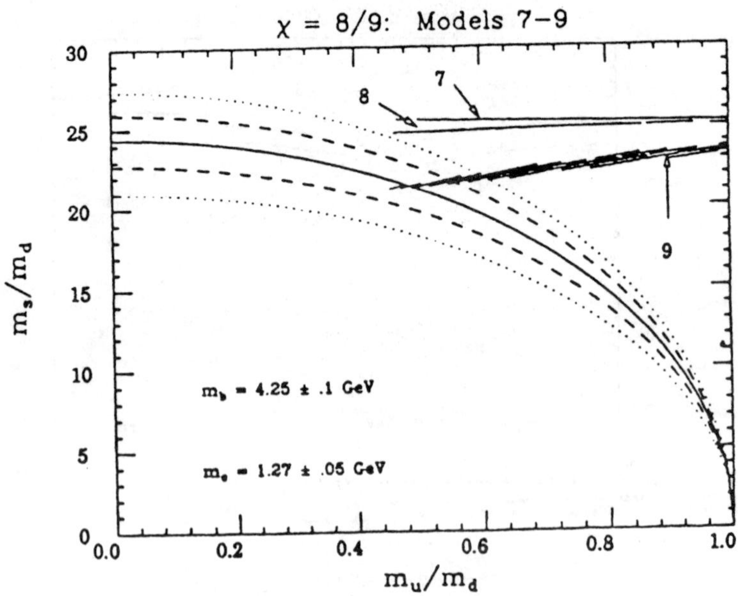

Figure 4c Prediction for $\frac{m_s}{m_d}$ and $\frac{m_u}{m_d}$.

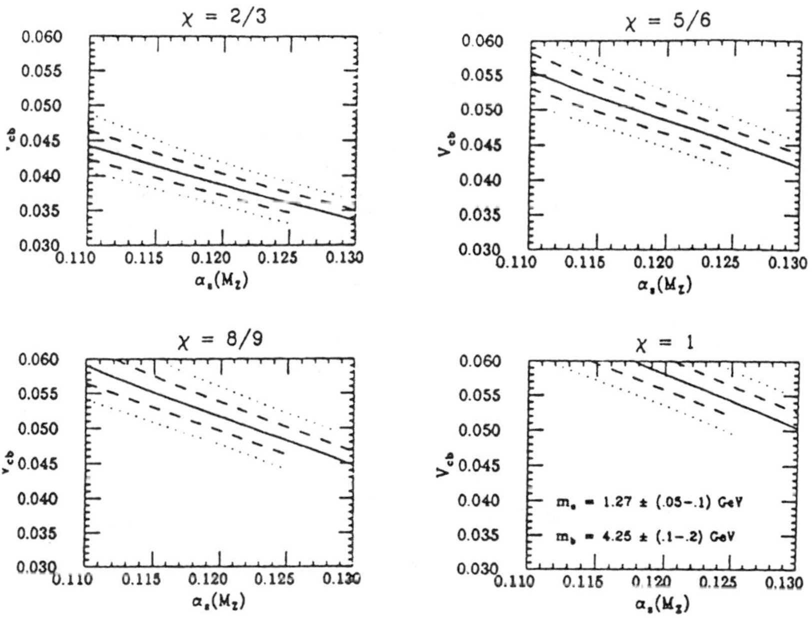

Figure 5 Prediction for V_{cb}

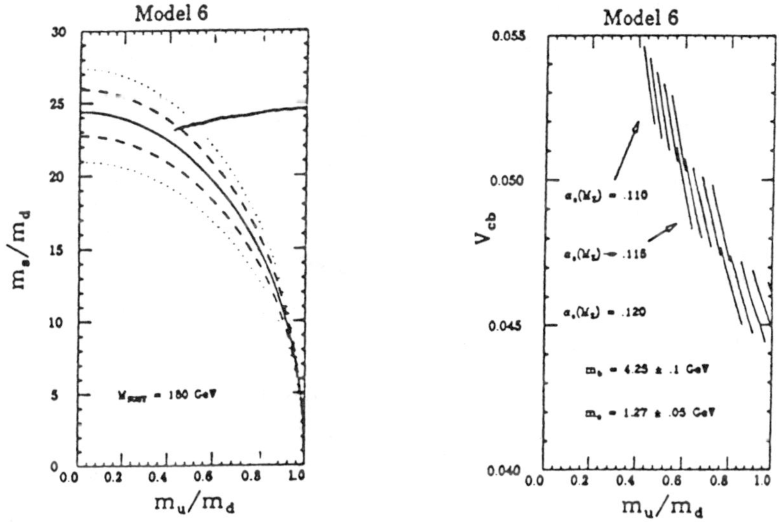

Figure 6

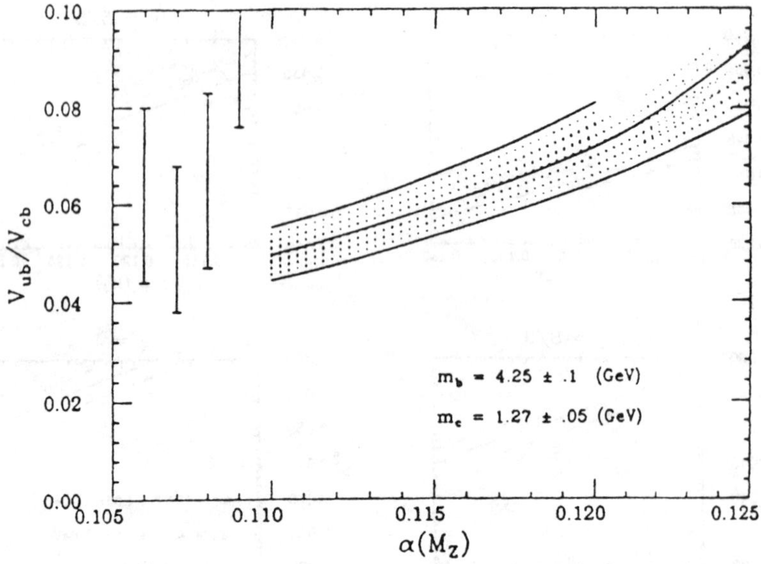

Figure 7 Prediction for $\frac{V_{ub}}{V_{cb}}$.

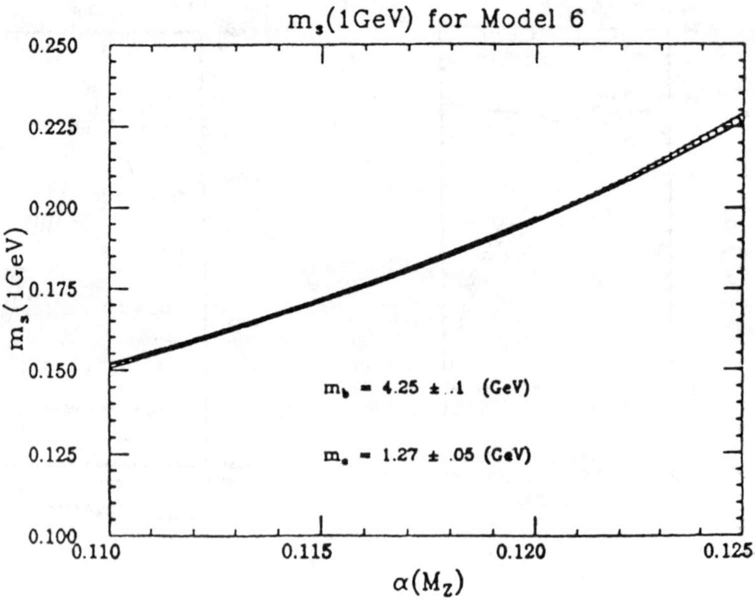

Figure 8 Prediction for m_s.

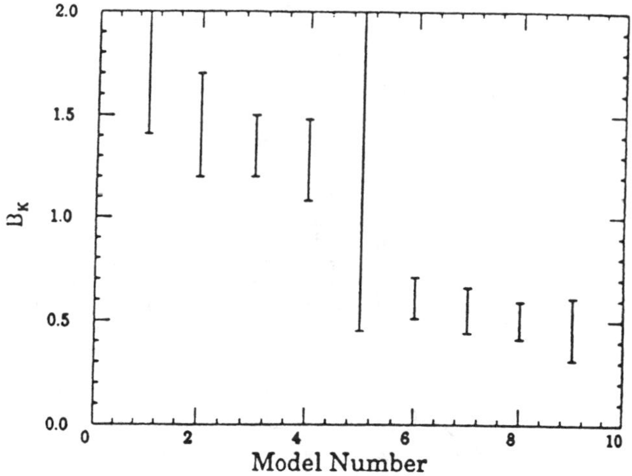

Figure 9

CP violation in B DECAY

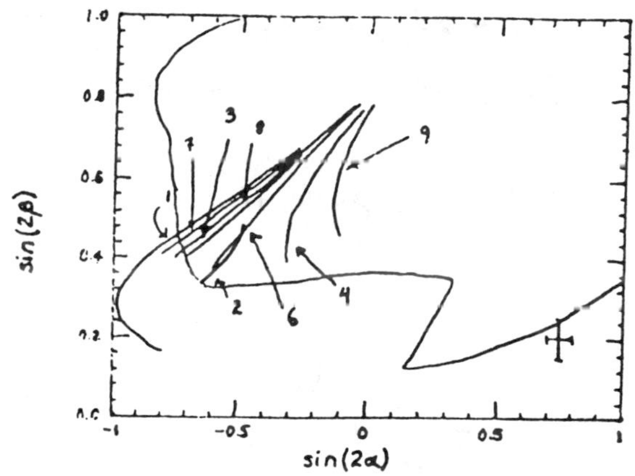

Figure 10

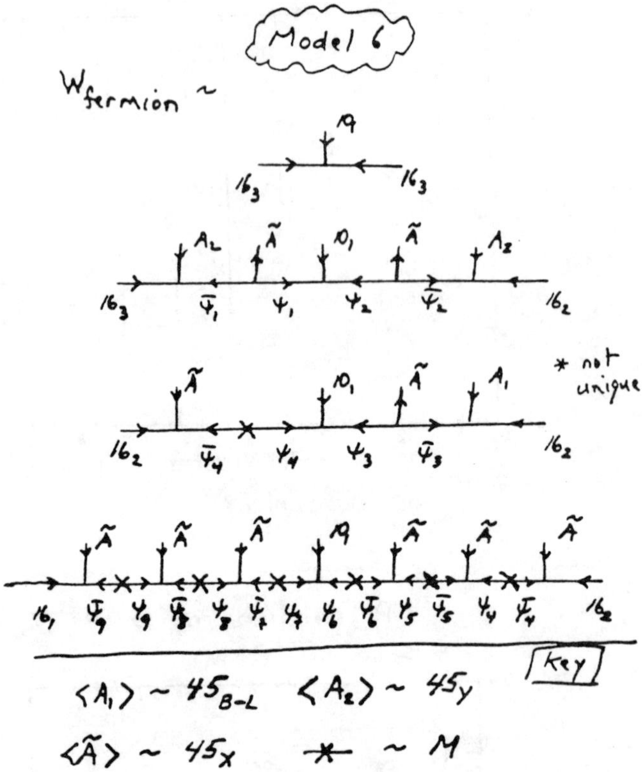

Figure 11

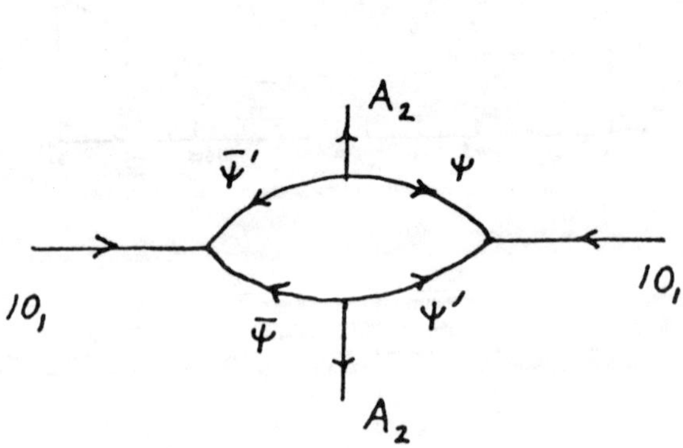

Figure 12

ELECTROWEAK SYMMETRY BREAKING WITH NON-UNIVERSAL SCALAR SOFT TERMS IN SUPERYMMETRIC SO(10) MODELS

M. OLECHOWSKI *

Institute of Theoretical Physics, Warsaw University
ul. Hoża 68, PL-00-681 Warsaw, Poland

ABSTRACT

We discuss radiative electroweak symmetry breaking in the case of large $\tan\beta$. We identify two classes of models with non-universal soft scalar mass terms which give solutions with $A_o \simeq B_o \simeq M_{1/2} \simeq O(M_Z)$ and with $\mu \gg M_Z$ or $\mu \simeq O(M_Z)$, respectively. It is shown that the minimal supersymmetric SO(10) model with universal soft mass terms at the GUT scale is ruled out by the combination of constraints from the $b \to s\gamma$ decay and the condition $\Omega h^2 < 1$ for the lightest (stable) neutralino. The constraints are, however, easily satisfied for certain class of supersymmetric SO(10) models with non-universal scalar masses which gives small supersymmetric corrections to the bottom quark mass and light higgsino-like neutralinos.

1. Introduction

The mechanism of radiative electroweak symmetry breaking in supersymmetric models[1] has been extensively studied in the literature. The $SU(2) \times U(1)$ gauge symmetry is broken by quantum corrections to the tree level potential which is postulated at the GUT scale. The quantum corrections are included by means of the renormalization group evolution of the parameters of the lagrangian from the GUT scale to the electroweak scale. Most of the work[2] has so far been done under the assumption that the soft supersymmetry breaking parameters: gaugino masses, scalar masses and trilinear couplings have at the GUT scale some universal values $M_{1/2}$, m_o and A_o, respectively. Then, with two additional parameters: B_o (the soft Higgs mixing term) and μ_o (the Higgs mixing in the superpotential), the model depends on five free parameters. However, exact universality of the soft mass terms at the GUT scale is neither phenomenologically necessary nor theoretically sound.

Radiative electroweak breaking is expected to be sensitive to the non-universal corrections to the soft terms in those regions of the parameter space where the breaking with universal terms requires a large degree of correlation among them. This is the case for the large $\tan\beta$ solutions[3,4] and we will study such solutions in the presence of non-universal scalar soft mass terms. We will investigate also the constraints due to the requirement that the relic abundance of the lightest neutralino, which is the lightest supersymmetric particle (LSP) in these models, does not overclose the Universe. Furthermore, we study the restrictions imposed by the recent observation

*Partially supported by the Polish Committee of Scientific Research.

of the inclusive decay $b \to s\gamma$ by the CLEO II Collaboration[5] which implies a 95% c.l. upper and lower limits on this branching ratio of $3.4 \cdot 10^{-4}$ and $1.2 \cdot 10^{-4}$, respectively[6]. Both types of constraints may turn out to be important for superymmetric SO(10) models, given the correlations present in their parameter spaces.

The issue of the "large $\tan\beta$" vacuum has been discussed for some time[3,7,8] as the possible explanation to the breaking of the $SU(2)_V$ symmetry of the quark masses with $\tan\beta \equiv v_2/v_1 \cong m_t/m_b$ and $h_t \simeq h_b$ where v_1, v_2 are the vacuum expectation values of the two Higgs fields which couple to the down and up quarks, respectively, and h_t and h_b are Yukawa couplings.

Special attention to the large $\tan\beta$ scenario has been given in the framework of the models with underlying SO(10) symmetry[8], which *predict* the equality of the top and bottom Yukawa couplings at the GUT scale.

2. Universal soft supersymmetry breaking terms

2.1. Electroweak symmetry breaking

In a recent paper[4] a detailed study has been performed of radiative electroweak breaking with complete gauge and Yukawa unification and universal soft mass terms at the GUT scale. It has been found that for large values of h_t (as required for the heavy top quark) solutions with large $\tan\beta$ are obtained only for strongly correlated values of the GUT scale parameters.

This can be easily understood in a qualitative way by studying the tree level scalar potential and the 1–loop RG evolution of its parameters from the M_Z scale to the GUT scale. For easy reference we recall the main points here. The Higgs potential

$$V = m_1^2 H_1^\dagger H_1 + m_2^2 H_2^\dagger H_2 - m_3^2 \left(H_1^\dagger \, i\tau_2 \, H_2 + h.c. \right) + \text{ quartic terms} \tag{1}$$

has, for large $\tan\beta$ values, two characteristic features. It follows from the minimization conditions that

$$m_2^2 \simeq -\frac{M_Z^2}{2} \tag{2}$$

and

$$m_3^2 \simeq \frac{M_A^2}{\tan\beta} \simeq 0 \tag{3}$$

with

$$M_A^2 \simeq m_1^2 + m_2^2. \tag{4}$$

Equations (2) and (4) combined together give a useful constraint on the low energy parameters

$$m_1^2 - m_2^2 \approx M_Z^2 + M_A^2 > M_Z^2. \tag{5}$$

Equations (2) and (3) are the two main constraints on the parameters of the scalar potential, which are characteristic for large $\tan\beta$ solutions.

The next information we need is the running of the parameters from the GUT scale to the M_Z scale. For universal soft supersymmetry breaking parameters at the GUT scale and for $Y_t = Y_b = Y$ ($Y_{t,b} = h^2_{t,b}/4\pi$) and large (such that $Y(M_Z)$ is close to its infrared quasi-fixed point value Y_f, as required by the heavy top quark), the approximate solutions to the RG equations read[4]:

$$m^2_{H_1} \simeq m^2_{H_2} = m^2_o + 0.5\ M^2_{1/2} - \frac{3}{7}\Delta m^2 \tag{6}$$

where $m^2_i = \mu^2 + m^2_{H_i}$,

$$\begin{aligned}\Delta m^2 \ &\simeq\ 3m^2_o\frac{Y}{Y_f} - 4.6\ A_o M_{1/2}\frac{Y}{Y_f}\left(1-\frac{Y}{Y_f}\right) \\ &+\ A^2_o\frac{Y}{Y_f}\left(1-\frac{Y}{Y_f}\right) + M^2_{1/2}\left[14\frac{Y}{Y_f} - 6\left(\frac{Y}{Y_f}\right)^2\right]\end{aligned} \tag{7}$$

and

$$\mu^2 = 2\mu^2_o\left(1-\frac{Y}{Y_f}\right)^{6/7}. \tag{8}$$

We observe that for large values of Y the masses $m^2_{H_1}$ and $m^2_{H_2}$ tend to become large and negative for increasing values of m_o and/or $M_{1/2}$. For the running of the soft supersymmetry breaking trilinear and bilinear couplings we have:

$$A_t \simeq A_o\left(1-\frac{Y}{Y_f}\right) - M_{1/2}\left(4.2 - 2.1\frac{Y}{Y_f}\right), \tag{9}$$

$$B \simeq B_o - \frac{6}{7}\frac{Y}{Y_f}\ A_o + M_{1/2}\left(2\frac{Y}{Y_f} - 0.6\right) \tag{10}$$

Finally, the squark masses read

$$m^2_U \simeq m^2_D = m^2_o + 6.7\ M^2_{1/2} - \frac{2}{7}\Delta m^2, \tag{11}$$

$$m^2_Q \simeq m^2_o + 7.2\ M^2_{1/2} - \frac{2}{7}\Delta m^2. \tag{12}$$

It is clear that in the approximation (6) the condition (5), necessary for the proper symmetry breaking, is not satisfied. However, eq. (6) neglects small differences in the running of the two Higgs masses which follow from the different hypercharges of the right top and bottom squarks, from the difference in the running of the bottom and top Yukawa couplings (equal at the GUT scale) and from the effects due to the τ lepton Yukawa. As can be inferred from the RG equations, the inclusion of those effects leads to the following result:

$$m^2_1 - m^2_2 - aM^2_{1/2} + bm^2_o \tag{13}$$

where $b < 0$ due to the Y_τ effects, $a > 0$ due to the other effects mentioned above and both are $O(0.1)$ in absolute values. With the condition (5) we see now that large $\tan\beta$ solutions must be driven by large values of $M_{1/2}$:

$$M_{1/2} > \frac{M_Z}{\sqrt{a}} \quad , \qquad M_{1/2} > \sqrt{\frac{|b|}{a}}m_o \tag{14}$$

Also, it follows from eqs. (2), (6) and (7) that the low energy superpotential parameter μ (or its GUT value μ_o; see eq. (8)) is strongly correlated with $M_{1/2}$. For large $M_{1/2}$, large values of μ^2 are needed to cancel large negative values of $m^2_{H_2}$ and to satisfy the condition (2). Indeed, in the limit $Y_t \to Y_f$ we get

$$\mu^2 \approx 3M^2_{1/2}. \tag{15}$$

In summary, in the model with universal soft supersymmetry breaking terms large $\tan\beta$ solutions to radiative electroweak symmetry breaking are characterized by large ($\gg M_Z$) and linearly correlated values of the parameters $M_{1/2}$, μ and $(7Y_f B_o - 6Y A_o)$, with the constraint $M_{1/2} > m_o$. Numerical calculations[4] based on the 2–loop RGE for the gauge and Yukawa couplings, and one–loop equations for the Higgs and supersymmetric mass parameters with all the leading supersymmetric threshold corrections included (bottom–up approach of ref.[9]) confirm the above qualitative considerations.

Another question one must address in the context of the large $\tan\beta$ scenario, with $Y_t = Y_b = Y_\tau$ at the GUT scale, is about the prediction for the bottom and top quark masses in the model with radiative electroweak breaking. As stressed in ref.[4] this prediction is sensitive to potentially large supersymmetric threshold corrections to the bottom mass (mainly gluino and higgsino exchange):

$$\frac{\delta m_b}{m_b} \simeq \left(\frac{\alpha_s}{3\pi}\frac{M_{\tilde{g}}\mu}{m^2_{\tilde{q}}} + \frac{Y_t}{8\pi}\frac{A_t\mu}{m^2_{\tilde{q}}}\right)\tan\beta \tag{16}$$

which in turn depend on the parameter space strongly constrained by the requirement of radiative electroweak breaking. In the model with universal soft terms the supersymmetric loop correction to the bottom mass are large, O(20–50%) and almost constant for fixed $\tan\beta$.

Since the final value of the pole mass should be in the experimental range $m_b = (4.9 \pm 0.3)$ GeV the bottom quark mass before supersymmetric corrections must be either small or large enough to accomodate those large corrections. The first option is inconsistent with Yukawa and gauge coupling unification[10,4] and, therefore, the supersymmetric correction (16) must be negative, i.e. the parameter space is further constrained by the requirement $\mu M_{1/2} < 0$ ($\mu A_t > 0$). As a representative example, we have taken $m_t = 160$ GeV and $\tan\beta = 40$ which give $m_b = 6.1$ GeV before susy corrections.

2.2. Constraints from the $b \to s\gamma$ decay and the relic abundance of the LSP

We can now discuss experimental constraints on the version of the model with universal boundary conditions coming from BR($b \to s\gamma$) and from the requirement that the neutralino relic abundance satisfies $\Omega h^2 < 1$. Our numerical calculation is based on the formalism developed in ref.[11,12] for $b \to s\gamma$ and in ref.[13] for the neutralino relic abundance.

The numerical results can be qualitatively understood in terms of the summarized above properties (9,15,16) of the parameter space and the spectra. For large $\tan\beta$, the dominant contribution to the $b \to s\gamma$ decay rate (additional to the standard model one) comes from the charged Higgs and chargino exchanges. We can estimate them by using the formulae of ref.[11] in the approximation of no mixing between the gaugino and higgsino and in the limit of large $\tan\beta$:

$$A_{H^+} \approx \frac{1}{2}\frac{m_t^2}{m_{H^+}^2} f^{(2)}\left(\frac{m_t^2}{m_{H^+}^2}\right), \tag{17}$$

$$A_{\widetilde{\chi}^+} \approx -\frac{\tan\beta\, m_t}{4\ \ \mu}\left[f^{(3)}\left(\frac{m_{\tilde{t}_1}^2}{\mu^2}\right) - f^{(3)}\left(\frac{m_{\tilde{t}_2}^2}{\mu^2}\right)\right] \tag{18}$$

where

$$f^{(2)}(x) = \frac{3-5x}{6(x-1)^2} + \frac{3x-2}{3(x-1)^3}\ln x, \tag{19}$$

$$f^{(3)}(x) = \frac{7x-5}{6(x-1)^2} - \frac{x(3x\ \ 2)}{3(x-1)^3}\ln x \tag{20}$$

and $m_{\tilde{t}_k}$ are the eigenvalues of the top squark mass matrix. The chargino contribution for the parameter space consistent with radiative electroweak breaking and the SO(10) unification is mainly due to the exchange of the heavier (higgsino–like) chargino. The relative sign of the H^+ and $\widetilde{\chi}^+$ contributions depends on the sign of the product μA_t and is positive for $\mu A_t > 0$. We recall that the latter is required for a proper correction to m_b.

Thus, the branching ratio BR($b \to s\gamma$) is bounded from below by the sum of the SM and the charged Higgs contributions (with negligible chargino contribution for heavy charginos) at the experimantal result puts a strong lower bound on the mass of the charged Higgs and, in consequence ($m_{H^\pm} \approx m_A$) also on m_A. In our numerical calculations of BR($b \to s\gamma$) we included the theoretical and experimental uncertainities following the method of ref.[12].

The annihilation of the lightest neutralino which, due to the relation (15), is strongly bino–like proceeds dominantly through s–channel CP odd Higgs exchange, whose coupling to $\tau\bar{\tau}$ and $b\bar{b}$ is strongly enhanced for large $\tan\beta$. The dominant part of the $\widetilde{\chi}_1^0\widetilde{\chi}_1^0 A$ coupling is proportional to the product $Z_{11}Z_{13}$, with $Z_{11} \sim 1$. Ωh^2 is an increasing function of $m_{\widetilde{\chi}^0}$ since the relic mass is proportional to $m_{\widetilde{\chi}^0}$ and the Z_{13} coupling behaves as $Z_{13} \sim 1/m_{\widetilde{\chi}^0}$. In addition, in the parameter space constrained by radiative electroweak breaking, for fixed $m_{\widetilde{\chi}^0}$, the rate of annihilation increases with m_A (since always $m_A < 2m_{\widetilde{\chi}^0}$ and for increasing m_A we approach closer the pole in the s-channel). Thus, for fixed $m_{\widetilde{\chi}^0}$, Ωh^2 increases when m_A decreases. Its lower bound corresponds to the upper bound on m_A for this value of $m_{\widetilde{\chi}^0}$. The net effect is shown in Fig. 1a; the broadness of the band is determined by the discussed above dependence on m_A for fixed $m_{\tilde{\chi}^0}$. Finally, we can now understand the dependence of Ωh^2 on m_A whose relevant feature is the lower bound on Ωh^2 which is increasing with m_A. For fixed m_A, the bound corresponds to the lower limit on $m_{\tilde{\chi}^0}$ which is increasing with increasing m_A. Thus, the rise of Ωh^2 with $m_{\tilde{\chi}^0}$ translates itself into the rise of the lower bound on Ωh^2 with m_A. Clearly, it remains to be checked if the lower bound on $m_{H^+} \approx m_A$ obtained from BR($b \to s\gamma$) is compatible with $\Omega h^2 < 1$. The result is shown in Fig. 1b: the combination of the two constraints rules out the minimal SO(10) model with universal soft supersymmetry breaking terms at the GUT scale.

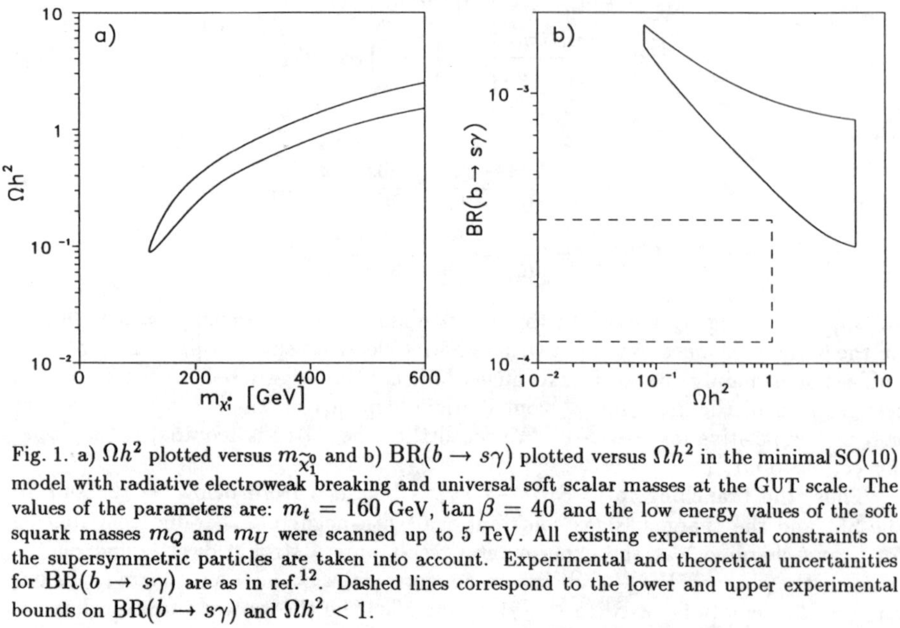

Fig. 1. a) Ωh^2 plotted versus $m_{\tilde{\chi}_1^0}$ and b) BR($b \to s\gamma$) plotted versus Ωh^2 in the minimal SO(10) model with radiative electroweak breaking and universal soft scalar masses at the GUT scale. The values of the parameters are: $m_t = 160$ GeV, $\tan\beta = 40$ and the low energy values of the soft squark masses m_Q and m_U were scanned up to 5 TeV. All existing experimental constraints on the supersymmetric particles are taken into account. Experimental and theoretical uncertainities for BR($b \to s\gamma$) are as in ref.[12]. Dashed lines correspond to the lower and upper experimental bounds on BR($b \to s\gamma$) and $\Omega h^2 < 1$.

3. Non–universal soft scalar masses

3.1. Electroweak symmetry breaking

Because of the above mentioned problems of the minimal SO(10), it is very interesting to investigate the version of the model with non–universal soft supersymmetry breaking terms[14,15]. In the minimal SO(10) model with $Y_t \approx Y_b \approx Y_\tau$ radiative electroweak breaking is very sensitive to small perturbations in the universal boundary conditions, if they can reverse the sign of b in eq. (14): for $b > 0$, qualitatively new solutions become possible[15], with $M_{1/2} \simeq 0$ and

$$m_o > \frac{M_Z}{\sqrt{b}} \tag{21}$$

i.e. contrary to the universal case, radiative electroweak breaking can be driven by soft scalar masses. From inspection of the RGE, it is easy to see that there are (at least) three ways to reverse the sign of b by perturbing slightly the universal boundary conditions for the third generation at the GUT scale:

$$\begin{array}{lrcl} & Y_t & > & Y_b \\ \text{and/or} & (m_o^U)^2 & > & (m_o^D)^2 \\ \text{and/or} & (m_o^{H_1})^2 & > & (m_o^{H_2})^2 \end{array}$$

Already O(20-30%) perturbation is sufficient to reverse the sign of b. The possibility of small deviation from the equality $Y_t = Y_b$ has already been discussed in ref.[4]. Here we focus on the non-universal scalar masses. For the boundary values of the scalar masses at the GUT scale we introduce the following notation

$$(m_o^{Q,D,U})^2 \equiv (1+\delta_{Q,U,D})m_o^2 \ , \qquad (m_o^{H_{1,2}})^2 \equiv (1+\delta_{1,2})m_o^2 \tag{22}$$

which defines the universal mass scale m_o and the departure from the universality.

In the presence of non-universal boundary values for the scalar masses the coefficients of the m_o^2 terms in eqs.(6) and (7) are appropriately modified. By solving the RGE in the same approximation we get[15]

$$\begin{aligned} m^2_{H_1,H_2} = \Bigg\{ 1 \ &- \ \frac{9Y}{7Y_f} + \delta_{1,2} - \frac{3Y}{14Y_f}(2\delta_Q + \delta_U + \delta_D + \delta_2 + \delta_1) \\ &\pm \ \frac{3}{10}\left[1 - \left(1 - \frac{Y}{Y_f}\right)^{5/7}\right](\delta_U - \delta_D + \delta_2 - \delta_1)\Bigg\} m_o^2 + \ldots \end{aligned} \tag{23}$$

where dots represent contributions proportional to A_o and $M_{1/2}$ which are the same as in eqs.(6) and (7) The change of parameter b is given by[15]

$$\Delta b = \left[\frac{2}{5} + \frac{3}{5}\left(1 - \frac{Y}{Y_f}\right)^{5/7}\right](\delta_1 - \delta_2) + \frac{3}{5}\left[1 - \left(1 - \frac{Y}{Y_f}\right)^{5/7}\right](\delta_U - \delta_D) \,. \tag{24}$$

It takes values O(0.1) for $(m_o^{H_1})^2 - (m_o^{H_2})^2$ or $(m_o^U)^2 - (m_o^D)^2$ of order $O(20\%)m_o^2$ and O(0.5) for $O(m_o^2)$, respectively.

It should be stressed that the pattern (21) of symmetry breaking is determined by the positive sign of the coefficient b and essentially does not depend separately on the values of $\delta_{1,2}$ and $\delta_{U,D}$ in eq.(24). However, as we shall see, there are important properties of the solutions which do depend in more detail on the character of the non-universality at the GUT scale, specified by the parameters δ_i.

Further properties of the solutions and certain phenomenological classification of non-universal boundary conditions follow from the equation (2). Since $m_2^2 = m_{H_2}^2 + \mu^2$, this equation can be interpreted as an equation for μ^2. As recalled earlier, in the universal case large values of μ^2 are needed to cancel large negative values of $m_{H_2}^2$. Now, with non-universal scalar terms it follows from eqs.(2), (23) and (24) that

$$\mu^2 = -\frac{M_Z^2}{2} + \frac{3}{7}\frac{Y}{Y_f}\left(1 - \frac{Y}{Y_f}\right) A_o \left(A_o - 4.6 M_{1/2}\right) + c_M M_{1/2}^2 + c_m m_o^2 \qquad (25)$$

where

$$c_M \simeq 6\frac{Y}{Y_f} - \frac{18}{7}\left(\frac{Y}{Y_f}\right)^2 - 0.5 \ , \qquad c_m \simeq \frac{9}{7}\frac{Y}{Y_f} - 1 + \Delta$$

and

$$\Delta = -\delta_2 + \frac{3Y}{14Y_f}\left(2\delta_Q + \delta_U + \delta_D + \delta_2 + \delta_1\right) + \frac{3}{10}\left[1 - \left(1 - \frac{Y}{Y_f}\right)^{5/7}\right]\left(\delta_U - \delta_D + \delta_2 - \delta_1\right)$$

is the term generated by departures from universality. Several observations follow from eq. (25): The strong $(M_{1/2},\ \mu)$ correlation triggered by eq. (25) in the limit of universal scalar terms disappears, due to the potential importance of both m_o and $M_{1/2}$ contributions, depending on the particular solution. Nevertheless, the values of μ^2 remain correlated with m_o or $M_{1/2}$ in the limit $m_0 \gg M_{1/2}$ and $M_{1/2} \gg m_o$, respectively. The values of μ^2 depend, contrary to the sign of b, on the pattern of the deviation from universality in the scalar masses. We obtain the following classification:

(A) $\mu^2 > c_M M_{1/2}^2$ for $c_m > 0$

(B) $\mu^2 < c_M M_{1/2}^2$ for $c_m < 0$

It is clear that with $b > 0$, as in eq. (24), the first option can be realized e.g. for $m_o^{H_2} < m_o$ or $m_o^U > m_o$, with the masses of the other scalars at m_o^2. Generically the values of μ remain then large $\mu \gg M_Z$ even in the limit $M_{1/2} \simeq 0$, due to the positive correlation with m_0^2.

The possibility (B) (with $b > 0$) can be realized e.g. when $m_o^D < m_o^U < m_o$ or/and $m_o < m_o^{H_2} < m_o^{H_1}$, for large enough deviations from universality. The parameter μ can then be arbitrarily small due to the cancellation between scalar and gaugino contributions in eq.(25). Note that for phenomenological reasons (experimental bounds) we are actually not interested in the strict limit $M_{1/2} = 0$. Thus, radiative breaking can be driven by $m_o \geq M_{1/2} \simeq O(M_Z)$, with $\mu \simeq O(M_Z)$ which is not correlated with the value of m_o. We stress that the above approximate considerations are only meant as a qualitative quideline to complete numerical calculations. Still, they provide a useful clasification of the non–universal boundary conditions for scalar masses which is based on the sign of their overall contribution to the equation for μ^2 and leads to very different solutions.

We have not addressed yet the question of consistency of radiative breaking with full unification of gauge and Yukawa couplings. As we know from ref.[4], this requirement is a strong constraint for the model with universal soft terms due to the large supersymmetric corrections to the bottom mass. With non–universal boundary conditions, in most of the parameter space consistent with radiative breaking, supersymmetric corrections to m_b are also very large. The procedure to check the consistency of radiative breaking with coupling unification in the minimal SO(10) is as follows. For chosen values of (m_t, α_s) we know the values of m_b predicted by the coupling unification, before correcting for supersymmetric loops. At the same time we can calculate those corrections at each point of the parameter space consistent with radiative electroweak breaking and check if the physical b mass is predicted to lie in the experimental range $m_b = (4.9 \pm 0.3)$ GeV. For given values of $(m_t, \tan\beta)$ we have therefore, the parameter space obtained after radiative breaking *and* constrained by the requirement of the right magnitude for the correction δm_b. Contrary to the universal case, it is possible to have solutions with small correction to the bottom mass with $\delta m_b / m_b \leq 10\%$, say. The condition

$$\left|\frac{\delta m_b}{m_b}\right| \approx 0.5 \left|\frac{M_{\tilde{g}}\mu}{m_{\tilde{q}}^2}\right| < 0.1 \tag{26}$$

can be fulfilled in case (A) (which gives $\mu \sim O(m_{\tilde{q}})$) for $m_{\tilde{q}} > 5M_{\tilde{g}}$. Taking into account the experimental limit $M_{\tilde{g}} > 150$ GeV we need $m_{\tilde{q}} > O(1-2)$ TeV and correspondingly $m_o \sim O(2-4)$ TeV (since $M_{\tilde{g}} \ll m_{\tilde{q}}$, the squark masses must be almost entirely determined by m_o).

In case (B), due to the lower limit on $M_{1/2}$ as a function of m_o (which follows from the experimental condition $\mu^2 \geq (50\ \text{GeV})^2$ and eq. (25)), the squark masses $m_{\tilde{q}}^2 \sim O(0.2)m_o^2 + O(4)M_{1/2}^2$ are determined by the gluino contribution and $M_{\tilde{g}}/m_{\tilde{q}} \sim O(1)$. The condition (26) is satisfied for $m_{\tilde{q}} > O(5\mu)$ which is achieved for $m_{\tilde{q}} > O(300\ \text{GeV})$, as μ can take small values. We shall illustrate those general considerations with several examples. Our numerical calculation is based on the method developed in ref.[9] and ref.[4] and our choice of parameters is as follows:
(a) $m_t = 170$ GeV, $\tan\beta = 48$ and with the GUT scale non–universal scalar masses $m_U^2 = 1.2m_o^2$, $m_D^2 = 0.8m_o^2$ for the third generation;
(b) $m_t = 180$ GeV, $\tan\beta = 53$ and $m_{H_2}^2 = 0.7m_o^2$;
(c) $m_t = 180$ GeV, $\tan\beta = 53$ and $m_{H_1}^2 = 2m_o^2$, $m_{H_2}^2 = 1.5m_o^2$;
m_o is the universal mass of the other scalars. The choice of the top quark mass and $\tan\beta$ values is dictated by our interest in consistent solutions with $|\delta m_b/m_b| < 10\%$. Therefore we pick up the values of m_t and $\tan\beta$ which correspond to the bottom mass values 5.1 and 5.0, respectively, (without susy corrections) and give acceptable final bottom quark mass with supersymmetric corrections in the range $|\delta m_b/m_b| \leq 0.1$ (at least for the negative sign of the correction). As stressed earlier the mechanism of radiative electroweak breaking and the pattern of solutions with non–universal scalar masses are totally insensitive to this particular choice of m_t and $\tan\beta$.

Our examples (a) and (b) illustrate the non–universality of type (A) and example (c) is a realization of case (B). First we refer to the overall regions in Fig.2a showing all solutions to radiative breaking: the only constraints which are taken into account are the present experimental limits on the sparticle masses and the requirement that the lightest neutralino is the lightest supersymmetric particle. For comparison the ($M_{1/2}$, m_o) region allowed in a universal case is also shown. In case (A) represented by our examples (a) and (b) the ($M_{1/2}$, m_o) region is bounded from below by the constraints (15) and (18). The values of (a, b) are generically $(O(0.1), O(0.01))$ and they depend, of course, on the assumed boundary values of the scalar masses. For large values of $M_{1/2}$ we have a lower bound on m_o from the requirement $m_{\tilde{l}} > m_{\chi_1^0}$ (the lightest neutralino). For large values of m_o the lower bound on $M_{1/2}$ is the experimental one obtained from the CDF limit on the gluino mass. In case (B) (represented by our example (c) with both a and b of order $O(0.1)$) additional constraints follow from the condition $\mu^2 \geq (50\text{ GeV})^2$ (given by the experimental limit on the chargino mass) which for a given m_o gives a lower bound on $M_{1/2}$. The absolute lower bound on $M_{1/2}$ is therefore higher than the experimental bound of $\sim$50 GeV. The plot in Fig.2b is also nicely consistent with our qualitative discussion. The lower limit on $M_{1/2}$ as a function of μ, in example (c), follows from eq. (25). The upper bound for $M_{1/2}$ as a function of μ in examples (a) and (b) corresponds to the linear correlation present in the universal case. Comparing examples (a) and (b), we see the effects of approaching the infrared fixed point limit $Y = Y_f$. The regions constrained by the condition (26) are marked in Fig. 2 and are characterized by upper bounds on $M_{1/2}$ for given m_o (i.e. $m_{\tilde{q}}$) or given μ. Again, comparing (a) and (b) we see the effect of approaching the limit $Y \to Y_f$.

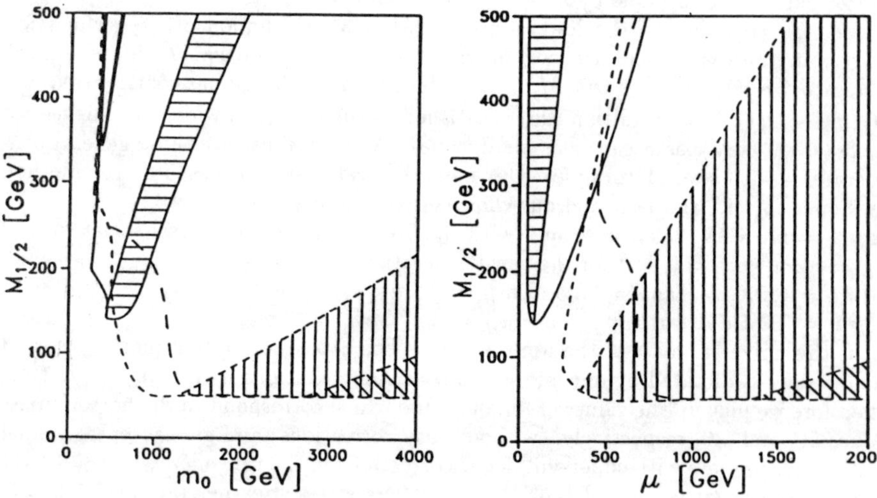

Fig. 2. The regions in a) ($M_{1/2}$, m_o) and b) ($M_{1/2}$, μ) which give radiative electroweak symmetry breaking and consistent with all experimental limits on sparticle masses.
Short dashed curves: $m_t = 170$ GeV, $\tan\beta = 48$, $m_{\bar{U}}^2(0) = 1.2m_o^2$, $m_{\bar{D}}^2(0) = 0.8m_o^2$.
Long dashed curves: $m_t = 180$ GeV, $\tan\beta = 53$, $m_{H_2}^2(0) = 0.7m_o^2$.
Solid curves: $m_t = 180$ GeV, $\tan\beta = 53$, $m_{H_1}^2(0) = 2.0m_o^2$, $m_{H_2}^2(0) = 1.5m_o^2$.
Heavy curve shows the region allowed in a universal case.

Morover, eq. (26) gives a lower bound on $M_{1/2}$ as a function of μ which is seen in Fig.2b. But at the same time μ is given by eq. (25) so this bound must be consistent with eq. (25). We obtain then a lower bound on m_o as a function of $M_{1/2}$, whereas the upper bound is given by the condition $\mu^2 \geq (50\ \mathrm{GeV})^2$. With $c_m < 0$, the allowed region is never empty. This explains the pattern seen in Fig.2a. One should stress the importance of the non–zero gaugino mass in satisfying simultaneously all the constraints in case (B). In particular it assures the positivity of the squark masses squared and of the μ^2 for a given deviation from universality in the scalar masses.

The superpartner mass spectra obtained with non–universal scalar masses show several new and interesting features as compared to the universal case. The most important aspects are (we discuss the spectra with eq. (26) taken into account):

1. Light charginos and neutralinos; they reflect very small values of $M_{1/2}$ (case (A)) or μ (case (B)). This latter case is particularly interesting as it revives our interest in light higgsino–like neutralinos.

2. Clear differences between sparticle spectra in case (A) and (B): they reflect the available parameter regions in both cases. We got, respectively
a) very heavy, $\geq$ O(1 TeV), or light, $\geq$ O(100 GeV), sleptons.
b) very heavy, $\geq$ O(1 TeV), or moderately light, $\geq$ O(300 GeV), squarks.
c) very light, $\geq$ O(150 GeV), or somewhat heavier, $\geq$ O(300 GeV), gluinos.

3. The Higgs sector is characterized by generically light pseudoscalar A. This follows from the equation $M_A^2 = aM_{1/2}^2 + bm_o^2 - M_Z^2$, with small coefficients a and b. In the allowed parameter range, both in cases (A) and (B), the mass M_A can reach the present experimental limit of about 50 GeV and increases with increasing m_o and/or $M_{1/2}$. The mass of the light scalar h is strongly correlated with M_A and changes from $M_h \sim 60$ GeV for $M_A \sim 50$ GeV to $M_h \sim (120-140)$ GeV for $M_A > 100$ GeV.

3.2. Experimental constraints

It is clear from our discussion of the universal case that the strong constraints from $b \rightarrow s\gamma$ and the neutralino relic abundance follow from the combination of the two properties: a) gaugino–like neutralinos b) negative corrections to m_b, which imply positive chargino contribution to the amplitude for $b \rightarrow s\gamma$ (the sign of μA_t is correlated with the sign of μM_2 by the RG running). A departure from universality of the soft terms can change both properties of the solutions. In both cases, (A) and (B), one can have solutions with small corrections $\delta m_b/m_b \leq 0.1$. Thus, Yukawa and gauge coupling unification is now consistent with both signs of the correction δm_b, while m_b remains in the acceptable range. Nevetherless, with gaugino–like neutralinos (A) the constraints from $b \rightarrow s\gamma$ and $\Omega h^2 < 1$ remain critical. The main annihilation channel is the one with CP-odd Higgs exchange. Since for the same reasons as in the universal case Ωh^2 is rising with $m_{\tilde{\chi}^0}$, and the annihilation amplitude is inversely proportional to $|m_A^2 - 4m_{\tilde{\chi}^0}^2|$ (typically, now, $m_A > 2m_{\tilde{\chi}^0}$), the condition $\Omega h^2 < 1$ gives us an upper bound on m_A. The corresponding contribution to $b \rightarrow s\gamma$ from the charged Higgs boson exchange is generically in conflict with the experimental result. The smallness of the correction $\delta m_b/m_b$ allows now, in principle, for both signs of this correction and, in consequence, for both relative signs of the H^+ and $\widetilde{\chi}^+$ contribution to $b \to s\gamma$. However, the latter turns out to be negligible and cannot cancel out too large corrections from the H^+-exchange: for this class of non-universal scalar masses (with $\mu \gg M_{1/2}$) the smallness of $\delta m_b/m_b$ can be achieved only at the expense of very heavy squarks, $m_{\tilde{q}} \gtrsim \mathcal{O}(2\,\mathrm{GeV})$ and the chargino contribution to $b \to s\gamma$ is strongly suppressed.

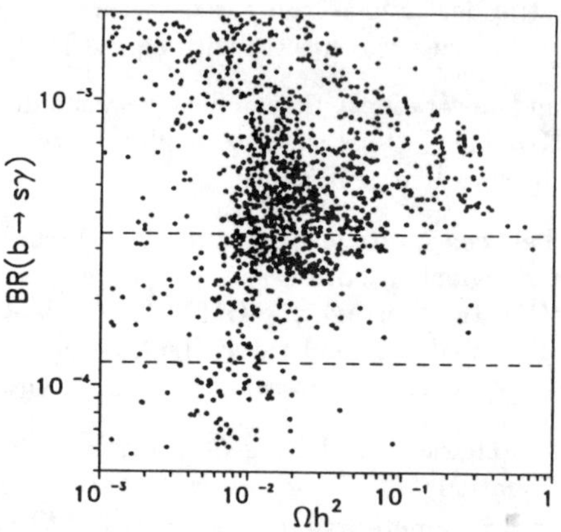

Fig. 3. Scatter plot of the central value (as defined in ref.[12]) of $\mathrm{BR}(b \to s\gamma)$ versus Ωh^2 in the minimal SO(10) model with radiative electroweak breaking and *non-universal* soft scalar masses at the GUT scale for $m_t = 180$ GeV, $\tan\beta = 53$, $m_{H_1}^2 = 2m_0^2$, $m_{H_1}^2 = 1.5m_0^2$ and the GUT scale values of the other scalar masses m_0^2; the low energy values of the m_Q and m_U were scanned up to 5 TeV. Dashed lines as in Fig. 1b.

The constraints discussed in this paper are easily satisfied for the class (B) of non–universal scalar masses which leads to a higgsino–like lightest neutralino. First of all, the neutralino annihilation can proceed now by Z exchange and essentially does not constrain the parameter space (apart from the known general limits on the neutralino mass). The relic abundance of the lightest neutralino is typically small, $\Omega h^2 \sim 10^{-2} - 10^{-1}$. Moreover, the supersymmetric contribution to the $\mathrm{BR}(b \to s\gamma)$ can be small due to the cancellation between the charged Higgs boson and higgsino amplitudes. In the limit of a pure higgsino–like lightest neutralino, ($m_{\tilde{\chi}^+} \approx \mu$), this can be seen from formula (18) taken in the limit $x = m_{\tilde{q}}^2/m_{\tilde{\chi}^+}^2 \to \infty$. For the considered class of non–universal scalar masses the limit is a good approximation: radiative electroweak breaking gives solutions[15] with $\mu < M_{1/2}$ and the condition (26) with $M_{\tilde{g}} \sim 2.7M_{1/2}$ and $m_{\tilde{q}}^2 \approx O(5)M_{1/2}^2$, gives $\mu \leq \frac{1}{3}M_{1/2}$, i.e. $\mu \leq \frac{1}{6}m_{\tilde{q}}$. In the limit $x \to \infty$

$$\Delta f^{(3)} = f^{(3)}(x) - f^{(3)}(x+\delta) \approx -\frac{\delta}{x}\frac{\ln x}{x} \tag{27}$$

where

$$x = \frac{m_{\tilde{t}_1}^2}{\mu^2} \ , \qquad \delta = \frac{\Delta m_{\tilde{t}_1}^2}{\mu^2} \approx \frac{2A_t m_t}{\mu^2} \ . \tag{28}$$

From eq. (9) we have $A_t \approx -2M_{1/2}$ and we indeed estimate that the contribution (17) and (18) can be of the same order of magnitude: For fixed $m_{\tilde{\chi}_2^+} \approx \mu$ the chargino contribution increases when the squark mass (i.e. $M_{1/2}$) decreases. However, the requirement $\delta m_b/m_b < 0.1$ gives a lower bound on $M_{1/2}$, i.e. a lower bound on $m_{\tilde{q}}$, for fixed μ and, therefore, an upper bound for the chargino contribution. Since squarks are aways much heavier than charginos, this upper bound remains strong (of the order of the H^+ contribution) even for light charginos. It decreases as μ is increasing because, for the maximal value of the ratio $\mu/M_{1/2}$ (fixed by $\delta m_b/m_b$), chargino contribution is proportional to $m_t/M_{1/2}$. We plot in Fig. 3 the value of the BR($b \to s\gamma$) versus Ωh^2 for our example (c) with non–universal terms of type (B).

4. Summary

In summary, non–universal scalar terms at the GUT scale open qualitatively new possibilities for radiative elecktroweak breaking with large $\tan\beta$ solutions. Two different patterns of non–universality can be identified, with distinct predictions for the particle spectra. Solutions of type (B) with $A_o \sim B_o \sim M_{1/2} \sim \mu \sim O(M_Z)$, $m_o \geq M_Z$ look particularly interesting. They can be consistent with experimental data on $b \to s\gamma$ and the relic abundance of the lightest neutralino $\Omega h^2 < 1$. On the other hand, the minimal SO(10) model with universal soft supersymmetry breaking terms at the GUT scale is ruled out by the combination of these two experimental constraints.

5. References

1. L.E. Ibanez and G.G. Ross, *Phys. Lett.* **B110** (1982) 215;
   K. Inoune et al., *Prog. Theor. Phys.* **68** (1982) 927;
   L. Alvarez–Gamumé, J. Polchinksy and M. Wise, *Nucl. Phys.* **B221** (1983) 495;
   J. Ellis, J. Hagelin, D. Nanopoulos and K. Tamvakis, *Phys. Lett.* **B125** (1983) 275;
   L.E. Ibánez and G. López, *Nucl. Phys.* **B233** (1984) 411.
2. For the most recent results see:
   R. Arnowitt and P. Nath, *Phys. Rev. Lett.* **69** (1992) 1014; *Phys. Lett.* **B287** (1992) 89 and **B289** (1992) 368;
   M. Olechowski and S. Pokorski, *Nucl. Phys.* **B404** (1993) 590;
   M. Carena, M. Olechowski, S. Pokorski and C. Wagner, *Nucl. Phys.* **B419** (1994) 213;
   V. Barger, M.S. Barger and P. Ohmann, preprint MAD/PH/801 (1994);
   G.L. Kane, C. Kolda, L. Roszkowski and J.D. Wells, UM–TH–94–03 (1994).

3. M. Olechowski and S. Pokorski, *Phys. Lett.* **B214** (1988) 393;
G.F. Giudice and G. Ridolfi, *Z. Phys.* **C41** (1988) 447;
H.P. Nilles, "Beyond the Standard Model", in Proceedings of the 1990 Theoretical Advanced Study Institute in Elementary Particle Physics, p. 633; Eds. M. Cvetic and P. Langacker, World Scientific;
P.H. Chankowski, Diploma Thesis (1990), University of Warsaw;
A. Seidl, Diploma Thesis (1990), Technical University, Munich;
W. Majerotto and B. Mösslacher, *Z. Phys.* **C48** (1990) 273;
S. Bertolini, F. Borzumati, A. Masiero and G. Ridolfi, *Nucl. Phys.* **B353** (1991) 591;
M. Drees and M.M. Nojiri, *Nucl. Phys.* **B369** (1992) 54;
B. Ananthanarayan, G. Lazarides and Q. Shafi, BA–92–29.
4. M. Carena, M. Olechowski, S. Pokorski and C. Wagner, *Nucl. Phys.* **B426** (1994) 269.
5. R. Ammar et. al (CLEO Collaboration), CLEO CONF 94-1.
6. F.M. Borzumati, M. Drees, M.M. Nojiri, *Phys. Rev.* **D51** (1994) 27.
7. P. Krawczyk and S. Pokorski, *Phys. Rev. Lett.* **60** (1988) 182.
8. B. Anantharayan, G. Lazarides and Q. Shafi, *Phys. Rev.* **D44** (1991) 1613
S. Dimopoulos, L.J. Hall and S. Raby, *Phys. Rev. Lett.* **68** (1992) 1984 and *Phys. Rev.* **D45** (1992) 4192
G.W. Anderson, S. Raby, S. Dimopoulos and L.J. Hall, *Phys. Rev.* **D47** (1993) 3702
G.W. Anderson, S. Raby, S. Dimopoulos, L.J. Hall and G.D. Starkman, *Phys. Rev.* **D49** (1994) 3660.
9. M. Olechowski and S. Pokorski, *Nucl. Phys.* **B404** (1993) 590.
10. L.J. Hall, R. Rattazzi and H. Sarid, LBL–33997 (1993),
R. Hampfling, *Phys. Rev.* **D49** (1994) 6168.
11. R. Barbieri and G.F. Giudice, *Phys. Lett.* **B309** (1993) 86.
12. A. J. Buras, M. Misiak, M. Münz and S. Pokorski, *Nucl. Phys.* **B424** (1994) 374.
13. G. Gelmini and P. Gondolo, *Nucl. Phys.* **B360** (1991) 145,
P. Gondolo, M. Olechowski and S. Pokorski, to appear.
14. D. Matalliotakis and H.P. Nilles, MPI–PhT/94-39, TUM-HEP-201-94 (1994)
N. Polonsky and A. Pomarol, University of Pennsylvania preprint UPR–0627T (1994), *Phys. Rev. Lett.* **73** (1994) 2292;
R. Hempfling, DESY preprint 94-078 (1994);
A. Leyda and C. Munoz, *Phys. Lett.* **B317** (1993) 82;
T. Kobayashi, D. Suematsu and Y. Yamagishi, KANAZAWA–94–06, hep–ph 9403330;
R. Rattazzi, U. Sardir and L.J. Hall, SU–ITP–94/15, hep–ph 9405313;
15. M. Olechowski and S. Pokorski, preprint MPI–Ph/94–40 (1994) to be published in Phys. Lett. B.

Superstring Inspired SO(10) and A Theory of Fermion Masses

K.S. Babu[1]
Bartol Research Institute
University of Delaware
Newark, DE 19716

and

R.N. Mohapatra[2]
Department of Physics and Astronomy
University of Maryland
College Park, MD 20742

Abstract

We give a general prescription for deriving predictive mass matrix ansatze for quarks and leptons in the framework of superstring inspired $SO(10)$ models. The key to our approach is a new way to naturally implement the doublet–triplet splitting which enables us to obtain symmetric quark and lepton mass matrices which have different structures in the up and the down quark sectors. We illustrate our method by deriving the Georgi–Jarlskog texture which has six predictions in the flavor sector, and then show how it generalizes to other symmetric texture models.

---

[1]Work supported by the Department of Energy Grant #DE-FG02-91ER406267

[2]Work supported by the National Science Foundation Grant #PHY-9119745

I. Introduction:

One of the major challenges in particle physics today is an understanding of the spectrum of quarks and leptons. In the standard model, all the fermion masses and the mixing angles are arbitrary, accounting for 13 of the 19 free parameters of the theory. This situation is clearly unsatisfactory and has been viewed by many as a hint for physics beyond the standard model. A more fundamental theory, it is hoped, will reduce this arbitrariness, by providing a common origin perhaps for several of these parameters. In the absence of a fundamental theory of fermion masses, it has been popular to assume certain restricted forms for the quark and lepton mass matrices that results in predictions for some of the observables. The hope in this approach is that if the predictions of a particular mass matrix ansatz are borne out by experiments, then a model that leads to a natural derivation of that ansatz may provide the next step in our search for the fundamental theory of nature.

The next step beyond the standard model may very well contain a supersymmetric grand unified theory (SUSY GUT). This speculation is supported by the dramatically accurate unification of gauge couplings that has been observed to occur [1] with supersymmetry around the TeV scale following the improved measurements of the low energy couplings at LEP and SLC. Superstring models, which are candidate theories of unifying gravity with the strong and electroweak forces, can lead to a variety of such SUSY GUT groups via an appropriate compactification scheme. It therefore appears to us to be a very desirable program to see if a successful mass matrix ansatz can receive a plausible derivation within a SUSY GUT model inspired by superstrings. This is what we take up in this paper.

There are a variety of mass matrix ansatze which can be characterized by their "texture" zeros. For increased predictivity, it is usually assumed that the matrices are symmetric. A well-known example is the Fritzsch ansatz [2] which has a texture similar in form for the up and the down quarks. Such matrices are easily derived from GUTs such as $SO(10)$, which has built-in left-right symmetry and up-down symmetry. If the recent CDF data [3] on the top quark, viz., $m_t = 174 \pm 17\ GeV$, holds up, it would appear that the Fritzsch ansatz would be ruled out even after renormalization group corrections are taken into account [4]. Another popular and predictive ansatz is the so-called Georgi-Jarlskog (GJ) texture [5], which assumes symmetric quark and lepton mass matrices, which however, have different forms in the up and down quark sectors. Such matrices, owing to their up-down asymmetry, are less trivial to derive in the context of SUSY GUTs such as $SO(10)$. In this letter, we provide a prescription for deriving such *up-down asymmetric* matrices in the context of superstring inspired $SO(10)$ models. We will illustrate our method by deriving the GJ texture, which has been the subject of substantial improvement and polishing including renormalization group effects in recent literature [6,7,8,9,10]. We will briefly summarize the six predictions of the ansatz to show its consistency with present data, especially with a heavy top as indicated by the recent CDF results [3]. Then we show how our method generalizes to other symmetric texture models. In particular, we show how it becomes possible to derive all of the five symmetric models listed in Ref. [11].

SUSY $SO(10)$ seems to us to be the ideal setting for addressing the quark and lepton masses. All fermions of a family are unified to a single irreducible representation of $SO(10)$, which facilitates the generation of symmetric mass matrix textures. (This is not the case in $SU(5)$ GUTs, where the down quark and the charged lepton mass matrices are not symmetric. Higher symmetries such as E_6 invariably brings in exotic particles which can mix with the known quarks and leptons, thus generally reducing the predictive power.) The emergence of the right-handed neutrino in $SO(10)$ leads to small but non-zero neutrino masses via the see-saw mechanism which may be highly desirable. The vexing problem of doublet-triplet splitting of SUSY GUTs also has an elegant resolution in $SO(10)$.

It has been known for some time that conventional SUSY GUTs such as $SO(10)$ with massless matter superfields belonging to the adjoint representation (needed for gauge symmetry breaking) can arise in the free fermionic formulation of superstrings [12]. It is well-known that superstring theories restrict considerably the number and the nature of gauge multiplets that survive to low energies, thereby reducing the arbitrariness present in a general GUT. They also provide naturally the discrete symmetries which are often needed in restricting the texture of quark and lepton mass matrices. Specifically, we are encouraged by the recent works of Choudhuri, Chung and Lykken [13] and Cleaver [14] who have constructed explicit $SO(10)$ models with adjoint scalars at the Kac-Moody level of two. These authors also classify the allowed representations that emerge as massless chiral multiplets below the Planck scale at the level two construction. While there can be any number of vectors (**10**), spinors ($\overline{\mathbf{16}} + \mathbf{16}$) and gauge singlets, the number of adjoints is restricted to be at most 2. Similarly, no more than one **54** can remain light, although no explicit example with any **54** has been constructed so far.

We shall be guided in our derivation of the mass matrix "textures" by the superstring constraints listed above. Specifically, the Higgs representations that we shall use for symmetry breaking and for fermion mass generation will be a spinorial $\overline{\mathbf{16}} + \mathbf{16}$, a pair of **10**, two adjoint **45** and a few singlets. In order to build a realistic and predictive model for quarks and leptons at low energies, we shall impose the following requirements: (i) there must be a consistent mechanism to break the SO(10) symmetry down to the standard model at the GUT scale; (ii) the light particle spectrum of the theory must be such that it preserves the successful prediction for

$\sin^2\theta_W$; (iii) the doublet–triplet splitting should be achieved naturally (i.e., without fine–tuning) in such a way that only one pair of Higgs doublet remains light (to be identified with H_u and H_d of MSSM); and finally, (iv) the same symmetry that helps satisfy the above requirements must provide an interesting texture for fermion mass matrices.

II. The Georgi-Jarlskog Texture and its predictions:

Let us first briefly summarize the predictions of the Georgi–Jarlskog ansatz. It assumes the following form for the up quark, down quark and the charged lepton mass matrices at the GUT scale:

$$M_u = \begin{pmatrix} 0 & a & 0 \\ a & 0 & b \\ 0 & b & c \end{pmatrix} ; \; M_d = \begin{pmatrix} 0 & de^{i\phi} & 0 \\ de^{-i\phi} & f & 0 \\ 0 & 0 & g \end{pmatrix} ; \; M_\ell = \begin{pmatrix} 0 & d & 0 \\ d & -3f & 0 \\ 0 & 0 & g \end{pmatrix} . \quad (1)$$

There are 7 parameters in all to fit the 13 observables (9 masses, 3 mixing angles and one CP phase), thereby resulting in six predictions. Three of these predictions are the b, s and d–quark masses:

$$m_b = \eta_{b/\tau}^{-1} m_\tau; \; \frac{m_d/m_s}{(1-m_d/m_s)^2} = 9\frac{m_e/m_\mu}{(1-m_e/m_\mu)^2}; \; (m_s - m_d) = \frac{1}{3}\eta_{s/\mu}^{-1}(m_\mu - m_e) \; . \quad (2)$$

The other three predictions are for the quark mixing angles and the CP phase J:

$$|V_{cb}| = \eta_{KM}^{-1}\eta_{u/t}^{1/2}\sqrt{\frac{m_c}{m_t}} ; \qquad \frac{|V_{ub}|}{|V_{cb}|} = \sqrt{\frac{m_u}{m_c}} ;$$

$$J = \eta_{KM}^{-2}\eta_{u/t}\sqrt{\frac{m_d}{m_s}}\sqrt{\frac{m_c}{m_t}}\sqrt{\frac{m_u}{m_t}}\left[1 - \frac{1}{4}\left(\sqrt{\frac{m_u}{m_c}}\sqrt{\frac{m_s}{m_d}} + \sqrt{\frac{m_c}{m_u}}\sqrt{\frac{m_d}{m_s}} - \sqrt{\frac{m_c}{m_u}}\sqrt{\frac{m_s}{m_d}}|V_{us}|^2\right)^2\right]^{\frac{1}{2}} \quad (3)$$

Here the η's are renormalization factors for the various parameters in going from the low energy scale to the GUT scale. If the bottom–quark Yukawa coupling h_b is much smaller than the top Yukawa coupling h_t, (corresponding to $\tan\beta \lesssim 10$ or so) these RGE factors can be expressed analytically as

$$\eta_{KM} = \eta_{d/b} = \left(1 - \frac{Y_t}{Y_f}\right)^{\frac{1}{12}} ; \; \eta_{u/t} = \left(1 - \frac{Y_t}{Y_f}\right)^{\frac{1}{4}} ; \eta_{s/\mu} = \left(\frac{\alpha_1}{\alpha_G}\right)^{-10/99}\left(\frac{\alpha_3}{\alpha_G}\right)^{-8/9} ;$$

$$\eta_{b/\tau} = \left(\frac{\alpha_1}{\alpha_G}\right)^{-10/99}\left(\frac{\alpha_3}{\alpha_G}\right)^{-8/9}\left(1-\frac{Y_t}{Y_f}\right)^{-1/12} . \tag{4}$$

Here α_G is the unified gauge coupling strength, $Y_t = h_t^2$ at the weak scale and Y_f is the fixed point value of Y_t. Y_t cannot strictly be equal to Y_f, since that would correspond to infinite Y_t at the GUT scale. If we demand that $Y_t \lesssim 4$ at GUT scale, Y_t/Y_f can at most be 0.98. The renormalization factors in Eq. (4) are all well–behaved even when Y_t differs from Y_f by only 2% due to the small exponents.

In the three Yukawa unification scheme ($h_t = h_b = h_\tau$ at the GUT scale), the exponents $(1/12, 1/12, 1/4)$ corresponding to $(\eta_{KM}, \eta_{d/b}, \eta_{u/t})$ in Eq. (4) will be replaced by $(1/7, 2/7, 2/7)$; $\eta_{s/\mu}$ will remain unchanged, while $\eta_{b/\tau}$ is modified to

$$\eta_{b/\tau} = \left(\frac{\alpha_1}{\alpha_G}\right)^{-19/376}\left(\frac{\alpha_3}{\alpha_G}\right)^{-2/9}\left(\frac{Y_t}{Y_\tau}\right)^{-3/8}\left(1-\frac{Y_t}{Y_f}\right)^{-1/14} . \tag{5}$$

The η factors given in Eqs. (2)-(5) correspond to running the parameters from the weak scale (taken here to be m_t which is also assumed to be the SUSY threshold) up to M_{GUT}. For the light fermion masses there is QCD and QED running factors below m_t as well. These are obtained numerically using three loop QCD and one–loop QED β and γ functions. Corresponding to $\alpha_s(M_Z) = 0.12$ and $\alpha^{-1}(M_Z) = 127.9$, these factors are $(\eta_u, \eta_{d,s}, \eta_c, \eta_b, \eta_{e,\mu}, \eta_\tau) = (0.401, 0.404, 0.460, 0.646, 0.982, 0.984)$. If $\alpha_s(M_Z) = 0.125$ is chosen these factors become $(0.356, 0.358, 0.422, 0.630, 0.982, 0.984)$. Using $\alpha_s(M_Z) = 0.12, m_c(m_c) = 1.27\ GeV, m_u(1\ GeV) = 5.1\ MeV$, $|V_{us}| = 0.22$, $m_t^{\rm phys} = 174\ GeV$ and $Y_t/Y_f = 0.98$ as input values, we obtain $m_d(1\ GeV) = 7.7\ MeV$, $m_s(1\ GeV) = 193\ MeV, m_b(m_b) = 4.26\ GeV, |V_{cb}| = 0.050$, $|V_{ub}|/|V_{cb}| = 0.059$, $J = 2.96\times10^{-5}$. The value of $|V_{cb}|$ corresponding to $\alpha_s(M_Z) = 0.125, m_c(m_c) = 1.22\ GeV, m_t^{\rm phys} = 190\ GeV$ is $|V_{cb}| = 0.045$. In the case of three Yukawa unification, $h_t = h_b = h_\tau$, $|V_{cb}|$ is unrenormalized and results in a larger value, which is therefore disfavored. Note that all the predictions of the GJ ansatz are presently in good agreement with experiments, especially for the case of small tanβ.

III. Superstring inspired SO(10) and doublet-triplet splitting:

We now turn to the derivation of eq. (1) in superstring inspired SO(10) models. First notice that the up and the down quark mass matrices in Eq. (1) have a very asymmetric structure. In generic SUSY SO(10) models, they tend to have similar form. The reason is that the Higgs multiplets that contributes to fermion masses such as **10** and $\overline{\mathbf{126}}$ contain both up and down type Higgs doublets of MSSM and unless care is taken, the both light Higgs doublets that survive the GUT symmetry breaking contain contributions from the same GUT multiplet, so that after symmetry breaking M_u and M_d have similar forms. So the first thing we have to show is how to generate such an up–down asymmetry within $SO(10)$.

Let us consider the mass matrix in Eq.(1). The effective **10** or $\overline{\mathbf{126}}$ operator that generates the (23) element of M_u must develop an electroweak VEV only along the up (and not along the down) direction. Similarly, the $\overline{\mathbf{126}}$ operator that induces the (22) element of M_d and M_l must develop an electroweak VEV only along the down direction. Now, if the successful prediction of $\sin^2\theta_W$ of SUSY GUTs is to be preserved, only one pair of Higgs(ino) doublets can remain light (H_u and H_d of MSSM). These are the only doublets that acquire electroweak VEVs. There could be several doublet Higgs(ino)s at the GUT scale which mix with one another, but in such a way that only one pair of them remains light [15]. The texture of the mass matrices in Eq. (1) then requires that the Higgs doublet coupling to the (23) entry in M_u should have an admixture of H_u, but not of H_d and similarly, the Higgs doublet generating (22) entry of M_d and M_l should have an admixture of H_d, but not H_u. This leads to the important observation that the texture zeros of the fermion mass matrices can be ensured only if the Higgs(ino) mass matrix has its own texture zeros. Furthermore, it requires that the Higgs(ino) mass matrix be asymmetric, otherwise an up–down asymmetry cannot be induced.

Which linear combination of up–type and down–type Higgs(ino) doublets remains light is of course related to the question of doublet–triplet splitting of SUSY GUTs. The simplest way to achieves a natural doublet–triplet splitting in $SO(10)$ is via the Dimopoulos–Wilczek mechanism[16]. This is achieved by the coupling H_1AH_2 where H_1 and H_2 belong to the **10** of $SO(10)$, while A is the adjoint **45**. This term gives masses to the color triplets in **10**'s but not to the $SU(2)$ doublets if the VEV of the adjoint A is chosen to be along the $(B-L)$ direction: $\langle A\rangle = diag.(a,a,a,0,0)\times i\tau_2$. In realistic SO(10) models, this pattern of vev's is not stable and new techniques are needed to make this method useful [17,18]. Furthermore, it leads to up-down symmetric mass matrix textures for the fermions, which are not useful for our purpose.

In order to generate an asymmetry in the Higgs(ino) mass matrix we propose to mix the **10**-plets (H_1 and H_2) with the spinorial $\mathbf{16}+\overline{\mathbf{16}}$ ($\psi_H+\overline{\psi}_H$), which are needed anyway for symmetry breaking. Each **10** contains in it one up–type and one down–type Higgs doublet, ψ_H contains a down–type doublet, while $\overline{\psi}_H$ contains an up–type doublet. We have therefore a total of three up–type and three down–type Higgs doublets. The **10**'s also contain a color triplet–antitriplet pair, ψ_H has a color antitriplet while $\overline{\psi}_H$ has the color triplet. The superpotential involving the $SU(2)$ doublets and the color triplets is assumed to be

$$W_{DT} = \lambda_1\psi_H\psi_H H_1 + \lambda_2\overline{\psi}_H\overline{\psi}_H H_2 + H_1AH_2\ . \tag{6}$$

This is the most general superpotential relevant for the doublet–triplet splitting compatible with the set of discrete symmetries given in Table. 1. Eq. (6) leads to the following mass matrices for the Higgs(ino) doublets and color triplets (written in the basis where the rows correspond to $(\psi_{H_d}, H_{2d}, H_{1d})$ and the columns stand for $(\overline{\psi}_{H_u}, H_{1u}, H_{2u})$ for the doublets and similarly for the color triplets)

$$M_D = \begin{pmatrix} 0 & \lambda_1 v_R & 0 \\ \lambda_2 v_R & 0 & 0 \\ 0 & 0 & 0 \end{pmatrix}; \quad M_T = \begin{pmatrix} 0 & \lambda_1 v_R & 0 \\ \lambda_2 v_R & \lambda_3 a & 0 \\ 0 & 0 & -\lambda_3 a \end{pmatrix}. \tag{7}$$

Here $\langle \psi_H \rangle = \langle \overline{\psi}_H \rangle = v_R$ and $\langle A \rangle = a$. This gives GUT scale Dirac masses to all three triplet–antitriplet pairs. Two pairs of Higgs(ino) doublets become superheavy, leaving one pair of doublets light. The H_u and H_d fields of MSSM are easily identified as $H_u = H_{2u}$ and $H_d = H_{1d}$. Note that the desired asymmetry has been achieved, H_1 will acquire a VEV only along the down direction ($\langle H_{1d} \rangle \neq 0, \langle H_{1u} \rangle = 0$), while H_2 will acquire a VEV only along the up direction ($\langle H_{2d} \rangle = 0, \langle H_{2u} \rangle \neq 0$). This serves as the first step in our derivation of the GJ ansatz. In this scheme, $\tan\beta \neq m_t/m_b$ owing to the (10-16) mixing.

IV. GUT symmetry breaking:

Let us now turn to the symmetry breaking sector. Nonrenormalizable operators suppressed by appropriate powers of inverse Planck mass, which are expected to arise in superstring theories, will play a crucial role in our analysis. In addition, we shall make use of a few gauge singlet superfields to generate the GUT scale starting from the Planck scale and the SUSY breaking scale $m_0 \sim 1\ TeV$. To see how this works, consider a nonrenormalizable superpotential involving a singlet superfield

$$W = \lambda \frac{\phi^n}{M_{Pl}^{n-3}} \tag{8}$$

where n is an integer. The corresponding scalar potential, including soft supersymmetry breaking terms is given by

$$V = m_0^2 |\phi|^2 + \left| \frac{n\lambda\phi^{n-1}}{M_{Pl}{}^{n-3}} \right|^2 + m_0 A_n \left[\frac{\lambda \phi^n}{M_{Pl}{}^{n-3}} + H.C. \right] \tag{9}$$

In the supersymmetric limit, the potential has a unique minimum given by $\langle \phi \rangle = 0$.

But including SUSY breaking, the potential develops two other minima given by

$$\langle|\phi|\rangle = M_{Pl}\left[\frac{1}{2}\left|\frac{A_n m_0}{\lambda M_{Pl} n(n-1)}\right|\left(1 \pm \sqrt{1-\frac{4(n-1)}{A_n^2}}\right)\right]^{\frac{1}{n-2}} . \tag{10}$$

For $n = 8, 9, 10$, this leads to $\langle|\phi|\rangle \sim 10^{16}\ GeV$ which is near the GUT scale. The existence of these symmetry breaking minima requires that $A_n^2 \geq 4(n-1)$ (see Eq. (10)). We have investigated if this constraint is satisfied automatically in $N = 1$ supergravity models with hidden sector SUSY breaking *a la* Polony. We found $|A_n| = |n - \sqrt{3}|$ in this scheme, which satisfies the constraint for $n \geq 7$.

Let us now display the part of the superpotential that breaks the gauge symmetry down to the standard model. A minimum of one adjoint **45** and a $\mathbf{16}+\overline{\mathbf{16}}$ pair Higgs superfields are required for this [17,18]. In order to generate the GJ texture, we shall also need a second adjoint A'. A, which has a VEV along the $(B-L)$ direction, $\langle A\rangle = diag.(a,a,a,0,0) \times i\tau_2$, is responsible for the doublet–triplet splitting. A' develops a VEV along the I_{3R} direction, $\langle A'\rangle = diag.(0,0,0,a',a') \times i\tau_2$. In addition, we shall need a few singlets, in order to generate the GUT scale as discussed above, as well as for the fermion mass matrix texture. These singlets are denoted by $X, Y, Z, \overline{Z}$ and $S_{1,2,3}$, all of which develop GUT scale VEVs except for the field X which has zero VEV. The superpotential which achieves the symmetry breaking and the desired VEV structure is given by

$$\begin{aligned} W_{sym} = X(\overline{\psi}_H\psi_H\ -\ Y^2)\ + A^2Z^2 + A^2\overline{Z}^2 + A^4 + A^2A'^2 + A'^2Z^2 + A'^2\overline{Z}^2 + \\ A'^4 + (Z\overline{Z})^2 + Z^4 + \overline{Z}^4 + M\overline{Z}Z + S_1^8 + S_2^8 + S_1^2S_2^4S_3^4 + X^8A^2 + \ldots \end{aligned} \tag{11}$$

This is the most general superpotential up to relevant orders consistent with the discrete symmetries given in Table 1. We should mention here that we have not tried to be economical with the choice of the discrete symmetry. Rather, we have identified the maximal symmetry of the desired superpotential in Table 1. The full symmetry of Table 1 may not be essential to achieve GJ texture. It is understood that in Eq. (11), each term is suppressed by the dimensionally appropriate powers of M_{Pl}. Terms such as $A^4, A^2A'^2$ in Eq. (11) stand for more than one possible contraction of indices. If the mass parameter M is chosen to be M_{GUT}^2/M_{Pl}, $\langle A\rangle, \langle A'\rangle \sim M_{GUT}$ will be generated. The superpotential gives rise to the VEVs of A and A' in the desired $(B-L)$ and I_{3R} directions respectively. Except for the question of linking the (A, A') sector to the $(\psi_H, \overline{\psi}_H)$ sector (this issue will be addressed shortly), which is necessary to avoid pseudogoldstone bosons, the symmetry breaking down to the standard model has now been achieved. As for the corrections to the doublet Higgs(ino) mass matrix, the leading order terms are $H_1H_2Y^4/M_{Pl}^3$ and $H_1H_2(\psi_H\overline{\psi}_H)^2/M_{Pl}^3$. These will induce an effective μ parameter for H_u and H_d of MSSM. Demanding that the μ parameter is $\lesssim 1\ TeV$, we find that $\langle\psi_H\rangle, \langle Y\rangle \sim 10^{15}\ GeV$, which is satisfactorily close to the GUT scale. Terms such as $\psi_H\psi_H H_2$ that can also contribute to the μ parameter do not arise until very high order.

V. Planck scale discrete symmetry and derivation of the G-J mass texture:

We now proceed to the derivation of the GJ texture. We assume that the underlying superstring theory leaves an appropriate discrete symmetry, which we assume for our purpose to be that displayed in table 1. The relevant superpotential involving the matter fields consistent with these discrete symmetries is

$$\begin{aligned} W_{Yuk} &= M^{-1}\left(h_{33}\psi_3\psi_3 H_1 S_1 + h'_{33}\psi_3\psi_3 H_2 S_2 + h_{23}\psi_2\psi_3 H_2 S_3\right) + \\ &\quad M^{-2} h_{22}\psi_2\psi_2 H_1 A A' + M^{-3}\left(h_{12}\psi_1\psi_2 H_1 S_2^3 + h'_{12}\psi_1\psi_2 H_2 S_1^3\right) . \quad (12) \end{aligned}$$

Here M is not necessarily the Planck mass, Eq. (12) could arise by integrating out vector–like fermions such as $\overline{\mathbf{16}} + \mathbf{16}$ which have an intermediate mass between the GUT and the Planck scale. All parameters in Eq. (12) can be made real by field redefinitions, but the VEVs of A, A' are complex in general, so one unremovable phase will reappear after symmetry breaking as in Eq. (1).

The h_{22} term in Eq. (13) has quite interesting properties, which needs a bit of explanation. In general, the $H_1 AA'$ can give rise to effective $\mathbf{10}$ as well as $\overline{\mathbf{126}}$ operators. The GJ texture requires that only the $\overline{\mathbf{126}}$ contributes. The interesting feature of this term with $\langle A\rangle$ along $(B-L)$ direction and $\langle A'\rangle$ along I_{3R} direction is that only the $\overline{\mathbf{126}}$ operator contributes to the mass matrices. Effective $\mathbf{10}$ can arise in three different ways, all of which vanish at the minimum, due to the orthogonality of $\langle A\rangle$ and $\langle A'\rangle$. To see this, we first observe that $\mathrm{Tr}(AA') = 0$, which sets one such contribution to zero. Other terms of the type $A_{ab}A'_{bc}H_{1c}$ and $A'_{ab}A_{bc}H_{1c}$ also vanish at the minimum. One is left with the term $A_{ab}A'_{cd}H_e$, which is precisely the $\overline{\mathbf{126}}$ contribution. This does not vanish and as a result, we are able to reproduce the relative factor of -3 in the (22) entry of M_d and M_l.

If all the Yukawa coupling parameters in Eq. (12) are chosen to be of order one, the hierarchy in the masses can be explained purely by the ratio $\langle S_i\rangle/M \equiv \epsilon$. Except for S_2, which generates the top quark mass, we will choose these ratios to be in the range 1/10 to 1/30. As noted already, the doublet–triplet splitting mechanism implies that $H_{1d} = H_d$ and $H_{2u} = H_u$. Making this identification, we see that below the GUT scale, Eq. (12) results in the following effective Yukawa superpotential (with redefined h_{ij}).

$$\begin{aligned} W^{eff}_{Yuk} &= h'_{33}Q_3 H_u u_3^c + \epsilon h_{33}(Q_3 H_d d_3^c + L_3 H_d e_3^c) + h_{23}\epsilon(Q_2 H_u u_3^c + Q_3 H_u u_2^c) + \\ &\quad h_{22}\epsilon^2(Q_2 H_d d_2^c - 3L_2 H_d e_2^c) + h'_{12}\epsilon^3(Q_1 H_u u_2^c + Q_2 H_u u_1^c) + \\ &\quad h_{12}\epsilon^3(Q_1 H_d d_2^c + Q_2 H_d d_1^c + L_1 H_d e_2^c + L_2 H_d e_1^c) . \quad (13) \end{aligned}$$

The desired GJ texture follows from this after electro-weak symmetry breaking. It is easy to see that the texture zeros of the mass matrices are protected to very high order by the discrete symmetry.

Let us now turn briefly to the neutrino sector. The Dirac neutrino matrix is identical to the up quark matrix in the model. As for the right–handed neutrino Majorana mass matrix, we have no more freedom to choose its form since its discrete symmetry assignment has already been fixed in the process of deriving the GJ ansatz. The allowed right–handed neutrino mass terms are

$$W_{\nu_R} = \left(M^{-3}\psi_3\psi_3 S_1 A + M^{-2}\psi_2\psi_2 A' + M^{-10}\psi_1\psi_1 A'^3 S_2^6\right)\overline{\psi}_H\overline{\psi}_H \; . \tag{14}$$

The resulting ν_R mass matrix is nonsingular, and thus generates small neutrino masses for ν_e, ν_μ and ν_τ. Note that although the last term in Eq. (14) has higher inverse powers of M, it is compensated largely by the VEV of S_2, which cannot be too much below M as it generates the top quark mass. Thus the neutrino spectrum is similar to what is expected in non–SUSY $SO(10)$ models with an intermediate scale. Such a spectrum is known to be capable of resolving the solar neutrino puzzle via the MSW mechanism while leaving the ν_τ mass in the cosmologically significant multi–eV range. It is interesting that the superstring constraint on the Higgs spectrum is what is responsible for the weakening of the see–saw suppressions somewhat.

Let us now turn to the question of linking the (A, A') and the $(\overline{\psi}_H, \psi_H)$ sectors. If the two sectors were not linked, there will be pseudogoldstones belonging to a **10** + $\overline{\mathbf{10}}$ of $SU(5)$. Although the successful prediction of $\sin^2\theta_W$ will be unaffected, since they form complete multiplets of $SU(5)$, they do upset the prediction for m_b. In order to make them superheavy, one cannot link the two sectors directly, as that would upset the VEV of A. The simplest way to achieve this is to assume a term $\mathrm{Tr}(AA'A'')$ along with $A''\overline{\psi}_H\psi_H$ in the superpotential. Here A'' is another adjoint. The $\mathrm{Tr}(AA'A'')$ term, due to its complete anti-symmetry, does not upset the VEV structure of A, A', yet it makes all the pseudogoldstones superheavy. The introduction of a third adjoint would appear to be in conflict with the superstring constraints, however, A'' need not survive below the Planck scale, in which case there is no contradiction. To be specific, let us add the following terms to the superpotential:

$$W' = \overline{\psi}_H\psi_H A''P + AA'A''Q + (A''Q)^2 + Q\overline{Q} + (Q\overline{Q})^2 + (P\overline{Q})^8 \tag{15}$$

where $P, Q, \overline{Q}$ are gauge singlets. These terms are clearly consistent with the discrete symmetries of Table 1. W' also has a Z_n symmetry under which $(A'', \overline{Q})$ and (P, Q) have opposite charges. This superpotential admits $\langle Q\rangle \sim M_{Pl}$, so that A'' has a mass of order M_{Pl}. P gets a VEV of order M_{GUT} from the last term in Eq. (15). The coupling of A'' to the other fields make all pseudogoldstones heavy, of order

M_{GUT}^2/M_{Pl}. A'' will receive an induced VEV, but it is of order M_{GUT}^3/M_{Pl}^2. The only effect of A'' on the fermion mass matrices is to give a tiny correction to the (22) element of $M_{d,l}$, via the term $\psi_2\psi_2 H_1 A'' Q$, which is about 10^{-3} times smaller than the leading term.

VI. Conclusions and Outlook:

Let us finally show how the method developed here facilitates the derivation of other symmetric mass matrix textures. Take for example, Model (4) of Ref. (11), which is obtained by adding a (22) entry in M_u of Eq. (1). Such a texture follows readily in our scheme by a new Yukawa term $\psi_2\psi_2 H_2 S$ with S a gauge singlet carrying discrete charge of $(4,4,0,2)$. All the five symmetric texture models of Ref. (11) can be derived in an analogous fashion. Our method can be applied to derive asymmetric texture models as well in the context of $SO(10)$ [19].

We conclude by observing some interesting variations of the doublet–triplet splitting scheme (Eqs. (6)-(7)). One could add to Eq. (6) direct mass terms $(M_\psi \psi_H \overline{\psi}_H + M_1 H_1^2 + M_2 H_2^2)$ with $M_\psi M_1 M_2 = 0$ to ensure a light doublet. If $M_1 = 0$, the MSSM fields H_u and H_d are $H_u \propto (M_\psi M_2 H_{1u} + \lambda_1\lambda_2 v_R^2 H_{2u} - \lambda_1 v_R M_2 \overline{\psi}_{H_u})$ and $H_d = H_{1d}$. Thus $\langle H_{1u}\rangle, \langle H_{1d}\rangle, \langle H_{2u}\rangle, \langle \overline{\psi}_{H_u}\rangle \neq 0$, while $\langle H_{2d}\rangle = \langle \overline{\psi}_{H_d}\rangle = 0$. Such a spectrum enables one to use Yukawa couplings to the H_1 field to induce some common elements in $M_{u,d,l}$ while generating an up–down asymmetry via the couplings of H_2 and $\overline{\psi}_{H_u}$. If $M_\psi = 0, M_1 M_2 \neq 0$, then $H_d = \cos\theta H_{1d} + \sin\theta \psi_{H_d}$ and $H_u = \cos\theta' H_{2u} + \sin\theta' \overline{\psi}_{H_u}$, with θ, θ' Higgs(ino) mixing angles. The up–down asymmetry is similar to Eq. (7), but now there is the freedom of involving the $\psi_{H_d}, \overline{\psi}_{H_u}$ fields in the fermion mass generation.

In summary, we have constructed the first consistent string inspired model that reproduces a predictive fermion mass matrix, which is stable under Planck-scale induced corrections. The key ingredients are (i) a new doublet-triplet splitting mechanism; (ii) a new way to simulate the effect of $\overline{\mathbf{126}}$ using only $\mathbf{45}$ dim. representations. We have indicated how this method can be used to generate other mass matrix textures. In our work, we had to assume a maximal form for the discrete symmetry to restrict the fermion mass matrices. As further work on string theories reveal more about the nature of the discrete symmetries, this gap between string theories and low energy physics can be closed leading to a more satisfactory grand unified framework for particle physics.

References

1. U. Amaldi, W. de Boer, and H. Furstenau, Phys. Lett. **B260**, 447 (1991); P. Langacker and M.X. Luo, Phys. Rev. **D 44**, 817 (1991), J. Ellis, S. Kelley,

and D.V. Nanopoulos, Phys. Lett. **B260**, 131 (1991).

2. H. Fritzsch, Phys. Lett. **70B**, 436 (1977); Nucl. Phys. **B155**, 189 (1979).
3. CDF Collaboration, F. Abe et. al., Phys. Rev. Lett. **73**, 225 (1994).
4. K.S. Babu and Q. Shafi, Phys. Rev. **D 47**, 5004 (1993); Y. Achiman and T. Greiner, Phys. Lett. **B329**, 33 (1994).
5. H. Georgi and C. Jarlskog, Phys. Lett. **86B**, 297 (1979).
6. J. Harvey, P. Ramond and D. Reiss, Phys. Lett. **B92**, 309 (1980); Nucl. Phys. **B199**, 223 (1982).
7. X.G. He and W.S. Hou, Phys. Rev. **D41**, 1517 (1990)
8. S. Dimopoulos, L.J. Hall and S. Raby, Phys. Rev. Lett. **68**, 752 (1992); Phys. Rev. **D 45**, 4192 (1992).
9. H. Arason, D. Castano, E.J. Pirad and P. Ramond, Phys. Rev. **D 47**, 232 (1993).
10. V. Barger, M.S. Berger, T. Han and M. Zralek, Phys. Rev. Lett. **68**, 3394 (1992).
11. P. Ramond, R.G. Roberts and G.G. Ross, Nucl. Phys. **B406**, 19 (1993).
12. D. Lewellen, Nucl. Phys. **B337**, 61 (1990); A. Font, L. Ibanez, and F. Quevedo, Nucl. Phys. **B345**, 389 (1990).
13. S. Chaudhuri, S. Chung and J. Lykken, FERMILAB-PUB-94/137-T (1994).
14. G. Cleaver, Ohio State Preprint OHSTPY-HEP-T-94-007 (1994).
15. D. Lee and R.N. Mohapatra, Phys. Rev. **D51**, (to appear) (1995).
16. S. Dimopoulos and F. Wilczek, Report No. NSF-ITP-82-07, August 1981 (unpublished); R.N. Cahn, I Hinchliffe and L. Hall, Phys. Lett. **109B**, 426 (1982).
17. K.S. Babu and S.M. Barr, Phys. Rev. **D 48**, 5354 (1993); Phys. Rev. **D 50**, 3529 (1994).
18. K.S. Babu and S.M. Barr, Bartol Preprint BA-94-45 (1994).
19. S.M. Barr, Phys. Rev. Lett. **64**, 353 (1990); Z. Berezhiani and R. Rattazzi, Nucl. Phys. **B 407**, 249 (1993); G. Anderson, S. Dimopoulos, L. Hall, S. Raby and G.D. Starkman, Phys. Rev. **D 49**, 3660 (1994).

Table. 1. The matter superfields along with their discrete charge assignments. The subscript in Z and $\overline{Z}$ stand for a Z_2 symmetry under which all other fields are even.

Superfield	$Z_8 \times Z_{16} \times Z_{16} \times Z_4$
$\psi_H(\mathbf{16})$	(1,0,0,0)
$\bar{\psi}_H(\overline{\mathbf{16}})$	(7,2,0,0)
$H_1(\mathbf{10})$	(6,0,0,0)
$H_2(\mathbf{10})$	(2,12,0,0)
$A(\mathbf{45})$	(0,4,0,0)
$A'(\mathbf{45})$	(0,12,0,2)
$Y(\mathbf{1})$	(0,1,0,0)
$X(\mathbf{1})$	(0,14,0,0)
$Z(\mathbf{1})$	$(0,4,0,0)_-$
$\bar{Z}(\mathbf{1})$	$(0,12,0,0)_-$
$\psi_1(\mathbf{16})$	(7,4,6,1)
$\psi_2(\mathbf{16})$	(1,0,0,1)
$\psi_3(\mathbf{16})$	(0,0,1,1)
$S_1(\mathbf{1})$	(2,0,14,2)
$S_2(\mathbf{1})$	(14,4,14,2)
$S_3(\mathbf{1})$	(13,4,15,2)

SOLVING SUSY GUT PROBLEMS: GAUGE HIERARCHY AND FERMION MASSES

ZURAB BEREZHIANI

Istituto Nazionale di Fisica Nucleare, Sezione Ferrara
44100 Ferrara, Italy

and

Institute of Physics, Georgian Academy of Sciences
38077 Tbilisi, Georgia

ABSTRACT

The supersymmetric $SU(6)$ model accompanied by flavour-blind discrete symmetry Z_3 can succesfully deal with such key problems of SUSY GUTs, as are gauge hierarchy/doublet-triplet splitting, μ-problem and flavour problem. The Higgs doublets arise as Goldstone modes of the spontaneously broken *accidental* global $SU(6) \times U(6)$ symmetry of the Higgs superpotential. Their couplings to fermions have peculiarities leading to consistent picture of the quark and lepton mass spectrum and mixing, without invoking any of horizontal symmetry/zero texture concepts. Namely, the only one particle that has the direct Yukawa coupling with Higgs doublet is top quark. Other fermion masses appear from higher order operators, with natural mass hierarchy. Specific mass formulas are also obtained.

1. Introduction

The experimental developments of the last few years have confronted us with the impressive phenomenon of gauge coupling crossing[1] in the framework of minimal supersymmetric standard model (MSSM): at the scale $M_G \simeq 10^{16}$ GeV, $SU(3)\times SU(2)\times U(1)$ can be consistently embedded in $SU(5)$, which in itself can be further extended to some larger group at some larger scale. At this point two ideas, elegant GUT and beautiful SUSY, already long time wanted each other, succesfully meet. It has become completely clear that GUT without SUSY is not viable:[1] non-supersymmetric $SU(5)$ is excluded while the larger GUTs require some intermediate scale as an extra parameter, whereas the SUSY $SU(5)$ prediction[2] for $\sin^2\theta_W$ well agrees with experiment. On the other hand, SUSY without GUT (i.e. MSSM directly from the string unification), gives too small $\sin^2\theta_W$. In this view, it is more attractive to think that at compactification scale $M \sim 10^{18}$ GeV, the string theory first reduces to some SUSY GUT containing $SU(5)$, which then breaks down to standard $SU(3) \times SU(2) \times U(1)$ at $M_G \simeq 10^{16}$ GeV.

Besides this experimental hint, grand unification needs supersymmetry for conceptual reasons[5], related to so-called gauge hierarchy problem. At the level of standard model this is essentially a problem of the Higgs mass ($\sim M_W$) stability against radiative corrections (quadratic divergences). It is removed as soon as one appeals to

SUSY, which links the masses of scalars with those of their fermion superpartners while the latter are protected by the chiral symmetry. However, in the context of grand unification the gauge hierarchy problem rather concerns the origin of scales: why the electroweak scale is so small as compared to the GUT scale M_G, which in itself is not far from the Planck scale. This question is inevitably connected with the puzzle of the so-called doublet-triplet (DT) splitting: in GUT supermultiplets the MSSM Higgs doublets are accompanied by the colour triplet partners, which mediate an unacceptably fast proton decay (especially via $d = 5$ operators[3]) unless their masses are very large ($\sim M_G$). In addition light triplets, even being decoupled from quarks and leptons by some means[4], would spoil the unification of gauge couplings.

For example, in minimal SUSY $SU(5)$ with fermions $(\bar{5}+10)_i$, where $i = 1,2,3$ is a family index, and Higgs sector consisting of superfields $\Phi(24)$ and $\phi(5)+\bar{\phi}(\bar{5})$, the $SU(3)\times SU(2)\times U(1)$ decomposition of the latter is $\phi(5) = T(3,1)+h_2(1,2)$ and $\bar{\phi}(\bar{5}) = \bar{T}(\bar{3},1)+h_1(1,\bar{2})$, where $h_{1,2}$ are the MSSM higgs doublets and $T, \bar{T}$ are their triplet partners. The only source of DT splitting can be the interaction of $\phi, \bar{\phi}$ with the 24-plet Φ. Indeed, the most general superpotential of these fields has the form:

$$W_{Higgs} = M\Phi^2 + m\bar{\phi}\phi + \lambda\Phi^3 + f\bar{\phi}\Phi\phi \tag{1}$$

The $SU(5)$ breaking to $SU(3)\times SU(2)\times U(1)$ is given by supersymmetric vacuum $\langle\Phi\rangle = V_G\mathrm{diag}(2,2,2,-3,-3)$, $\langle\phi\rangle, \langle\bar{\phi}\rangle = 0$, with $V_G = 2M/3\lambda$. Then the masses of the T and h fragments are respectively $m+2fV_G$ and $m-3fV_G$, so that massless doublets require the relation

$$3\lambda m = 2fM \tag{2}$$

and mass of triplet is unavoidably $O(M)$ in this case. Non-renormalization theorem guarantees that in the exact supersymmetry limit this constraint is stable against radiative corrections. However, this so-called technical solution[5] is nothing but *fine tuning* of different unrelated parameters in the superpotential.

The actual task is to achieve the DT-splitting in a natural way (without fine tuning) due to symmetry reasons, which implies a certain choice of GUT symmetry and its field content. Several attempts have been done, but non of them seems satisfactory. The sliding singlet scenario[6] was shown to be unstable under radiative corrections[7]. The group-theoretical trick known as a missing partner mechanism[8] needs rather complicated Higgs sector when implemented in $SU(5)$. In a most economic way it works in the *flipped* $SU(5)$,[9] which is not however a *genuine* GUT unifying the gauge couplings. The missing VEV mechanism[10] implemented in SUSY $SO(10)$ also requires rather artificial Higgs sector if one attempts[11] to support it by some symmetry. In addition, the 'missing' (partner, VEV) mechanisms once being motivated by symmetry reasons so that the Higgs doublets are strictly massless in the exact supersymmetry limit, miss also a solution of so-called μ-problem[12]. The μ-term of the order of soft SUSY breaking terms, in fact should be introduced by hand.

Another theoretical weakness of SUSY GUTs is a lack in the understanding of flavour. Although GUTs can potentially unify the Yukawa couplings within one family, the origin of inter-family hierarchy and weak mixing pattern remains open.

Moreover, in the light families the Yukawa unification simply contradicts to observed mass pattern, though the $b-\tau$ Yukawa unification[13] may constitute a case of partial but significant success. In order to deal with the flavour problem in GUT frameworks, some additional ideas (horizontal symmetry, zero textures) are required[14,15].

2. GIFT – Goldstones Instead of Fine Tuning

An attractive possibility to solve the gauge hierarchy problem and the related problem of the DT splitting, suggested in[16,17,18], can be simply phrased as follows: Higgs doublets can appear as Goldstone modes of a spontaneously broken global symmetry, which is larger than the local symmetry of the GUT. These doublets, being strictly massless in the exact SUSY limit, acquire nonzero masses after supersymmetry breaking and thereby triger the electroweak symmetry breaking. In ref.[17] this mechanism was elegantly named as GIFT (*Goldstones Instead of Fine Tuning*).

In refs.[16,17] GIFT mechanism was implemented in $SU(5)$ model, by *ad hoc* assumption that the Higgs superpotential has larger global symmetry $SU(6)$. This was done by adding a singlet superfield I to the minimal Higgs sector of $SU(5)$: $\Phi(24)+\bar{\phi}(\bar{5})+\phi(5)+I(1)$ is just the $SU(5)$ decomposition of the $SU(6)$ adjoint representation $\Sigma(35)$. If one assumes that the Higgs superpotential has the simple form $W_{Higgs}=M\Sigma^2+\lambda\Sigma^3$, then it possesses the $SU(6)$ global symmetry. The supersymmetric ground state $\langle\Sigma\rangle = V_G\mathrm{diag}(1,1,1,1,-2,-2)$ (one among the other discretely degenerated vacua), breaking $SU(6)$ down to $SU(4)\times SU(2)\times U(1)$, gives rise to Goldstone supermultiplets in fragments $(4,\bar{2})+(\bar{4},2)$. At the same time the gauged part $SU(5)\subset SU(6)$ breaks down to $SU(3)\times SU(2)\times U(1)$, so that the fragments $(3,\bar{2})+(\bar{3},2)$ are eaten up by Higgs mechanism. The remaining Goldstone fragments $h_1=(1,2)$ and $h_2=(1,\bar{2})$ stay massless until supersymmetry (and thus also larger global symmetry of Higgs Potential) remains unbroken. However, the global $SU(6)$ symmetry in the Higgs sector of $SU(5)$ seems rather artificial. In general the Higgs superpotential of the fields involved should be

$$W_{Higgs}=\mu^2 I+M\Phi^2+M'I^2+m\bar{\phi}\phi+\lambda\Phi^3+\lambda' I^3+\lambda''\Phi^2 I+f\bar{\phi}\Phi\phi+f'I\bar{\phi}\phi \quad (3)$$

while the $SU(6)$ invariance is equivalent to imposing the following constraints

$$\mu=0, \quad M=M'=m/2, \quad \lambda=-\sqrt{15/8}\lambda'=\sqrt{10/3}\lambda''=f/3=-\sqrt{5/6}f' \quad (4)$$

Without valid dynamical or symmetry reasons these constraints look as *several* unnatural fine tunings instead of *one* tuning (2) needed in minimal SUSY $SU(5)$. Thus, if one remains within the SUSY $SU(5)$ frames, GIFT is essentially LOST![1)]

A much more attractive scenario is that the larger global symmetry arises in an accidental way[18]. In other words, it should be an automatic consequence of the gauge symmetry and the field content of the model. Obviously, this requires extension of the $SU(5)$ gauge symmetry, say to $SU(6)$,[18] with the anomaly-free fermion

[1)]Lots Of Strange Tunings

sector consisting of chiral superfields $(\bar{6}+\bar{6}'+15)_i$, where $i = 1,2,3$ is a family index. The Higgs sector contains supermultiplets Σ and $H+\bar{H}$, respectively in adjoint 35 and fundamental $6+\bar{6}$ representations. If the Higgs superpotential has a structure $W = M\Sigma^2 + \lambda\Sigma^3 + S(\bar{H}H - V_H^2)$, where S is an auxiliary singlet, then it acquires an extra global symmetry $SU(6)_\Sigma \times U(6)_H$ [2)] In the exact SUSY limit the vacuum state has continuous degeneration: $\langle H\rangle = \langle\bar{H}\rangle = V_H(1,0,0,0,0,0)$, $\langle\Sigma\rangle = V_G U^\dagger \mathrm{diag}(1,1,1,1,-2,-2)U$, where U is arbitrary 6×6 unitary matrix.[3)] If the true vacuum state corresponds to $U=1$ (as it can appear after SUSY breaking), then these VEVs break the $SU(6)$ gauge symmetry down to the standard $SU(3)\times SU(2)\times U(1)$ symmetry. At the same time, the global symmetry $SU(6)_\Sigma \times U(6)_H$ is spontaneously broken to $[SU(4)\times SU(2)\times U(1)]_\Sigma \times U(5)_H$, and the corresponding Goldstone modes are presented by fragments $[(4,\bar{2})+(\bar{4},2)]_\Sigma \to [\bar{T}'(3,\bar{2})+T'(\bar{3},2)+\bar{h}(1,\bar{2})+h(1,2)]_\Sigma$ and $\bar{5}_{\bar{H}}+5_H \to [\bar{T}(\bar{3},1)+\bar{h}(1,\bar{2})]_{\bar{H}} + [T(3,1)+h(1,2)]_H$. Clearly, all these are eaten up by corresponding fragments in vector superfields of the gauge $SU(6)$ symmetry, except the combinations $h_1 \propto V_H\bar{h}_\Sigma - 3V_G\bar{h}_{\bar{H}}$, $h_2 \propto V_H h_\Sigma - 3V_G h_H$, which remain massless and can be be identified with the MSSM Higgs doublets.

In order to maintain the gauge coupling unification, we have to assume $V_H \gtrsim V_G$, so that $SU(6)$ is first broken by $H,\bar{H}$ down to $SU(5)$. Then, at the scale $V_G \simeq 10^{16}$ GeV, the VEV of Σ breaks $SU(5)$ down to $SU(3)\times SU(2)\times U(1)$. As far as the superpotential of Σ does not feel the local symmetry breaking by the $H,\bar{H}$ VEVs and continues to carry the global $SU(6)_\Sigma$ symmetry, in the limit $V_H \gg V_G$ the theory is automatically reduced to the $SU(5)$ GIFT model of refs.[16,17].

After the SUSY breaking enters the game (presumably through the hidden supergravity sector), the Higgs potential, in addition to the (supersymmetric) squared F and D terms, includes also the soft SUSY breaking terms[25]. These are given by $V_{SB} = AmW_3 + BmW_2 + m^2|\phi_k|^2$, where ϕ_k imply all scalar fields involved, $W_{3,2}$ are terms in superpotential respectively trilinear and bilinear in ϕ_k, and A, B, m are soft breaking parameters. Due to these terms the VEV of Σ is shifted, as compared to the one calculated in the exact SUSY limit, by an amount of $\sim m$. Through the Σ^3 term in the superpotential, this shift gives rise to term $\mu h_1 h_2$. Thus, the GIFT scenario automatically solves the μ-problem: the (supersymmetric) μ-term for the resulting MSSM in fact arises as a result of SUSY breaking, with $\mu \sim m$.

The Higgs doublets acquire also the soft SUSY breaking mass terms, but not all of them immediately. Clearly, V_{SB} also respects the larger global symmetry $SU(6)_\Sigma \times U(6)_H$, so that one combination of the scalars h_1 and h_2, namely $\tilde{h} = h_1 + h_2^*$ remains massless as a truly Goldstone boson. However, as far as SUSY breaking relaxes radiative corrections, the latter will remove the vacuum degeneracy and provide non-zero mass to $\tilde{h}$ (situation, much similar to the case of axion). The effects of radiative

[2)]Notice, however, that the crossing term $H\Sigma H$ is put to zero by hand

[3)] In fact, $SU(6)_\Sigma \times SU(6)_H$ is not a global symmetry of a whole Higgs Lagrangian, but only of the Higgs superpotential. In particular, the Yukawa as well as the gauge couplings (D-terms) do not respect it. However, in the exact supersymmetry limit (i) it is effective for the field configurations on the vacuum valley, where $D=0$, (ii) the Yukawa interactions cannot spoil the extra global symmetry of superpotential.

corrections, which lift vacuum degeneracy and lead to the electroweak symmetry breaking, were studied in ref.[19]. It was shown that GIFT scenario does not imply any upper bound on the top quark mass, in spite of earlier claims[17,20] and it can go up up to its infrared fixed limit.

In fact, the $SU(6)$ model[18] is a minimal extension of the standard $SU(5)$ model. At the scale V_H the fermion content is reduced to the minimal fermion content of standard $SU(5)$. Indeed, the $SU(5)$ decomposition of various supermultiplets reads

$$\begin{array}{lll} H = (5+1)_H, & \bar{H} = (\bar{5}+1)_{\bar{H}}, & \Sigma = (24+5+\bar{5}+1)_\Sigma \\ \bar{6}_i = (\bar{5}+1)_i, & \bar{6}'_i = (\bar{5}+1)'_i, & 15_i = (10+5)_i \qquad (i=1,2,3) \end{array} \tag{5}$$

so that the fermion sector at this scale consists of six $\bar{5}$-plets, three 5-plets, three 10-plets and six singlets. According to the survival hypothesis[21], after the breaking $SU(6) \to SU(5)$ the extra fermions $(5+\bar{5}')_i$ become heavy (with masses $\sim V_H$) owing to the Yukawa couplings $\Gamma_{ij}15_i\bar{H}\bar{6}'_j$, and decouple from from the light states which remain in $(\bar{5}+10)_i$ and singlets. Thus, below the scale V_H we are left with minimal $SU(5)$ GUT with three standard fermion families.

The $SU(6)$ theory[18] drastically differs from any other GUT approaches. Usually, in GUTs the Higgs sector consists of two different sets: one is for the GUT symmetry breaking (e.g. 24-plet in $SU(5)$), while another is just for the electroweak symmetry breaking and fermion mass generation (like $5+\bar{5}$ in $SU(5)$). In contrast, in the $SU(6)$ theory no special superfields are indroduced for the second function. The Higgs sector consisting of the 35-plet and $6+\bar{6}$, is a minimal one needed for the $SU(6)$ breaking down to the $SU(3)\times SU(2)\times U(1)$. As for the Higgs doublets, they arise as Goldstone modes, from the $SU(2)\times U(1)$ doublet fragments in Σ and $H, \bar{H}$.

Due to this reason, their couplings with the fermion fields have some peculiarities, which could provide new possibilities towards the understanding of flavour. Namely, if by chance the Yukawa superpotential also respects the $SU(6)_\Sigma \times U(6)_H$ global symmetry, then the Higgs doublets, as Goldstone fields, have *vanishing* Yukawa couplings with the fermions that remain massless after the $SU(6)$ symmetry breaking. In fact, these are the chiral families $(\bar{5}+10)_i$ of ordinary quarks and leptons. Yukawa couplings $15\,\bar{H}\,\bar{6}$ respect extra global symmetry and cannot generate their masses, so that one has to invoke the higher order operators scaled by inverse powers of some large mass $M \gg V_H$. These could appear due to nonperturbative quantum gravity effects, with $M \sim M_{Pl}$. Alternatively, they can arise by integrating out some heavy states with masses above the $SU(6)$ breaking scale. Indeed, such states in vectorlike (real) representations can naturally present in SUSY (stringy) GUTs. According to survival hypothesis, they should acquire maximal allowed masses M if there are no symmetry reasons to keep them light. Then masses of ordinary light fermions can appear as a result of 'seesaw' mixing with these heavy states[22]. In model[18], operators relevant for down quarks appear as $\frac{1}{M}15\Sigma\bar{H}\bar{6}$ while the operators relevant for upper quarks are $\frac{1}{M^2}15H\Sigma H15$. So, it seems that model leads to unacceptable case $m_b \gg m_t$.

As it was shown in ref.[23], the problem can be resolved by introducing a fermion 20-plet, which $SU(5)$ content is $20 = 10+\overline{10}$. Since 20 is a pseudo-real representation (the tensor product 20×20 contains singlet only in an antisymmetric combination), the survival hypothesis does not apply to it. More generally, if in original theory 20-plets present in odd number then *one* of them should inevitably stay massless. Then its Yukawa couplings $g20\Sigma20$ and $G_i 20 H 15_i$ *explicitly* violate the global $SU(6)_\Sigma \times U(6)_H$ symmetry. As a result, the *only* particle which gets direct Yukawa coupling with the 'Goldstone' Higgs doublet is an upper quark contained in 20, that is top quark. Other fermions stay massless at this level, and for generating their masses one has to appeal to higher order operators. In order to achieve a proper operator structure, additional symmetries are needed. On the other hand, consistency of the GIFT scenario also requires some extra symmetry in order to forbid the crossing term $\bar{H}\Sigma H$ – otherwise the Higgs superpotential has no accidental global symmetry.

Below we describe a consistent SUSY $SU(6)$ model with flavour-blind discrete Z_3 symmetry[24]. The role of the latter is important: first, it guarantees that Higgs superpotential has automatic global $SU(6)_\Sigma \times U(1)_H$ symmetry without putting to zero of some terms by hand, and second, it provides proper structure of the higher order operators generating realistic mass and mixing pattern for *all* fermion families.

3. $SU(6) \times Z_3$ model

Consider the supersymmetric model with $SU(6)$ gauge symmetry, with the set of chiral superfilds consisting of two sectors:

(i) The 'Higgs' sector: vectorlike supermultiplets $\Sigma_1(35)$, $\Sigma_2(35)$, $H(6)$, $\bar{H}(\bar{6})$ and an auxiliary singlet S.

(ii) The 'fermion' sector: chiral, anomaly free supermultiplets $(\bar{6}+\bar{6}')_i$, 15_i ($i = 1,2,3$ is a family index) and 20.

We introduce also two flavour-blind discrete symmetries. One is usual matter parity Z_2, under which the fermion superfields change the sign while the Higgs ones stay invariant. Such a matter parity, equivalent to R parity, ensures the proton stability. Another discrete symmetry is Z_3 acting in the following way:

$$\begin{aligned} &\Sigma_1 \to \omega\,\Sigma_1, \quad \Sigma_2 \to \bar{\omega}\,\Sigma_2, \quad H,\bar{H} \to H,\bar{H}, \quad S \to S \\ &\bar{6}^i_{1,2} \to \omega\,\bar{6}^i_{1,2}, \quad 15^i \to \bar{\omega}\,15^i, \quad 20 \to \omega\,20 \qquad (\omega = e^{i\frac{2\pi}{3}}) \end{aligned} \tag{6}$$

Let us consider first the Higgs sector. The most general renormalizable superpotential compatible with the $SU(6) \times Z_3$ symmetry is $W_{Higgs} = W_\Sigma + W_H + W(S)$, where $W(S)$ is a polynomial containing linear, quadratic and cubic terms, and

$$W_\Sigma = M\Sigma_1\Sigma_2 + \lambda_0 S\Sigma_1\Sigma_2 + \lambda_1\Sigma_1^3 + \lambda_2\Sigma_2^3\,, \qquad W_H = M'\bar{H}H + \lambda S\bar{H}H \tag{7}$$

This superpotential automatically has larger global symmetry $SU(6)_\Sigma \times U(6)_H$, related to independent transformations of the Σ and H fields. In the exact supersymmetry limit, the condition of vanishing F and D terms allows, among other discretely (and continuously) degenerated vacua, the VEV configuration

$$\begin{aligned} &\langle\Sigma_{1,2}\rangle = V_{1,2}\,\mathrm{diag}(1,1,1,1,-2,-2) \\ &\langle H\rangle = \langle\bar{H}\rangle = V_H\,(1,0,0,0,0,0) \end{aligned} \tag{8}$$

For a proper parameter range, these configuration can appear as a true vacuum state afterthat the vacuum degeneracy is lifted by soft SUSY breaking and subsequent radiative corrections[19].

The VEVs (8) lead to the needed pattern of gauge symmetry breaking: $H, \bar{H}$ break $SU(6)$ down to $SU(5)$, while $\Sigma_{1,2}$ break $SU(6)$ down to $SU(4) \times SU(2) \times U(1)$. Both channels together break the local symmetry down to $SU(3) \times SU(2) \times U(1)$. At the same time, the global symmetry $SU(6)_\Sigma \times U(6)_H$ is broken down to $[SU(4) \times SU(2) \times U(1)]_\Sigma \times U(5)_H$. The Goldstone degrees which survive from being eaten by Higgs mechanism constitute the couple $h_1 + h_2$ of the MSSM Higgs doublets, which in terms of the doublet (anti-doublet) fragments in $\Sigma_{1,2}$ and $H, \bar{H}$ are given as

$$
\begin{aligned}
h_2 &= \cos\alpha(\cos\gamma\, h_{\Sigma_1} + \sin\gamma\, h_{\Sigma_2}) + \sin\alpha\, h_H \\
h_1 &= \cos\alpha(\cos\gamma\, \bar{h}_{\Sigma_1} + \sin\gamma\, \bar{h}_{\Sigma_2}) + \sin\alpha\, \bar{h}_{\bar{H}}
\end{aligned} \tag{9}
$$

where $\tan\gamma = V_2/V_1$ and $\tan\alpha = 3V_G/V_H$, $V_G = (V_1^2 + V_2^2)^{1/2}$. In the following we adopt the case $V_H \gg V_1 \gg V_2$. Thus, in this case the Higgs doublets dominantly come from Σ_1, while in Σ_2 and $H, \bar{H}$ they are contained with small weights $\varepsilon_2/\varepsilon_1$ and $3\varepsilon_1$ respectively, where $\varepsilon_{1,2} = V_{1,2}/V_H$.

4. Fermion masses

The most general Yukawa superpotential compatible with the $SU(6) \times Z_3$ symmetry has the form

$$
W_{Yuk} = g\,20\Sigma_1 20 \;+\; G\,20H15_3 \;+\; \Gamma_{ij}15_i\bar{H}\bar{6}'_j \qquad i,j = 1,2,3 \tag{10}
$$

where all Yukawa coupling constants are assumed to be $O(1)$ (for comparison, we remind that the gauge coupling constant at GUT scale is $g_{GUT} \simeq 0.7$). Without loss of generality, one can always redefine the basis in 15-plets so that only the 15_3 state couples to 20-plet in (10). Also, among six $\bar{6}$-plets one can always choose the basis when only three of them (denoted in eq. (10) as $\bar{6}'_{1,2,3}$) couple to $15_{1,2,3}$, while other three states $\bar{6}_{1,2,3}$ have no Yukawa couplings.

For $V_H \gg V_G$, already at the breaking $SU(6) \to SU(5)$, the light fermion states are identified from this superpotential, whereas the extra fermion states become superheavy. Indeed, the $SU(5) \supset SU(3) \times SU(2) \times U(1)$ decomposition of the fermion multiplets under consideration reads

$$
\begin{aligned}
20 &= 10 + \overline{10} = (q + u^c + e^c)_{10} + (Q^c + U + E)_{\overline{10}} \\
15_i &= (10 + 5)_i = (q_i + u_i^c + e_i^c)_{10} + (D_i + L_i^c)_5 \\
\bar{6}_i &= (\bar{5} + 1)_i = (d_i^c + l_i)_{\bar{5}} + N_i \\
\bar{6}'_i &= (\bar{5} + 1)'_i = (D_i^c + L_i)_{\bar{5}'} + N'_i
\end{aligned} \tag{11}
$$

According to eq. (10), the extra fermion pieces with non-standard $SU(5)$ content, namely $\overline{10}$ and $5_{1,2,3}$, form massive particles being coupled with 10_3 and $\bar{5}'_{1,2,3}$

$$
G\,V_H\,\overline{10}\,10_3 \;+\; \Gamma_{ij}V_H\,5_i\,\bar{5}'_j \;+\; g\,V_1\,(U\,u^c - 2E\,e^c) \tag{12}
$$

and thereby decouple from the light states which remain in $\bar{5}_{1,2,3}$, $10_{1,2}$ and 10 (if neglect the small, $\sim \varepsilon_1$ mixing between the $u^c - u^c_3$ and $e^c - e^c_3$ states). On the other hand, since the couplings of 20-plet explicitly violate the global $SU(6)_\Sigma \times U(6)_H$ symmetry, the Higgs doublet h_2 has *non-vanishing* couplings with up-type quarks from 20- and 15_3-plets. Indeed, it follows from eq. (10) that *only* Yukawa coupling relevant for light states is contained in $g20\Sigma_1 20 \to g10\, 5_{\Sigma_1} 10$. Therefore, only one up-type quark (to be identified with top quark), dominantly contained in 20-plet, has a direct Yukawa coupling $g\, qu^c\, h_2$, so that its mass has to be in the 100 GeV range. Other fermions stay massless at this level, unless we invoke the higher order operators.

Higher order operators scaled by inverse powers of some large mass $M \gg V_H$ could appear due to quantum gravity effects, with $M \sim M_{Pl}$. Alternatively, they can arise by integrating out some heavy states with masses $M \gg V_H$. In the Sect. 5 we adopt the second viewpoint, namely that these operators appear from the exchange of some heavy fermion superfields[22]. The reason is twofold: first, as we see shortly, the fermion mass pattern favours the scale $M \sim 10^{18}$ GeV (string scale?) rather than M_{Pl}, and second, the mechanism of heavy fermion exchange is rather instructive for obtaining the realistic fermion mass pattern.

Before addressing the concrete scheme of heavy fermion exchanges, let us start with the general operator analysis. Z_3 symmetry forbids any 'Yukawa' operator in the superpotential at $1/M$ order.[4)] However, operators at the next $(1/M^2)$ order are allowed and they are the following:

$$\mathcal{A} = \frac{a}{M^2} 20\bar{H}\Sigma_1\bar{H}\bar{6}_3\,, \quad \mathcal{B} = \frac{b_{ij}}{M^2} 15_i H\Sigma_2 H 15_j\,, \quad \mathcal{C} = \frac{c_{ik}}{M^2} 15_i(\Sigma_1\Sigma_2\bar{H})\bar{6}_k \tag{13}$$

where a, b, c are $O(1)$ 'Yukawa' constants. Analogous operators involving heavy $\bar{6}'_i$ states are irrelevant for the light fermion masses. According to eq. (12) the state $10_3 \subset 15_3$ is also heavy, and it is decoupled from the light particle spectrum. Therefore, operators (13) are relevant only for $10 \subset 20$, $10_i \subset 15_i$ $(i = 1,2)$ and $\bar{5}_k \subset \bar{6}_k$ $(k = 1,2,3)$ states. Without loss of generality, we redefine the basis of $\bar{6}$-plets so that only the $\bar{6}_3$ state couples to 20-plet in eq. (13). It is worth to note that in fact $\mathcal{C}$ can contain two relevant combinations with different convolutions of the $SU(6)$ indices indicated by brackets in an obvious way:

$$C_1 = 15(\Sigma_1\Sigma_2\bar{H})\bar{6}, \quad C_2 = 15(\Sigma_1\bar{H})(\Sigma_2\bar{6}), \quad C_3 = (15\bar{H})(\Sigma_1\Sigma_2\bar{6}), \quad C_4 = (15\bar{H}\bar{6})(\Sigma_1\Sigma_2) \tag{14}$$

C_1 and C_2 provide different Clebsch coefficients for the down quark and charged lepton mass terms, while C_3 and C_4 are irrelevant for the mass generation and they lead only to some minor rotation of the heavy fermion states.

Let us analyse now the impact of these operators on the fermion masses. Obviously, the operator $\mathcal{A}$ is responsible for the b quark and τ lepton masses, and at the MSSM level it reduces to Yukawa couplings $a\varepsilon_H^2\,(qd^c_3 + e^c l_3)\, h_1$, where $\varepsilon_H = V_H/M$.

---

4) Operators involving an odd number of fermion superfields are forbidden by matter parity.

Hence, though b and τ belong to the same family as t (namely, to 20-plet), their Yukawa couplings are substantially (by factor $\sim \varepsilon_H^2$) smaller than $\lambda_t \approx g$. Moreover, we automatically have almost precise $b-\tau$ Yukawa unification at the GUT scale:

$$\lambda_b = a\varepsilon_H^2 = \lambda_\tau[1 + (\varepsilon_1 g/G)^2] \tag{15}$$

where the $\sim \varepsilon_1^2$ correction is due to mixing between the e^c and e_3^c states (see eq. (12)).

As far as the fermions of the third family are already defined as the states belonging to $10 \subset 20$ and $\bar{5}_3 \subset \bar{6}_3$, the operators $\mathcal{B}$ and $\mathcal{C}$ generate mass terms for the fermions of the first two families, which in general case are expected to be of the same order. In order to achieve mass splitting between the second and first families, one can assume that the 'Yukawa' matrices b_{ij} and c_{ik} are *rank-1* matrices, so that each of operators $\mathcal{B}$ and $\mathcal{C}$ can provide only one non-zero mass eigenvalue (i.e. c and s quark masses). Then, without loss of generality, we can redefine the basis of $15_{1,2}$ and $\bar{6}_{1,2}$ states so that these matrices have the form

$$b_{ij} = (0,\beta)^T \bullet (0,\beta) = \begin{pmatrix} 0 & 0 \\ 0 & b \end{pmatrix}$$

$$c_{ik} = (\gamma_1,\gamma_2)^T \bullet (0,\delta_2,\delta_3) = \begin{pmatrix} 0 & c_2\sin\theta & c_3\sin\theta \\ 0 & c_2\cos\theta & c_3\cos\theta \end{pmatrix} \tag{16}$$

where $\tan\theta = \gamma_1/\gamma_2$. Hence, in this basis only $b_{22} = b$ component of the symmetric matrix b_{ij} is nonzero, and c quark should be identified as an up-quark state from $q_2, u_2^c \subset 10_2 \subset 15_2$. Then s quark state is the down quark state in $q_2' \subset 10_2' \subset 15_2'$ and $d_2^c \subset \bar{5}_2 \subset \bar{6}_2$, where $15_2' = \sin\theta \cdot 15_1 + \cos\theta \cdot 15_2$ is an effective combination of the $15_{1,2}$ states which couples $\bar{6}_2$ and $\bar{6}_3$ states (it is not difficult to recognize that in fact θ is the Cabibbo angle, which in general tends to be $O(1)$). Clearly, μ-lepton is also contained in $15_2' + \bar{6}_2$. Thus, operators $\mathcal{B}$ and $\mathcal{C}$ reduce to MSSM Yukawa couplings for the second family quarks and leptons

$$b(\varepsilon_2/\varepsilon_1)\varepsilon_H^2\, q_2 u_2^c\, h_2, \qquad c_{2,3}\varepsilon_2\varepsilon_H^2\,(q_2' d_{2,3}^c + K e_2'^c l_{2,3})\, h_1 \tag{17}$$

where K is the Clebsch coefficient depending on weights of operators (14) entering $\mathcal{C}$ (for example, $K=1$ if $\mathcal{C} \propto C_1$ and $K=-2$ if $\mathcal{C} \propto C_2$).

For the first family fermion masses one can appeal to $1/M^3$ operators which can be presented as following:

$$\mathcal{D} = \frac{d_{ik}}{M^3} 15_i' \Sigma_1^3 \bar{H} \bar{6}_k\,, \qquad \mathcal{E} = \frac{e_{ij}}{M^3} 15_i H \Sigma_1^2 H 15_j \tag{18}$$

where $15_1'$ is a state orthogonal to $15_2'$. As in the case of operator $\mathcal{C}$, these operators can have different convolutions of the $SU(6)$ indices. For $\mathcal{D}$ the relevant combinations, giving different relative Clebsch factors P for the d quark and electron masses, are

$$D_1 = 15(\Sigma_1^3\bar{H})\bar{6}, \quad D_2 = 15(\Sigma_1^2\bar{H})(\Sigma_1\bar{6}), \quad D_3 = 15(\Sigma_1\bar{H})(\Sigma_1^2\bar{6}), \quad D_4 = 15(\Sigma_1\bar{H})\bar{6}(\Sigma_1^2) \tag{19}$$

Operator $\mathcal{D}$, with arbitrary couplings d_{ij} of the order of 1, will provide d-quark and electron masses in the correct range. As for the operator $\mathcal{E}$, for $e_{11} \sim 1$, it leads to somewhat large value of u-quark mass. Therefore, it is more suggestive to think that only e_{12} couplings are non-zero, while $e_{11} = 0$. As we show in Sect. 5, this is really the case in the context of heavy fermion exchange model.

Keeping only the leading contributions to each component, for the quark and lepton Yukawa couplings at the GUT scale we obtain

$$\begin{array}{c} \\ q_1 \\ q_2 \\ q \end{array} \begin{array}{c} \begin{array}{ccc} u_1^c & u_2^c & u^c \end{array} \\ \begin{pmatrix} 0 & e_{12}\varepsilon_1\varepsilon_H^3 & 0 \\ e_{12}\varepsilon_1\varepsilon_H^3 & b(\varepsilon_2/\varepsilon_1)\varepsilon_H^2 & 0 \\ 0 & 0 & g \end{pmatrix} \end{array} \cdot h_2 \tag{20a}$$

$$\begin{array}{c} \\ q_1' \\ q_2' \\ q \end{array} \begin{array}{c} \begin{array}{ccc} d_1^c & d_2^c & d_3^c \end{array} \\ \begin{pmatrix} d_{11}\varepsilon_1^2\varepsilon_H^3 & d_{12}\varepsilon_1^2\varepsilon_H^3 & d_{13}\varepsilon_1^2\varepsilon_H^3 \\ d_{21}\varepsilon_1^2\varepsilon_H^3 & c_2\varepsilon_2\varepsilon_H^2 & c_3\varepsilon_2\varepsilon_H^2 \\ 0 & 0 & a\varepsilon_H^2 \end{pmatrix} \end{array} \cdot h_1 \tag{20b}$$

$$\begin{array}{c} \\ e_1'^c \\ e_2'^c \\ e^c \end{array} \begin{array}{c} \begin{array}{ccc} l_1 & l_2 & l_3 \end{array} \\ \begin{pmatrix} Pd_{11}\varepsilon_1^2\varepsilon_H^3 & Pd_{12}\varepsilon_1^2\varepsilon_H^3 & Pd_{13}\varepsilon_1^2\varepsilon_H^3 \\ Pd_{21}\varepsilon_1^2\varepsilon_H^3 & Kc_2\varepsilon_2\varepsilon_H^2 & Kc_3\varepsilon_2\varepsilon_H^2 \\ 0 & 0 & a\varepsilon_H^2 \end{pmatrix} \end{array} \cdot h_1 \tag{20c}$$

Thus, the Yukawa coupling eigenvalues at the GUT scale are $\lambda_t = g \sim 1$ and

$$\begin{aligned} \lambda_b = \lambda_\tau = a\varepsilon_H^2 \qquad & \Rightarrow \quad \lambda_b/\lambda_t \sim \varepsilon_H^2 \\ \lambda_c = b(\varepsilon_2/\varepsilon_1)\varepsilon_H^2 \qquad & \Rightarrow \quad \lambda_c/\lambda_b \sim \frac{\varepsilon_2}{\varepsilon_1} \\ \lambda_s = \lambda_\mu/K = c_2\varepsilon_2\varepsilon_H^2 \qquad & \Rightarrow \quad \lambda_s/\lambda_b \sim \varepsilon_2 \\ \lambda_d = \lambda_e/P = d_{11}\varepsilon_1^2\varepsilon_H^3 \qquad & \Rightarrow \quad \lambda_d/\lambda_s \sim \varepsilon_2\left(\frac{\varepsilon_1}{\varepsilon_2}\sqrt{\varepsilon_H}\right)^2 \\ \lambda_u = \tfrac{e_{12}^2}{b}(\varepsilon_1^3/\varepsilon_2)\varepsilon_H^4 \qquad & \Rightarrow \quad \lambda_u/\lambda_d \sim \frac{\varepsilon_2}{\varepsilon_1}\left(\frac{\varepsilon_1}{\varepsilon_2}\sqrt{\varepsilon_H}\right)^2 \end{aligned} \tag{21}$$

In order to connect these Yukawa constants to the physical masses of the quarks and leptons, the renormalization group (RG) running has to be considered[15,26]. We have:

$$\begin{aligned} m_t &= \lambda_t A_u \eta_t y^6 v \sin\beta, & m_b &= \lambda_b A_d \eta_b y v \cos\beta, & m_\tau &= \lambda_\tau A_e \eta_\tau v \cos\beta \\ m_c &= \lambda_c A_u \eta_c y^3 v \sin\beta, & m_s &= \lambda_s A_d \eta_d v \cos\beta, & m_\mu &= \lambda_\mu A_e \eta_l v \cos\beta \\ m_u &= \lambda_u A_u \eta_u y^3 v \sin\beta, & m_d &= \lambda_d A_d \eta_d v \cos\beta, & m_e &= \lambda_e A_e \eta_l v \cos\beta \end{aligned} \tag{22}$$

where $v = 174$ GeV and $\tan\beta = v_2/v_1$ is a famous ratio of the h_2 and h_1 VEVs. The factors A_f account for the running induced by gauge couplings from the GUT scale M_G to the SUSY breaking scale M_S (for the definiteness we take $M_S \simeq m_t$), and y

includes the running induced by the top quark Yukawa coupling:

$$y = \exp\left[-\frac{1}{16\pi^2}\int_{\ln M_S}^{\ln M_G} \lambda_t^2(\mu)\mathrm{d}(\ln\mu)\right] \tag{23}$$

The factors η_f encapsulate the running from M_S down to m_f for heavy quarks $f = t, b, c$ or $\mu = 1\,\mathrm{GeV}$ for light quarks $f = u, d, s$. Taking all these into the account, we see that qualitatively correct description of *all* fermion masses can be achieved with $\varepsilon_H = V_H/M \sim 0.1$, $\varepsilon_1 = V_1/V_H \sim 0.1$ and $\varepsilon_2/\varepsilon_1 = V_2/V_1 \sim 0.3$ (so that $\frac{\varepsilon_1}{\varepsilon_2}\sqrt{\varepsilon_H} \sim 1$), provided that $\tan\beta = 1 - 1.5$. Interestingly, this region of $\tan\beta$ is favoured by the electroweak symmetry radiative breaking picture in the presence of $b - \tau$ Yukawa unification (see eq. (15)). In addition, $b - \tau$ unification and small $\tan\beta$ require substantially large λ_t, actually close to its infrared fixed point[27], which implies for the physical top mass $M_t \approx \sin\beta(190 - 210)\,\mathrm{GeV}$. As far as the scale $V_1 \simeq 10^{16}\,\mathrm{GeV}$ is fixed by the $SU(5)$ unification of gauge couplings, these relations in turn imply that $V_H \sim 10^{17}\,\mathrm{GeV}$ and $M \sim 10^{18}\,\mathrm{GeV}$.

Obtained mass matrices give rise to a quolitatively correct picture of quark mixing. In particular, one obtains the CKM matrix at the unification scale as

$$V_{\mathrm{CKM}} \approx \begin{pmatrix} 1 & s_{12} & s_{12}s_{13} - s_{13}e^{-i\delta} \\ -s_{12} & 1 & s_{23} + s_{12}s_{13}e^{-i\delta} \\ s_{13}e^{i\delta} & -s_{23} & 1 \end{pmatrix} \tag{24}$$

where δ is a CP-phase and

$$\begin{aligned} s_{12} &\approx \gamma_1/\gamma_2 \sim 1 & &(|V_{us}| = 0.220 \pm 0.002) \\ s_{23} &\approx c_3\lambda_s/c_2\lambda_b \sim \varepsilon_2 & &(|V_{cb}| = 0.04 \pm 0.01) \\ s_{13} &\approx d_{13}\lambda_d/d_{11}\lambda_b \sim \varepsilon_2^2 & &(|V_{td}| = 0.03 - 0.1) \end{aligned} \tag{25}$$

for comparison, in the brackets the 'experimental' values of mixing angles are shown. The questions of the neutrino mass pattern and the proton decay features due to $d = 5$ Higgsino mediated operators are considered in refs.[23,24].

5. Yukawa couplings generated by heavy particle exchanges

¿From the previous section, we are left with two problems: the difficulty in splitting the masses of the first two families (in Sect. 4 the form (16) for the coupling constants in operators $\mathcal{B}$ and $\mathcal{C}$ was assumed by hand), and the need to suppress the coupling e_{11} in operator $\mathcal{E}$, which leads to unacceptably large u quark mass.

Here we show how both problems can be solved, still without appealing to any flavour symmetry, by assuming that all higher order operators are generated by the exchanges of some heavy superfields with $\sim M$ masses. As we see below, this mechanism provides also specific predictions for the Clebsch coefficients K and P distinguishing down quark and charged lepton masses.

Z_3:	Higgs	fermions	F-fermions
ω	Σ_1	$\bar{6}_i$, $\bar{6}'_i$, 20	15_F, $\overline{15}_F^{1,2}$, 20_F, 84_F
$\bar{\omega}$	Σ_2	15_i	$\overline{15}_F$, $15_F^{1,2}$, $\overline{20}_F$, $\overline{84}_F$
inv.	S, H, $\bar{H}$	-	$\overline{15}'_F$, $15'_F$, $20_F^{1,2}$, $\overline{105}_F$, 105_F, $\overline{210}_F$, 210_F

Table 1. Z_3-transformations of various supermultiplets.

Let us introduce the set of heavy vectorlike fermions (in the following referred as F-fermions) with masses $O(M)$ and transformation properties under $SU(6)\times Z_3$ given in Table 1. Certainly, we prescribe negative matter parity to all of them.

The operators $\mathcal{A},\mathcal{B},\mathcal{C}$ obtained by the proper exchange chains are shown in Fig. 1. We see that these operators generate only the third and second family masses. Indeed, coupling $15_F\Sigma_1\overline{15}_F^1$ defines $\overline{15}_F^1$ state while the corresponding 15_F^1 state defines $\bar{6}_3$ through the coupling $15_F^1\bar{H}\bar{6}_3$. The operator $\mathcal{A}$ is unambiguosly built in this way. On the other hand, coupling 15_2H20_F defines 15_2 state, so that the operator $\mathcal{B}$ contributes only the c quark mass. The coupling $(\gamma_1 15_1+\gamma_2 15_2)\Sigma_1\overline{15}'_F$ defines $15'_2$ state, which in general does not coincide with 15_2, and the coupling $15_F^2\bar{H}\bar{6}_2$ defines $\bar{6}_2$ state. Therefore, the operator $\mathcal{C}$ providing only the s quark and μ-lepton masses, in general implies the large Cabibbo angle, $\tan\theta=\gamma_1/\gamma_2$. In addition, operator $\mathcal{C}$ built in this way, acts as combination $\mathcal{C}\propto C_1+2C_2$ of operators (14), which leads to specific Clebsch coefficient $K=-5$ in eq. (17).

Exchanges generating operators $\mathcal{D},\mathcal{E}$ are shown in Fig. 2. In reproducing these operators, we have taken into account the following restriction: $\mathcal{D}$ built by F-fermion exchange should be irreducible to lower $(1/M^2)$ order operator, in order to guarantee the mass hierarchy between first and second families. In other words, the exchange chain should not allow to replace $\Sigma_1\Sigma_1$ by Σ_2. This condition requires large representations of $SU(6)$ involved into the exchange. Then the operator $\mathcal{D}$ built as shown in Fig. 1D acts in combination $\mathcal{D}\propto D_1+D_3-D_4$, which gives relative Clebsch coefficient $P=5/8$ for the d quark and electron masses. On the other hand, the operator $\mathcal{E}$ built as in Fig. 2E, can only mix 15_1 state containing u quark, with 15_2 state containing c quark, but cannot provide direct mass term for the former.

As a result, higher order operators obtained by exchange of F-fermions defined in Table 1, consistently reproduce the mass matrix ansatz given in Sect. 4. Moreover, specific Clebsch coefficients are obtained, leading to relations $\lambda_d=\frac{1}{5}\lambda_\mu$ and $\lambda_d=\frac{8}{5}\lambda_e$ (small ($\sim\varepsilon_1$) corrections to these can arise from the interference of the operators $\mathcal{C}$ and $\mathcal{D}$). According to eqs. (22), this relations imply

$$\frac{m_d}{m_s}\simeq 8\,\frac{m_e}{m_\mu}\approx\frac{1}{25} \tag{26}$$

In addition, by taking into account the uncertainties of renormalization factors (22), mainly due to uncertainty in $\alpha_3(M_Z)$, for the quark running masses at $\mu=1\,\text{GeV}$ we obtain

$$m_s=90-150\,\text{MeV},\qquad m_d=4-7\,\text{MeV} \tag{27}$$

in agreement with the experimental values.

Let us conclude with following remark. As we have seen, the fermion mass pattern requires that scales M, V_H and V_G are related as $V_G/V_H \sim V_H/M \sim 0.1$. The superpotential (7) includes mass parameters, which are not related to M. Therefore, it cannot explain why the scales should be arranged in this way. Bearing in mind the possibility that considered $SU(6)$ theory in principle could be stringy SUSY GUT, one can assume that the superfields $H, \bar{H}$ and $\Sigma_{1,2}$ are zero modes, and their superpotential has the form not containing mass terms:

$$W = S[\bar{H}H - (\varepsilon_H M)^2] + \lambda_1 \Sigma_1^3 + \lambda_2 \Sigma_2^3 + \frac{(\bar{H}H)}{M}(\Sigma_1 \Sigma_2) \tag{28}$$

The last term can be effectively obtained by exchange of the singlet superfield Z with a large mass term MZ^2, as shown in Fig. 3. Then the relation $V_G/V_H \sim V_H/M$ follows naturally. Certainly, the origin of small linear term ($\varepsilon_H \sim 0.1$) in (28) remains unclear. It may arise due to some hidden sector outside the GUT.

Non-perturbative effects in principle could induce the higher order operators scaled by inverse powers of the Planck mass. If all such operators unavoidably occur, this would completely spoil the GIFT picture. For example, already operator $\frac{1}{M_{Pl}}(\bar{H}\Sigma_1)(\Sigma_2 H)$ would provide unacceptably large mass to the Higgs doublets. One may hope, however, that not all possible structures appear in higher order terms. Alternatively, one could try to suppress dangerous high order operators by symmetry reasons, in order to achieve consistent 'all order' solution[28].

References

1. P. Langacker and M. Luo, *Phys. Rev.* **D44** (1991) 817; U. Amaldi, W. de Boer and H. Furstenau, *Phys. Lett.* **B260** (1991) 447; J. Ellis, S. Kelley and D.V. Nanopoulos, *Phys. Lett.* **B260** (1991) 131.
2. S. Dimopoulos, S. Raby and F. Wilczek, *Phys. Rev.* **D24** (1981) 1681; S. Dimopoulos and H. Georgi, *Nucl. Phys.* **B193** (1981) 150; L. Ibañez and G.G. Ross, *Phys. Lett.* **B105** (1981) 439.
3. S. Weinberg, *Phys. Rev.* **D26** (1982) 287; N. Sakai and T. Yanagida, *Nucl. Phys.* **B197** (1982) 533.
4. G.R. Dvali, *Phys. Lett.* **287B** (1992) 101; Z. Berezhiani, Ferrara preprint INFN-FE 13-93 (1993).
5. L. Maiani, in *Comptes Rendus de l'Ecole d'Etè de Physique des Particules* (Gif-sur-Yvette, 1979), IN2P3, Paris, 1980, p. 3; S. Dimopoulos and H. Georgi, *Nucl. Phys.* **B150** (1981) 193; M. Sakai, *Z. Phys.* C **11** (1981) 153; E. Witten, *Nucl. Phys.* **B188** (1981) 573.
6. E. Witten, *Phys. Lett.* **B105** (1981) 267; D.V. Nanopoulos and K. Tamvakis, *ibid.* **B113** (1982) 151.
7. H.P. Nilles, M. Srednicki and D. Wyler, *Phys. Lett.* **B124** (1982) 337; A. Lahanas, *ibid.,* 341.

8. S. Dimopoulos and F. Wilczek, in *Erice Summer Lectures* (Plenum, New York, 1981); B. Grinstein, *Nucl. Phys.* **B206** (1982) 387; H. Georgi, *Phys. Lett.* **B108** (1982) 283; A. Masiero *et al., ibid.* **B115** (1982) 380.
9. I. Antoniadis *et al., Phys. Lett.* **B194** (1987) 231.
10. S. Dimopoulos and F. Wilczek, preprint NSF-ITP-82-07 (*unpublished*); M. Srednicki, *Nucl. Phys.* **B202** (1982) 327.
11. K.S. Babu and S.M. Barr, *Phys. Rev.* **D48** (1993) 5354; **D50** (1994) 3529.
12. J.E. Kim and H.P. Nilles, *Phys. Lett.* **B138** (1984) 150.
13. M.S. Chanowitz, J. Ellis and M.K. Gaillard, *Nucl. Phys.* **B128** (1977) 506; A.J. Buras, J. Ellis, M.K. Gaillard and D.V. Nanopoulos, *Nucl. Phys.* **B135** (1978) 66; D.V. Nanopoulos and D.A. Ross, *Nucl. Phys.* **B157** (1979) 273.
14. H. Fritzsch, *Nucl. Phys.* **B155** (1979) 189; H. Georgi and C. Jarlskog, *Phys. Lett.* **B86** (1979) 297; J. Harvey, P. Ramond and D. Reiss, *Nucl. Phys.* **B199** (1982) 223; P. Ramond, R.G. Roberts and G.G. Ross, *Nucl. Phys.* **B406** (1993) 18.
15. S. Dimopoulos, L. Hall and S. Raby, *Phys. Rev.* **D45** (1992) 4195; G. Anderson, S. Dimopoulos, S. Raby, L. Hall and G. Starkman, *Phys. Rev.* **D49** (1994) 3660.
16. K. Inoue, A. Kakuto and T. Takano, *Progr. Theor. Phys.* **75** (1986) 664.
17. A. Anselm and A. Johansen, *Phys. Lett.* **B200** (1988) 331; A. Anselm, *Sov. Phys. JETP* **67** (1988) 663.
18. Z. Berezhiani and G. Dvali, *Sov. Phys. Lebedev Inst. Reports* **5** (1989) 55.
19. R. Barbieri, G. Dvali and M. Moretti, *Phys. Lett.* **B312** (1993) 137.
20. T. Goto, K. Inoue, Y. Okada and T. Yanagida, *Phys. Rev.* **D46** (1992) 4808.
21. H. Georgi, *Nucl. Phys.* **B156** (1979) 126.
22. C.D. Froggatt and H.B. Nielsen, *Nucl. Phys.* **B147** (1979) 277; Z.G. Berezhiani, *Phys. Lett.* **B129** (1983) 99; *Phys. Lett.* **B150** (1985) 177; S. Dimopoulos, *Phys. Lett.* **B129** (1983) 417.
23. R. Barbieri, G. Dvali, A. Strumia, Z. Berezhiani and L. Hall, preprint IFUP-TH.7/94 (1994), hep-ph/9405428.
24. R. Barbieri and Z. Berezhiani, preprint INFN FE 13-94 (1994).
25. R. Barbieri, S. Ferrara and C. Savoy, *Phys. Lett.* **B110** (1982) 343; P. Nath, R. Arnowitt and A. Chamseddine, *Phys. Rev. Lett.* **49** (1982) 970.
26. V. Barger, M. Berger and P. Ohmann, *Phys. Rev.* **D47** (1993) 1093.
27. J. Giveon, L. Hall and U. Sarid, *Phys. Lett.* **B271** (1991) 138; P. Langaker and N. Polonsky, *Phys. Rev.* **D47** (1993) 4028; M. Carena, S. Pokorsky and C. Wagner, *Nucl. Phys.* **B406** (1993) 140; W. Bardeen, M. Carena, S. Pokorski and C. Wagner, *Phys. Lett.* **B320** (1994) 110.
28. Z. Berezhiani, C. Csaki and L. Randall, in preparation.

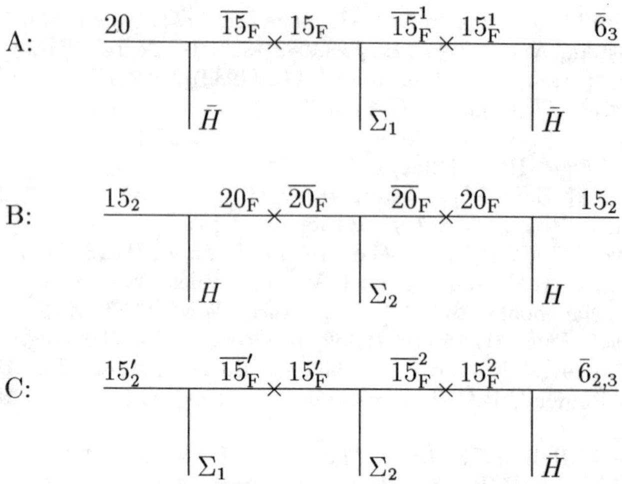

Fig. 1. diagrams giving rise to the operators $\mathcal{A}$, $\mathcal{B}$, $\mathcal{C}$ respectively.

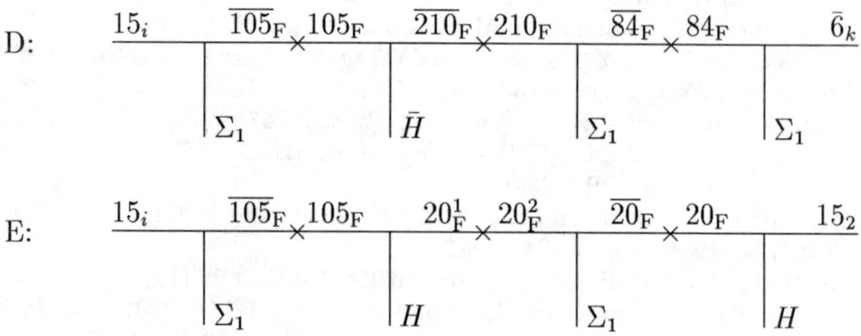

Fig. 2. diagrams giving rise to the operators $\mathcal{D}$ and $\mathcal{E}$ respectively.

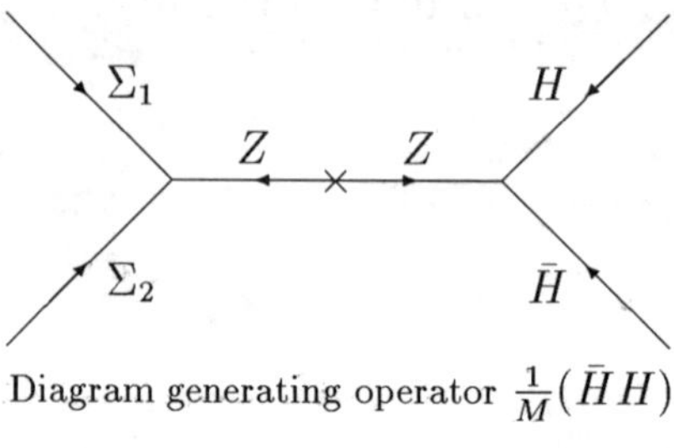

Fig. 3. Diagram generating operator $\frac{1}{M}(\bar{H}H)(\Sigma_1\Sigma_2)$.

THRESHOLD EFFECTS, EFFECTIVE COUPLINGS AND SUSY UNIFICATION

M. BASTERO–GIL
Laboratorio de Astrofísica y Física Fundamental
Apartado 50727, 28080 Madrid, Spain

ABSTRACT

We are interested in how threshold effects due to the presence of massive degrees of freedom in the theory modify the evolution with the scale of the gauge coupling constants, and the implications that these modifications have for the perturbative unification of the gauge couplings in the MSSM.

1. Introduction

When studying the unification picture in the Minimal Supersymmetric Standard Model (MSSM), we search for the convergence of its three gauge couplings, α_3, α_2 and α_1, defined at the electroweak scale, m_Z, towards a common value, α_X, associated with the unification group G, defined at high energies, beyond a scale M_X.

The scale evolution of the couplings is controlled by their β_i functions, whose explicit expressions depend on the renormalization scheme we choose to compute them, and more specifically, on the subtraction procedure. We can choose between two main classes of procedures: (i) mass independent subtraction procedures, such as modified minimal subtraction, which lead to a β_i function that depends only on the number of the degree of freedom present at scale (μ) at which one computes β_i; (ii) and mass dependent subtraction procedures[1], which yield β_i functions which depend not only on the number, but also on the particle masses (m_i).

Thus, with the second class of procedures one automatically includes threshold effects and the possible decoupling of the massive particles. This is incorporated via "threshold functions", $f(m_i/\mu)$, one for each massive particle. As we can see in Fig.(1) the behavior of the threshold function is very similar to a Heaviside "step–function" $\theta(m_i^2 - \mu^2)$. But, unlike the step–function, it is not strictly zero for scales below the particle's mass. This is a quantum–mechanical effect, reflecting the Heisenberg uncertainty principle: we have a non–zero probability of producing a particle even for momenta be-

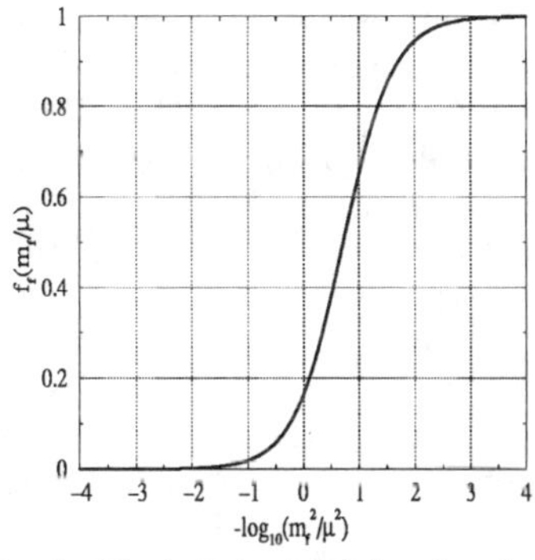

Fig. 1. Typical threshold function due to a loop of massive fermions.

low the mass scale of the particle; furthermore, the contribution of the degree of freedom to the β–function spreads over a few orders of magnitude in momentum.

Hence, the threshold function gives a more accurate description of the threshold crossing than the step–function approximation. We want to stress this point, because a change in the derivative (β_i function) when crossing the thresholds, affects the value of the coupling even for momenta well away from the threshold region.

Susy unification theories are a good theoretical laboratory to study how threshold effects impact on the evolution of the coupling constants. On the one hand, we have a high number of thresholds (massive particles) contributing to the running of the couplings; on the other, we run the couplings over a long distance in momentum scale, so that all the thresholds are crossed and their effects can accumulate.

2. Susy Unification

To study the perturbative unification of the gauge couplings in the MSSM, we consider the simplest unification scenario, neglecting the contribution of the heavy degrees of freedom introduced by the unification group (we choose $G = SU(5)$).

About the procedure, it can be shown that the renormalized couplings using a mass dependent subtraction procedure are completely analogous to effective charges[2], in the sense of the effective charge defined by Gell–Mann and Low, Stuckelberg and Peterman in QED. The advantages of working with effective charges are that (a) they are finite, (b) they are independent of renormalization scheme, because they are given only in terms of bare couplings, and (c) they are universal, i.e., process independent.

We assume about the supersymmetric model that susy is broken at the Planck scale by a hidden sector, so that the majority of susy masses, defined at the weak scale, can be determined by a small number of mass parameters, defined at the unification scale. For the sake of simplicity, we neglect the mixing between winos and higgsino, and stops left and right. Thus, we take six mass parameters: $m_{1/2}$ (gauginos), m_0 (common scalar mass), m_μ (higgsinos), and m_H (heaviest higgses) for the susy partners, and also m_h (lightest Higgs) and* m_t (top). These mass parameters are not completely arbitrary. Besides experimental lower bounds[3] ($m_{1/2} > 45\,GeV$, $m_t > 91\,GeV$ and $m_h > 60\,GeV$), we also have a theoretical upper bound on susy masses of about $1\,TeV$, which is commonly known as the "naturalness bound".

To study the unification of the gauge couplings, as a first approach we simply plot the evolution with scale of the three gauge couplings, at one loop–order, and look if they meet at a common value at same point of the high energy region, high enough (above $10^{16}\,GeV$) to fulfill the experimental lower bounds on proton lifetime. We can see in Fig (2.a) that, when using the step–function approximation there is crossing of the three gauge couplings at a common value, while the situation with effective couplings is not so pristine. This result is related to the presence of gauge

*The work presented here was carried out before the announcement of the CDF Collaboration about the top, so that for us the top mass was arbitrary. Indeed, the conclusions reached here are not modified if we fix the top mass.

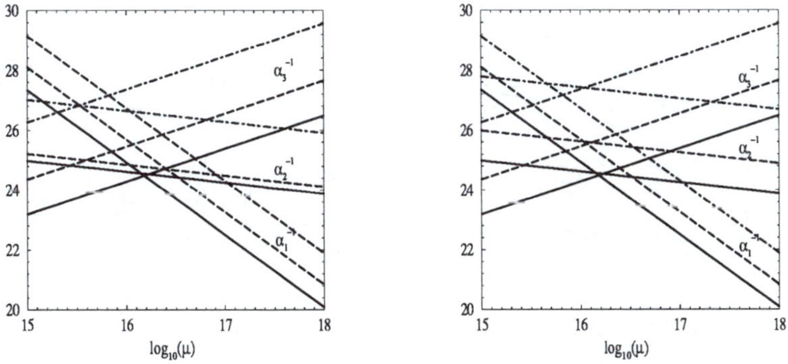

(a) (b)

Fig. 2. Evolution of the three couplings of the MSSM at 1–loop order calculated with (i) step–function with $M_{susy} = m_Z$ (solid lines); (ii) effective couplings with $m_{1/2} = 45\ GeV$ and $m_0 = m_Z$ (dashed lines), and $m_{1/2} = m_0 = 1\ TeV$ (dot–dashed lines); $m_\mu = m_H = m_0$ and $m_t = m_h = 200\ GeV$. (a) All thresholds are included, (b) without gauge thresholds in α_2^{-1}.

boson thresholds[4] in the evolution of α_2. The opposite effects from the susy thresholds (increasing) and the gauge boson thresholds (decreasing) cancel between them, mainly for susy masses of order m_Z, forcing α_2 to shift away from the unification point. If we do not take into account these gauge thresholds, we can recover the unification of the effective couplings, in the sense that they meet at a common value (Fig. (2.b)). The susy threshold effects raise the value of the unification coupling, and decrease slightly the unification scale, with respect to that obtained with the step–function.

Since gauge thresholds are always present, to get unification with effective couplings we have to increase the values of the susy mass parameters beyond their naturalness bound of $1\ TeV$. In order to avoid this unwanted result, another way to recover unification would be to increase the initial value of the QCD gauge coupling, α_3. For the previous plot, we have taken $\alpha_3(m_Z) = 0.113$[3]. This is a good value for getting unification with the step–function *at one–loop order*. But indeed, this coupling is the one worst determined by experiments, with an assortment of experimental data which ranges from 0.108 to 0.125. Next step will be to derive the value of $\alpha_3(m_Z)$ from the unification condition, as a function of the susy mass parameters[5].

In Fig. (3) we have plotted the results obtained with both, the effective couplings and the step–function approximation. The trend is the same for both procedures: the lower the susy masses, the higher the value of $\alpha_3(m_Z)$. Since the step–function does not distinguish values of the susy mass parameters lower than m_Z, this gives us an upper bound: $\alpha_3(m_Z) < 0.116$. On the contrary, using effective couplings we can get unification for any value of $\alpha_3(m_Z)$, if we increase enough the susy masses. But if we want to maintain the naturalness bound for the susy masses, we need a value of the QCD coupling dangerously close to its upper experimental bound.

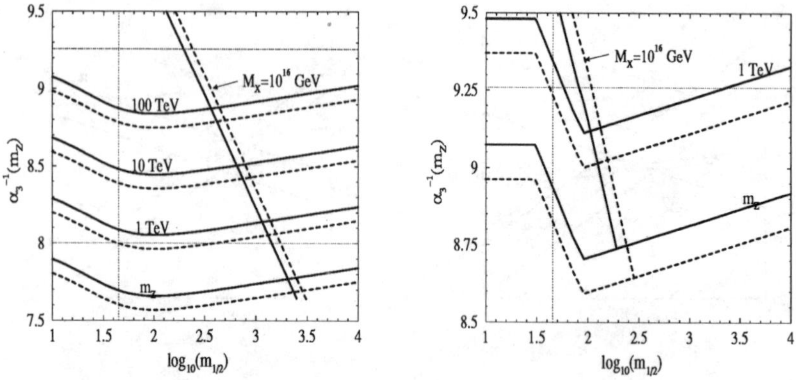

(a) (b)

Fig. 3. Values of $\alpha_3^{-1}(m_Z)$, compatible with the unification condition, for different values of $m_\mu = m_H$. Solid lines are for $m_t = m_h = 200\ GeV$, and dashed lines for $m_t = 91\ GeV$ and $m_h = 60\ GeV$. The straight lines are the upper limit obtained for $\alpha_3^{-1}(m_Z)$ when imposing $M_X = 10^{16}\ GeV$. (a) Effective couplings. (b) Step–function.

We conclude by pointed out that the results for susy unification depend on the procedure chosen to compute the thresholds. The positive results achieved when using a step–function approximation are not always maintained when using a more accurate description of threshold crossing, as it is given by the effective couplings. This we have seen to be the case in the simplest model at one–loop order. Finally, we remark that not only are the susy thresholds important, but also the gauge boson thresholds, present in any extension of the Standard Model.

3. Acknowledgments

I would like to thank Dr. J. Pérez–Mercader for introducing me to this problem and for his useful comments about it.

4. References

1. H. Georgi and H. D. Politzer, *Phys. Rev.* **D14** (1976) 1829; O. Nachtmann and W. Wetzel, *Nucl. Phys.* **B146** (1979) 272.
2. M. Bastero–Gil, V. Manías and J. Pérez–Mercader, *to be published in Int. J. of Mod. Phys. A* (1994); D. C. Kennedy and B. Lynn, *Nucl. Phys.* **B322** (1989) 1.
3. Particle Data Group, *Phys. Rev.* **D45 S1** (1992) 1.
4. D. A. Ross, *Nucl. Phys.* **B140** (1978) 1.
5. M. Bastero–Gil and J. Pérez–Mercader, *Phys. Lett.* **B322** (1994) 355.

SUPERSYMMETRIC LEFT-RIGHT MODEL AND ITS PHENOMENOLOGICAL IMPLICATIONS

KATRI HUITU, MARTTI RAIDAL
Research Institute for High Energy Physics
P.O.Box 9, FIN-00014 University of Helsinki, Finland

and

JUKKA MAALAMPI
Department of Theoretical Physics,
P.O.Box 9, FIN-00014 University of Helsinki, Finland

ABSTRACT

We review here our study of a supersymmetric left-right model (SLRM). In the model the R-parity is spontaneously broken. Phenomenologically novel feature of the model is the occurrance of the doubly charged particles in the Higgs sector, which are possibly light enough to be seen in the next linear collider. Detection of the doubly charged higgsinos in the next linear collider is discussed.

1. Introduction

The left-right symmetric electroweak model based on the $SU(2)_L \times SU(2)_R \times U(1)_{B-L}$ symmetry has many attractive features. In particular, in the see-saw mechanism it offers a beautiful and very natural explanation for the lightness of the ordinary neutrinos. On the other hand, like in the Standard Model it has a hierarchy problem in the scalar sector, which can be solved by making the theory supersymmetric.

The left-right models are especially interesting, if the experiments on solar [2] and atmospheric [3] neutrinos continue to show deviation from the standard model, as well as the existence of the hot dark matter component [4] explaining some features of the power spectrum of density fluctuations of the Universe persists. All these results seem to indicate that neutrinos indeed have a small mass.

To achieve the see-saw mechanism, the $SU(2)_L \times SU(2)_R \times U(1)_{B-L}$ symmetry has to be broken by scalar triplets of $SU(2)_R$. A novel feature of the model is that the triplet superfields contain among others also doubly charged particles.

2. The model

The model is described by the superpotential

$$\begin{aligned} W &= h_{\phi Q}\widehat{Q}_L^T i\tau_2\widehat{\phi}\widehat{Q}_R^c + h_{\chi Q}\widehat{Q}_L^T i\tau_2\widehat{\chi}\widehat{Q}_R^c \\ &\quad + h_{\phi L}\widehat{L}_L^T i\tau_2\widehat{\phi}\widehat{L}_R^c + h_{\chi L}\widehat{L}_L^T i\tau_2\widehat{\chi}\widehat{L}_R^c + h_\Delta \widehat{L}_R^{cT} i\tau_2\widehat{\Delta}\widehat{L}_R^c \\ &\quad + \mu_1 \mathrm{Tr}(i\tau_2\widehat{\phi}^T i\tau_2\widehat{\chi}) + \mu_2 \mathrm{Tr}(\widehat{\Delta}\widehat{\delta}), \end{aligned} \tag{1}$$

where $\widehat{Q}_{L(R)}$ denote the left (right) handed quark superfield doublets and similarly for the leptons $\widehat{L}_{L(R)}$. The triplet and the bidoublet Higgs superfields of $SU(2)_L \times SU(2)_R \times U(1)_{B-L}$ are given by

$$\widehat{\Delta} = \begin{pmatrix} \widehat{\Delta}^-/\sqrt{2} & \widehat{\Delta}^0 \\ \widehat{\Delta}^{--} & -\widehat{\Delta}^-/\sqrt{2} \end{pmatrix}, \qquad \widehat{\delta} = \begin{pmatrix} \widehat{\delta}^+/\sqrt{2} & \widehat{\delta}^{++} \\ \widehat{\delta}^0 & -\widehat{\delta}^+/\sqrt{2} \end{pmatrix}$$

$$\widehat{\phi} = \begin{pmatrix} \widehat{\phi}_1^0 & \widehat{\phi}_1^+ \\ \widehat{\phi}_2^- & \widehat{\phi}_2^0 \end{pmatrix}, \qquad \widehat{\chi} = \begin{pmatrix} \widehat{\chi}_1^0 & \widehat{\chi}_1^+ \\ \widehat{\chi}_2^- & \widehat{\chi}_2^0 \end{pmatrix}, \tag{2}$$

where the different fields transform as $\widehat{\Delta} \sim (\mathbf{1},\mathbf{3},-2)$, $\widehat{\delta} \sim (\mathbf{1},\mathbf{3},2)$, $\widehat{\phi} \sim (\mathbf{2},\mathbf{2},0)$, and $\widehat{\chi} \sim (\mathbf{2},\mathbf{2},0)$. Corresponding to each scalar multiplet with non-zero $U(1)$ quantum number, one has to include another multiplet with an opposite $U(1)$ quantum number in order to avoid chiral anomalies for the fermionic superpartners. Also another bidoublet Higgs superfield is added to get a nontrivial Kobayashi-Maskawa matrix.

We find a region in the parameter space for which the scalar fields in the minimum have the following vacuum expectation values [6]:

$$\langle \Delta^0 \rangle = v_\Delta, \;\; \langle \delta^0 \rangle = v_\delta, \;\; \langle \phi_1^0 \rangle = \kappa_1, \;\; \langle \chi_2^0 \rangle = \kappa_2, \;\; \langle \tilde{\nu} \rangle = \sigma_R. \tag{3}$$

Applying the minimization conditions $\partial V/\partial \kappa_1 = \partial V/\partial \kappa_2 = \partial V/\partial v_\Delta = \partial V/\partial v_\delta = \partial V/\partial \sigma_R = 0$ one can find the scalar masses. In the minimum the $mass^2$ of all the scalars in the Higgs sector must be positive. This requirement has fundamental consequences for the R-parity, $R = (-1)^{3(B-L)+2s}$. The R-parity is automatically conserved in Lagrangian in this type of models [5], but it may be broken spontaneously if $\langle \tilde{\nu} \rangle \neq 0$. In the case of conserved R-parity, *i.e.* $\langle \tilde{\nu}_{R,L} \rangle = 0$, the pseudoscalar mass matrix is given by four two by two blocks. One of the blocks contains the sneutrinos and we need not consider it here. Two of the blocks contain the Goldstone bosons which make two of the neutral gauge bosons massive. The physical pseudoscalar particles have the masses [6]

$$\begin{aligned} m_{A_1}^2 &= m_{\phi\chi}^2 \left(\frac{\kappa_1}{\kappa_2} + \frac{\kappa_2}{\kappa_1} \right), \qquad m_{A_2}^2 = m_{\Delta\delta}^2 \left(\frac{v_\delta}{v_\Delta} + \frac{v_\Delta}{v_\delta} \right), \\ m_{A_{3,4}} &= \frac{1}{2} \Big\{ m_{A_1}^2 \pm [\, m_{A_1}^4 + 4(m_{W_R}^2 \cos 2\gamma - m_{W_L}^2 \cos 2\beta)^2 \\ &\quad -4(m_{W_R}^2 \cos 2\gamma - m_{W_L}^2 \cos 2\beta) m_{A_1}^2 \cos 2\beta \,]^{1/2} \Big\}, \end{aligned} \tag{4}$$

where it is defined $\tan^2\gamma = (v_\delta^2 + \frac{1}{2}\kappa_1^2)/(v_\Delta^2 + \frac{1}{2}\kappa_2^2)$, $\tan\beta = \kappa_2/\kappa_1$, and $\tan\delta = v_\delta/v_\Delta$.

On the other hand the masses of the doubly charged scalars are given by

$$m_{H_{1,2}^{++}}^2 = \frac{1}{2} \left\{ m_{A_2}^2 \pm \sqrt{m_{A_2}^4 + 8 m_{W_R}^2 \cos 2\gamma [m_{A_2}^2 \cos 2\delta + 2 m_{W_R}^2 \cos 2\gamma]} \right\}. \tag{5}$$

It is easily seen that trying to make both pseudoscalar and doubly charged $mass^2$ positive one ends in contradiction. Necessarily at least one of the $\langle \tilde{\nu} \rangle \neq 0$.

However, if the right-sneutrino has a vev, it is found that ranges of parameters exist, where all the squares of the Higgs masses are positive. Furthermore, one of the doubly charged Higgses turns out to be lighter than 500 GeV for $h_\Delta \leq 0.8$. The majority of the scalars are heavier than 1 TeV.

3. Testing SLRM in the colliders

Another particle of interest in the Higgs sector is the supersymmetric counterpart of the doubly charged Higgs. This particle is very suitable for experimental search for many reasons. It is doubly charged, which means that it does not mix with other particles. Consequently its mass is given by a single parameter, the susy Higgs mixing parameter μ_2.

The next generation linear electron colliders will, besides the usual e^+e^- reactions, be able to work also in e^-e^-, $e^-\gamma$ and $\gamma\gamma$ modes. The high energy photon beams can be obtained by back-scattering an intensive laser beam on high energy electrons. The doubly charged higgsinos can be produced in any of these operation modes [7]:

$$e^+e^- \to \tilde{\Delta}^{++}\tilde{\Delta}^{--}, \tag{6}$$

$$e^-e^- \to \tilde{\Delta}^{--}\tilde{\chi}^0, \tag{7}$$

$$\gamma e^- \to \tilde{l}^+\tilde{\Delta}^{--}, \tag{8}$$

$$\gamma\gamma \to \tilde{\Delta}^{--}\tilde{\Delta}^{++}. \tag{9}$$

We have chosen these reactions for investigation because they all have a clean experimental signature: a few hard leptons and missing energy. Furthermore, they all have very small background from other processes.

In large regions of the parameter space, the kinematically favoured decay mode of the triplet higgsino is $\tilde{\Delta}^{++} \to \tilde{l}^+l^+$. Which of the slepton decay modes is dominant, depends on kinematics, but one possibility is the decay to a lepton and the lightest neutralino: $\tilde{l} \to l\tilde{\chi}^0$. The experimental signature of the doubly charged higgsino could be then

$$\tilde{\Delta}^{--} \to \tilde{l}^-l^- \to l^-l^-\tilde{\chi}^0, \tag{10}$$

where l can be any of the e, μ, and τ with practically equal probabilities. The experimental signal of reactions (6) and (9) would be then four leptons and missing energy. The total cross section of reaction (6) for the collision energy $\sqrt{s}$ =1 TeV and the slepton and higgsino masses in the range of 100–400 GeV is about 0.5 pb. The reaction (9) is a model independent way to produce doubly charged higgsinos, since it depends only on the parameter μ_2. For $\mu_2 \lesssim 300$ GeV its cross section is larger than 1 pb.

The slepton pair production

$$e^+e^- \to \tilde{l}^+\tilde{l}^- \tag{11}$$

tests also the Higgs sector of the theory since the process is mediated among others by the doubly charged higgsino [8].

The selectron pair production in supersymmetric LR-model has a larger cross section than the corresponding process in the MSSM by about an order of magnitude [8]. This is due to two factors, firstly the number of gauginos in t-channel is larger and secondly the triplet higgsino contribution in u-channel is large, though dependent on the unknown triplet higgsino coupling to the electron and selectron.

The u-channel exchange of the doubly charged higgsino occurs only for a right-handed electron and a left-handed positron (P_{+-} polarization), whereas in s- and t-channel processes also other chirality combinations may enter. Use of polarized beams could therefore give us more information of the triplet higgsino contribution. For P_{+-} polarization there is a peak in the backward direction in the angular distribution of the final state electron. Whether there is a forward peak, depends on the neutralino content. For gaugino dominated neutralinos, the peak exists, but for higgsino dominated neutralinos it is suppressed by the lepton Yukawa couplings.

The cross section of the pair production of smuons and staus are in general expected to be smaller than that of selectron pair production, since the neutralinos do not contribute. On the other hand the cross sections are in general larger than in the case of the MSSM because of the nondiagonal couplings of the triplet higgsinos.

The work has been supported by the Academy of Finland.

1. M. Gell-Mann, P. Ramond and R. Slansky, in *Supergravity*, eds. P. van Niewenhuizen and D.Z. Freedman (North Holland 1979); T. Yanagida, in Proceedings of *Workshop on Unified Theory and Baryon Number in the Universe*, eds. O. Sawada and A. Sugamoto (KEK 1979).
2. K. Lande et al., in Proc. XXVth Int. Conf. on High Energy Physics, eds. K.K. Phua and Y. Yamaguchi (World Scientific, Singapore, 1991); K.S. Hirata et al., Phys. Rev. Lett. **66** (1990) 1301; K. Nakamura, Nucl. Phys. **B 31** (Proc. Suppl.) (1993); A.I. Abazov et al., Phys. Rev. Lett. **67** (1991) 3332; V.N. Gavrin, in Proc. XXVIth Int. Conf. in High Energy Physics (Dallas 1992), to appear.
3. K.S. Hirata et al., Phys. Lett. **B 280** (1992) 146; D. Casper et al., Phys. Rev. Lett. **66** (1993) 2561.
4. M. Davis, F.J. Summers, D. Schlegel, Nature 359 (1992) 393; A. N. Taylor, M. Rowan-Robinson, ibid. 396.
5. R.N. Mohapatra, Phys. Rev. **D34** (1986) 3457; A. Font, L.E. Ibanez, F. Quevedo, Phys. Lett. **B 228** (1989) 79; S.P. Martin, Phys. Rev. **D46** (1992) 2769.
6. K. Huitu, J. Maalampi, to appear in Phys. Lett. **B** .
7. K. Huitu, J. Maalampi, M. Raidal, Nucl. Phys. **B 420** (1994) 449.
8. K. Huitu, J. Maalampi, M. Raidal, Phys. Lett. **B 328** (1994) 60.

THE LINEAR MULTIPLET AND QUANTUM STRING EFFECTIVE ACTIONS

J.-P. DERENDINGER
Physics Institute, University of Neuchâtel, Breguet 1
CH-2000 Neuchâtel, Switzerland

ABSTRACT

Quantum symmetries of four-dimensional superstrings are frequently realized in an anomaly-cancellation mode in the effective low-energy supergravity. The massless antisymmetric tensor plays an important rôle in this mechanism and the choice of its supersymmetric description, using either a chiral or a linear multiplet, appears to introduce significant conceptual and practical differences at the string loop level. This paper reviews the construction of loop-corrected string effective supergravities with the dilaton and antisymmetric tensor embedded in a linear multiplet. Using anomaly cancellation and the linear multiplet allows to obtain an all-order renormalization-group invariant effective lagrangian for a pure gauge sector with field-dependent gauge coupling constant.

1. The linear multiplet

A massless antisymmetric tensor $b_{\mu\nu}$ plays an important rôle in the low-energy effective supergravity of superstrings. Its existence is a universal property: the gravity sector of four-dimensional superstrings contains, together with the graviton, $b_{\mu\nu}$ and a real scalar, the dilaton. It is then natural to take advantage of the particular properties of the supermultiplet generalizing the antisymmetric tensor when discussing the structure of the effective supergravity theory.

This section will review some aspects of the supersymmetric description of the antisymmetric tensor. For simplicity, the discussion will first concentrate on global supersymmetry. The case of supergravity which will be of importance for the superstring case will be introduced at the end of the section.

In global supersymmetry, the description of the antisymmetric tensor uses two superfields[1], the prepotential ψ^α and the linear multiplet L. The prepotential is a chiral, spinor (the index α) superfield with the gauge transformation

$$\delta\psi^\alpha = i\overline{\mathcal{D}}\overline{\mathcal{D}}\mathcal{D}^\alpha K, \tag{1}$$

where K is a real vector superfield. This transformation can be used to choose a gauge in which some of the components of ψ^α are eliminated. In this gauge, ψ^α contains the antisymmetric tensor, a real scalar C and a spinor χ. And (1) reduces to the residual (bosonic) gauge symmetry

$$\delta b_{\mu\nu} = \partial_\mu k_\nu - \partial_\nu k_\mu, \tag{2}$$

which preserves the gauge choice.

The real linear multiplet[1] is the supersymmetrization of the field strength

$$v_\mu = \frac{1}{\sqrt{2}}\epsilon_{\mu\nu\rho\sigma}\partial^\nu b^{\rho\sigma} \tag{3}$$

of $b_{\mu\nu}$. In terms of the prepotential, it is defined by

$$L = \mathcal{D}^\alpha\psi_\alpha + \overline{\mathcal{D}}_{\dot\alpha}\overline{\psi}^{\dot\alpha}, \tag{4}$$

which is invariant under the transformation (1). Alternatively, it can be obtained by imposing the constraints

$$\mathcal{D}\mathcal{D}L = \overline{\mathcal{D}\mathcal{D}}L = 0, \tag{5}$$

on a real vector superfield. The component expansion of L is

$$L = C + i\theta\chi - i\overline{\theta}\overline{\chi} + \theta\sigma^\mu\overline{\theta}v_\mu - \frac{1}{2}\theta\theta\overline{\theta}(\partial_\mu\chi\sigma^\mu) - \frac{1}{2}\overline{\theta\theta}\theta(\sigma^\mu\partial_\mu\overline{\chi}) - \frac{1}{4}\theta\theta\overline{\theta\theta}\Box C.$$

It is interesting to remark that L does not contain any auxiliary field. A lagrangian involving the linear multiplet L only does not include a scalar potential and supersymmetry cannot spontaneously break.

A generic lagrangian for L and chiral matter denoted collectively by Σ is of the form

$$\mathcal{L}_{global} = 2\int d^2\theta d^2\overline{\theta}\,\Phi(L,\Sigma,\overline{\Sigma}) + \int d^2\theta\, w(\Sigma) + \int d^2\overline{\theta}\,\overline{w}(\overline{\Sigma}),$$

with an arbitrary real function Φ and a superpotential w independent from L. Coupling this theory to gauge fields leads to

$$\mathcal{L}_{global} = 2\int d^2\theta d^2\overline{\theta}\,\Phi(\hat{L},\Sigma,\overline{\Sigma}e^V) + \int d^2\theta\, w(\Sigma) + \int d^2\overline{\theta}\,\overline{w}(\overline{\Sigma}), \tag{6}$$

where $V = V^aT^a$ is the Lie-algebra-valued vector superfield of gauge potentials, the superpotential is a gauge-invariant analytic function,

$$\hat{L} = L - 2\Omega, \tag{7}$$

and Ω is the Chern-Simons superfield. This last object can be defined by its relation with the chiral superfield of gauge curvatures,

$$W^\alpha = -\frac{1}{4}\overline{\mathcal{D}\mathcal{D}}e^{-V}\mathcal{D}^\alpha e^V = W^{\alpha\,a}T^a,$$

which reads

$$\overline{\mathcal{D}\mathcal{D}}\Omega = W^aW^a, \qquad \mathcal{D}\mathcal{D}\Omega = \overline{W}^a\overline{W}^a. \tag{8}$$

Since theory (6) only depends on $b_{\mu\nu}$ through its curl v_μ [eq. (3)], a duality transformation can be performed to replace $b_{\mu\nu}$ by a (pseudo)scalar field. With supersymmetry, the duality transformation replaces L by a chiral superfield, an operation which will be discussed in detail in the next section.

For reasons to be explained below, conformal supergravity[2] offers the appropriate formalism to extend the global lagrangian (6) to local supersymmetry. One needs to introduce a chiral compensating supermultiplet S_0 with conformal and chiral weights equal to one.* Since the conformal weight of L is two, the supergravity lagrangian generalizing (6) is

$$\mathcal{L}_{local} = \left[S_0\overline{S}_0\Phi(\frac{\hat{L}}{S_0\overline{S}_0}, \Sigma, \overline{\Sigma}e^V)\right]_D + \left[S_0^3 w(\Sigma)\right]_F. \tag{9}$$

The real and chiral density formula are denoted respectively by $[\ldots]_D$ and $[\ldots]_F$. The Poincaré supergravity theory is obtained by imposing in (9) gauge-fixing conditions for the superconformal symmetries not contained in the super-Poincaré algebra. These gauge choices use in particular the scalar and fermion components z_0 and ψ_0 of S_0, and leave the auxiliary component f_0 and the gauge field A_μ of the chiral, internal $U(1)$ of the superconformal algebra as supergravity auxiliary fields.

It is worth recalling that the linear multiplet has made at least two appearances in past developments of supergravity theories. Firstly, 'new minimal supergravity' is obtained in the conformal framework using a linear compensating multiplet[4,3], with lagrangian $[L \log L]_D$. And secondly, early investigations[5] of the tree-level effective supergravity of superstrings suggested the relevance of '16+16 supergravity'[6], which corresponds in the conformal approach to a theory with a linear and a chiral multiplet, with some particular gauge fixing[7].

2. Gauge kinetic lagrangian and gauge coupling constants

Since field-dependent gauge coupling constants in superstring effective supergravities will be at the center of the discussion, we briefly consider[8] the sector of gauge kinetic terms in the global lagrangian (6) and in its supergravity extension (9).

In global supersymmetry, the expansion of the Chern-Simons superfield in the Wess-Zumino gauge contains the following terms:

$$\begin{aligned}\Omega &= -\tfrac{1}{2}\theta\theta\overline{\theta\theta}\left(\tfrac{1}{2}D^aD^a - \tfrac{1}{2}i\lambda^a\sigma^\mu\partial_\mu\overline{\lambda}^a + \tfrac{1}{2}i\partial_\mu\lambda^a\sigma^\mu\overline{\lambda}^a - \tfrac{1}{4}F^{a\,\mu\nu}F^a_{\mu\nu}\right)\\ &\quad +\tfrac{1}{2}(\theta\sigma^\mu\overline{\theta})\epsilon^{\mu\nu\rho\sigma}\omega_{\nu\rho\sigma} + \tfrac{1}{4}\theta\theta\overline{\lambda}^a\overline{\lambda}^a + \tfrac{1}{4}\overline{\theta\theta}\lambda^a\lambda^a + \ldots,\end{aligned}$$

where $\omega_{\nu\rho\sigma}$ is the gauge Chern-Simons form, normalised by

$$\partial_\mu\epsilon^{\mu\nu\rho\sigma}\omega_{\nu\rho\sigma} = \frac{1}{4}\epsilon^{\mu\nu\rho\sigma}F^a_{\mu\nu}F^a_{\rho\sigma} = \frac{1}{2}\tilde{F}^a_{\mu\nu}F^{a\,\mu\nu}.$$

Its highest component is the super-Yang-Mills lagrangian (multiplied by $-1/2$) and it does not contain any $\tilde{F}^a_{\mu\nu}F^{a\,\mu\nu}$ contribution. Gauge kinetic terms in lagrangian (6)

*This choice of compensator leads to the old minimal set of supergravity auxiliary fields, which allows the most general matter couplings[3].

will then clearly be of the form

$$2\Phi_C\left(\frac{1}{2}D^aD^a-\frac{1}{2}i\lambda^a\sigma^\mu\partial_\mu\overline{\lambda}^a+\frac{1}{2}i\partial_\mu\lambda^a\sigma^\mu\overline{\lambda}^a-\frac{1}{4}F^{a\,\mu\nu}F^a_{\mu\nu}\right),$$

with $\Phi_C = \frac{\partial}{\partial C}\Phi(C,z,\overline{z}) = \left[\frac{\partial}{\partial L}\Phi(\hat{L},\Sigma,\overline{\Sigma}e^V)\right]_{\theta=\overline{\theta}=0}$. The field-dependent gauge coupling constant is then[9,8]

$$\frac{1}{g^2}=2\Phi_C. \tag{10}$$

With the linear multiplet, the gauge coupling constant is a real function which is not constrained to be harmonic. It is obtained in a real D-term, contrary to the chiral case where it is related to a holomorphic function. This illustrates the fact that holomorphicity properties of gauge couplings do not follow from supersymmetry only but require also a selection of supersymmetric multiplets[9].

Lagrangian (6) also contains a term

$$-\left[\Phi_{CC}v_\mu - i\Phi_{Cz}\partial_\mu z + i\Phi_{C\overline{z}}\partial_\mu\overline{z}\right]\epsilon^{\mu\nu\rho\sigma}\omega_{\nu\rho\sigma}. \tag{11}$$

This contribution will only give rise, by partial integration, to an $\tilde{F}^a_{\mu\nu}F^{a\,\mu\nu}$ with the very special choice

$$\Phi=\hat{L}[\phi(\Sigma)+\overline{\phi}(\overline{\Sigma})]. \tag{12}$$

In this case, $\Phi_{CC}=0$, $\Phi_{Cz}\partial_\mu z=\partial_\mu\phi(z)$, $\Phi_{C\overline{z}}\partial_\mu\overline{z}=\partial_\mu\overline{\phi}(\overline{z})$, and (11) becomes

$$\begin{aligned} i\partial_\mu[\phi(z)-\overline{\phi}(\overline{z})]\epsilon^{\mu\nu\rho\sigma}\omega_{\nu\rho\sigma} &= 2\mathrm{Im}\,\phi(z)\partial_\mu\epsilon^{\mu\nu\rho\sigma}\omega_{\nu\rho\sigma}+\text{total derivative} \\ &= \mathrm{Im}\,\phi(z)\tilde{F}^a_{\mu\nu}F^{a\,\mu\nu}+\text{total derivative.} \end{aligned}$$

In this case, the gauge coupling constant is proportional to the real part of $\phi(z)$:

$$\frac{1}{g^2}=4\mathrm{Re}\,\phi(z).$$

Notice however that since $\int d^2\theta d^2\overline{\theta}\,L[\phi(\Sigma)+\overline{\phi}(\overline{\Sigma})]=$ total derivative, the choice (12) completely eliminates the linear multiplet. Then,

$$\mathcal{L}=-4\int d^2\theta d^2\overline{\theta}\,\Omega[\phi(\Sigma)+\overline{\phi}(\overline{\Sigma})]=\int d^2\theta\,\phi(\Sigma)W^aW^a+\text{h.c.},$$

which is the usual super-Yang-Mills theory with gauge coupling depending on chiral matter. The fact that the gauge coupling is the real part of a holomorphic function appears then to depend on the description of matter with chiral multiplets only.

3. Supersymmetric duality

Any theory with a linear multiplet can be in principle transformed into a theory with chiral multiplets only, using a supersymmetric duality transformation. Suppose we want to transform (6) into a theory with a chiral superfield S replacing L. The starting point is to write a classically equivalent theory, of the form:

$$\mathcal{L}_1 = 2\int d^2\theta d^2\overline{\theta}\,\Phi(\hat{L}-U,\Sigma,\overline{\Sigma}e^V) + \left\{\int d^2\theta\left[-\frac{1}{2}S\overline{\mathcal{D}\mathcal{D}}U + w(\Sigma)\right] + \text{h.c.}\right\}. \qquad (13)$$

This lagrangian possesses a new gauge invariance under

$$U \longrightarrow U+\mathcal{U}, \qquad L \longrightarrow L+\mathcal{U}, \qquad (14)$$

where the gauge parameter is a linear superfield $\mathcal{U}$: $\mathcal{D}\mathcal{D}\mathcal{U} = \overline{\mathcal{D}\mathcal{D}}\mathcal{U} = 0$. The equation of motion for S implies that U is a linear multiplet and (6) is then simply the gauge choice $U=0$ in (13), after solving for S. Notice that a superfield U submitted to the gauge transformation (14) can be regarded as the supersymmetric extension of an antisymmetric tensor $a_{\mu\nu\rho}$ with gauge variation $\delta a_{\mu\nu\rho} = \partial_{[\mu}\mathcal{U}_{\nu\rho]}$. The expansion of the superfield $L-U$ contains

$$\frac{1}{\sqrt{2}}(\theta\sigma^\mu\overline{\theta})\epsilon_{\mu\nu\rho\sigma}D^{[\nu}b^{\rho\sigma]},$$

with a derivative

$$D^{[\nu}b^{\rho\sigma]} = \partial^{[\nu}b^{\rho\sigma]} - a^{\nu\rho\sigma}$$

which is invariant under $\delta b_{\mu\nu} = \mathcal{U}_{\mu\nu}$. Theory (13) can be rewritten

$$\begin{aligned}
\mathcal{L}_2 &= 2\int d^2\theta d^2\overline{\theta}\left[\Phi(\hat{L}-U,\Sigma,\overline{\Sigma}e^V) - (S+\overline{S})(\hat{L}-U) - 2(S+\overline{S})\Omega\right] \\
&\quad + \left\{\int d^2\theta\, w(\Sigma) + \text{h.c.}\right\} \\
&= 2\int d^2\theta d^2\overline{\theta}\left[\Phi(\hat{L}-U,\Sigma,\overline{\Sigma}e^V) - (S+\overline{S})(\hat{L}-U)\right] \\
&\quad + \left\{\int d^2\theta\,[SW^aW^a + w(\Sigma)] + \text{h.c.}\right\}.
\end{aligned} \qquad (15)$$

The equation of motion for U (or $\hat{L}-U$) can in principle be used to express $\hat{L}-U$ as a function of $S+\overline{S}$, to finally obtain the dual theory

$$\mathcal{L}_S = \int d^2\theta d^2\overline{\theta}\,K(S+\overline{S},\Sigma,\overline{\Sigma}e^V) + \left\{\int d^2\theta\,[SW^aW^a + w(\Sigma)] + \text{h.c.}\right\}. \qquad (16)$$

In this expression, which possesses the symmetry $S \longrightarrow S + i(\text{real constant})$, the gauge coupling constant is always

$$\frac{1}{g^2} = 4\text{Re}\, s, \qquad (17)$$

s being the scalar lowest component of S. And there is a term of the form

$$\text{Im}\, sF^a_{\mu\nu}F^{a\,\mu\nu} = -2(\partial_\mu \text{Im}\, s)\epsilon^{\mu\nu\rho\sigma}\omega_{\nu\rho\sigma} + \text{total derivative}. \qquad (18)$$

The duality transformation is performed by solving the equations of motion for the components $(c, \tilde{\chi}, m, n, \tilde{v}_\mu, \tilde{\lambda}, d)$ of U, for instance in the gauge $L = 0$. In particular, these equations indicate that[8]

$$\begin{aligned} \mathrm{Re}\, s &= \tfrac{1}{2}\Phi_c, \\ \partial_\mu \mathrm{Im}\, s &= \tfrac{1}{2}\Phi_{cc}\tilde{v}_\mu - \tfrac{i}{2}\Phi_{cz}\partial_\mu z + \tfrac{i}{2}\Phi_{c\bar{z}}\partial_\mu \bar{z} + \text{fermionic terms}. \end{aligned} \tag{19}$$

These relations show the equivalences of (17) and (18) with respectively (10) and (11).

These results have a straightforward extension to conformal and Poincaré supergravities[8]. Equation (17), in particular, remains true in the local case. It provides an important information on the description of superstring effective supergravities. The two dual theories have a different physical interpretation. Working with the chiral theory, with S, implies that the gauge coupling constant appearing in the lagrangian is the expectation value of $\mathrm{Re}\, s$. This field is then directly related to a bare, unphysical quantity. The physical (renormalized) gauge coupling constant, which appears in the effective action (the generating functional of one-particle irreducible Green's functions), is a function of $\mathrm{Re}\, s$ (as well as other fields, in general). It is certainly not given by the expectation value of $\mathrm{Re}\, s$. Then, since $\mathrm{Re}\, s$ is related to a bare quantity of the effective field theory, it is not necessarily in simple and natural relation with string parameters at the (string) quantum level. And there is no reason to expect that quantum symmetries of the superstring have a natural action on this field.

4. Superstring effective lagrangians

The effective supergravity of superstrings describes the low-energy† physics of string massless modes. It then contains an ultraviolet physical cutoff, above which 'string physics has been integrated'. The effective supergravity can be defined using a Wilson lagrangian $\mathcal{L}_W$. At the level of superstring perturbation theory, in which supersymmetry does not break, the Wilson effective lagrangian has a string loop expansion

$$\mathcal{L}_W = \sum_{N\geq 0} \mathcal{L}_W^{(N)}. \tag{20}$$

The Wilson lagrangian and the corresponding action $S_W = \int d^4x\, \mathcal{L}_W$ should not be confused with the more familiar effective action of quantum field theory,

$$S_\Gamma = \int d^4x\, \mathcal{L}_\Gamma,$$

which is the generating functional of one-particle irreducible Green's functions. While $\mathcal{L}_W$ is a quantized lagrangian, $\mathcal{L}_\Gamma$ is a functional of classical fields.

†With respect to the inverse string length and compactification radius.

For large classes of superstrings, the tree-level lagrangian $\mathcal{L}_W^{(0)}$ is well understood. Since the present discussion focusses on the rôle of the antisymmetric tensor described by a linear multiplet L, or on the dual theory with a chiral S, it will be sufficient to use

$$\mathcal{L}_W^{(0)} = -\frac{1}{\sqrt{2}}\left[\hat{L}\left(\frac{\hat{L}e^{\hat{K}/3}}{S_0\overline{S}_0}\right)^{-3/2}\right]_D + [S_0^3 w]_F, \tag{21}$$

as tree-level Wilson lagrangian[10]. The Kähler function $\hat{K}$ and the superpotential w depend in a gauge invariant way on the chiral multiplets describing moduli and charged matter. Using (16), theory (21) is, by duality, equivalent to the chiral lagrangian

$$\mathcal{L}_S^{(0)} = -\frac{3}{2}\left[S_0\overline{S}_0 e^{-\mathcal{K}/3}\right]_D + [SW^aW^a + S_0^3 w]_F, \qquad \mathcal{K} = -\log(S+\overline{S}) + \hat{K}. \tag{22}$$

The tree-level Wilson lagrangian possesses the (formal) Kähler symmetry

$$\begin{aligned} S_0 &\longrightarrow e^{\varphi/3}S_0, & w &\longrightarrow e^{-\varphi}w, \\ \hat{K} &\longrightarrow \hat{K}+\varphi+\overline{\varphi}, & \hat{L} &\longrightarrow \hat{L}, \end{aligned} \tag{23}$$

where φ is an arbitrary holomorphic function of chiral multiplets.

The lagrangian (21) is superconformal invariant. Two different choices of compensator fixings are of interest. Imposing canonical normalisation of the Einstein term leads to

$$e^{-1}\mathcal{L}_W^{(0)} = -\frac{1}{2\kappa^2}R - \frac{1}{4}(\kappa^2 C)^{-1}F^a_{\mu\nu}F^{a\,\mu\nu} + \frac{1}{4}\kappa^2(\kappa^2 C)^{-2}v_\mu v^\mu + \ldots,$$

and the gauge coupling constant is

$$g^{-2}_{tree} = (\kappa^2 C)^{-1} = 4\mathrm{Re}\, s. \tag{24}$$

The second choice is the 'string frame' where the same bosonic terms become

$$e^{-1}\mathcal{L}_W^{(0)} = \frac{\varphi^{-1/2}}{\kappa^2}\left[-\frac{1}{2}R - \frac{1}{4}M_s^{-2}F^a_{\mu\nu}F^{a\,\mu\nu} + \frac{1}{4}M_s^{-4}v_\mu v^\mu\right] + \ldots$$

where $\varphi = \kappa^2 C$ is the string 'loop-counting parameter' and $M_s^2 = \langle C\rangle$ is the string scale. With this choice,

$$g^{-2}_{tree} = (\kappa^2 C)^{-1} = (4\mathrm{Re}\, s)^{2/3},$$

since the relation between $\mathrm{Re}\, s$ and C depends on the superconformal gauge fixing.

5. One loop: a string symmetry, anomaly cancellation

Some contributions to $\mathcal{L}_W^{(1)}$, in eq. (20), have been revealed by the effective field-theory description of string one-loop calculations of gauge kinetic terms in (2,2) symmetric orbifolds[11,12,13]. The moduli dependence of these corrections is in large part dictated by target-space duality, a perturbative string symmetry. In the effective supergravity, the crucial observation[9] is that since target-space duality is an anomalous symmetry of $\mathcal{L}^{(0)}$, invariance is restored by an anomaly-cancellation mechanism due to specific contributions in $\mathcal{L}^{(1)}$. This mechanism is similar to the cancellation of gauge and gravitational anomalies in ten-dimensional superstrings[14]. Target-space duality acts on the tree-level lagrangian (21) like a Kähler transformation (23). The cancellation of target-space duality anomaly is then a mechanism of Kähler anomaly cancellation in supergravity[15].

This mechanism will be reviewed here in its simplest version only. Specifically, we will consider the case of vanishing threshold corrections[‡] We will also for simplicity restrict the discussion to a simple gauge group, which could be the hidden E_8 group present in symmetric (2,2) orbifolds. The discussion of the general case can be found in the original literature[9,16].

The nature of Kähler symmetry is easily elucidated. Clearly, it is a sigma-model symmetry: it acts on $\hat{K}$ and $\hat{K}$ is the Kähler potential of the supersymmetric sigma-model defining the couplings of chiral multiplets. But in supergravity, it also acts as an R-symmetry which, in particular, rotates all fermions in the theory. This is due to the presence of the supergravity auxiliary field A_μ, which in the superconformal theory is the gauge field of the internal chiral $U(1)$. This chiral group does not commute with supersymmetry transformations and hence acts like an R-symmetry. The equation of motion of its gauge field is

$$A_\mu = \frac{i}{3}\left[\frac{\partial\hat{K}}{\partial z}(\partial_\mu z) - \frac{\partial\hat{K}}{\partial\overline{z}}(\partial_\mu\overline{z})\right] + \ \dots \ ,$$

for an arbitrary chiral scalar component z. The Kähler transformation (23) is then a chiral $U(1)$ gauge transformation of A_μ, once the equation of motion for A_μ has been solved, and $\hat{K}$ can be assimilated to the (superfield) Kähler connection.

The existence of chiral couplings of the Kähler connection to fermions implies that a non-zero anomalous triangle graph with, for instance, two external gauge fields and one external Kähler connection will in general arise at one-loop. This diagram corresponds to a non-local contribution to the effective lagrangian $\mathcal{L}_\Gamma$ (not to $\mathcal{L}_W$), which, in the absence of chiral charged matter as assumed here, is of the form

$$\Delta_{triangle} = -\frac{A}{4}\left[\frac{1}{3}W^aW^a\mathcal{P}_C\hat{K}\right]_F , \tag{25}$$

[‡]This means that anomaly cancellation also cancels the one-loop modulus-dependent correction to physical gauge coupling constants.

where $\mathcal{P}_C$ is the non-local chiral projector. Gauginos only contribute to the anomaly and its coefficient is $A = 3C(G)/(8\pi^2)$, $C(G)$ being the quadratic Casimir of the gauge group G. The transformation of $\Delta_{triangle}$ under (23) is

$$\delta(\Delta_{triangle}) = -\frac{A}{4}\left[\frac{1}{3}W^a W^a \varphi\right]_F, \tag{26}$$

a local expression.

The one-loop Wilson lagrangian $\mathcal{L}_W^{(1)}$ should then contain a Kähler-variant term able to cancel (26) by its variation. This contribution is[9]

$$\mathcal{L}_W^{(1)} = \frac{A}{3}\frac{1}{4}\left[\hat{L}\hat{K}\right]_D, \tag{27}$$

which includes a coupling of the antisymmetric tensor to the Kähler connection[§]. With this one-loop correction, the Wilson lagrangian becomes

$$\mathcal{L}_W^{(0)} + \mathcal{L}_W^{(1)} = \left[-\frac{1}{\sqrt{2}}\hat{L}\left(\frac{Le^{\hat{K}/3}}{S_0\overline{S}_0}\right)^{-3/2} + \frac{A}{12}\hat{L}\hat{K}\right]_D + [S_0^3 w]_F. \tag{28}$$

Repeating the steps (15) and (16) to perform the duality transformation leads to the substitution

$$S + \overline{S} \longrightarrow S + \overline{S} - \frac{A}{3}\hat{K} \tag{29}$$

in the tree-level Kähler function $\mathcal{K}$ given in eq. (22). This result is reminiscent of the mechanism cancelling gauge anomalies for $U(1)$ factors of the gauge group, where the substitution is

$$S + \overline{S} \longrightarrow S + \overline{S} - \alpha V,$$

V being the vector superfield of the $U(1)$ gauge potential and α the coefficient of the anomaly[17].

6. An effective all-order gauge sector

In the previous section, the information that target-space duality is a quantum string symmetry has been used to derive a one-loop contribution to the Wilson effective lagrangian, using anomaly cancellation as a basic principle. The one-loop lagrangian (28) has been written in the superconformal formalism. It is then also scale invariant as long as a superPoincaré invariant gauge choice has not been taken. It is tempting to extend[18] the method applied to restore the anomalous target-space duality also to the dilatation symmetry contained in the conformal algebra. Since the gauge beta-function corresponds[19] to the dilatation anomaly obtained when varying the compensator field S_0, the superconformal formalism used here is certainly appropriate.

§The existence of this coupling has been demonstrated by direct string calculations[12].

The discussion given in this section only applies to a super-Yang-Mills theory without chiral charged matter. This situation corresponds, for instance, to the hidden E_8 sector of $(2,2)$ superstrings. The reason for this limitation is not only simplicity. The generalization to a string gauge sector with chiral charged matter and a non simple gauge group is not known at present. Furthermore, the all-order field theory results on beta-functions, which are necessary to check the results of the construction, are not available with arbitrary chiral matter.

To proceed, consider the effect of a scale change on the gauge coupling constant. Under

$$M \longrightarrow \lambda M,$$

the variation of the one-loop gauge coupling constant is

$$\delta(g^{-2}) = A \log \lambda,$$

where $A = 3C(G)/(8\pi^2)$ is the coefficient of the one-loop beta function which already appeared in the Kähler/target-space duality anomaly (25). In an effective lagrangian formalism, this anomalous variation reads

$$\delta\left(-\frac{1}{4}\frac{1}{g^2}F^a_{\mu\nu}F^{a\,\mu\nu}\right) = -\frac{1}{4}A \log \lambda F^a_{\mu\nu}F^{a\,\mu\nu}. \tag{30}$$

The next step is to construct a supersymmetric expression such that its variation compensates in particular (30). This should be done without introducing any new Kähler anomaly. Using the same notation as in the preceding section, there exists[18,8] a unique supersymmetric expression with a variation including (30):

$$\frac{A}{4}\left[\hat{L}\log(\hat{L}/\mu^2)\right]_D. \tag{31}$$

Its anomaly behaviour follows from the conformal weight two of $\hat{L}$. The quantity μ is an arbitrary number. It is not a scale since a constant supermultiplet must have zero conformal dimension. It is only after conformal gauge fixing that μ will acquire the rôle of a reference scale, as will be seen below.

Collecting its tree, one-loop and running contributions, the complete Wilson lagrangian is

$$\mathcal{L}_W = -\frac{1}{\sqrt{2}}\left[\hat{L}\left(\frac{\hat{L}e^{\hat{K}/3}}{S_0\overline{S}_0}\right)^{-3/2} + \frac{A}{4}\hat{L}\log\left(\frac{\hat{L}e^{\hat{K}/3}}{\mu^2}\right)\right]_D + \left[S_0^3 w\right]_F. \tag{32}$$

Notice that the anomaly term can also be written:

$$\frac{A}{4}\left[\hat{L}\log\left(\frac{\hat{L}e^{\hat{K}/3}}{S_0\overline{S}_0}\right) + \hat{L}\log\left(\frac{S_0\overline{S}_0}{\mu^2}\right)\right]_D.$$

This last expression shows that both Kähler and dilatation anomalies are conveniently carried by the chiral compensator S_0, a natural observation since S_0 has chiral *and* Weyl weights equal to one and Kähler symmetry finds its superconformal origin in the chiral internal $U(1)$.

The effective lagrangian $\mathcal{L}_\Gamma$ can be easily obtained from the Wilson theory (32). It differs from $\mathcal{L}_W$ (viewed as a functional of classical fields instead of an operator-valued lagrangian) by contributions arising from loop diagrams, which are partially non-local:

$$\mathcal{L}_\Gamma = \mathcal{L}_W - \frac{A}{4}\left[W^a W^a \mathcal{P}_C \log\left(\frac{\hat{L}e^{\hat{K}/3}}{\mu^2}\right)\right]_F + \dots, \tag{33}$$

the dots indicating contributions unrelated to gauge kinetic terms which are not of interest here. As it should, it is invariant under both Kähler and scale (conformal) transformations.

The Poincaré invariant theory is obtained by a superconformal gauge fixing which uses in particular the scalar z_0 and fermionic ψ_0 components of the compensating multiplet S_0. The gravitational lagrangian will take the canonical form $-\frac{1}{2\kappa^2}eR$ if one chooses

$$U \equiv \frac{1}{\sqrt{2}}\left(\frac{Ce^{\hat{K}/3}}{z_0\overline{z}_0}\right)^{-3/2} = \frac{1}{\kappa^2 C} - \frac{A}{6}, \tag{34}$$

which defines $|z_0|$ and introduces the Planck scale via $\kappa = \sqrt{8\pi G_N} = (M_P/8\pi)^{-1}$. Notice that the quantity U is Kähler and scale invariant.

The field-dependent Wilson gauge coupling constant is defined as the coefficient of gauge kinetic terms in $\mathcal{L}_W$:

$$\mathcal{L}_W = -\frac{1}{4}\frac{1}{g_W^2}F^a_{\mu\nu}F^{a\,\mu\nu} + \dots \tag{35}$$

It is a bare, unphysical quantity, like all parameters appearing in $\mathcal{L}_W$. In theory (32), one finds,

$$\frac{1}{g_W^2(z_0\overline{z}_0)} = U - \frac{A}{3}\log U + \frac{A}{2}\log\left(\frac{z_o\overline{z}_0}{\mu^2}\right) + c, \tag{36}$$

where the constant $c = \frac{A}{2}(1 - \frac{1}{3}\log 2)$ can be absorbed in μ. The Wilson coupling is neither Kähler nor scale invariant. Notice also that the Poincaré gauge fixing promotes the quantity μ to a scale, once $|z_0|$ is measured in units of the Planck scale.

The field-dependent effective gauge coupling constant is defined as the coefficient of gauge kinetic terms in the effective lagrangian:

$$\mathcal{L}_\Gamma = -\frac{1}{4}\frac{1}{g_\Gamma^2}F^a_{\mu\nu}F^{a\,\mu\nu} + \dots \tag{37}$$

The non-local contributions to $\mathcal{L}_\Gamma$ in eq. (33) include local gauge kinetic terms, and one simply obtains

$$\frac{1}{g_\Gamma^2(\mu)} = U, \tag{38}$$

a Kähler and target-space duality invariant result. According to (34), the physical effective gauge coupling constant in the loop-corrected theory is specified by the expectation value of the real, dimensionless field $\kappa^2 C$.¶ And, as discussed in section 2, the Wilson lagrangian (32) does not contain any $F^a_{\mu\nu}\tilde{F}^{a\,\mu\nu}$ term.

The dependence on the arguments $z_0\overline{z}_0$ and μ of g_W and g_Γ follows from the following remark. Using eqs. (36) and (38), one finds:

$$\mu\frac{\partial}{\partial\mu}g_\Gamma = -\frac{3C(G)}{16\pi^2}g_\Gamma^3\left[1-\frac{C(G)}{8\pi^2}g_\Gamma^2\right]^{-1}, \tag{39}$$

$$|z_0|\frac{\partial}{\partial|z_0|}g_W = -\frac{3C(G)}{16\pi^2}g_W^3. \tag{40}$$

The scale dependence of the physical coupling g_Γ coincides with the all-order renormalisation-group equation in pure super-Yang-Mills theory[20]. And the Wilson gauge coupling is only affected by one-loop running. Relation (40) leads to

$$\frac{1}{g_W^2(\mu^2)} = \frac{1}{g_\Gamma^2(\mu)} + \frac{A}{3}\log g_\Gamma^2(\mu). \tag{41}$$

Equations (40) and (41) have been established in pure super-Yang-Mills theories by Shifman and Vainshtein [21]. They have been obtained here from anomaly cancellation in an effective lagrangian construction.

Finally, the dual Wilson lagrangian with L replaced by the chiral multiplet S has a gauge coupling constant given by

$$\frac{1}{g_W^2(z_0\overline{z}_0)} = 4\mathrm{Re}\, s, \tag{42}$$

as in (17). The duality transformation is performed by expressing U as a function of $\mathrm{Re}\, s$, using eqs. (36) and (42). The solution can however only be found in a perturbative expansion: while the linear multiplet theory (32) is known exactly, with all-order gauge couplings (36) and (38), the dual chiral theory can only be expressed as an infinite series.

7. Conclusions

Formally, a supergravity theory with a linear multiplet can always be transformed into a theory with chiral multiplets only, using duality. String effective supergravities indicate however that this formal equivalence does not imply conceptual equivalence, besides the fact that duality can be hard or impossible to perform analytically. The duality transformation is applied to the Wilson lagrangian, with the loop expansion (20). Since the scalar component C of the linear multiplet appears to be the string

¶The field $\kappa^2 C$ is the lowest component of $\kappa^2\hat{L}$ in the Wess-Zumino gauge.

loop-counting parameter, it seems natural to formulate the loop expansion with L. The relation between S and L changes order by order. Duality will then induce a rearrangement of perturbation theory which can obscure the identification of string parameters and amplitudes with the fields of the effective supergravity. This is exemplified by the effective description of thereshold corrections in orbifolds[11,9], and also by substitution (29), at one loop.

The construction of the effective gauge sector provides a number of interesting results. The all-order renormalization-group behaviour is obtained from anomaly cancellation. In some sense, there is a non-renormalization property similar to an Adler-Bardeen theorem. The field-dependent gauge coupling, a physical parameter, is the expectation value of the real field C. The chiral version of the theory uses a complex scalar s, which gives the bare unphysical Wilson coupling. And the linear theory does not include any $F\tilde{F}$ term.

Finally, the strength of the gauge interaction and the scale at which this force becomes confining are characterized by the renormalization-group invariant scale Λ, given by the real superfield

$$\Lambda^3 = \mu^3 U e^{\ 3U/A},$$

which is also target-space duality invariant. This is also the scale at which dynamical supersymmetry breaking is expected to be induced by gaugino condensates. The gauge sector defined by the Wilson lagrangian (32) can be used for an effective lagrangian treatment of this phenomenon[22].

8. References

1. S. Ferrara, J. Wess and B. Zumino, *Phys. Lett.* **B51** (1974) 239;
   W. Siegel, *Phys. Lett.* **B85** (1979) 333.
2. M. Kaku, P.K. Townsend and P. van Nieuwenhuizen, *Phys. Rev.* **D17** (1978) 3179;
   T. Kugo and S. Uehara, *Nucl. Phys.* **B222** (1983) 125; **B226** (1983) 49.
3. S. Ferrara, L. Girardello, T. Kugo and A. Van Proeyen, *Nucl. Phys.* **B223** (1983) 191.
4. B. de Wit and M. Roček, *Phys. Lett.* **B109** (1982) 439.
5. W. Lang, J. Louis and B. A. Ovrut, *Phys. Lett.* **B158** (1985) 40.
6. G. Girardi, R. Grimm, M. Müller and J. Wess, *Phys. Lett.* **B147** (1984) 81.
7. W. Siegel, *Class. Quantum Grav.* **3** (1986) L47;
   C. S. Aulakh, J.-P. Derendinger and S. Ouvry, *Phys. Lett.* **B169** (1986) 201.
8. J.-P. Derendinger, F. Quevedo and M. Quirós, *Nucl. Phys.* **B428** (1994) 282.
9. J.-P. Derendinger, S. Ferrara, C. Kounnas and F. Zwirner, *Nucl. Phys.* **B372** (1992) 145.
10. S. Cecotti, S. Ferrara and M. Villasante, *Int. J. Mod. Phys.* **A2** (1987) 1839.
11. L. J. Dixon, V. S. Kaplunovsky and J. Louis, *Nucl. Phys.* **B355**(1991) 649.
12. I. Antoniadis, K. S. Narain and T. R. Taylor, *Phys. Lett.* **B267** (1991) 37;

I. Antoniadis, E. Gava and K. S. Narain, *Phys. Lett.* **B283** (1992) 209; *Nucl. Phys.* **B383** (1992) 93.
13. P. Mayr and S. Stieberger, *Nucl. Phys.* **B407** (1993) 725; **B412** (1994) 502.
14. M. B. Green and J. H. Schwarz, *Phys. Lett.* **B149** (1984) 117.
15. G. L. Cardoso and B. A. Ovrut, *Nucl. Phys.* **B369** (1992) 351; **B392** (1993) 315.
16. J.-P. Derendinger, in *Particles, Strings and Cosmology*, edited by P. Nath and S. Reucroft (World Scientific, Singapore, 1992), p. 766;
J. Louis, in *Particles, Strings and Cosmology*, edited by P. Nath and S. Reucroft (World Scientific, Singapore, 1992), p. 751;
D. Lüst and L. E. Ibáñez, *Nucl. Phys.* **B382** (1992) 305.
17. M. Dine, N. Seiberg and E. Witten, *Nucl. Phys.* **B289** (1987) 589.
18. J.-P. Derendinger, S. Ferrara, C. Kounnas and F. Zwirner, *Phys. Lett.* **B271** (1991) 307.
19. See for instance: M. T. Grisaru and P. C. West, *Nucl. Phys.* **B254** (1985) 249, or: S. J. Gates Jr., M. T. Grisaru, M. Roček and W. Siegel, *Superspace* (Benjamin/Cummings, Reading, 1983).
20. D. R. T. Jones, *Phys. Lett.* **B123** (1983) 45;
V. A. Novikov, M. A. Shifman, A. I. Vainshtein and V. I. Zakharov, *Nucl. Phys.* **B229** (1983) 381; *Phys. Lett.* **B166** (1986) 329.
21. M.A. Shifman and A.I. Vainshtein, *Nucl. Phys.* **B277** (1986) 456; *Nucl. Phys.* **B359** (1991) 571.
22. C. P. Burgess, J.-P. Derendinger, F. Quevedo and M. Quirós, in preparation.

SUPERGRAVITY, LINEAR MULTIPLETS,
AND CHERN-SIMONS FORMS

R. GRIMM
Centre de Physique Théorique - CNRS - Luminy, Case 907
F-13288 Marseille Cedex 9, France

ABSTRACT

Some general features of locally supersymmetric theories (N=1 in four dimensions) involving Chern-Simons forms and antisymmetric tensors are sketched out. The relevance of the three-form multiplet both for the description of Chern-Simons forms and the supersymmetry properties of the gaugino condensate is pointed out.

1. Introduction

The supersymmetric standard model has a rather simple structure concerning the basic building blocks involved in its construction: only chiral and Yang-Mills multiplets are needed. On the other hand, when it comes to the discussion of low energy effective theories of superstrings one has to deal with more sophisticated ingredients. First of all the general Kähler structure, in particular the Kähler phase transformations, should be taken into account. This can be done in $U_K(1)$ superspace[1,2], a geometric framework which embodies at the same time supergravity and matter, with Kähler transformations appearing ab initio in the structural group, on the same footing as the local Lorentz transformations of supergravity.

Moreover, in superstring inspired scenarios, less familiar multiplet structures and new couplings appear, as for instance linear multiplets and Chern-Simons forms induced from the Green-Schwarz anomaly cancellation in the superstring[3,4]. Again, $U_K(1)$ superspace provides an appropriate background for a geometric description of these new structures[5,6]. In turn, supersymmetric Chern-Simons forms are closely related to the three-form multiplet[7]: the three-form superspace geometry provides the relevant framework for the Chern-Simons forms, Yang-Mills as well as gravitational[8,9]. In a similar line of reasoning it will be pointed out here that the three-form multiplet is the relevant object which accounts for the supersymmetry properties of the gaugino condensate supermultiplet. Finally, it might be conceivable that three-form multiplets enter the stage of supersymmetric theories in their own right.

2. $U_K(1)$ Superspace: Unification of Supergravity and Matter

The basic object in this geometric formulation is the frame $E^A = dz^M E_M{}^A(z)$, a differential one-form in superspace, which, as an extension of traditional general relativity, introduces the usual moving frame $e_m{}^a(x)$ together with the Rarita-Schwinger field $\psi_m{}^\alpha$, its supersymmetry partner, in a unified manner. In distinction to the traditional approach[10], the frame is not only subject to local Lorentz transformations, but also to local chiral transformations, with chiral weights defined as

$$w(E^a) \;=\; 0, \qquad w(E^\alpha) \;=\; +1, \qquad w(E_{\dot\alpha}) \;=\; -1. \tag{1}$$

Correspondingly, in the definition of the torsion a new term appears, in addition to the familiar spin connection:

$$T^A \;=\; dE^A + E^B \Phi_B{}^A + w(E^A) E^A \mathrm{A}. \tag{2}$$

The crucial point is that the $U_K(1)$ gauge potential $\mathrm{A} = E^A \mathrm{A}_A$ is given in terms of the Kähler potential $K(\phi, \bar\phi)$ of the matter sector - the traditional unconstrained prepotential of supersymmetric gauge theory has been consistently replaced by the Kähler potential, which is a function of the chiral and antichiral matter superfields, giving rise to a unified geometric description of the supergravity-matter system. In this construction Kähler transformations are built in from the very beginning, on the same footing as local Lorentz transformations. The transformation

$$K(\phi, \bar\phi) \;\mapsto\; K(\phi, \bar\phi) + F(\phi) + \bar F(\bar\phi), \tag{3}$$

of the Kähler potential itself induces, by construction[2], a transformation

$$\mathrm{A} \;\mapsto\; \mathrm{A} + \frac{i}{2} d\, Im\, F, \tag{4}$$

on the $U_K(1)$ gauge potential which in turn serves to covariantize the Kähler phase transformations

$$E^A \;\mapsto\; E^A \exp\left(-\frac{i}{2} w(E^A) Im\, F\right), \tag{5}$$

of the frame in superspace. Observe that these assignements determine completely the Kähler properties of the supergravity superfields. This holds in particular for the superfields R, $R^\dagger$ and G_a of weights $w(R) = +2$, $w(R^\dagger) = -2$ and $w(G_a) = 0$, which are subject to the conditions

$$\mathcal{D}^{\dot\alpha} R \;=\; 0, \qquad \mathcal{D}_\alpha R^\dagger \;=\; 0, \tag{6}$$

and

$$\mathcal{D}^\alpha \mathcal{D}_\alpha R - \mathcal{D}_{\dot\alpha} \mathcal{D}^{\dot\alpha} R^\dagger \;=\; 4i \mathcal{D}_a G^a, \tag{7}$$

with Lorentz and Kähler covariant derivatives. For a more exhaustive presentation of $U_K(1)$ superspace see the literature cited so far.

3. Yang-Mills in $U_K(1)$ Superspace and Chern-Simons Forms

Having set up the geometrical framework for the supergravity-matter system we shall now include gauge interactions. To this end we consider matter fields ϕ and $\bar{\phi}$ in some representation of the gauge group and with gauge transformations

$$ {}^g\phi \;=\; g^{-1}\phi, \qquad {}^g\bar{\phi} \;=\; \bar{\phi}g, \tag{8} $$

where g is a superspace dependent group element, like $g(z) = \exp\, i\alpha^{(r)}(z)T_{(r)}$, with superfield gauge parameters $\alpha^{(r)}(z)$ and generators $T_{(r)}$. Exterior covariant derivatives

$$ d\phi - \mathcal{A}\phi \;=\; D\phi \;=\; E^A\mathcal{D}_A\phi, \qquad d\bar{\phi} + \bar{\phi}\mathcal{A} \;=\; D\bar{\phi} \;=\; E^A\mathcal{D}_A\bar{\phi}, \tag{9} $$

are defined in terms of the superspace one-form Lie algebra valued gauge potential

$$ \mathcal{A} \;=\; E^A\mathcal{A}_A^{(r)}T_{(r)}, \qquad {}^g\mathcal{A} \;=\; g^{-1}\mathcal{A}g - g^{-1}dg, \tag{10} $$

and the matter fields are now constrained to be covariantly chiral, resp. antichiral:

$$ \mathcal{D}^{\dot{\alpha}}\phi \;=\; 0, \qquad \mathcal{D}_\alpha\bar{\phi} \;=\; 0. \tag{11} $$

As is well known, the covariant field strength

$$ d\mathcal{A} + \mathcal{A}\mathcal{A} \;=\; \mathcal{F} \;=\; \frac{1}{2}E^AE^B\mathcal{F}_{BA}, \tag{12} $$

is subject to constraints. Correspondingly, the gaugino superfields $\mathcal{W}_\alpha$ and $\mathcal{W}^{\dot{\alpha}}$ of Kähler weights $w(\mathcal{W}_\alpha) = +1$ and $w(\mathcal{W}^{\dot{\alpha}}) = -1$, respectively, are subject to

$$ \mathcal{D}_\alpha\mathcal{W}^{\dot{\alpha}} \;=\; 0, \qquad \mathcal{D}^{\dot{\alpha}}\mathcal{W}_\alpha \;=\; 0, \tag{13} $$

$$ \mathcal{D}^\alpha\mathcal{W}_\alpha \;=\; \mathcal{D}_{\dot{\alpha}}\mathcal{W}^{\dot{\alpha}}. \tag{14} $$

In turn, as a consequence of this constrained superspace geometry, the gaugino-squared superfield $\operatorname{tr}\mathcal{W}^2$ and its complex conjugate are chiral, resp. antichiral superfields with the additional property

$$ \left(\mathcal{D}^\alpha\mathcal{D}_\alpha - 24R^\dagger\right)\operatorname{tr}\mathcal{W}^2 - \left(\mathcal{D}_{\dot{\alpha}}\mathcal{D}^{\dot{\alpha}} - 24R\right)\operatorname{tr}\overline{\mathcal{W}}^2 \;=\; -2i\varepsilon^{dcba}\operatorname{tr}\left(\mathcal{F}_{dc}\mathcal{F}_{ba}\right). \tag{15} $$

This last equation has a very natural interpretation in relation with supersymmetric Chern-Simons forms built from the superspace gauge potential $\mathcal{A}$ defined as

$$ \operatorname{tr}\left(\mathcal{A}d\mathcal{A} + \frac{2}{3}\mathcal{A}\mathcal{A}\mathcal{A}\right) \;=\; \mathcal{Q} \;=\; \frac{1}{3}E^AE^BE^C\mathcal{Q}_{CBA}, \tag{16} $$

and satisfying $d\mathcal{Q} \;=\; \operatorname{tr}(\mathcal{F}\mathcal{F})$. We will come back to this issue in a short while after having introduced the linear multiplet and its superspace geometry.

4. Superfield Actions

Given the geometric structures of the previous two sections, we turn now to the invariant actions. In the geometric approach, invariance under general coordinate transformations and under supersymmetry means nothing else than invariance under general reparametrizations in superspace. In addition, the actions should be invariant under Lorentz, Kähler and Yang-Mills transformations. It turns out that already the simplest invariant, namely the basic volume density, is quite non-trivial: it provides at the same time the kinetic terms for supergravity and for the matter sector:

$$\mathcal{L}_{supergravity\,+\,matter} \;=\; -3\int E. \tag{17}$$

Here the integration is performed over full superspace, the commuting and the anti-commuting directions. In the kinetic terms for the Yang-Mills sector,

$$\mathcal{L}_{Yang-Mills} \;=\; \frac{1}{8}\int \frac{E}{R}\, f_{(r)(s)}(\phi)\, \mathcal{W}^{\alpha(r)}\mathcal{W}^{(s)}_{\alpha} + \frac{1}{8}\int \frac{E}{R^{\dagger}}\, \bar{f}_{(r)(s)}(\bar{\phi})\, \mathcal{W}^{(r)}_{\dot{\alpha}}\mathcal{W}^{\dot{\alpha}(s)}, \tag{18}$$

chiral and antichiral volume elements of non-trivial Kähler weights are used. In general, holomorphic gauge coupling functions can be present. In the potential terms,

$$\mathcal{L}_{pot} \;=\; \frac{1}{2}\int \frac{E}{R} e^{K/2} W(\phi) + \frac{1}{2}\int \frac{E}{R^{\dagger}} e^{K/2} \bar{W}(\bar{\phi}), \tag{19}$$

the superpotential transforms as

$$W(\phi) \;\mapsto\; e^{-F(\phi)} W(\phi), \qquad \bar{W}(\bar{\phi}) \;\mapsto\; e^{-\bar{F}(\bar{\phi})} \bar{W}(\bar{\phi}), \tag{20}$$

and Kähler invariance is established due to the factors $e^{K/2}$. In other words, the combinations

$$e^{K/2} W(\phi), \qquad e^{K/2} \bar{W}(\bar{\phi}), \tag{21}$$

have well-defined Kähler weights

$$w\left(e^{K/2} W(\phi)\right) \;=\; +2, \qquad w\left(e^{K/2} \bar{W}(\bar{\phi})\right) \;=\; -2, \tag{22}$$

which are compensated by those of R and $R^{\dagger}$, respectively.

The sum of these three separately invariant actions provides the dynamics of the complete traditional supergravity-matter and Yang-Mills system. In its explicit expansion in terms of component fields, which still requires some computational and book-keeping effort[2], the combinations

$$\mathcal{D}^{\alpha}\mathcal{D}_{\alpha} R + \mathcal{D}_{\dot{\alpha}}\mathcal{D}^{\dot{\alpha}} R^{\dagger}, \tag{23}$$

$$\mathcal{D}^{\alpha}\mathcal{D}_{\alpha}\, \mathrm{tr}\, \mathcal{W}^2 + \mathcal{D}_{\dot{\alpha}}\mathcal{D}^{\dot{\alpha}}\, \mathrm{tr}\, \overline{\mathcal{W}}^2, \tag{24}$$

orthogonal to those of Eq. (7) and Eq. (15), are of crucial importance.

5. Linear Multiplet and Chern-Simons Forms

The linear multiplet is another supermultiplet of helicity content $(0, 1/2)$. In distinction to the chiral multiplet which is composed of a complex scalar, a Majorana spinor and a complex scalar auxiliary field, the linear multiplet contains an antisymmetric tensor gauge potential, a real scalar and a Majorana spinor, no auxiliary field. The presence of the antisymmetric tensor

$$b_{mn} = -b_{nm}, \qquad b_{mn} \mapsto b_{mn} + \partial_m \xi_n - \partial_n \xi_m, \tag{25}$$

makes the linear multiplet a gauge multiplet. As an antisymmetric tensor should be viewed as a two-form gauge potential, the linear multiplet has a well-established geometric formulation based on the superspace two-form

$$B = \frac{1}{2} E^A E^B B_{BA}(z). \tag{26}$$

As the corresponding field strength, obtained from applying the exterior derivative, is a three-form, it may be combined with the Chern-Simons form $\mathcal{Q}$ encountered before in the Yang-Mills case, such that

$$dB + k\mathcal{Q} = H = \frac{1}{3!} E^A E^B E^C H_{CBA}, \tag{27}$$

with some constant k. The Bianchi identity is then simply

$$dH = k\,\mathrm{tr}(\mathcal{F}\mathcal{F}). \tag{28}$$

At this point a detailed anlysis of the superspace structure reveals that this provides indeed a compatible geometric system and the Bianchi identities correspond to the modified linearity conditions

$$\left(\mathcal{D}^2 - 8R^\dagger\right) L = 2k\,\mathrm{tr}\,\overline{\mathcal{W}}^2, \tag{29}$$

$$\left(\overline{\mathcal{D}}^2 - 8R\right) L = 2k\,\mathrm{tr}\,\mathcal{W}^2. \tag{30}$$

These equations characterize the properties of the linear multiplet, coupled to Chern-Simons forms, in analogy to the chirality constraints of the usual matter superfields. They will become crucial in the evaluation of component field actions and the discussion of the ensuing interaction terms. Another information from the superspace Bianchi identities is the equation

$$\left([\mathcal{D}_\alpha, \mathcal{D}_{\dot\alpha}] - 4\sigma^a_{\alpha\dot\alpha} G_a\right) L = -\frac{1}{3} \sigma_{d\alpha\dot\alpha} \varepsilon^{dcba} H_{cba} - 4k\,\mathrm{tr}\left(\mathcal{W}_\alpha \mathcal{W}_{\dot\alpha}\right), \tag{31}$$

which identifies the field strength tensor H_{cba} in the superfield expansion of the linear superfield.

6. Invariant Actions with Linear Multiplets

The coupling of the linear multiplet (for simplicity, only one linear multiplet is considered here, in general an arbitrary number is possible) can be incorporated in promoting the superfield Kähler potential, which so far was a function of the chiral and antichiral matter superfields only, to a more general function

$$K(\phi, \bar{\phi}, L), \tag{32}$$

depending on the linear superfield L as well. In this case, the action obtained from the volume of superspace, Eq. (17), will now describe the kinetic terms for the linear multiplet as well. In addition, due to the modified linearity conditions, *i.e.* the gaugino superfield squared terms, it will also give rise to gauge kinetic terms with a gauge coupling function proportional to $\partial K/\partial L$. Hence, the simple volume term of superspace, when expanded in terms of component fields reveals quite a lot of information.

On the other hand, in the explicit evaluation of the component field action one finds that the normalization of the curvature scalar acquires a field dependend contribution proportional to

$$\left(1 - \frac{1}{3} L \frac{\partial K}{\partial L}\right)^{-1}. \tag{33}$$

In the geometric approach used here, the normalization of the Einstein term can easily be modified making use of the super-Weyl transformations adopted to $U_K(1)$ superspace with linear superfield (whose proper rescalings must be taken into account too). Parametrizing the rescaling in terms of a superfield $F(\phi, \bar{\phi}, L)$ (not to be confused with Kähler transformations denoted similarly), and requiring, for instance, canonical normalization for the curvature scalar, one finds that the corresponding action is given as

$$\mathcal{L}_{supergravity\,+\,chiral\ matter\,+\,linear\,+\,Yang-Mills} = -3 \int E\, F(\phi, \bar{\phi}, L), \tag{34}$$

with the function F related to the modified Kähler potential through the equation[5,14,6]

$$F - L\frac{\partial F}{\partial L} = 1 - \frac{1}{3} L \frac{\partial K}{\partial L}, \tag{35}$$

implying that F should be of the general form

$$F(\phi, \bar{\phi}, L) = 1 + L\, V(\phi, \bar{\phi}) + \frac{1}{3} L \int \frac{dL}{L} \frac{\partial K}{\partial L}. \tag{36}$$

Here, $V(\phi, \bar{\phi})$ is a new arbitrary function, relevant for supersymmetric theories with non-holomorphic gauge coupling functions[11]. Also, it is precisely this superfield which appears in effective theories with Kähler anomaly cancellation mechanism[12,13,14,6].

The superfield appearance of the explicit Yang-Mills action, Eq. (18), and of $\mathcal{L}_{pot}$, Eq. (19), remain unchanged.

7. Linear Multiplet vs. Chiral Multiplet

The modified linearity conditions, Eqs. (29,30), can be obtained from a variational principle in superspace. This goes as follows. First of all we express the gaugino-squared terms as derivatives of the Chern-Simons superfield,

$$\mathrm{tr}\,\overline{\mathcal{W}}^2 = \frac{1}{2}\left(\mathcal{D}^2 - 8R^\dagger\right)\Omega, \tag{37}$$

$$\mathrm{tr}\,\mathcal{W}^2 = \frac{1}{2}\left(\overline{\mathcal{D}}^2 - 8R\right)\Omega. \tag{38}$$

Then one writes down the first order action

$$\mathcal{L}_{(1)} = -3\int \left(F(\phi,\bar{\phi},L) + (L-k\Omega)(S+\bar{S})\right). \tag{39}$$

In this action L is understood to be unconstrained, while S and $\bar{S}$ are chiral resp. antichiral. Variation of this action with respect to those chiral superfields, taking into account the chirality constraints and integration by parts in superspace, results in the modified linearity equations (29,30) and we get back the theory described in the previous section. On the other hand, variation of the first order action with respect to L yields[14]

$$(S+\bar{S})\left(1-\frac{1}{3}L\frac{\partial K}{\partial L}\right) = \frac{1}{3}F\frac{\partial K}{\partial L} - \frac{\partial F}{\partial L}. \tag{40}$$

This equation should be read as an equation which serves to express L in terms of ϕ, $\bar{\phi}$ and of $S+\bar{S}$, giving rise to a theory entirely in terms of chiral multiplets with the special property that S and $\bar{S}$ appear only through their sum.

One should, however, again keep track of the normalization of the curvature scalar: canonical normalization is established if one requires the condition

$$F(\phi,\bar{\phi},L) + L(S+\bar{S}) = 1, \tag{41}$$

where L is solution of the above equation. Combination of these two equations leads again to Eq. (35) of the previous section.

It is in this sense that a linear multiplet is said to be dual to a chiral multiplet. In this dual theory, K is a true Kähler potential, it depends only on chiral superfields, and one might use this correspondance to give a geometric interpretation to the function $K(\phi,\bar{\phi},L)$ in the previous theory. As an example consider[14]

$$K(\phi,\bar{\phi},L) = K_0(\phi,\bar{\phi}) + \alpha\log L, \tag{42}$$

$$F(\phi,\bar{\phi},L) = 1-\frac{1}{3}\alpha + L\,V(\phi,\bar{\phi}). \tag{43}$$

Eq. (40) is solved by $\alpha/3L = S + \bar{S} + V(\phi, \bar{\phi})$ yielding

$$K(\phi, \bar{\phi}, S+\bar{S}) = K_0(\phi, \bar{\phi}) + \alpha \log \alpha/3 - \alpha \log\left(S + \bar{S} + V(\phi, \bar{\phi})\right). \quad (44)$$

8. Three-Form Multiplet

The three-form multiplet is yet another helicity $(0, 1/2)$ supermultiplet. It consists of a complex scalar, a Majorana spinor, an antisymmetric three-index tensor gauge potential and a real scalar auxiliary field. Again, it is a gauge multiplet, and its properties can be determined in superspace geometry[7]. As a starting point consider

$$C = \frac{1}{3!} E^A E^B E^C C_{CBA}(z), \quad (45)$$

a three-form in $U_K(1)$ superspace with field strength

$$dC = \Sigma = \frac{1}{4!} E^A E^B E^C E^D \Sigma_{DCBA}, \quad (46)$$

and Bianchi identity $d\Sigma = 0$. A detailed superspace analysis allows to identify the three form multiplet in terms of a chiral superfield T,

$$\mathcal{D}^{\dot{\alpha}} T = 0, \qquad \mathcal{D}_\alpha \bar{T} = 0, \quad (47)$$

with an additional constraint

$$\left(\mathcal{D}^\alpha \mathcal{D}_\alpha - 24 R^\dagger\right) T - \left(\mathcal{D}_{\dot{\alpha}} \mathcal{D}^{\dot{\alpha}} - 24 R\right) \bar{T} = \frac{8i}{3} \varepsilon^{dcba} \Sigma_{dcba}, \quad (48)$$

where the vectorial fieldstrength superfield

$$\varepsilon^{dcba} \Sigma_{dcba} = \varepsilon^{dcba} \left(4\, \mathcal{D}_d C_{cba} + 6\, T_{dc}{}^{\alpha} C_{\alpha\, ba} + 6\, T_{dc\dot{\alpha}} C^{\dot{\alpha}}{}_{ba}\right), \quad (49)$$

appears. The solution of these constraints is given in terms of an unconstrained prepotential superfield Ω such that

$$\bar{T} = \left(\mathcal{D}^2 - 8R^\dagger\right) \Omega, \qquad T = \left(\bar{\mathcal{D}}^2 - 8R\right) \Omega. \quad (50)$$

As we are working in $U_K(1)$ superspace the Kähler chiral weights are read off to be $w(T) = +2$ and $w(\bar{T}) = -2$. In coupling the three-form multiplet one may perfectly well include the superfields T and $\bar{T}$ into the Kähler potential and the function F, with appropriate care, however, to the normalization of the Einstein term. As to the superpotential, one should maintain the Kähler transformations defined in Eq. (20). Expanding the superpotential $\mathcal{W}(\phi, T)$ in powers of T, this may be achieved with suitable insertions of exponentials of the Kähler potential

$$\mathcal{W}(\phi, T) = \sum_{n \geq 0} W_n(\phi) \left(e^{-K/2} T\right)^n, \quad (51)$$

and Kähler transformations

$$W_n \mapsto e^{(n-1)F} W_n. \tag{52}$$

of the coefficient functions, thus establishing $\mathcal{W} \mapsto e^{-F}\mathcal{W}$.

9. Comments

The investigation of the three-form multiplet is less academic as it might appear. First of all it seems that this multiplet structure applies to the description of the supercurrent multiplet and its anomaly structure. On the other hand it has been advantageously employed in the study of supersymmetric Chern-Simons forms, both Yang-Mills and gravitational[8,9]: four-dimensional Chern-Simons forms are three-forms which change under gauge-transformations by the the exterior derivative of a two-form.

In a similar line of reasoning the three-form multiplet should be relevant in gluino condensation: if the supersymmetry properties of the condensate are supposed to reflect those of the superfield $tr\,\mathcal{W}^2$ and its complex conjugate, the natural multiplet structure for an effective description of the gluino condensate is the three-form. This is related to the fact that the gaugino superfield is not just chiral, Eq. (13), but subject to the additional condition Eq.(14). As a consequence, the superfield $tr\,\mathcal{W}^2$ is not just chiral either, but subject to the same additional restrictions as the three-form (cf. Eq. (15) and Eq. (48)).

Apart from this possibility of parametrizing the gaugino condensate, the couplings of generic three-form multiplets may of course be investigated in their own right. Let me close with the description of a mechanism paraphrasing the Chern-Simons - antisymmetric tensor coupling on the level of the three form. To this end consider abelian one- and two-form gauge potentials A and B with fieldstrength $F = dA$ and $H = dB$. In addition, consider the three form gauge potential C with field strength Σ defined as

$$\Sigma = dC + \tau H\, A. \tag{53}$$

Inspection of this geometric structure in superspace shows that the one-, two- and three-form geometries are indeed compatible with this definition of Σ. Whereas in the one- and two-form sectors one has the usual equations for the gaugino and linear superfields, the chiral, antichiral superfields T, $\bar{T}$ are now subject to a modified additional constraint

$$\begin{aligned}&\left(\mathcal{D}^\alpha \mathcal{D}_\alpha - 24R^\dagger\right) T - \left(\mathcal{D}_{\dot{\alpha}} \mathcal{D}^{\dot{\alpha}} - 24R\right) \bar{T} - \frac{8i}{3}\varepsilon^{dcba}\Sigma_{dcba} \\ &\quad + 16\tau \left(\mathcal{D}^\alpha \mathcal{W}_\alpha + \mathcal{D}_{\dot{\alpha}} \mathcal{W}^{\dot{\alpha}}\right) L + 64\tau \left(\mathcal{W}^\alpha \mathcal{D}_\alpha L + \mathcal{W}_{\dot{\alpha}} \mathcal{D}^{\dot{\alpha}} L\right) = 0, \end{aligned} \tag{54}$$

reflecting the properties of the modified Bianchi-identities $d\Sigma = \tau H\, F$. It is then a straightforward task to implement this geometrical structure in a supersymmetric dynamical context, which will be presented in a forthcoming publication[15].

10. Acknowledgements

The material presented here is based on work carried out over the years in collaboration with P. Adamietz, P. Binétruy, G. Girardi and M. Müller.

11. References

This list of references is by no means intended to be exhaustive, I apologize in advance for any undue omission.

1. P. Binétruy, G. Girardi, R. Grimm and M. Müller, *Phys. Lett.* **B189** (1987) 83.
2. P. Binétruy, G. Girardi and R. Grimm, *Supergravity and matter: a geometric description*, preprint LAPP-TH-275/90, August 1990.
3. S. Ferrara and M. Villasante, *Phys. Lett.* **B186** (1987) 85.
4. S. Cecotti, S. Ferrara and M. Villasante, *Int. J. Mod. Phys.* **A2** (1987) 1839.
5. P. Binétruy, G. Girardi, R. Grimm and M. Müller, *Phys. Lett.* **B195** (1987) 389.
6. P. Adamietz, P. Binétruy, G. Girardi and R. Grimm, *Nucl. Phys* **B401** (1993) 257.
7. S.J. Gates, Jr., *Nucl. Phys.* **B184** (1981) 381.
8. G. Girardi and R. Grimm, *Phys. Lett.* **B260** (1991) 365.
9. G. Girardi and R. Grimm, *The structure of supersymmetric theories with gravitational Chern-Simons forms*, forthcoming paper.
10. J. Wess and J. Bagger, *Supersymmetry and Supergravity*, second edition (Princeton University Press, Princeton, 1992).
11. J. Louis, in *Particles,Strings and Cosmology*, ed. P. Nath and S. Reucroft (World Scientific,1991).
12. G. L. Cardoso and B. Ovrut, *Nucl. Phys* **B369** (1992) 351.
13. J.-P. Derendinger, S. Ferrara, C. Kounnas and F. Zwirner, *Nucl. Phys* **B372** (1992) 145.
14. P. Binétruy, G. Girardi and R. Grimm, *Phys. Lett.* **B265** (1991) 111.
15. P. Binétruy, G. Girardi and R. Grimm, in preparation.

SOME RESULTS ON $(0,2)$ Z_N ORBIFOLD THEORIES WITH CONTINUOUS WILSON LINES

GABRIEL LOPES CARDOSO, DIETER LÜST[†]
Humboldt Universität zu Berlin
Institut für Physik
D-10115 Berlin, Germany

and

THOMAS MOHAUPT[‡]
DESY-IfH Zeuthen
Platanenallee 6
D-15738 Zeuthen, Germany

ABSTRACT

We discuss recent results on orbifold compactifications with (0,2) world sheet supersymmetry and continuous Wilson lines, emphasizing the role of modular symmetries.

1. Introduction

Orbifold compactifications [1,2,3] of the heterotic string are interesting to study, because they are both exactly solvable and give rise to a $N = 1$ supersymmetric spectra with interesting gauge groups and chiral matter. Whereas for a general compactification one needs $(2,2)$ world sheet (WS) supersymmetry in order to compute some quantities of interest, one can consider here models with $(0,2)$ WS supersymmetry which is sufficent in order to have $N = 1$ space time (ST) supersymmetry. If the model is constructed such that some components of the Wilson line can still be varied continuously, then the rank of the gauge group can be smaller then 16 [4] and its level can be bigger then 1 [5], thus allowing reasonable gauge groups and a realistic Higgs sector for GUT models. The physical properties can be worked out using the effective supergravity action valid below the Planck or string scale. Here one uses the fact there is a one to one correspondence between the moduli parametrizing deformations of the string theory and scalars with flat potential in the effective action. Thus the generalized kinetic term of these scalars is fixed by the Kähler potential of the moduli metric. The global structure of the moduli space is also reflected by the effective action which must be invariant under the modular group of the internal sector. This can be used [6] once the Kähler potential K is known in order to restrict

[†]e-mail addresses: GCARDOSO@QFT2.PHYSIK.HU-BERLIN.DE, LUEST@QFT1.PHYSIK.HU-BERLIN.DE
[‡]e-mail address: MOHAUPT@HADES.IFH.DE

the form of the superpotential W, because for example the effective potential does only depend on the combination $G = K + \log |W|^2$.

2. Results

2.1. Local structure of moduli space, Kähler potentials

The moduli of an orbifold model fall into two classes: the untwisted and the twisted moduli. The untwisted moduli are those moduli of the original model, which are compatible with the twist and therefore are still moduli of the twisted model. The twisted moduli, which we will not study here, live in the the so called twisted sectors that one has to add in order to preserve WS modular invariance.

In the case of Narain compactifications the moduli space is locally isomorhic to $SO(22,6)/SO(22) \otimes SO(6)$ [7]. $SO(22,6)$ is the group of allowed deformations of the Narain lattice (which parametrizes the moduli dependent part of the theory) and the subgroup of Euclidean rotations is modded out, because its action is physically trivial. If one now constructs an asymmetric orbifold by modding out a rotation $\Theta = \Theta_L \otimes \Theta_R \in SO(22) \otimes SO(6)$, then the space of untwisted moduli is locally given by the subgroup of $SO(22,6)$ commuting with Θ modulo the subgroup of pure rotations. As shown in [8] the result is

$$\mathcal{M}_O(\Theta) = \bigotimes_{i=1}^{n} \frac{SU(p_i,q_i)}{SU(p_i) \otimes SU(q_i) \otimes U(1)} \otimes \frac{SO(r,s)}{SO(r) \otimes SO(s)} \otimes \frac{SO(u,v)}{SO(u) \otimes SO(v)} \tag{1}$$

where p_i, r, u and q_i, s, v are the multiplicities of the eigenvalues $e^{2\pi i \phi_i}, -1, 1$ of Θ_L and Θ_R respectively. If Θ_R fulfills the constraint $SU(2) \not\ni \Theta_R \in SU(3)$ imposed by $N = 1$ ST supersymmetry [2], then $v = 0$ and either $s = 0$ or $s = 2$ and the moduli space is a complex Kähler manifold. The Kähler potential can now be found along the lines of [8,9]. In the limit of vanishing Wilson lines the results of [10,11] are reproduced. Let us now look at the effect of continuous Wilson lines in an concrete example and assume that the internal torus factorizes as $\mathbf{T}^6 = \mathbf{T}^2 \otimes \mathbf{T}^4$ such that the internal twist acts as $-I_2$ on $\mathbf{T}^2$.

In this case there is a T and a U modulus as usual [12,11], which are now supplemented by two continous Wilson line associated with the inequivalent directions on the torus, taking values in that part of the Cartan subalgebra of the gauge group on which the gauge twist acts as $-I$. Assuming for defineteness that this subspace has also dimension two, we have four real Wilson moduli which can be rearranged into two complex ones, B and C. According to the general result the moduli space is now

$$\mathcal{M}_{4,2} = \frac{SO(4,2)}{SO(4) \otimes SO(2)} \text{ and } K = -\log\left((T+\overline{T})(U+\overline{U}) - \frac{1}{2}(B+\overline{C})(C+\overline{B})\right) \tag{2}$$

is the Kähler potential as shown in [8]. For vanishing Wilson lines the moduli space factorizes into two separate $SU(1,1)/U(1)$ cosets corresponding to complex and to Kähler deformations. Thus in this limit, where the $(0,2)$ WS supersymmetry is enhanced to $(2,2)$ special geometry is restored. The holomorphic mixing terms are physically interesting because terms of this type can generate once local supersymmetry is spontanously broken a Higgs mass term of order $m_{3/2}$ thus offering a solution to the μ problem [13]. As observed in [14,15] such terms appear naturally through matter fields in $(2,2)$ compactifications.

2.2. Modular symmetries

Moduli spaces of string compactifications have a very non–trivial global structure because some large deformations are actualy automorphisms [16,12,17]. This is analog to moduli spaces of Riemann surfaces. The modular groups of $SO(r,2)$ and $SU(m,n)$ cosets are denoted by $SO(r,2,\mathbf{Z})$ and $SU(m,n,\mathbf{Z})$, but note that they depend on the details of the underlying lattices and therefore some care is required in their definition. For compactifications on $\mathbf{T}^2/\mathbf{Z}_2$ with two complex Wilson line moduli we find that the well known subgroup $SL(2,\mathbf{Z})_T$ of the modular group acts by

$$T \to \frac{aT-ib}{icT+d}, \quad U \to U - \frac{ic}{2}\frac{BC}{icT+d}, \quad B \to \frac{B}{icT+d}, \quad C \to \frac{C}{icT+d} \tag{3}$$

with $a,b,c,d \in \mathbf{Z}$, $ad-bc=1$. Note that the same transformation law also applies to matter field $(2,2)$ compactifications [15]. As is evident from the so called mirror symmetry $T \leftrightarrow U$, there exists another subgroup $SL(2,\mathbf{Z})_U$ with an obvious action on the moduli.

There are also modular transformations which are only non–trivial when the Wilson lines are switched on. These contain translations of the Wilson lines by lattice vectors [18] which have the following action on the complex moduli [8]:

$$T \to T + pU + \frac{p}{2}\sqrt{2}(B+C), \quad U \to U, \quad B \to B + p\sqrt{2}U, \quad C \to C + p\sqrt{2}U, \tag{4}$$

with $p \in \mathbf{Z}$. All these transformations leave the Kähler potential invariant up to a Kähler transformation.

2.3. Modular Forms and Superpotentials

Finally we want to use the transformation properties of the Kähler potential to construct some candidates for non perturbative superpotentials along the lines of [19]. The starting point is the mass formula for string states in a $\mathbf{Z}_2$ sector of an $N=1$ orbifold, which we will choose to be the one related to an $SO(4,2)$ coset and with Wilson lines taking values in the CSA of a $SU(2)^2$ [9]:

$$\alpha' M^2 = \frac{|\mathcal{M}|^2}{Y} + 2(N - 1 + l^T l + m^T n) \tag{5}$$

The first term on the right hand side is moduli dependent and contains the holomorpic mass $\mathcal{M}$ [19] and a real analytic function Y which is related to the Kähler potential by $K = -\log Y$. The second term is moduli independent and contains the number N of left moving excitations together with the vectors $l=(l_i)$, $n=(n_i)$ and $m=(m_i)$, $i=1,2$ of charge, winding and momentum quantum numbers. The term $l^T l + m^T n$

is proportional to the $SO(4,2)$ invariant scalar product. Setting the second term to zero gives an $SO(4,2)$ invariant constraint which restricts the quantum numbers to live in a certain orbit $\mathcal{O}$ of $SO(4,2,\mathbf{Z})$. Comparing the mass formula to the form of the G function, one learns that $W_{\mathcal{O}}$, where $\log W_{\mathcal{O}} = \sum_{\mathcal{O}} \log \mathcal{M}$, is a candidate for a non–perturbative superpotential because it transforms formally in the right way. Since the sums in general diverge some regularization is needed. Two interesting orbits are given by $\mathcal{O}_1 : N = 1 \Rightarrow l^T l + m^T n = 0$ and $\mathcal{O}_0 : N = 0 \Rightarrow l^T l + m^T n = 1$. Here, we will only consider the following simplest $SO(2,2,\mathbf{Z})$- invariant suborbits, namely $\mathcal{O}_{1,0}$: $l = 0, m^T n = 0$ and $\mathcal{O}_{0,1}, \ldots \mathcal{O}_{0,4}$: $l^T l = 1$, $m = n = 0$, which give rise to three inequivalent functions,

$$W_1 = \eta^{-2}(U)\eta^{-2}(T)\left[1 - \frac{BC}{2}\partial_U \log \eta^2(U)\partial_T \log \eta^2(T) + O(B^2C^2)\right], \tag{6}$$

$W_2 \simeq B+C$ and $W_3 \simeq C-B$. As expected these functions are (in case of W_1 to leading order in BC) only covariant with respect to $SO(2,2,\mathbf{Z})$ but fail to transform correctly under shifts of the Wilson lines. Note that W_1 is the same superpotential as was postulated in [15] in the context of $(2,2)$ orbifolds because of its correct transformation properties under $SO(2,2,\mathbf{Z})$. We will give a more complete discussion of these issues in [9]. Note that the BC-terms in (6) yield an alternative solution to the the μ- problem [20].

3. References

1. L. Dixon, J. Harvey, C. Vafa, and E. Witten, *Nucl. Phys.* **B 261** (1985) 678.
2. L. Dixon, J. Harvey, C. Vafa, and E. Witten, *Nucl. Phys.* **B 274** (1986) 285.
3. K. Narain, M. Sarmadi and C. Vafa, *Nucl. Phys.* **B 288** (1987) 551.
4. L. E. Ibáñez, H. P. Nilles and F. Quevedo, *Phys Lett.* **B 192** (1987) 332.
5. A. Font, L. E. Ibáñez and F. Quevedo, *Nucl. Phys.* **B 345** (1990) 389.
6. S. Ferrara, D. Lüst, A. Shapere and S. Theisen, *Phys. Lett.* **255** (1989) 363.
7. K. Narain, *Phys. Lett.* **B 169** (1986) 41.
8. G. Lopes Cardoso, D. Lüst and T. Mohaupt, hep-th-9405002.
9. G. Lopes Cardoso, D. Lüst and T. Mohaupt, in preparation.
10. S. Ferrara, C. Kounnas and M. Porrati, *Phys. Lett.* **B 181** (1986) 263.
11. M. Cvetic, J. Louis and B. A. Ovrut, *Phys. Lett. B* **206** (1988) 227.
12. R. Dijkgraaf, E. Verlinde, H. Verlinde, THU-87/30.
13. G.F. Giudice and A. Masiero, *Phys. Lett.* **B 206** (1988) 480.
14. J. Louis and V. Kaplunovsky, *Phys. Lett.* **B 306** (1993) 269.
15. I. Antoniadis, E. Gava, K.S. Narain and T.R. Taylor, hep-th 9405024.
16. A. Giveon, E. Rabinovici, and G. Veneziano, *Nucl. Phys.* **B 322** (1989) 167.
17. M. Spalinski, *Phys. Lett.* **B 275** (1992) 47.
18. J. Lauer, unpublished.
19. S. Ferrara, C. Kounnas, D. Lüst, F. Zwirner, *Nucl. Phys.* **B 365** (1991) 431.
20. J. A. Casas and C. Munoz, *Phys. Lett.* **B 306** (1993) 288.

FERMION MASSES IN SUPERSTRING THEORY*

I. ANTONIADIS
Ecole Polytechnique, Centre de Physique Théorique,†
91128 Palaiseau, France

and

T.R. TAYLOR
Department of Physics, Northeastern University,
Boston, MA 02115, U.S.A.

ABSTRACT

We give a model-independent discussion of fermion masses in four-dimensional heterotic superstring theories. We discuss the tree level contributions and quantum corrections, including one-loop threshold effects and masses generated as a result of non perturbative supersymmetry breaking. We also point out that superstring models give rise to a generic μ-term in the effective low energy Lagrangian.

1. OVERVIEW

The basic assumption of superstring phenomenology is that all light particles originate from superstring excitations that are massless at the superstring unification scale. The hierarchy between this high energy scale and small masses is created by the vacuum expectations of Higgs fields and by the supersymmetry breaking scale. Quarks and leptons acquire masses as a result of direct Yukawa couplings to Higgs scalars. The masses of squarks and sleptons receive contributions from the so-called soft terms generated by supersymmetry breaking. What is less known, and will be discussed later in this review, is that once local supersymmetry is broken, additional supersymmetric mass terms can also be generated for bosons and fermions, including the two Higgsinos of the minimal supersymmetric standard model (MSSM) and other fermions.

According to yet another standard lore of superstring phenomenology, supersymmetry breaking occurs as a result of non-perturbative effects in some "hidden sectors" which couple to the "observable" world with gravitational-strength, non-renormalizable interactions only; the soft masses are then of order of the gravitino mass, $m_{3/2} \propto M^3_{\rm HID}/M^2_{\rm PLANCK}$. In order for these masses to be of order 1TeV, the scale

*Research supported in part by the National Science Foundation under grant PHY-93-06906, in part by the EEC contracts SC1-CT92-0792 and CHRX-CT93-0340, and in part by a CNRS-NSF collaborative grant INT-92-16146.
†Laboratoire Propre du CNRS UPR A.0014

of non-perturbative effects must be $M_{\rm HID} \propto 10^{14}$GeV. In typical superstring models, supersymmetry breaking is due to gaugino condensation of a non-abelian gauge group which becomes strong at such energy scales.

We begin by discussing fermion mass generation due to standard Yukawa interactions that originate from explicit superpotential terms

$$W = W_{ijk} Q^i Q^j h^k + s_{kl} h^k h^l + \ldots, \tag{1}$$

where $Q_{i,j}$ represent quarks or leptons, and $h^{k,l}$ the Higgs superfields. The "Yukawa couplings" W_{ijk} are gauge singlet superfield functions. They may depend on the moduli and other fields whose vacuum expectation values (VEVs) are completely arbitrary to all orders in string perturbation theory. On the other hand, since Higgs particles are assumed to be massless at the string level, the functions s_{kl} must have zero VEVs at the string unification scale.

When computing the quark and lepton masses, special care must be taken with the wave function normalization factors. In fact, the Kähler potential can be expanded in powers of matter fields:

$$K = G + Z_{i\bar{\imath}} Q^i \bar{Q}^{\bar{\imath}} + Z_{k\bar{k}} h^k \bar{h}^{\bar{k}} + (H_{kl} h^k h^l + c.c.) + \ldots \tag{2}$$

The matter-independent function G corresponds to the Kähler potential for the moduli and other gauge singlet fields. The wave function normalization factors for the matter fields are determined by the functions Z. The fermion mass matrix is

$$M_{ij} = \lambda_{ijk} \langle h^k \rangle \tag{3}$$

where the physical Yukawa couplings

$$\lambda_{ijk} = e^{G/2} (Z_{i\bar{\imath}} Z_{j\bar{\jmath}} Z_{k\bar{k}})^{-1/2} W_{ijk} \tag{4}$$

and $\langle h^k \rangle$ are the VEVs of canonically normalized Higgs fields.

The problem of computing quark and lepton masses in a given superstring model consists of two parts. The first part is to compute the functions G, W_{ijk} and Z that enter into the physical Yukawa couplings. The most important problem at this point is to determine how these functions depend on the moduli and other gauge singlet fields with flat potentials, since VEVs of all these fields are *a priori* unknown. After solving this problem, what is generally called determination of the moduli-dependence of physical couplings, one should compute all relevant VEVs. Whereas the first part is of a purely kinematical nature, the second part involves some real superstring dynamics, like gaugino condensation and other non-perturbative phenomena.

The strategy followed in order to compute the Kähler potential and the superpotential in the effective supergravity theory of massless string excitations, is to consider the appropriate scattering amplitudes. In this way, the tree level quantities and one-loop corrections can be determined, as discussed later in this review. The tree-level superpotential (1) does not receive loop corrections, in agreement with the standard non-renormalization theorems. On the other hand, the loop expansion of the Kähler potential (2) takes the form:

$$\begin{aligned} G &= -\ln(S+\bar{S}) + G^{(0)} + \frac{2}{S+\bar{S}}G^{(1)} + \cdots \\ Z &= Z^{(0)} + \frac{2}{S+\bar{S}}Z^{(1)} + \cdots \\ H &= H^{(0)} + \frac{2}{S+\bar{S}}H^{(1)} + \cdots \end{aligned} \tag{5}$$

Here, S is the dilaton superfield which contains the dilaton as the real part of its scalar component and the universal axion as the imaginary part.[‡] The dilaton VEV determines the four-dimensional string coupling constant g: $\mathrm{Re}\langle S\rangle = 1/g^2$

Note that the function H, which mixes Higgs particles in the Kähler potential (2), does not enter in the fermion mass formula (3). However, once local supersymmetry is broken by non-vanishing VEVs of some auxiliary fields, $\langle F^\alpha\rangle \propto m_{3/2}$, this mixing gives rise to Higgsino masses

$$M_{kl} = (Z_{k\bar{k}}Z_{l\bar{l}})^{-1/2}\mu_{kl}, \tag{6}$$

where

$$\mu_{kl} = m_{3/2}H_{kl} - \langle \bar{F}^{\bar{\alpha}}\rangle\partial_{\bar{\alpha}}H_{kl}. \tag{7}$$

These masses should be interpreted as originating from the effective low-energy superpotential term $\mu_{kl}h^k h^l$ – the so-called μ-term.

The μ-term plays important role in the minimal supersymmetric standard model. MSSM requires the existence of two Higgs doublets, h^1 and h^2, carrying opposite hypercharges. h^1 provides with masses the down quarks and leptons, while h^2 gives masses to up quarks. A superpotential term $\mu h^1 h^2$ is necessary in the low energy Lagrangian in order to generate masses for Higgsinos and for undesirable electroweak axions. In MSSM, one usually introduces by hand a parameter μ of order of the weak scale, creating a hierarchy problem.

A class of solutions to this problem extends the MSSM to include light singlets s with Yukawa couplings to Higgs fields, as in the second term of eq.(1). This singlet could acquire a non-vanishing VEV at the electro-weak scale, driven by the soft supersymmetry breaking, generating an effective μ-term. Similarly, a non-renormalizable effective superpotential term of the form $M_{\mathrm{PLANCK}}^{1-n}s^n h^1 h^2$ could be present, in which case $\mu \propto M_{\mathrm{PLANCK}}^{1-n}\langle s\rangle^n$, with $\langle s\rangle$ now of the order of some intermediate scale, such as the decay constant of an invisible axion.[1] Another solution is to introduce Kähler mixing of the form $Hh^1h^2 + c.c.$, as in the last term of eq.(2). In this case, the singlets are only gravitationally coupled and acquire in general Planck scale VEVs. A μ-term can be then induced by local supersymmetry breaking,[2] see eq.(7). Superstring theory provides a natural setting for the latter mechanism[3] since H is generically a non-vanishing function of moduli fields, as discussed later in this review.

[‡]This axion is dual to the two-index antisymmetric tensor.

As it is clear from eqs.(3) and (7), the computation of fermion masses requires determination not only of the moduli and other scalar VEVs, but also of their auxiliary components.[§] This brings us back to the supersymmetry breaking problem. It is well known that gaugino condensation generates non-perturbative potential for the dilaton and moduli. Below, we give a simple physical explanation of the origin of such potentials.

In the effective low-energy Lagrangian, the gauge kinetic terms are of the form $\sum_{\mathcal{G}} f_{\mathcal{G}} \mathcal{W}_{\mathcal{G}}^2$, where $\mathcal{W}_{\mathcal{G}}$ are the gauge field strength superfields and the functions $f_{\mathcal{G}}$ are field-dependent gauge couplings associated with group $\mathcal{G}$: $1/g_{\mathcal{G}}^2 = \langle f_{\mathcal{G}} \rangle$. At the tree level, these functions are universal: $f_{\mathcal{G}} = S$. This universality is violated already at the one loop level by the threshold corrections which depend on the moduli as well as on the matter fields. As an example, consider gauginos of a pure gauge hidden sector. Non-perturbative effects give rise to a superpotential whose magnitude is determined by the gaugino condensate $\langle \lambda\lambda \rangle \propto M_{\text{HID}}^3 \propto \exp(3/2bg_{\text{HID}}^2) M_{\text{PLANCK}}^3$, where b is the one-loop beta function coefficient and g_{HID} is the coupling constant, $1/g_{\text{HID}}^2 = \langle f_{\text{HID}} \rangle$, of the hidden gauge group. The fact that a non-perturbative superpotential is generated for the dilaton and moduli is due to the dependence of the gauge coupling function f_{HID} on these fields. This superpotential can be derived rigorously from the effective Lagrangian describing non-perturbative gaugino condensation .[4] In some simple cases, like orbifold compactifications, explicit expressions can be obtained for the superpotential by using symmetry arguments based on large-small compactification radius duality.

The gauge coupling function f_{HID} depends also on the matter fields. In particular, it contains terms of the form $M_{\text{PLANCK}}^{-2} \partial_k \partial_l f_{\text{HID}} h^k h^l$, which depend on Higgs fields. In this way, gaugino condensation, which gives rise to $\langle \mathcal{W}_{\text{HID}}^2 \rangle \propto M_{\text{HID}}^3$, generates an additional contribution to the μ-term, of order of $m_{3/2}$. In other words, an explicit, supersymmetric mass term is generated for Higgsinos by non-perturbative effects in hidden sectors .[5,6] This fact is illustrated later in this review on an orbifold example.

To summarize, there exist three basic sources of fermion masses in superstring theory: 1) tree-level superpotential Yukawa couplings, 2) mixed terms in the Kähler potential which give rise to masses once local supersymmetry is spontaneously broken, and 3) explicit mass terms generated by non-perturbative effects like gaugino condensation. In addition, fermion masses can be generated by some higher weight interactions which are briefly mentioned in section 2.

This review is organized as follows. In the next two sections, we discuss tree-level computations of the quantities that enter into determination of fermion masses, and illustrate them on orbifold examples. The last two sections are devoted to quantum corrections, illustrated again on similar orbifold examples.

2. TREE-LEVEL RESULTS

We will restrict our discussion to Calabi-Yau compactifications of the heterotic

[§]Gaugino masses can be then computed by using standard formulas.

superstring, in which case the underlying internal superconformal field theory has $N = (2,2)$ world-sheet supersymmetry. In this case, the gauge group is $E_6 \times E_8$ and the matter fields transform as **27** or $\overline{\mathbf{27}}$ under E_6 and they are in one-to-one correspondence with the moduli: **27**'s are related to (1,1) moduli and $\overline{\mathbf{27}}$'s to (1,2) moduli. In models with E_6 grand unified group, all known particles are usually assigned to **27** representations, and Yukawa couplings originate from the superpotential terms of the form $\mathbf{27}^3$. E_6 can be however broken to $SO(10)$ or to another subgroup at the string scale; non-trivial Yukawa couplings can be then generated between particles contained in $\overline{\mathbf{27}}$'s and **27**'s. In the following discussion we will assume that Higgs fields can originate from both **27** and $\overline{\mathbf{27}}$. The knowledge of low-energy effective Lagrangians is then absolutely crucial in order to identify quarks, leptons and Higgs particles in this class of models.

The Kähler potential has the following power expansion in the matter fields:

$$K = G + A^\alpha A^{\bar\alpha} Z^{(1,1)}_{\alpha\bar\alpha} + B^\nu B^{\bar\nu} Z^{(1,2)}_{\nu\bar\nu} + (A^\alpha B^\nu H_{\alpha\nu} + c.c.) + \dots , \tag{8}$$

where A and B refer to **27**'s and $\overline{\mathbf{27}}$'s, respectively. The function G defines the moduli metric which at the tree-level is block-diagonal in (1,1) and (1,2) moduli: $G^{(0)} = G^{(1,1)} + G^{(1,2)}$. The moduli metrics as well as the matter metrics $Z^{(1,1)}$ and $Z^{(1,2)}$ and the Kähler mixing H have been studied in the literature up to one-loop level.[7,8,5]

At the tree-level the various quantities which determine the effective low energy $N = 1$ supergravity are not independent. They are related because of the $N =$ 2 world-sheet supersymmetry in the right-moving (bosonic) sector of the heterotic superstring. An interesting consequence of the corresponding Ward-identities is the so-called special geometry, which relates the tree-level moduli metric to the Yukawa couplings:[7,9]

$$R^{(0)}_{a\bar a b\bar d} = G^{(0)}_{a\bar c} G^{(0)}_{b\bar d} + G^{(0)}_{a\bar d} G^{(0)}_{b\bar c} - e^{2G^{(0)}} W_{abe} \overline{W}_{\bar c\bar d\bar f} G^{(0)e\bar f} , \tag{9}$$

where $R^{(0)}_{a\bar c b\bar d}$ is the Riemann tensor of the moduli Kähler geometry $G^{(0)}$, and the above equation holds separately for $(1,1)$ and $(1,2)$ moduli. Eq.(9) can be understood as a differential equation which determines the moduli metric in terms of the analytic superpotential and it can be solved in several examples.[10] On the other hand, the tree-level matter metrics are proportional to the moduli metrics:

$$\begin{aligned} Z^{(1,1)}_{\alpha\bar\alpha} &= G^{(1,1)}_{\alpha\bar\alpha} \exp(G^{(1,2)} - G^{(1,1)})/3 , \\ Z^{(1,2)}_{\nu\bar\nu} &= G^{(1,2)}_{\nu\bar\nu} \exp(G^{(1,1)} - G^{(1,2)})/3 . \end{aligned} \tag{10}$$

Another consequence of $N = 2$ Ward-identities is that the Kähler mixing function H satisfies the differential equation:[5] ¶

$$\partial_{\bar\beta} \partial_{\bar\mu} H^{(0)}_{\alpha\nu} = G^{(0)}_{\alpha\bar\beta} G^{(0)}_{\nu\bar\mu} . \tag{11}$$

¶This solution may not be appropriate at enhanced symmetry points, for instance in orbifold compactifications.

The above equation can be used to identify representations containing candidates for Higgs fields with non-trivial Kähler mixings that can result in a μ-term.

There is a complication which arises in the presence of Yukawa couplings of the charged fields A, B with (non-moduli) singlets. The interactions induced by H_{AB} mix with some other interactions, which are not described by the standard two-derivative supergravity, corresponding to higher dimensional F-terms.[5] These new interactions lead to additional contribution to the effective μ-term. In the globally supersymmetric limit, they have the form:

$$\int d^2\theta(\bar{D}^2 f^1)(\bar{D}^2 f^2)\,, \tag{12}$$

where $f^{1,2}$ are arbitrary functions of moduli (and singlets) and $\bar{D}^2$ is the chiral projection. These interactions also appear as basic building blocks in the holomorphic anomaly equations of topological amplitudes in the heterotic case.[11] We should note however that they vanish in the orbifold limit.

In the presence of higher weight interactions and singlets which couple to Higgs fields in the superpotential, the complete mass formula becomes rather complicated. Furthermore, Yukawa couplings of Higgs fields with singlets produce in general a direct superpotential mass since the singlets can acquire non-vanishing expectation values at the scale of supersymmetry breaking. In the case of compactifications which give rise to the particle content of the MSSM at low energies, there are no massless singlets coupled to Higgs particles and the above complication does not arise. In this case, the induced μ-term depends entirely on the Kähler function H satisfying eq.(11), as well as on eventual non-perturbative superpotential generated at the supersymmetry breaking scale through the matter field dependence of threshold corrections to gauge couplings.

3. TREE-LEVEL ORBIFOLD EXAMPLES

Symmetric orbifolds are flat compactifications on the cotient of a six dimensional torus over a discrete subgroup of $SU(3)$ so that one space-time supersymmetry remains unbroken. They correspond to singular points of Calabi-Yau manifolds with enhanced gauge symmetry $U(1)^2$, or larger. An important property of these models is space-time duality symmetry which contains a transformation exchanging large with small compactification radii.

For instance, for two compactified dimensions, there are four independent parameters corresponding to the three components of the metric G_{IJ} and one component of the antisymmetric tensor $B_{IJ} = b\epsilon_{IJ}$. They form two complex fields $T = 2(\sqrt{G} + ib)$ and $U = (\sqrt{G} + iG_{12})/G_{11}$ corresponding to (1,1) and (1,2) moduli, respectively. The duality symmetry in this case forms the group of $SL(2,Z) \times SL(2,Z)/Z_2$ transformations,

$$T \to \frac{aT - ib}{icT + d}, \qquad U \to \frac{a'U - ib'}{ic'U + d'}, \qquad T \leftrightarrow U, \tag{13}$$

where a, b, c, d are integers with $ad - bc = 1$ (similarly for primed parameters). The matter fields φ transform under this transformations as $SL(2,Z)$ modular forms of weight n_φ:

$$\varphi \to (icT+d)^{-n_\varphi}\varphi\,, \tag{14}$$

and similarly under U transformations. Moreover, $SL(2,Z)$ duality induces a Kähler transformation under which the superpotential W transforms as a form of weight 1:

$$\begin{aligned} K &\to K + \ln(icT+d) + \ln(-ic\overline{T}+d)\,, \\ W &\to (icT+d)^{-1}W\,. \end{aligned} \tag{15}$$

The massless states in orbifold models fall into two sectors: (a) The untwisted sector which contains the (1,1) and (1,2) moduli T_α and U_β, in correspondence with the **27**'s A_α and $\overline{\mathbf{27}}$'s B_β. Here, α,β label the internal complex planes: $\alpha = 1,2,3$ while β refers only to the Z_2-twisted planes, otherwise U is fixed to some background value. Also, in orbifolds with non abelian enhanced symmetries, T becomes a matrix and sums should be replaced by traces in the subsequent formulae. (b) The twisted sector which contains matter fields C's and $\overline{C}$'s in correspondence with the blowing-up moduli which allow to deform orbifolds into regular Calabi-Yau manifolds.

The tree-level moduli metric is:[7]

$$G^{(0)} = -\sum_{\alpha=1}^{3}\ln(T_\alpha+\overline{T}_\alpha) - \sum_\beta \ln(U_\beta+\overline{U}_\beta)\,, \tag{16}$$

while the matter metrics are:

$$\begin{aligned} &U \text{ fixed:} \\ &\quad Z^{(0)}_{A\overline{A}} = \frac{1}{T+\overline{T}}, \\ &Z_2\text{-twisted planes:} \\ &\quad Z^{(0)}_{A\overline{A}} = Z^{(0)}_{B\overline{B}} = \frac{1}{(T+\overline{T})(U+\overline{U})}, \\ &\quad Z^{(0)}_{C\overline{C}} = \prod_{\alpha=1}^{3}\frac{1}{(T_\alpha+\overline{T}_\alpha)^{n^C_\alpha}}, \end{aligned} \tag{17}$$

where T, U are the moduli associated to the matter fields A, B, and n^C_α are the modular weights of C. Moreover, the Kähler mixing H is nonvanishing only when the matter fields belong to the untwisted sector and are associated with a Z_2-twisted internal plane;[5] in addition, it depends on the moduli of this plane only:

$$\begin{aligned} H^{(0)}_{AB} &= \frac{1}{(T+\overline{T})(U+\overline{U})}\,, \\ H_{CC} &= H_{CA} = H_{CB} = 0\,. \end{aligned} \tag{18}$$

Note that the above results are consistent with the $SL(2,Z)$ large-small compactification radius T-duality (13-15) with the untwisted matter fields A and B having modular weights 1, provided the (1,2) modulus U is not inert in the presence of matter fields:

$$U \to U - \frac{ic}{icT + d} AB \,. \tag{19}$$

Similarly, T transforms under $SL(2,Z)$ U-duality.

Once supersymmetry is broken, the induced μ-term, eq.(7), and a possible non-perturbative superpotential term $W_{AB}AB$ yield the Higgsino mass:

$$m_{AB} \; = \; m_{3/2} + (T + \overline{T})F_T + (U + \overline{U})F_U + (T + \overline{T})(U + \overline{U})e^{G/2}W_{AB} \,, \tag{20}$$

where we used the tree-level expressions (16-18) for the moduli and matter metrics.

The most general superpotential, including non-perturbative contributions, has the following expansion in powers of matter fields:

$$W = W_0 + W_{AB}AB + \ldots \tag{21}$$

The gravitino mass is then $m_{3/2} = e^{G/2}W_0$. It follows from (15) that $SL(2,Z)$ invariance of the effective action under the transformations (13) and (19) requires that W_{AB} transforms as:

$$W_{AB} \to (icT + d)W_{AB} + ic\,\partial_U W_0 \,. \tag{22}$$

It is remarkable this transformation property automatically implies a non-vanishing mass term, $W_{AB} \neq 0$, if a moduli-dependent superpotential W_0 is generated. It is easy to check that the physical masses (20) transform with unobservable phase factors under the duality transformations (13) and (19).

Finally, the non vanishing superpotential terms at the trilinear level are of the form:

$$W \sim A_1A_2A_3,\; B_1B_2B_3,\; ACC,\; BCC,\; CCC. \tag{23}$$

The corresponding physical Yukawa couplings between three untwisted or one untwisted and two twisted **27**'s are constants,

$$\lambda^{(0)}_{123} = \lambda^{(0)}_{ACC} = \lambda^{(0)}_{BCC} = g\sqrt{2} \,, \tag{24}$$

while $\lambda^{(0)}_{CCC}$ are in general non trivial functions of the moduli T_α.[12]

It follows that in the context of orbifold models with Higgs fields h^1 and h^2 contained in **27** and $\overline{\mathbf{27}}$ of the untwisted sector, there is an automatic generation of a μ-term induced by the breaking of local supersymmetry through their mixing in the Kähler potential. The resulting Higgsino mass is given in equation (20). Furthermore, if the top quark gets mass as a result of trilinear superpotential couplings, its Yukawa coupling is unified with the gauge couplings at the string scale and is given in eq.(24). This implies by the renormalization group evolution that the top is in general heavy with a mass close to the fixed point value. An interesting possibility is when the whole third generation receives mass from the trilinear superpotential. In this case, a strict prediction is obtained since all three Yukawa couplings are equal at the unification scale: $\lambda_t = \lambda_b = \lambda_\tau = g\sqrt{2}$. This leads to $m_t \sim 180$ GeV together with a successful prediction for the bottom and tau masses in the region of large $\tan\beta$ (~ 50).[13]

4. ONE-LOOP RESULTS

Computing the loop corrections to superstring scattering amplitudes, one integrates not only over heavy particles, but over massless particles as well. This integration gives rise to on-shell infrared divergences, associated with the running of low energy couplings. In the analogous field-theoretical computations, such logarithmic divergences are usually regulated by going off-shell, to momentum $p^2 \neq 0$. It is then important to realize that in string theory, as well as in quantum field theory, the momentum-dependence of coupling constants is a purely infrared effect, and therefore the corresponding β-function coefficients of the $p^2 \to 0$ divergence depend on the massless particle content only.[14,15]

Consider for instance the one-loop case. A generic on-shell amplitude $\mathcal{A}$ corresponding to some physical coupling of the low-energy theory is written as an integral over the complex Teichmüller parameter $\tau = \tau_1 + i\tau_2$ of the world sheet torus inside its fundamental domain $\Gamma \equiv \{|\tau_1| \leq \frac{1}{2},\ |\tau| \geq 1\}$:

$$\mathcal{A} = \int_\Gamma \frac{d^2\tau}{\tau_2} B(\tau,\bar{\tau}). \tag{25}$$

The presence of massless particles propagating in the loop implies that the integrand B goes to a constant b as $\tau_2 \to \infty$, and the integral over the Teichmüller parameter diverges in the infrared. When the logarithmic divergence is regularized and compared to the field-theoretical $\overline{DR}$ scheme, it is converted to $b\ln(M_{st}^2/p^2)$, where $M_{st} \simeq 5 \times g \times 10^{17}$ GeV is the string unification scale. b can then be identified with the corresponding field-theoretical β-function, while the remaining finite part of the integral yields the moduli-dependent string threshold corrections:[15]

$$\mathcal{A} = b\ln\frac{M_{st}^2}{\mu^2} + \int_\Gamma \frac{d^2\tau}{\tau_2}[B(\tau,\bar{\tau}) - b]. \tag{26}$$

In particular, the one-loop corrections to the Kähler metric $K^{(1)}$ can be obtained by the computation of the one-loop three-point amplitude involving two complex scalars and the antisymmetric tensor field: $\mathcal{A}(b^{\mu\nu}\varphi(p_1)\overline{\varphi}(p_2)) \sim \varepsilon^{\mu\nu\lambda\rho}p_{1\lambda}p_{2\rho}K^{(1)}_{\varphi\overline{\varphi}}$. The result is:[8]

$$K^{(1)} = \frac{i}{16(2\pi)^3}\int_\Gamma \frac{d^2\tau}{\tau_2^2}\bar{\eta}^{-2}\mathrm{Tr}_R F(-1)^F\,, \tag{27}$$

where η is the Dedekind eta function and the trace is over the Ramond sector of the internal $N = 2$ (left-moving) superconformal field theory with $U(1)$-charge operator F.

Considering the Kähler metric corresponding to eq.(27), one obtains a finite correction for the moduli metric $G^{(1)}$ (at a generic point $T \neq U$), while the integral for the matter metric $Z^{(1)}$ is infrared divergent. The infinite part can be identified with the one-loop anomalous dimensions of a generic $N = 1$ supersymmetric field theory, in a gauge in which the superpotential remains unrenormalized. The remaining finite part gives the string threshold corrections to wave function factors.[8] These corrections determine the boundary conditions for the physical Yukawa couplings λ_{ijk} at the string unification scale:

$$\lambda_{ijk}(M_{st}) = \lambda_{ijk}^{(0)} \left[1 + g^2 \left(Y_i + Y_j + Y_k\right)\right]^{-1/2}, \tag{28}$$

where Y_i is defined as the finite part of $Z_{i\bar{i}}^{(1)}/Z_{i\bar{i}}^{(0)}$. The one-loop corrections to the Kähler mixing H_{AB} can be obtained similarly by expanding eq.(27) to first order in AB.

Finally, we discuss the matter field dependence of threshold corrections to gauge couplings which lead to an additional source of Higgsino mass via gaugino condensation. They can be obtained by the computation of the 4-point amplitude involving two gauge bosons and two matter fields A and B. It has been shown[8] that, to the leading order in matter fields, the one-loop threshold corrections $\Delta^{\text{1-loop}}$ satisfy

$$\partial_{\bar{\imath}}\partial_j \Delta^{\text{1-loop}} = \tilde{b} K_{\bar{\imath}j}^{(0)} + K_{\bar{\imath}j}^{(1)} , \tag{29}$$

where the indices $\bar{\imath}$, j represent fields which are neutral under the gauge group associated with $\Delta^{\text{1-loop}}$, and $-\tilde{b}$ is the quadratic Casimir of the adjoint representation. This result can be anticipated on purely field-theoretical grounds:[16,17] the first term on the r.h.s. of (29) is related to anomalous graphs involving the coupling of the Kähler current to gauginos, whereas the second term is due to the Green-Schwarz term which contributes to both Kähler potential and gauge couplings.[8] It turns out that in the case of a pure gauge group with no massless matter field representations (like E_8), equation (29) remains valid to higher orders in matter fields, as well.[5]

5. ONE-LOOP ORBIFOLD EXAMPLES

In orbifold models, the one-loop corrections to the relevant quantities which determine the effective field theory can be explicitly calculated. For instance, the moduli dependence of the one-loop threshold corrections to the wave function renormalization factors of untwisted matter fields are non vanishing only for fields associated to a Z_2-twisted plane, A and B:[8]

$$\begin{aligned} Y_A = Y_B \;&=\; -\tilde{b}_A\{\Delta(T) + \Delta(U)\} + G^{(1)} \\ \Delta(T) \;&\equiv\; \ln(T + \overline{T})|\eta(iT)|^4 \end{aligned} \tag{30}$$

Here, $\tilde{b}_A = \hat{b}_A/ind$, where $\hat{b}_A$ is equal to the β-function coefficient of the gauge group that transforms A and B non-trivially in the corresponding N=2 supersymmetric orbifold, and *ind* is the index of the little subgroup of the untwisted plane in the full orbifold group.[18] In eq.(30) we neglected additive constants which do not depend on the moduli T and U. It follows that the boundary relation (24) between the untwisted Yukawa couplings and the E_6 gauge coupling at the unification scale does not receive any moduli-dependent corrections at the one-loop level:

$$\lambda_{123}(M_{st}) = g_{E_6}(M_{st})\sqrt{2}\,, \tag{31}$$

up to moduli-independent constants. In eq.(31), $g_{E_6}(M_{st})$ is the one-loop E_6 gauge coupling constant at the unification scale.

The one-loop correction to the function H_{AB} is:[5]

$$H^{(1)}_{AB} = \tilde{b}_A\Delta_T(T)\Delta_U(U) - \frac{1}{U+\overline{U}}G^{(1)}_T - \frac{1}{T+\overline{T}}G^{(1)}_U\,, \tag{32}$$

where subscripts on the functions Δ and $G^{(1)}$ denote partial derivatives: $\Delta_T(T) \equiv \partial_T\Delta(T)$ *etc.* The E_8 threshold corrections, up to first order in the untwisted matter fields AB, read:

$$\begin{aligned}\frac{1}{g^2_{E_8}(M_{st})} &= \frac{1}{g^2} - \tilde{b}_{E_8}[\Delta(T)+\Delta(U)] + G^{(1)} + (H^{(1)}_{AB}AB + c.c.)\\ &+ \tilde{b}_{E_8}\{[\frac{1}{(T+\overline{T})(U+\overline{U})} - 4\partial_T\ln\eta(iT)\partial_U\ln\eta(iU)]AB + c.c.\}\end{aligned} \tag{33}$$

with $\tilde{b}_{E_8} = -30$. Note that since $G^{(1)}$ is invariant under the transformation (13), the above expressions are consistent with the invariance of the one loop correction to the Kähler potential, $G^{(1)} + (H^{(1)}_{AB}AB + c.c.)$, as well as of the gauge coupling $g^2_{E_8}(M_{st})$, under the full set of $SL(2,Z)$ duality transformations (13-15) and (19).

We conclude by presenting an explicit expression for the Higgsino mass (20) in the context of gaugino condensation. A non-perturbative superpotential that depends on the dilaton in the right way, and satisfies the symmetry requirements (15), (22), up to first order in the expansion of matter fields AB, is:[5]

$$W = e^{S/2\tilde{b}_{E_8}}\,\eta^{-2}(iT)\eta^{-2}(iU)\,[1 - 4AB\,\partial_T\ln\eta(iT)\partial_U\ln\eta(iU)]\,\widetilde{W}\,, \tag{34}$$

where $\widetilde{W}$ may depend on the moduli of the two other planes. The second term inside the bracket, which from the point of view of non-perturbative dynamics originates from matter field-dependent threshold corrections (33), gives rise to a direct mass for Higgs particles.

Using auxiliary field equations and eq.(20), we obtain $F_T = -m_{3/2}\Delta_T(T) + \mathcal{O}(g^2)$, $F_U = -m_{3/2}\Delta_U(U) + \mathcal{O}(g^2)$, and the Higgsino mass

$$m = -m_{3/2}(T+\overline{T})(U+\overline{U})\Delta_T(T)\Delta_U(U)\,, \tag{35}$$

where we neglected terms of order $\mathcal{O}(g^2)$.

REFERENCES

1. E.J. Chun, J.E. Kim and H.P. Nilles, Nucl. Phys. B 370 (1992) 105.
2. G.F. Giudice and A. Masiero, Phys. Lett. B 206 (1988) 480.
3. V.S. Kaplunovsky and J. Louis, Phys. Lett. B 306 (1993) 269.
4. S. Ferrara, N. Magnoli, T.R. Taylor and G. Veneziano, Phys. Lett. B 245 (1990) 409; A. Font, L.E. Ibáñez, D. Lüst and F. Quevedo, Phys. Lett. B 245 (1990) 401; H.P. Nilles and M. Olechowski, Phys. Lett. B 248 (1990) 268; P. Binétruy and M.K. Gaillard, Phys. Lett. B 253 (1991) 119.
5. I. Antoniadis, E. Gava and K.S. Narain and T.R. Taylor, preprint hep-th 9405024.
6. See also: J.A. Casas and C. Muñoz, Phys. Lett. B 306 (1993) 288.
7. L.J. Dixon, V.S. Kaplunovsky and J. Louis, Nucl. Phys. B 329 (1990) 27.
8. I. Antoniadis, E. Gava, K.S. Narain and T.R. Taylor, Nucl. Phys. B 407 (1993) 706.
9. S. Ferrara, Proceedings of STRINGS 1992, Roma, September 1992 (World Scientific P. C.); P. Candelas, Proceedings of 26th Erice Workshop From Superstrings to Supergravity, December 1992 (World Scientific P. C.); and references therein.
10. P. Candelas, X.C. de la Ossa, P.S. Green and L. Parkes, in *Proceedings of the Second International Symposium on Particles, Strings and Cosmology*, P. Nath and S. Reucroft eds. (World Scientific, 1992), p. 667.
11. I. Antoniadis, E. Gava, K.S. Narain, T.R. Taylor and C. Vafa, in preparation.
12. S. Hamidi and C. Vafa, Nucl. Phys. B 279 (1987) 465; L. Dixon, D. Friedan, E. Martinec, S. Shenker, Nucl. Phys. B 282 (1987) 13.
13. W.A. Bardeen, M. Carena, S. Pokorski and C. Wagner, Phys. Lett. B 320 (1994) 110; M. Carena, M. Olechowski, S. Pokorski and C. Wagner, preprint hep-ph 9402253; S. Pokorski, preprint hep-ph 9407276.
14. T.R. Taylor and G. Veneziano, Phys. Lett. B 212 (1988) 147.
15. V.S. Kaplunovsky, Nucl. Phys. B 307 (1988) 145 and Erratum, preprint hep-th 9205070.
16. G.L. Cardoso and B.A. Ovrut, Nucl. Phys. B 369 (1992) 351; J.-P. Derendinger, S. Ferrara, C. Kounnas and F. Zwirner, Nucl. Phys. B 372 (1992) 145.
17. I. Antoniadis, K.S. Narain and T.R. Taylor, Phys. Lett. B 267 (1991) 37; J. Louis, in *Proceedings of the Second International Symposium on Particles, Strings and Cosmology*, P. Nath and S. Reucroft eds. (World Scientific, 1992), p. 751.
18. L.J. Dixon, V.S. Kaplunovsky and J. Louis, Nucl. Phys. B 355 (1991) 649.

A SYSTEMATIC ANALYSIS OF NON-PERTURBATIVE EFFECTS IN MATRIX MODELS AND STRING EFFECTIVE LAGRANGIANS

RAM BRUSTEIN
Theory Division, CERN
CH-1211 Geneva 23, Switzerland

MICHAEL FAUX and BURT A. OVRUT
Department of Physics
University of Pennsylvania
Philadelphia, Pa 19104-6396

ABSTRACT

We present the results of an explicit and systematic calculation of non-perturbative interactions, induced by single eigenvalue instantons, in $d = 1$ bosonic matrix models and corresponding string effective Lagrangians. We briefly discuss the possibility that similar non-perturbative induced operators in a supersymmetric version of the theory may break supersymmetry dynamically.

Matrix models [1] allow the construction of space-time Lagrangians valid to all orders in the string coupling parameter, at least for noncritical strings propagating in $d = 2$ dimensions. These Lagrangians are derived using the techniques of collective field theory [2, 3]. All order Lagrangians have been constructed for both the $d = 1$ bosonic matrix model [4] (a special case was discussed in [5]) and also for the $d = 1, \mathcal{N} = 2$ supersymmetric matrix model [6]. There are two important and related features of these constructions. First, when interactions are included to all orders, the induced coupling blows up at finite points in space and delineates a zone of strong coupling. This is to be contrasted with the lowest order theory, where the coupling only diverges at spatial infinity. Secondly, since these all-order Lagrangians are derived from matrix models, they contain additional non-perturbative information which is directly accessible and computable (see, however, [7] for some possible caveats).

The existence of these new non-perturbative aspects of the theory relies on the observation that the matrix models contain two distinct sectors. The first of these is the so-called continuous sector, which consists of a continuous distribution of matrix eigenvalues. The second sector consists of discrete eigenvalues, which are distinguishable from the continuum eigenvalues. The classical configurations of the matrix model include time-dependent instanton solutions in which the discrete eigenvalues tunnel between two continuous eigenvalue sectors. We perform an explicit calculation of the leading order effects of such single eigenvalue instantons on the effective theory derived from a $d = 1$ bosonic matrix model. These consists of a set of induced operators, whose exact form we compute and exhibit. The results presented here are a summary of the results contained in [8].

We conjecture that, in the supersymmetric case, the same instantons described in this talk, and their associated bosonic and fermionic zero modes, provide a mechanism for supersymmetry breaking in the associated $d = 2$ effective superstring theory. The calculation we present here is a necessary preliminary ingredient to the explicit calculation of this effect. We hope that similar techniques will allow us to obtain conclusive results in the supersymmetric case as well. The observation on which we base our conjecture is that the supersymmetry of the D=2 superstring continuous theory can be shown to be precisely related to the existence of D=1 supersymmetry in the underlying matrix model. However, it is well known that this D=1 supersymmetry is broken by instanton tunneling processes of the type mentioned above. It follows that instantons tunneling across the strong coupling region break the D=1 supersymmetry of the matrix model and, hence, must also break the induced D=2 supersymmetry of the associated continuous superstring theory. This symmetry breaking should be manifest in the induced operators alluded to above.

Non-perturbative effects due to single eigenvalue instantons and their implications were discussed by Shenker [9], who put forward compelling arguments for their universal nature and were also studied in [10, 11, 12]. Recently, an interesting approach based on 2-dimensional world-sheet arguments was suggested [13]. The possible role of non-perturbative interactions of similar strength in 4-dimensional string theory and effective field theories were discussed in [14]. Since we obtain corrections to the effective action that contain only derivatives, our explicit calculations mildly support the argument [15, 14] that these effects appear only in the Kahler potential.

A $d = 1$ bosonic matrix model has a time-dependent $N \times N$ Hermitian matrix, $M(t)$, as its fundamental variable. Its dynamics are described by the Lagrangian

$$L(\dot{M}, M) = \frac{1}{2} Tr \dot{M}^2 - V(M). \tag{1}$$

The potential is taken to be polynomial,

$$V(M) = \sum_{n=0}^{\omega} a_n Tr M^n, \tag{2}$$

As $N \to \infty$, if the a_n are tuned simultaneously and appropriately, the associated partition function describes an ensemble of oriented two-dimensional Riemann surfaces, including contributions at all genus. It is argued that, in this limit, the model describes a string propagating in two space-time dimensions. In the large N limit, the potential may be written as

$$V(M) = Tr(NV_0 \cdot \mathbf{1} - \frac{1}{2}\omega^2 M^2), \tag{3}$$

where $\mathbf{1}$ is the $N \times N$ unit matrix. The parameters V_0 and ω each have mass dimension one, and are arbitrary. In (3) the scaling behavior of the coefficients has been made explicit. The Lagrangian, (1), is invariant under the global $U(N)$ transformation $M \to \mathcal{U}^\dagger M \mathcal{U}$, where $\mathcal{U}$ is an arbitrary $N \times N$ unitary matrix. The set of states which do not transform under $\mathcal{U}$ comprise the $U(N)$-singlet sector of the quantized theory. It can be shown that the physics of this singlet sector is described equivalently by a theory involving only the N eigenvalues, $\lambda_i(t)$, of the matrix $M(t)$ with the following Lagrangian,

$$L[\lambda] = \sum_{i=1}^{N} \{ \frac{1}{2}\dot{\lambda}_i^2 - (V_0 - \frac{1}{2}\omega^2\lambda_i^2) - \frac{1}{2}\sum_{j \neq i} \frac{1}{(\lambda_i - \lambda_j)^2} \}. \tag{4}$$

The eigenvalues are first restricted to lie in the interval $-\frac{L}{2} \leq \lambda_i \leq \frac{L}{2}$ for any i. When we take the limit $N \to \infty$, we will simultaneously take $L \to \infty$. In this limit, over a given range, l, to be made explicit below, there exist two possibilities. If n represents the number of eigenvalues within this range, then the average density is given by $\rho = n/l$. In the limit $N \to \infty, L \to \infty$, ρ can remain small, and the eigenvalues populate the region sparsely. We refer to this situation as a "low density" or "discrete" distribution of eigenvalues over the region l. In the second case, ρ becomes very large, and the eigenvalues populate the region densely. In this case, the eigenvalues can be aggregated into a "collective field" which describes their collective motion. We refer to this second case as a "high density" or"continuous" distribution of eigenvalues. We begin by studying the continuous case.

We introduce a continuous real parameter, x, constrained to lie in the interval $-\frac{L}{2} \leq x \leq \frac{L}{2}$, and over this line segment define a collective field,

$$\partial_x \varphi(x,t) = \sum_{i=1}^{N} \delta(x - \lambda_i(t)). \tag{5}$$

It follows from (5) that

$$\int_{x_0}^{x_0+l} dx \partial_x \varphi(x,t) = n, \tag{6}$$

where n is the number of eigenvalues in the range l. Thus, $\varphi' = \partial_x \varphi$ is the eigenvalue density. In the range l, φ' has n degrees of freedom. Provided that $n/l \to \infty$ as $N \to \infty, L \to \infty$, the average density of eigenvalues then becomes infinite, and, modulo some technical subtleties irrelevant to this discussion, the field φ becomes an unconstrained, ordinary two dimensional field. In effect, φ' ceases to be a sum over delta functions and becomes a continuous eigenvalue density. It can be shown, in this case, that the eigenvalue Lagrangian, (4), may be rewritten in terms of the collective field as follows,

$$L[\varphi] = \int dx \{ \frac{\dot{\varphi}^2}{2\varphi'} - \frac{\pi^2}{6} \varphi'^3 - (V_0 - \frac{\omega^2}{2} x^2) \varphi' \}. \tag{7}$$

The associated action is given by $S[\varphi] = \int dt L[\varphi]$. This expression describes the physics over all ranges of x where the eigenvalue density is large. The limits on the $\int dx$ integral are set accordingly. Since our interest is in the quantum theory, henceforth we will consider only the Euclidean version of the action, which is given by

$$S_E[\varphi] = \int dx dt \{ \frac{\dot{\varphi}^2}{2\varphi'} + \frac{\pi^2}{6} \varphi'^3 + (V_0 - \frac{\omega^2}{2} x^2) \varphi' \}. \tag{8}$$

The equation of motion, obtained by varying (8) is

$$\partial_t (\frac{\dot{\varphi}}{\varphi'}) - \frac{1}{2} \partial_x \left\{ \frac{\dot{\varphi}^2}{\varphi'^2} + \pi^2 \varphi'^2 - \omega^2 x^2 \right\} = 0. \tag{9}$$

The static solution is obtained by taking $\dot{\varphi} = 0$, so that (9) reduces to

$$\partial_x \left\{ \pi^2 \varphi'^2 - \omega^2 x^2 \right\} = 0. \tag{10}$$

The solution to this equation is the following,

$$\tilde{\varphi}_0'(x) = \frac{\omega}{\pi}\sqrt{x^2 - A^2}, \tag{11}$$

where A^2 is a positive constant. Additional analysis reveals that

$$A^2 = 2V_0/\omega^2. \tag{12}$$

Since φ' is now a continuous density of eigenvalues, we may use (6) to determine the approximate location of the first eigenvalues in the continuum; that is, those two eigenvalues closest to $x = \pm A$. We focus on the region $x \geq A$. There is an identical discussion regarding the opposite region, $x \leq -A$. Given (11), the first eigenvalue must live somewhere in the region $A \leq x \leq A + \epsilon_x$, where ϵ_x is determined by the following relation,

$$\begin{aligned} 1 &= \frac{\omega}{\pi}\int_A^{A+\epsilon_x} dx\sqrt{x^2 - A^2} \\ &= \frac{\omega A^2}{2\pi}\left\{ \frac{x}{A}\sqrt{(\frac{x}{A})^2 - 1} - \ln(\frac{x}{A} + \sqrt{(\frac{x}{A})^2 - 1}) \right\}\Bigg|_{x=A}^{x=A+\epsilon_x}. \end{aligned} \tag{13}$$

We make the important assumption that $\epsilon_x << A$. After some algebra, Eq.(13) then becomes

$$\frac{1}{2}(\frac{3\pi}{\omega A^2})^{2/3} = \frac{\epsilon_x}{A} + \mathcal{O}\left((\frac{\epsilon_x}{A})^2 \right). \tag{14}$$

For consistency, this requires that $(\omega A^2)^{-1} << 1$. This small dimensionless number will be central to much of the ensuing analysis, so we give it a special name,

$$g = \frac{1}{\omega A^2} << 1. \tag{15}$$

It is clear that the first eigenvalue does not live precisely at the value $x = A$. This distinction will prove a necessary and important regulator on quantities which we will encounter. For definiteness, we assume henceforth that the first eigenvalue in the static continuum has a value $x = A + \epsilon_x$, where

$$\epsilon_x = \frac{1}{2}(3\pi g)^{2/3} A \tag{16}$$

and g is a small, dimensionless number, which, in the present context, parameterizes the width of the discrete region as well as our ignorance regarding the "graininess" of eigenvalues near the edge of the continuous distribution, when we adopt a collective field point of view. We now turn our attention to the region $|x| \le A$. We assume, in addition to a continuum of eigenvalues λ_i for $i = 1$ to N, that there exists an additional discrete eigenvalue, which we denote λ_0. There are then $N+1$ total eigenvalues, and the Euclidean version of Lagrangian (4) now reads

$$L_E = \sum_{i=0}^{N}\{\frac{1}{2}\dot{\lambda}_i^2 + (V_0 - \frac{1}{2}\omega^2\lambda_i^2) + \frac{1}{2}\sum_{j\neq i}\frac{1}{(\lambda_i - \lambda_j)^2}\}. \tag{17}$$

Note that the index i now runs over the $N+1$ values from 0 to N. What do we mean by a discrete eigenvalue? The separation of the continuum eigenvalues nearest to $\pm A$ is of order ϵ_x. As long as $-A \le \lambda_0 \le A$, and

$$A - |\lambda_0| >> \epsilon_x, \tag{18}$$

the eigenvalue λ_0 is truly distinct from the continuum and, hence, discrete. Assuming that λ_0 satisfies (18), it is useful to rewrite this Lagrangian by separating the λ_0 contribution from the contribution due to the continuum eigenvalues, as follows,

$$\begin{aligned} L_E &= \frac{1}{2}\dot{\lambda}_0^2 + (V_0 - \frac{1}{2}\omega^2\lambda_0^2) + \sum_{i\neq 0}\frac{1}{(\lambda_0 - \lambda_i)^2} \\ &+ \sum_{i=1}^{N}\{\frac{1}{2}\dot{\lambda}_i^2 + (V_0 - \frac{1}{2}\omega^2\lambda_i^2) + \frac{1}{2}\sum_{j\neq i}\frac{1}{(\lambda_i - \lambda_j)^2}\}. \end{aligned} \tag{19}$$

As above, we may now rewrite this expression using the definition (5). We thus obtain

$$\begin{aligned} L_E[\lambda_0;\varphi] &= \frac{1}{2}\dot{\lambda}_0^2 + \frac{1}{2}\omega^2(A^2 - \lambda_0^2) + \int dx \frac{\varphi'}{(x - \lambda_0)^2} \\ &+ \int dx\{\frac{\dot{\varphi}^2}{2\varphi'} + \frac{\pi^2}{6}\varphi'^3 + \frac{1}{2}\omega^2(A^2 - x^2)\varphi'\}. \end{aligned} \tag{20}$$

The third term in this expression represents the mutual interaction of the discrete eigenvalue with the continuum eigenvalues, which are collectively described using the field φ. We obtain the Euclidean equations of motion for λ_0 and for φ by variation of (20). Respectively, these are found to be

$$\ddot{\lambda}_0 + \omega^2\lambda_0 + \int dx \frac{\varphi'}{(\lambda_0 - x)^3} = 0 \tag{21}$$

$$\partial_t\left(\frac{\dot\varphi}{\varphi'}\right)-\frac{1}{2}\partial_x\left\{\frac{\dot\varphi^2}{\varphi'^2}+\pi^2\varphi'^2-\omega^2x^2+\frac{2}{(\lambda_0-x)^2}\right\}=0. \tag{22}$$

We consider first the φ equation. It is possible to show, even in the presence of a non-trivial, but discrete, $\lambda_0(t)$, that the static background, $\tilde{\varphi}_0'$, derived above is still a valid solution to leading order in ϵ_x.

Next, we turn our attention to the λ_0 equation, (21). This is the Euclidean equation of motion,

$$\ddot\lambda_0-V'_{eff}(\lambda_0)=0, \tag{23}$$

where

$$V_{eff}(\lambda_0)=\frac{\omega}{2g}\left\{-\left(\frac{\lambda_0}{A}\right)^2+4g\frac{(\lambda_0/A)}{\sqrt{1-(\lambda_0/A)^2}}\tan^{-1}\left(\frac{(\lambda_0/A)}{\sqrt{1-(\lambda_0/A)^2}}\right)\right\}. \tag{24}$$

The effect of the second term in (24), is to turn the potential over near $\lambda_0=\pm A$, where it adds infinite confining walls. The eigenvalue, λ_0 can be treated as discrete, and $V_{eff}(\lambda_0)$ is well defined, for λ_0 sufficiently far from $\pm A$. When λ_0 approaches $\pm A$ to within order ϵ_x it is absorbed into the continuum, and disappears as a discrete entity. Of course, this process can be reversed. It is possible for the first eigenvalue of the continuum to "leak" out and become a discrete eigenvalue λ_0. We will return to such processes below.

This being said, we would like to find both static and time-dependent solutions for the Euclidean λ_0 equation of motion (23). In the small g limit we can replace (23) by

$$\begin{aligned}\ddot\lambda_0+\omega^2\lambda_0&=0 \quad ; \quad -A<\lambda_0<A\\ \ddot\lambda_0&=0 \quad ; \quad \lambda_0=\pm A.\end{aligned} \tag{25}$$

We also impose the following boundary conditions, $\lambda_0(t\to-\infty)=\pm A$ and, independently, $\lambda_0(t\to+\infty)=\pm A$. There are two static solutions to (25) which satisfy this boundary condition,

$$\widehat{\lambda}_{0\pm}=\pm A. \tag{26}$$

A simple time-dependent solution is given by

$$\widehat{\lambda}_0^{(+)}(t;t_1)=\begin{cases}-A & ; \quad t<t_1-\frac{\pi}{2\omega}\\ +A\sin\omega(t-t_1) & ; \quad t_1-\frac{\pi}{2\omega}\le t\le t_1+\frac{\pi}{2\omega}\\ +A & ; \quad t>t_1+\frac{\pi}{2\omega}\end{cases}, \tag{27}$$

where t_1 is arbitrary. The solution (27) describes an eigenvalue which rolls (tunnels) from $-A$ to $+A$ over a time interval of duration $\frac{\pi}{\omega}$, centered at an arbitrary time t_1. We refer to this solution as a "kink". Its mirror image is also a valid solution,

$$\widehat{\lambda}_0^{(-)}(t;t_1) = \begin{cases} +A & ; \quad t < t_1 - \frac{\pi}{2\omega} \\ -A\sin\omega(t-t_1) & ; \quad t_1 - \frac{\pi}{2\omega} \le t \le t_1 + \frac{\pi}{2\omega} \\ -A & ; \quad t > t_1 + \frac{\pi}{2\omega} \end{cases}, \tag{28}$$

It describes an eigenvalue which rolls from $+A$ to $-A$. It is referred to as an "anti-kink".

Taking into account the fact that, when at $\pm A$, the discrete eigenvalue gets reabsorbed in the continuum, we may rewrite the kink and antikink solutions as follows,

$$\lambda_0^{(\pm)} = \pm A\sin\omega(t-t_1) \quad ; \; t_1 - \frac{\pi}{2\omega} \le t \le t_1 + \frac{\pi}{2\omega}, \tag{29}$$

There exist more general solutions than those which we have already discussed, in which the identity of λ_0 is more complex. It is possible, for example, that a kink, which ends with eigenvalue λ_0 attaching to the continuum at $+A$, could be followed, at some later time, by an antikink, in which the eigenvalue λ_0 separates from the continuum at $+A$, rolls to $-A$ and then reattaches there. Such a kink-antikink sequence, which we denote $\lambda_0^{(+-)}$, would satisfy the Euclidean equation of motion, (25). It is also possible, however, that a kink, which ends with the eigenvalue λ_0 attaching to the continuum at $+A$, could be followed, at some later time, by another kink in which a different eigenvalue detaches from the continuum at $-A$, traverses the region between $-A$ and $+A$, and then reattaches to the continuum at $+A$ immediately next to the eigenvalue involved in the first kink. This kink-kink sequence, which we denote $\lambda_0^{(++)}$, also satisfies (25). There are thus $2^2 = 4$ solutions which involve two distinct kinks,

$$\begin{aligned} \lambda_0^{(++)} &= \begin{cases} +A\sin\omega(t-t_1) & ; \quad t_1 - \frac{\pi}{2\omega} \le t \le t_1 + \frac{\pi}{2\omega} \\ +A\sin\omega(t-t_2) & ; \quad t_2 - \frac{\pi}{2\omega} \le t \le t_2 + \frac{\pi}{2\omega} \end{cases} \\ \lambda_0^{(+-)} &= \begin{cases} +A\sin\omega(t-t_1) & ; \quad t_1 - \frac{\pi}{2\omega} \le t \le t_1 + \frac{\pi}{2\omega} \\ -A\sin\omega(t-t_2) & ; \quad t_2 - \frac{\pi}{2\omega} \le t \le t_2 + \frac{\pi}{2\omega} \end{cases} \\ \lambda_0^{(-+)} &= \begin{cases} -A\sin\omega(t-t_1) & ; \quad t_1 - \frac{\pi}{2\omega} \le t \le t_1 + \frac{\pi}{2\omega} \\ +A\sin\omega(t-t_2) & ; \quad t_2 - \frac{\pi}{2\omega} \le t \le t_2 + \frac{\pi}{2\omega} \end{cases} \\ \lambda_0^{(--)} &= \begin{cases} -A\sin\omega(t-t_1) & ; \quad t_1 - \frac{\pi}{2\omega} \le t \le t_1 + \frac{\pi}{2\omega} \\ -A\sin\omega(t-t_2) & ; \quad t_2 - \frac{\pi}{2\omega} \le t \le t_2 + \frac{\pi}{2\omega} \end{cases} \end{aligned} \tag{30}$$

In all four cases $t_2 \ge t_1 + \frac{\pi}{\omega}$, but both t_1 and t_2 are otherwise arbitrary. An arbitrary solution consists of q events which are randomly distributed between kinks and antikinks, where $0 \leq q < \infty$. For a given q there are 2^q distinct instanton configurations. Generically, we denote the 2^q instantons as $\lambda_0^{(q)}$. There are q zero modes associated with each $\lambda_0^{(q)}$. These correspond to the arbitrary times $t_1, ..., t_q$, where $t_q \geq t_{q-1} \cdots \geq t_1$, when the kinks or antikinks occur. We ignore all cases where several eigenvalues are simultaneously discrete, since the effect of these solutions is negligible.

The partition function associated with the theory discussed above can be written as a sum over different instanton sectors,

$$Z = \sum_{q=0}^{\infty} Z_q \tag{31}$$

where, schematically,

$$Z_q = \int [d\varphi] \int [d\lambda_o]_q e^{-S[\lambda_0;\varphi]}. \tag{32}$$

In this expression the symbol $[d\lambda_0]_q$ indicates that λ_0 is expanded around $\lambda_0^{(q)}$. For notational convenience we have suppressed a subscript E on the action, but it is assumed throughout this section that we are in euclidean space. We proceed to define equation (32) in more precise terms. First of all, remember that $\lambda_0^{(q)}$ generically represents all the 2^q instanton solutions which each have q single eigenvalue kinks-antikinks. Therefore, more specifically,

$$Z_q = \sum_{\{k_i\}} Z_{k_1 \cdots k_q}, \tag{33}$$

where $k_i = \pm$, the summation is over all 2^q possible sets $\{k_1 \cdots k_q\}$, and

$$Z_{k_1 \cdots k_q} = \int [d\varphi] \int [d\lambda_0]_{k_1 \cdots k_q} e^{-S[\lambda_0;\varphi]}. \tag{34}$$

The symbol $[d\lambda_0]_{k_1 \cdots k_q}$ indicates that λ_0 is expanded around $\lambda_0^{(k_1 \cdots k_q)}$. Thus, $Z_2 = Z_{++} + Z_{+-} + Z_{-+} + Z_{--}$, and so on. After some lengthy analysis, using a dilute gas approximation, we arrive at the following general result

$$\begin{aligned} Z &= \int [d\varphi] e^{-S_\varphi[\varphi]} \sum_{q=0}^{\infty} \frac{1}{q!} \mathcal{M}^q \prod_{i=1}^{q} \int dt_i \sum_{\{k_i\}} \prod_{j=1}^{q} e^{-S_I^{(k_j)}[\varphi;t_j]} \\ &= \int [d\varphi] e^{-S_\varphi[\varphi]} \sum_{q=0}^{\infty} \frac{1}{q!} \left\{ \mathcal{M} \int dt_1 \left(e^{-S_I^{(+)}[\varphi;t_1]} + e^{-S_I^{(-)}[\varphi;t_1]} \right) \right\}^q . \end{aligned} \tag{35}$$

The sum over q is now an exponential, so that

$$Z = \int [d\varphi] e^{-S_{eff}[\varphi]}, \tag{36}$$

where

$$S_{eff}[\varphi] = S_\varphi[\varphi] + \Delta S[\varphi] \tag{37}$$

is the effective action with the instanton effects systematically incorporated, and

$$\Delta S[\varphi] = \mathcal{M} \int dt_1 \left\{ e^{-S_I^{(+)}[\varphi;t_1]} + e^{-S_I^{(-)}[\varphi;t_1]} \right\} \tag{38}$$

is the associated change in the action. The action $S_I^{(\pm)}$ is given by

$$S_I^{(\pm)}[\varphi;t_j] = \int_{t_j-\frac{\pi}{2\omega}}^{t_j+\frac{\pi}{2\omega}} dt \int dx \left\{ \frac{\varphi'(x,t)}{(x-\lambda_0^{(\pm)}(t-t_j))^2} - \frac{\varphi'(x,t)}{(x-\lambda_\emptyset^{(\pm)}(t-t_j))^2} \right\}. \tag{39}$$

where

$$\lambda_\emptyset^{(\pm)}(t;t_1) = \left\{ \begin{array}{lcl} \mp A & ; & t_1 - \frac{\pi}{2\omega} \le t < t_1 \\ \pm A & ; & t_1 < t \le t_1 + \frac{\pi}{2\omega} \end{array} \right. . \tag{40}$$

The quantity $\mathcal{M}$ is a dimensionful parameter that sets the basic strength for induced non-perturbative interactions

$$\mathcal{M} = \omega \sqrt{\frac{\pi}{2g}} e^{-\frac{\pi}{2g}}. \tag{41}$$

So far, we have studied the collective field theory expressed in terms of the field φ. By examining equation (8), however, we discover that φ does not have a canonically normalized kinetic energy. We also find that the collective field Lagrangian is neither Lorentz invariant nor translation invariant. The first of these problems is solved, in part, by expanding φ around the solution to the euclidean field equation $\tilde{\varphi}_0$ given in (11). Thus, we define

$$\varphi(x,t) = \tilde{\varphi}_0(x) + \frac{1}{\sqrt{\pi}} \zeta(x,t). \tag{42}$$

As discussed at length elsewhere, a canonical kinetic energy is obtained by expressing the Lagrangian in terms of a new spatial coordinate τ defined by the following relation,

$$\tau'(x) = \frac{1}{\pi} (\tilde{\varphi}_0'(x))^{-1}. \tag{43}$$

Note that τ has mass dimension -1, which is the appropriate mass dimension for a spatial coordinate, whereas x has mass dimension $-\frac{1}{2}$. Expressing the euclidean collective field action (8) in terms of $\zeta(\tau,t)$, we find, in the absence of instanton effects, that

$$S_\zeta[\zeta] = \int dt \int d\tau \left\{ \frac{1}{2}(\dot{\zeta}^2 + \zeta'^2) - \frac{1}{2} \frac{g(\tau)\dot{\zeta}^2\zeta'}{1 + g(\tau)\zeta'} + \frac{1}{6} g(\tau)\zeta'^3 - \frac{1}{3}\frac{1}{g(\tau)^2} \right\}, \tag{44}$$

where $g(\tau)$ is a space dependent coupling parameter, which we define below, and the τ integration is over the limits $-\infty < \tau \leq \tau_0 + \frac{\sigma}{2}$ and $\tau_0 + \frac{\sigma}{2} \leq \tau < \infty$, where τ_0 and σ are independent integration constants which arise when solving (43). The reason why there are two integration constants rather than one, given that (43) is a first-order differential equation, is that we must solve (43) independently over the two separate regions $-\infty < x \leq A$ and $A \leq x < \infty$. The region $-A < x < A$, where there is no continuous collective field theory, is the low density region. In τ space, this region is given by $\tau_0 - \frac{\sigma}{2} < \tau < \tau_0 + \frac{\sigma}{2}$, so that τ_0 is the center of the low density region and σ is the width. The coupling parameter, defined over $-\infty < \tau \leq \tau_0 - \frac{\sigma}{2}$ and $\tau_0 + \frac{\sigma}{2} \leq \tau < \infty$, is given by $g(\tau) = (\pi^{3/2}\tilde{\varphi}_0(x))^{-1}$, and is found to be

$$g(\tau) = 4\sqrt{\pi}\frac{g}{\omega}\frac{\frac{1}{\kappa}e^{-2\omega|\tau-\tau_0|}}{(1-\frac{1}{\kappa}e^{-2\omega|\tau-\tau_0|})^2}, \tag{45}$$

where κ is a dimensionless number,

$$\kappa = \exp(-\omega\sigma), \tag{46}$$

which relates the width, σ, of the low density region in τ space to the natural length scale in the matrix model, $1/\omega$. Notice that the coupling parameter blows up as $\tau \to \tau_0 \pm \frac{\sigma}{2}$; that is, at the boundaries of the low density region.

We would now like to express the change in the effective action due to the instanton effects, equation (38), in terms of the canonical variable $\zeta(\tau, t)$. Since $S_I^{(\pm)}$ is linear in φ, it follows that

$$S_I^{(\pm)}[\varphi; t_1] = S_I^{(\pm)}[\tilde{\varphi}_0] + \frac{1}{\sqrt{\pi}}S_I^{(\pm)}[\zeta; \tau_0, t_1]. \tag{47}$$

The τ_0 dependence in the last term of this equation will be made clear presently. From (39), we find

$$S_I^{(\pm)}[\zeta; \tau_0, t_1] = \int_{t_1-\frac{\pi}{2\omega}}^{t_1+\frac{\pi}{2\omega}} dt \int d\tau \left\{ \frac{\zeta'(\tau,t)}{(x(\tau) - \lambda_0^{(\pm)}(t-t_1))^2} - \frac{\zeta'(\tau,t)}{(x(\tau) - \lambda_\emptyset^{(\pm)}(t-t_1))^2} \right\}, \tag{48}$$

where the prime now means differentiation with respect to τ, and where

$$x(\tau) = \begin{cases} -A\cosh\{\omega(\tau - \tau_0 + \sigma/2)\} & ; \tau \leq \tau_0 - \sigma/2 \\ +A\cosh\{\omega(\tau - \tau_0 - \sigma/2)\} & ; \tau \geq \tau_0 - \sigma/2 \end{cases}. \tag{49}$$

This last expression is found by integrating (43) to obtain $\tau(x)$ and then inverting the result to obtain $x(\tau)$. This function depends explicitly on τ_0. This explains why there is an explicit

τ_0 in equations (47) and (48). It is straightforward to compute $S_I^{(\pm)}[\tilde{\varphi}_0]$ and we find

$$S_I^{(\pm)}[\tilde{\varphi}_0] = -2^{3/2}\sqrt{\frac{A}{\epsilon_x}} + \ln\sqrt{\frac{A}{\epsilon_x}} + \mathcal{O}(\frac{\epsilon_x}{A}). \tag{50}$$

As discussed above, ϵ_x is the size of the inter-eigenvalue separation near the edge of the continuum and so provides the natural regulator for expressions such as (50). From (16) it follows that, to lowest order in g

$$e^{-S_I^{(\pm)}[\tilde{\varphi}_0]} = g^{1/3}e^{\mathcal{O}(g^{1/3})}. \tag{51}$$

Since all x-space integrations are cut-off at a distance ϵ_x from the edge of the low density region; that is, at $|x| = A + \epsilon_x$, it follows that all τ space integrals must be cut-off as well at a value ϵ_τ. Specifically, in (48) and in all other expressions which include a $\int d\tau$ integration, the following is implied,

$$\int d\tau = \int_{-\infty}^{\tau_0-\frac{\sigma}{2}-\epsilon_\tau} d\tau + \int_{\tau_0+\frac{\sigma}{2}+\epsilon_\tau}^{\infty} d\tau. \tag{52}$$

The value of ϵ_τ is simple to obtain. We require that

$$\begin{aligned} x(\tau - \frac{\sigma}{2} - \epsilon_\tau) &= -A - \epsilon_x \\ x(\tau + \frac{\sigma}{2} + \epsilon_\tau) &= A + \epsilon_x. \end{aligned} \tag{53}$$

Using (49) and (16) it follows, to leading order in g, that

$$\epsilon_\tau = \frac{1}{\omega\sqrt{2}}(3\pi g)^{1/3}. \tag{54}$$

Now, using (51), substituting (47) into (38), and using (41), we find that

$$\Delta S[\zeta] = \omega g^{-1/6}e^{-\frac{\pi}{2g}}\int dt_1\left\{e^{-S_I^{(+)}[\zeta;\tau_0,t_1]} + e^{-S_I^{(-)}[\zeta;\tau_0,t_1]}\right\}. \tag{55}$$

Equation (55) is a significant result. Concisely, it is the induced change in the canonical collective field theory which results from the systematic inclusion of instanton effects. A lengthy analysis allows us to calculate from Eq.(55) the induced action as an integral over a local density. Skipping a lot of details we simply state the results

$$\begin{aligned} S_I^{(+)} &= \frac{1}{\omega}h_{00}(\zeta_-' + \zeta_+') + \frac{1}{\omega^2}h_{01}(\zeta_-'' - \zeta_+'') + \frac{1}{\omega^2}h_{10}(\dot{\zeta}_-' - \dot{\zeta}_+') + \frac{1}{\omega^3}h_{11}(\dot{\zeta}_-'' + \dot{\zeta}_+'') + \cdots \\ S_I^{(-)} &= \frac{1}{\omega}h_{00}(\zeta_-' + \zeta_+') - \frac{1}{\omega^2}h_{01}(\zeta_-'' - \zeta_+'') - \frac{1}{\omega^2}h_{10}(\dot{\zeta}_-' - \dot{\zeta}_+') + \frac{1}{\omega^3}h_{11}(\dot{\zeta}_-'' + \dot{\zeta}_+'') + \cdots. \end{aligned} \tag{56}$$

where

$$h_{mn} = \frac{\omega^{m+n+1}}{m!n!} \int_{-\frac{\pi}{2}}^{\frac{\pi}{2}} dt \int_{-\infty}^{-\epsilon_\tau} d\tau \mathcal{J}(\tau,t)\tau^m t^n, \tag{57}$$

$$\zeta_\pm \equiv \zeta(\tau_0 \pm \frac{\sigma}{2}, t_1) \tag{58}$$

and

$$\mathcal{J}(\tau-\tau_0+\frac{\sigma}{2}, t-t_1) = \frac{1}{(x(\tau-\tau_0+\frac{\sigma}{2}) - \lambda_0^{(\pm)}(t-t_1))^2} - \frac{1}{(x(\tau-\tau_0+\frac{\sigma}{2}) - \lambda_\emptyset^{(\pm)}(t-t_1))^2} \tag{59}$$

It is straightforward to compute the coefficients h_{mn}. We find, for instance, to leading order in g, that

$$\begin{aligned} h_{00} &= -\frac{4\sqrt{2}}{9} \\ h_{10} &= -(\frac{8\pi g}{9})^{1/3} \\ h_{01} &= -\frac{\pi\sqrt{2}}{9}. \end{aligned} \tag{60}$$

In general, the h_{mn} are found to have the following g dependence,

$$h_{mn} \sim \begin{cases} g^{m/3} & ; \quad m \leq 3 \\ g & ; \quad m > 3 \end{cases} \tag{61}$$

Note, from (56) and (61), that, as the first index of h_{mn} increases, that the corresponding terms in $S_I^{(\pm)}$ depend on higher powers of g. However, none of h_{0n} have g dependence for any value of n. We proceed to analyze the relative impact of these terms on generic N-point functions. By putting (56) back into (55) we can find all relevant interaction vertices. These are obtained by Taylor expanding the exponentials in (55). For instance, we obtain the quadratic vertices $\frac{1}{\omega^2}h_{00}^2\zeta'_-\zeta'_-$ and $\frac{1}{\omega^3}h_{00}h_{10}\zeta'_-\zeta''_-$ where, as discussed above, $h_{00} \sim 1$ and $h_{10} \sim g^{1/3}$. It is clear that the effect of the second vertex, containing $h_{00}h_{10}$, on any N-point function, is suppressed by a factor $g^{1/3}p/w$, where p is a characteristic momentum, when compared with effects arising solely from the first vertex containing h_{00}^2. This is true at tree level. At the quantum level, there may be some subtleties to this argument which we will not discuss. Similar considerations apply to all other induced operators, involving higher h_{mn}. It can thus be shown, provided

$$p \lesssim \omega, \tag{62}$$

that, when working to leading order in g, we can consistently drop all but the h_{0n} terms in (56). Now, of the terms which remain, as n increases, the corresponding terms in $S_I^{(\pm)}$

depend on higher derivatives of ζ. Thus, the effect of any vertex, containing h_{0n}, on any N-point function, is suppressed by a factor $(p/\omega)^n$, relative to effects arising from vertices containing only h_{00}. If we further restrict momenta, such that

$$p << \omega, \tag{63}$$

we can then consistently neglect all but the h_{00} terms in (56). This results in a vast simplification of the final result, so we will assume this approximation. It would be completely straightforward, however, to lift the restriction (63), and only require (62). One would then have to keep all h_{0n} terms in (56). It follows from (56), that, to leading order in g,

$$\Delta S[\zeta] = 2\omega g^{-1/6} e^{-\frac{\pi}{2g}} \int dt_1 \exp\left\{\frac{4\sqrt{2}}{3\omega}\left(\zeta'(\tau_0 + \frac{\sigma}{2}, t_1) + \zeta'(\tau_0 - \frac{\sigma}{2}, t_1)\right)\right\}. \tag{64}$$

Note however that equation (64) includes nonlocal interactions, since it involves contributions coming from ζ' evaluated simultaneously at $\tau_0 - \frac{\sigma}{2}$ and also at $\tau_0 + \frac{\sigma}{2}$. This is not surprising though, since we have arrived at this result by integrating over single eigenvalue instantons, which link effects on the left-hand side of the low-density region with effects on the the right-hand side of this region, and because there is a finite separation between these two sectors. One may wish to find some further approximation which would render the effective theory local. This can be done as follows. Provided we consider momenta which satisfy (63), and provided also that $\omega \lesssim \frac{1}{\sigma}$, the effective width of the low density region as seen by any field will be essentially zero. We therefore Taylor expand $\zeta'(\tau_0 \pm \frac{\sigma}{2}, t_1)$ around the point (τ_0, t_1), thereby taking

$$\frac{1}{\omega}\zeta'(\tau_0 \pm \frac{\sigma}{2}, t_1) = \frac{1}{\omega}\zeta'(\tau_0, t_1) \pm \frac{\sigma\omega}{2\omega}\zeta''(\tau_0, t_1) + \cdots. \tag{65}$$

Then, in a manner identical to the previous discussion, we find that the contributions coming from vertices which involve σ are always suppressed by $(\sigma\omega)p/\omega$, where p is a characteristic momentum. Note that, since we now assume $\omega \lesssim \frac{1}{\sigma}$, the factor $(\sigma\omega)$ is $\lesssim \mathcal{O}(1)$. So, provided that

$$p << \omega \lesssim \frac{1}{\sigma}, \tag{66}$$

we may write the lowest order instanton-induced change in the collective field action approximately, in local form, as follows,

$$\Delta S[\zeta] = 2\omega g^{-1/6} e^{-\frac{\pi}{2g}} \int dt e^{-\frac{2\sqrt{2}}{3\omega}\zeta'(\tau_0, t)}. \tag{67}$$

We have dropped the subscript "1" on t_1 because it is now superfluous. This result can be written as a two-dimensional integral over a density $\Delta S = \int dt d\tau \Delta\mathcal{L}$, where

$$\Delta\mathcal{L} = 2\omega g^{-1/6} e^{-\frac{\pi}{2g}} \delta(\tau - \tau_0) e^{-\frac{2\sqrt{2}}{3\omega}\zeta'(\tau,t)}. \tag{68}$$

This is the final result of our calculation.

ACKNOWLEDGEMENT

Based on a talk given by B. O. at the Joint U.S.-Polish Workshop on Physics from Planck Scale to Electroweak Scale, Warsaw, Poland, 21-24 September 1994. The work of M. F. and B. O. was supported in part by DOE under Contract No. DOE-AC02-76-ERO-3071.

References

[1] D. J. Gross and N. Miljkovic, *Phys.Lett.* **B238** (1990) 217;
P. Ginsparg and J. Zinn-Justin, *Phys. Lett.* **B240** (1990) 333 ;
E. Brezin, V. Kazakov, Al. Zamolodchikov, *Nucl. Phys.* **B338** (1990) 673.

[2] S. R. Das and A. Jevicki, *Mod. Phys. Lett.* **A5** (1990) 1639.

[3] A. Jevicki, Brown preprint, BROWN-HET-918 (1993).

[4] R. Brustein and B. Ovrut, *Phys. Lett.* **B309** (1993) 45;
R. Brustein and B. Ovrut, preprint, UPR-523T (1992);
R. Brustein and B. Ovrut, Talk presented at 26th International Conference on High Energy Physics (ICHEP 92), Dallas, TX, 6-12 Aug 1992.

[5] R. Brustein and S. deAlwis, *Phys. Lett.* **B272** (1991) 285.

[6] R. Brustein, M. Faux and B. Ovrut, CERN preprint, Talk presented at the international Europhysics Conference on High Energy Physics, Marseille, France, July 22-28, 1993;
R. Brustein, M. Faux and B. Ovrut, *Nucl. Phys.* **B421** (1994) 293;
R. Brustein, M. Faux and B. Ovrut,Talk presented at the International Workshop on Supersymmetry and Unification of Fundamental Interactions (SUSY 93), Boston, MA, Mar. 29 - Apr. 1, 1993.

[7] J. Polchinski, Santa Barbara preprint, NSF-ITP-94-96 (1994).

[8] R. Brustein, M. Faux and B. Ovrut, CERN/Pennsylvania preprint, CERN-TH.7301/94 / UPR-608T, to appear in it Nucl. Phys. **B** (1994); R. Brustein, M. Faux and B. Ovrut, in proceedings of International Workshop on Supersymmetry and Unification of Fundamental Interactions SUSY 94, Univ. of Michigan, Ann Arbor, Michigan, USA May 14-17, 1994, C. Kolda and J. D. Wells Eds. pp.395-402.

[9] S. H. Shenker, talk presented at the Cargese Workshop on Random Surfaces, Quantum Gravity and Strings, Cargese, France, May 28 - Jun 1, 1990.

[10] A. Dabholkar, *Nucl. Phys.* **B368** (1992) 293.

[11] J. Lee and P. Mende, *Phys. Lett.* **B312** (1993) 433.

[12] A. Dahr, G. Mandal and S. R. Wadia, *Int. J. Mod. Phys.* **A8** (1993) 3811.

[13] J. Polchinski, Santa Barbara preprint, NSF-ITP-94-73 (1994); see also J. Polchinski, Santa Barbara preprint, NSF-ITP-94-97 (1994).

[14] T. banks and M. Dine, Rutgers preprint, RU-50-94 (1994).

[15] Soo-Jung Rey, Private communication.

CANCELLATION OF FOLDS IN A 2D STRING THEORY*

JACEK PAWEŁCZYK
Institute of Theoretical Physics, Warsaw University, Hoża 69
Warsaw PL-00-681, Poland.

ABSTRACT

A string model for cancellation of folded configurations of YM_2 is proposed. The model consists of two terms: the Nambu-Goto term and a topological invariant.

It is strongly believed that the dynamics of gauge fields can be described in terms of a string theory. The $1/N_c$ expansion[1] applied in 2-dimensional (2D) models[2,3] gave several hints how the QCD_2 (or YM_2) string theory may look like. It appeared that the string theory should contain the Nambu-Goto action and should have built-in mechanism of cancelation of folds.

In this note we propose such a mechanism. We shall supplement the Nambu-Goto action by a topological term which then will lead to cancellation of folds. The topological term is well defined in a target space-time of dimension 4 so we shall introduce two additional variables with values in R^2. The new variables will enter only the topological term - in this sense they will not have any dynamics.

We start with a similar problem for the 1-dimensional non-backtracking particle. This is a purely educational example which presents basic ideas in the simplified situation. The string model we propose will be a straightforward generalization of the 1D particle case.

In the first quantized language the dynamics of a particle moving in 1D target space-time is described by the ordinary path integral with the action $S[X] = \mu \int_0^1 dt\sqrt{\dot{X}^2}$, where $X = X(t)$ is a map from the unit interval $[0,1]$ into the real ax $X \in R$. The set of paths contributing to the path integral contains backtracking paths i.e. i.e. paths for which the velocity vanishes $\dot{X} = 0$ (what generically means that the velocity changes sign). Points for which this happens will be called singular and denoted by P_i (i enumerate such points) We assume that points P_i are isolated, what is the generic situation. We want to modify the action in such a way that the backtracking paths will not contribute. The idea is to introduce another field $Y \in R$, which will be interpreted as an additional coordinate (farther called also *vertical* coordinate). The transition from the map $X(t)$ to $(X(t),\ Y(t))$ is a lift. Although the map $X(t)$ is singular its lift is not, in general. By that we mean that $(X(t),\ Y(t))$ is an immersion[4] so e.g. the induced metric $\sqrt{\dot{X}^2 + \dot{Y}^2}$ is non-zero.

We supplement the action by a topological term:

$$S[X,Y] = \int_0^1 dt\mu\sqrt{\dot{X}^2} + i\pi R \tag{1}$$

*Work supported, in part, by Polish Government Research Grant KBN.

where $R = \frac{i}{2\pi}\int_0^1(\dot{X}\ddot{Y} - \dot{Y}\ddot{X})/(\dot{X}^2 + \dot{Y}^2)$ for flat space-times. The second term in (1) is the rotation number R of the vector tangent to the path $(X(t), Y(t))$. In the parameterization $\dot{X} = v\cos\alpha$, $\dot{Y} = v\sin\alpha$, $R = \int_0^1 dt\dot{\alpha}(t)$. This formula gives a unique answer for any lift if boundary values of α are fixed. The model described by (1) has an enormous group of local symmetries. These are local shifts of the vertical field $\delta_y Y(t) = \epsilon(t)$, where ϵ is such that $(X, Y + \epsilon)$ has the same R. This kind of variation we shall call the v-regular homotopy[4]. The space of immersions is divided into topological sectors by the equivalence relation induced by the symmetry δ_y. It seems that one can fix the gauge for this symmetry in such a way that each topological sector contributes only one representant to the function integral. We shall discuss this point in the future publication. Thus, after the gauge fixing of the v-regular homotopy the path integral for the model is

$$\int \mathcal{D}Xe^{-S[X]}\sum_{top} e^{i\pi R[X,Y]} \tag{2}$$

where the sum runs over all topological sectors of the model.

Below we describe topological sectors of the model (2). Let us consider a path $X(t)$ with singular points P_i. Because we work with immersions, $\dot{Y}(P_i) \neq 0$ necessarily. Thus $\dot{Y}(P_i) > 0$ or $\dot{Y}(P_i) < 0$ and the v-regular homotopy can not change the sign of $\dot{Y}(P_i)$. We can define the following map $f: P_i \to \mathrm{sign}(\dot{Y}(P_i))$ (all i). We show that any two lifts are v-regularly homotopic *if and only if* they correspond to the same map f. Let us consider two lifts $(X(t), Y_1(t))$, $(X(t), Y_2(t))$. If they are v-regularly homotopic then we can set $Y_1 = Y_2$ what defines f uniquely. One the other hand let us assume that two immersions are characterized by the same map f. Using the shift symmetry one can set $Y_1(P_i) = Y_2(P_i)$. f gives equality of tangents to both immersions at P_i's. Thus, both immersions are equal in the infinitesimally small neighborhood of P_i's (up to second power of an infinitesimal quantity). Away from the singular points $X(t)$ is an immersion. In this case the shift parameter $\epsilon(t)$ can be arbitrary so one can set $Y_1(t) = Y_2(t)$ there. Thus one can do it everywhere. We conclude that there is a one-to-one correspondence between topological sectors of the model and maps f. The sum over different topological sectors in (1) is just sum over different maps f. Moreover by direct inspection one can see that for two maps f and f' the difference between rotation numbers of the appropriate lifts is $\Delta R = 1/2\sum_i(f(P_i) - f'(P_i))$. . Thus the functional integral (2) will vanish if there are any singular points because for given singular point P_i, $f = 1$ and $f' = -1$ (or vice versa) so $|\Delta R| = 1$ and the factor $e^{i\pi R}$ has opposite sign for both lifts.

Now we go to string theory. The construction of the model will be a direct generalization of the particle case so will be rather sketchy[5]. The string theory functional integral for 2D targets is a sum over surface-to-surface maps $\Sigma \to M$, where Σ denotes the string world-sheet and M the 2D target space-time. It is known that generic surface-to-surface maps contain singularities which are folds and cusps[6]. In order to remove these singularities we introduce two additional, (vertical) world-sheet fields: $(X^3, X^4) \in R^2$ and consider all lifts of the (X^1, X^2) map i.e. maps of the form

$(X^1,\ X^2,\ X^3,\ X^4) \in M \times R^2$, where fields $(X^3,\ X^4)$ take arbitrary values. A lift is an immersion, in general.

In order to remove these singularities we introduce two additional, (vertical) world-sheet fields: $(X^3,\ X^4) \in R^2$ and consider all lifts of the $(X^1,\ X^2)$ map i.e. maps of the form $(X^1,\ X^2,\ X^3,\ X^4) \in M \times R^2$, where fields $(X^3,\ X^4)$ take arbitrary values. The proposed string action is

$$S[X] = \mu \int_M d^2\sigma \sqrt{g} + i\theta I[X], \tag{3}$$

where I is the self-intersection number of the surface immersed in the 4D space[1] The action (3) is invariant under arbitrary v-regular homotopy of vertical fields[4]: $\delta X^\mu(\xi) = \epsilon^\mu(\xi)$, $(\mu = 3,\ 4)$. A v-regular homotopy is a homotopy (i.e. just mentioned shift symmetry) which is an immersion for each homotopy parameter. The functional integral is just direct generalization of (2). Topological sectors of this model has a bit more complicated characteristic. Let us consider a fold (it is topologically S^1). At the fold the map X has one null tangent vector w i.e. there exists such a w that $w^a \partial_a X = 0$. To any lift we can assign a number $f \in \pi_1(S^1) = Z$ defined as the rotation number of w in the vertical plane $(X^3,\ X^4)$. For maps with many folds we assign the numbers $f_i \in \pi_1(S^1)$ to i-th connected component of the set of folds. The numbers $\{f_i\}$ are invariant under the v-regular homotopy. The classification of lifts gives the following[5]: *v-regular homotopy classes of lifts (i.e. topological sectors of the model) are in one-to-one correspondence with sets* $\{f_i\}$. Moreover the self-intersection number is $I[f] = \sum_{folds} \pm f_i$. We want to show that the originally folded configurations (X^1, X^2) will cancel out from the partition function. Fixing the shift symmetry in such a way that the gauge slice picks only one representant (or at least equal number of them) of each topological sector we get the following expression for the functional integral:

$$\int \mathcal{D}X^1 \mathcal{D}X^2\ e^{-S[X]} \sum_{\{f_i\}} e^{i\theta I[f]} \tag{4}$$

The sum over f_i's can be performed independently for each i because $I[f] = \sum_{folds} \pm f_i$. For one fold we get $\sum_{f \in Z} e^{+i\theta f} = 2\pi\delta(\theta)$. Thus all folded configurations vanish from the path integral for non-zero θ. Of course $\theta = \pi$ is the preferred value because then the model does not break parity.

Maps contributing to the vacuum-to-vacuum amplitude of the closed string necessarily have folds for the target space R^2. According to the above discussion the amplitude vanishes. This also holds for any correlation function of any finite set of local operators. Thus the final conclusion of this part of the paper is that the model (3) is trivial for the R^2 space-time.

In the end let us discuss shortly the Nambu-Goto action for maps without folds which may occur for topologically non-trivial space-times. In this case one can distinguish two classes of maps: orientation preserving (+) and orientation reversing maps

[1]The application of I in string theory was previously considered in various papers[7,10].

(−). $\int_M d^2\sigma\sqrt{g} = \pm\frac{1}{2}\int_M d^2\sigma\epsilon^{\mu\nu}\epsilon^{ab}\partial_a X^\mu \partial_b X^\nu$, $(\mu,\, \nu = 1, 2)$. The sign is chosen in such a way that the r.h.s. is positive. To some extend both sectors can be considered separately. In fact, this kind of theory has been considered recently[8] yielding very interesting results.

If one views (3) as a compactification of a 4D string then the additional dimensions should have topology of a 2D torus. We believe that the cancellation of folds holds also in this case although we have not considered it. We must also stress that we can say very little about non-generic surface-to-surface maps. Various subsets of these maps may be relevant for the description of various 2D gauge theories[9]. It is clear that the model (3) has straightforward extensions to 3 and 4 dimensional space-times. It is enough to make the additional dimensions dynamical i.e. add them to the Nambu-Goto action. Higher dimensional string models may require more terms e.g. the extrinsic curvature term[10].

Acknowledgment. I would like to thank T. Mostowski for discussions concerning geometrical aspects of the paper, K.Gawędzki and R.Dijkgraaf for reading the manuscript and valuable comments.

1. G.'t Hooft, *Nucl. Phys.* **B72** (1974) 461; *Nucl. Phys.* **B75** (1974) 461.
2. W.A.Bardeen, I,Bars, A.J.Hanson and R.D.Peccei, *Phys. Rev.* **D13** 2364 (1976); I.Bars and A.J.Hanson, *Phys. Rev.* **D13**, 1744 (1976).
3. D.J. Gross, *Nucl. Phys.* **B400** (1993) 161; D.J. Gross and W. Taylor, IV, *Nucl. Phys.* **B400** (1993) 181, *Nucl. Phys.* **B403** (1993) 395.
4. H.Whitney, *Ann.Math.* **45** (1944) 220.
5. J.Pawełczyk, hap-th/9403175.
6. H.Whitney, *Ann.Math.* **62** (55) 374.
   M.Golubitsky, V.Guillemin, *Stable Mappings and Their Singularities* (Springer-Verlag, 1973).
7. A.P.Balachandran, F.Lizzi and G.Sparano, *Nucl. Phys.* **B263** (1986) 608. P.O.Mazur and V.P.Nair, *Nucl. Phys.* **B 284** (1986) 146; J.Pawełczyk, *Phys. Lett.* **311B** (1993) 98.
8. S.Corder, G.Moore and S.Ramgoolam, Yale preprint YCTP-P23-93, hep-th/9402107.
9. E. Witten, *J.Geom.Phys.*9 (1992) 303; O. Ganor , J. Sonnenschein and S. Yankielowicz, Tel-Aviv Univ. preprint TAUP-2182-94, hep-th/9407114.
10. A.M.Polyakov, *Nucl. Phys.* **B268** (1986) 406. L. Peliti and S. Leibler, *Phys. Rev. Lett.* **54** (1985) 1690; D. Foerster, *Phys. Lett.* **114A** (1986) 115; H. Kleinert, *Phys. Lett.* **114A** (1986) 263, *Phys. Lett.* **174B** (1987) 335.

SEARCH FOR SUPERSYMMETRIC PARTICLES AND THE TOP QUARK WITH COLLIDER DETECTOR AT FERMILAB

Krzysztof Sliwa
Department of Physics and Astronomy, Tufts University
Medford, Massachusetts 02155, USA

ABSTRACT

A search for tri-lepton events, predicted by sypersymmetric (SUSY) models, at Collider Detector at Fermilab (CDF), is presented. No candidates were found in the data sample collected in the 1992/93 run, which corresponds to an integrated luminosity of 19 pb^{-1}. Within the framework of the Minimal Supersymmetric Standard Model (MSSM) this preliminary null result excludes chargino mass $M(\tilde{\chi}_1^{\pm}) < 46$ GeV/c^2 for -500< μ <-400 GeV, 2<tanβ <15, and $M(\tilde{q})=1.2\times M\tilde{g}$. Sensitivity of this and other searches in view of the expected increase of available statistics to more than 100 pb^{-1} by next fall, is discussed. A summary of the recently published CDF paper on a search for top quark is also presented. An excess of events is observed in the di-lepton final state, and in the lepton+multijet events with at least one of the jets with characteristics expected of a b-quark jet. The probability that the observed events can be simply a fluctuation of background is 0.26%, which for Gaussian errors would mean 2.8 σ. The excess is not significant enough to claim any discovery. Interpreting these events as due to $t\bar{t}$ production and decay, and assuming the Standard Model branching fractions, the number of observed events above background translates into the measurement of the $t\bar{t}$ cross section: $\sigma(t\bar{t}) = 16.8\pm^{7.4}_{5.9}$ pb for M_t=140 GeV/c^2, and $\sigma(t\bar{t}) = 13.7\pm^{6.0}_{4.7}$ pb for M_t=180 GeV/c^2. Prospects for future analyses of a larger data set, $\approx$ 100 pb^{-1}, expected to be available within a year, and clearly very much needed, are discussed.

1. SUSY search at CDF

1.1. Introduction

Supersymmetry[1] is a fundamental symmetry of space-time relating fermions and bosons. The new loop diagrams with SUSY particles introduced in SUSY models exactly cancel the quadratic divergencies of quantum field theory. This also leads, in many SUSY models, to a very small value of the cosmological constant, which is considered preferable given our present knowledge of the Universe. It also solves the gauge hierarchy problem. Supersymmetry explains the electroweak symmetry breaking dynamically through radiative corrections. Two-dimensional SUSY is a required element in string theories, SUSY in four-dimensions appears in many string theories.

Recently, extrapolations of the Standard Model gauge couplings from their values measured at low energies were found to converge at large energy Grand Unification(GUT) scale in SUSY GUT models, but not in non-supersymmetric GUT models.

If nothing more, Supersymmetry is a very attractive theoretical idea, which actually predicts phenomena which could very well be accessible to experimental tests very soon, if not now.

SUSY models predict the existence of a light Higgs boson, $M_h \approx$130-150 GeV, together with a large number of new ($\geq$20-30) particles, the "supersymmetric partners" of the Standard Model particles. Supersymmetric particles - squarks ($\tilde{q}$), sgluons ($\tilde{g}$), charginos ($\tilde{\chi}^{\pm}$), neutralinos ($\tilde{\chi}^{o}$), are expected to be produced in pairs in $p\bar{p}$ collisions at FLAB. Their cascade decays will invariably involve large missing transverse energy, $\not{E}_t$, carried away by the lightest supersymmetric particle, believed to be stable and neutral.

1.2. Signature and analysis cuts

Experimental signatures of production and decay of SUSY particles are quite distinctive:

i. multijet events with large $\not{E}_t$
ii. same sign di-lepton events
iii. tri-lepton events.

Multijets events with large $\not{E}_t$ ($\not{E}_t$ >40 GeV, or so) are sensitive to $\tilde{g}\tilde{g}$, $\tilde{q}\tilde{q}$ and $\tilde{g}\tilde{q}$ production, followed by cascade decays. The expected cross section are sizable, unfortunately $\not{E}_t$ is the poorest measured quantity in CDF. More work may bring understanding of $\not{E}_t$ to the level which would make this very promising channel accessible for SUSY searches, however, no results are available at present.

Same-sign di-lepton events may indicate $\tilde{g}\tilde{q}$ production and corresponding cascade decay, however, the predicted cross sections for producing such events are very small, and significantly more data than available is needed to reach any sensitivity.

The most promising signature for discovering SUSY particles at a hadron collider is that of a three lepton event. A chargino-neutralino pair,$\tilde{\chi}_1^{\pm}\tilde{\chi}_2^{o}$, may be produced with a reasonable cross section either via a virtual W (s-channel) or by a virtual squark (t-channel). With a subsequent leptonic decay ($\tilde{\chi}_1^{\pm} \rightarrow l\nu\tilde{\chi}_1^{o}$ and $\tilde{\chi}_2^{o} \rightarrow l\bar{l}\tilde{\chi}_1^{o}$), the resulting final state is that of three isolated leptons, not all the same sign, with very little jet activity expected in an event, and large $\not{E}_t$.

The total integrated luminosity of the data set from 1992/93 run is about 19 pb^{-1}. From the expected cross sections, CDF should be sensitive to chargino and neutralino masses in the range 40-50 GeV. This means that SUSY search must allow leptons of lower transverse energy (momentum) than required for leptons from W and Z decays. The experimental strategy adopted in the CDF SUSY search is outlined below:

i) define "gold" leptons:
$\mu^{\pm}$:P_T >11 GeV, $|\eta|$ <0.6 (η-pseudorapidity)
$e^{\pm}$:E_T >11 GeV, $|\eta|$ <1.1
ii) define "ordinary" leptons:
$\mu^{\pm}$:P_T > GeV, $|\eta|$ <1.2
$e^{\pm}$:E_T >5 GeV, $|\eta|$ <2.4
iii) require at least one "gold" lepton

iv) require the presence of e^+e^- or $\mu^+\mu^-$ pair

In both "gold" and "ordinary" lepton definitions the total energy within a cone of $\Delta R < 0.4$ was required to be less than ISO<2 GeV; $\Delta R = \sqrt{(\Delta\eta)^2 + (\Delta\phi)^2}$. Leptons should be separated by at least $\Delta R_{ll} > 0.4$ from any other lepton. For unlike sign leptons a di-lepton mass cuts are imposed to remove contributions from Z^o,J/ψ and Υ production. Vertex position is required to be within 60 cm from the nominal position of $Z_{vertex} = 0$, and a cut on the azimuth opening angle between any pair of leptons was imposed, $\Delta\phi_{ll} < 170^o$.

In Table I, below, the numbers of events passing analysis cuts are summarized. These results are preliminary.

cut	$\mu^\pm$ channel	$e^\pm$ channel
original sample	2 707 852	3 677 903
dilepton data sample	28 474	34 055
dilepton data sample with bad runs removed	26 069	31 081
dilepton selection, no ISO cuts	7 120	5 709
trilepton selection, no ISO cuts	148	86
trilepton selection, ISO<4 GeV	36	14
trilepton analysis cuts:		
ISO<2 GeV	22	4
$\lvert Z_{vertex}\rvert$ <60 cm	22	4
$\Delta R_{ll} > 0.4$	2	3
$\Delta\phi_{ll} < 170^o$	2	1
$\lvert\Sigma q_i\rvert$ <3 (electric charge)	2	1
require e^+e^- or $\mu^+\mu^-$	2	1
Z^o removal (75-105 GeV/c^2)	1	0
J/ψ removal (2.9-3.3 GeV/c^2)	1	0
Υ removal (9-11 GeV/c^2)	0	0
integrated luminosity (pb^{-1})	18.09±0.72	19.11±0.76

Table 1: Numbers of events passing selection cuts in the CDF SUSY tri-lepton analysis

1.3. Detection efficiencies and backgrounds

To translate the number of observed events into the cross section limit, or the mass limits of SUSY particles, one needs to know the detection efficiencies and backgrounds. To calculate the acceptances and detection efficiencies we used in our full detector simulation Monte Carlo a MSSM model[7] with the following set of parameters:

μ=-400 and μ=-500
M_{top}=170 GeV/c^2
M($\tilde{q}$)=1.2×M($\tilde{g}$)
M(H_A)=500 GeV/c^2
A_t=0
α_s=0.12

The detection efficiencies were found to be relatively insensitive to the values of the MSSM parameters used, and have been parametrized as (M($\tilde{\chi}_1^{\pm}$) in GeV/c^2):

ϵ_{eee}=0.191·M($\tilde{\chi}_1^{\pm}$)-6.09
$\epsilon_{ee\mu}$=0.204·M($\tilde{\chi}_1^{\pm}$)-5.99
$\epsilon_{e\mu\mu}$=0.256·M($\tilde{\chi}_1^{\pm}$)-7.60
$\epsilon_{\mu\mu\mu}$=0.276·M($\tilde{\chi}_1^{\pm}$)-8.41

The principal source of background events is the Drell-Yan process, others are $b\bar{b}$, Z+X and di-boson (WZ,ZZ) events. The sum of all backgrounds in our sample of 19 pb^{-1} is estimated to be 0.75 events (0.61 events from Drell-Yan).

1.4. Interpretation of results - Mass limits

Taking 3.2 events, which is a 95% condidence upper limit using Poisson statistics for zero observed events, detection efficiencies and luminosity, and convoluting the total systematic uncertainty of 19% (taken as a Gaussian smearing) the upper limit on the cross section× branching fraction can be calculated. Figure 1 shows CDF's upper limit on σ·BF($\tilde{\chi}_1^{\pm}\tilde{\chi}_2^{o} \rightarrow 3lX$) as a smooth function of M($\tilde{\chi}_1^{\pm}$). Branching fraction of all four trilepton final states were assumed equal. There are several families of points shown for several values of tanβ values at μ=-400 (taking μ=-500 yields virtually identical results). It is clear that M($\tilde{\chi}_1^{\pm}$) < 46 GeV/c^2 is excluded by CDF data, for a set of MSSM model parameters assumed in our analysis. The corresponding sgluino mass values, M($\tilde{g}$), for M($\tilde{\chi}_1^{\pm}$) = 46 GeV/c^2 are 131,149,161 and 167 GeV/c^2 for tanβ=2,4,8 and 15, respectively.

1.5. Prospects for future analyses

CDF is presently taking new data. It is expected that by the end of the 1994/95 run the total integrated luminosity will exceed 100 pb^{-1}, optimists hope that 150-175 pb^{-1} might be possible. Simple extrapolation of the results from 1992/93 run, assuming that no new tri-lepton events are found, would raise the lower limit on a chargino mass, M($\tilde{\chi}_1^{\pm}$) >75 GeV/c^2. Other decay modes, for example same-sign di-leptons and di-leptons (for which $t\bar{t}$ production is a background !) are being analysed, and with more data the results could be interesting.

Obviously the most startling prediction of SUSY models is that Higgs may have a relatively low mass, M_H ≈130-150 GeV/c^2. The search for its signature is a primary challenge for the future analyses. In the same category is the prediction by SUSY models[3] of a light stop quark, a supersymmetric partner of a top quark, M($\tilde{t}$) ≈130 GeV/c^2. The signatures are expected to be similar to the $t\bar{t}$ events, with one important difference - the final state b-quark jets transverse energies could be significantly lower than of those coming from $t\bar{t}$ production a decay. It is likely that the presently adopted strategies for top searches have low acceptances for light stop production. A reanalysis of the lepton+jets data with lower jet E_T cuts may be required to search for the light stop. One should bear in mind, however, that relaxing the jet E_T cuts will increase contributions from QCD W+jets process which constitutes the dominant source of background. Coping with those may require an interesting but difficult analysis.

2. Top quark search at CDF

2.1. Introduction

At Fermilab, top quarks are expected to be produced in pairs, and for $M_t > 120$ GeV/c^2 the dominant production mechanism is $q\bar{q}$ annihilation, with contribution from gg fusion decreasing fast with increasing top quark mass due to the very different structure functions for quarks and gluons. Assuming Standard Model decays, the events due to $t\bar{t}$ production and decay have three distinct classes of signatures: di-leptons+2 jets (2 b-quark jets); lepton+4 jets (2 b-quark jets); and 6 jets, two of which are b-quark jets. The corresponding branching fractions are:4/81, 24/81 and 36/81 respectively, with only electron and muon leptonic decays of W taken into account.

The all hadronic jets mode is the most abundant, however, QCD backgrounds are formidable. With progress in b-tagging techniques this mode should yield the largest statistics of top candidates, however, with less favourable signal/background ratio than for the other two classes of top events, no results are available in this mode at present.

CDF searches for top quark in the di-lepton and lepton+jets channels observe a marginally significant excess of events above background expected from the known sources. After long deliberations (whether to publish or not, what title...) CDF decided to publish its findings, although the observed excess is not yet significant enough to claim a discovery. The $t\bar{t}$ hypothesis is the simpliest explanation of the excess events, because it is the one expected in the Standard Model. The analyses described in the published CDF paper[4] are summarized here.

2.2. Search for top quark in di-lepton events

If both top quarks decays leptonically, into $t \to bl^+\nu_l$, and $\bar{t} \to \bar{b}l^-\bar{\nu}_l$, the final state will be that of a lepton+lepton+$\not{E}_t$ event. The branching fractions for decay modes in this cetegory (we shall call it a di-lepton mode) are 1/81 for e^+e^- and $\mu^+\mu^-$ and 2/81 for $e^\pm\mu^\mp$ mode.

The event selection used the following cuts:

i) lepton $E_T > 20$ GeV/c, for both leptons
ii) $|\eta_l| < 1.0$
iii) at least one lepton must pass "tight cuts"; efficiency ϵ=87±1%
iv) second lepton may pass "looser cuts"; efficiency ϵ=94±1%
v) leptons must have opposite sign of their electric charge
vi) $M(e^+e^-)$ and $M(\mu^+\mu^-)$ outside of Z^o mass region, 75 GeV/c^2-105 GeV/c^2
vii) $\not{E}_t > 25$ GeV
viii) $\Delta\phi(\not{E}_t, l) > 20^o$ when $\not{E}_t < 50$ GeV
ix) $\Delta\phi(\not{E}_t, jet) > 20^o$ when $\not{E}_t < 50$ GeV
x) require two jets with $|\eta_{jet}| < 2.4$ and $E_T > 10$ GeV

The numbers of events accepted in three di-lepton modes are listed in Table 2 below. The detection efficiencies, based on detailed Monte Carlo simulation of the

Cut	$e\mu$	ee	$\mu\mu$
P_t	8	702	588
Opposite-Charge	6	695	583
Isolation	5	685	571
Invariant Mass	5	58	62
$\not{E}_t$ magnitude	2	0	1
$\not{E}_t$ direction	2	0	0
Two-jet	2	0	0

Table 2: Number of data events surviving consecutive requirements.

CDF detector, and the expected number of events in 1992/1993 data sample are listed in Table 3.

M_{top} GeV/c^2	ϵ_{DIL}	$\sigma_{t\bar{t}}$ pb	$N_{e\mu}$ events	$N_{e\mu,ee,\mu\mu}$ events
120	0.0049	38.9	2.2	3.7
140	0.0066	16.9	1.3	2.2
160	0.0078	8.2	0.8	1.3
180	0.0086	4.2	0.4	0.7

Table 3: Detection efficiencies, $\epsilon_{DIL} = BF \cdot \epsilon_{total}$, the predicted central value of $t\bar{t}$ production cross section from Ref. 5, and the number of events expected in 19 pb^{-1}, as functions of top mass.

The total background for a sample of 19 pb^{-1} is estimated to be $0.31^{+0.24}_{-0.10}$ events, the largest contributions being from WW and Z→ $\tau\tau$ production. The two $e\mu$ events observed in the data thus constitute an excess. The probability that the two candidate events are due to the background fluctuation is 12%.

2.3. Search for top in W+jets events with b-quark tag

If one of the top quarks decays leptonically, into $t \to bl^+\nu_l$, and the other hadronically, $\bar{t} \to \bar{b}q\bar{q}$, the final state will be that of a lepton+multijet+$\not{E}_t$ event. The branching fraction for this, semileptonic, decay mode is 6 times the branching fraction for the leptonic mode, - $t \to l^+\nu_l b$ and $\bar{t} \to l^-\bar{\nu}_l\bar{b}$, where l is either an electron or a muon. The backgrounds to the possible $t\bar{t}$ signal coming from the W+jets process of QCD have to be addressed. The results of analytic calculations[6] showed that in a wide range of top quark mass above $M_t \approx$120 GeV, the signal in the semileptonic W+4jet mode may actually be (depending on jet E_T cuts used) comparable to the QCD W+4jet backgrounds. In addition, CDF used in its top quark search the requirement that at least one of the jets has one of the signature expected of a b-quark jets. Two tagging techniques were used - precision vertex detector (SVX) to detect the presence of a secondary vertex in an event, and soft-lepton technique (SLT), which identifies leptons within jets.

The event selection used the following cuts:

i) lepton E_T > 20 GeV/c
ii) $\not{E}_t$ >20 GeV
iii) require at least 3 jets $|\eta_{jet}|$ <2.0 and E_T >15 GeV

There are three sources of events in the final samples: top production, W+jets QCD production, and backgrounds from other processes. The numbers of events passing the cuts are listed in Table 4 below. Acceptance for top events is slowly

Jet Multiplicity	Electron Events	Muon Events
0 Jet	10,663	6,264
1 Jet	1058	655
2 Jets	191	90
3 Jets	30	13
$\geq$ 4 Jets	7	2

Table 4: Summary of W candidate event yields as a function of observed jet multiplicity for electron+jets and muon+jets events.

increasing with top mass. Table 5 (in the end of this paper) summarizes the expected number of events from top production, as a function of top mass. Cross section calculations taken from Ref. 5.

There are 52 events in the data passing the lepton W+3 or more jets requirements. The estimates of W+jets QCD bacground, as calculated using VECBOS Monte Carlo program which incorporates results of Ref.5, vary significantly (by a factor of 2) with the choice of $< Q^2 >$ scale. To improve signal/background ratio the SVX and SLT tagging techniques were used. The SVX tagging efficiency is 0.20±0.06 (for 140 GeV/c^2 <M_t <180 GeV/c^2). In the SLT case, the corresponding efficiency is about

0.16±0.02 in the same M_t range. The probability of mis-tagging has been estimated in several ways, using data and Monte Carlo. A summary of the candidate events, together with the numbers of expected tags and background events for several values of M_t, is presented in Table 6. There are 6 SVX tagged events, and 7 SLT tagged events. Three events appear in both the SVX tagged and SLT tagged samples. The probability that the observed excess due to a background fluctuation is 3.2% and 3.8% for SVX and SLT analyses, respectively.

2.4. Interpretation of results - Cross sections

CDF observes an excess of events above the expected backgrounds. The probability that the observed excess in all three top analyses combined (di-lepton, SVX amd SLT) is merely due to a background fluctuation is 0.26%. Assuming that the excess is due to top production, one can translate the above numbers of candidates, background estimates, efficiencies and acceptances into the measurement of the $t\bar{t}$ cross section. In Table 7, the measured cross sections are listed separately for each of the three analyses, together with the combined values. In Figure 2, the combined measured cross section is compared to the theoretical calculation. The CDF measurement and the theoretical prediction cross at $M_t \approx 140$ GeV/c^2, although one should remember that the errors are very large.

2.5. Kinematical fits to W+4jet events

In a $t\bar{t}$ event with one of the W decaying leptonically and the other hadronically there is only one parton, a neutrino, which cannot be measured. However, if one measures the lepton and all four jets in an event, there are enough kinematical constraints to perform a fit to a hypothesis that a given W+4jet event is due to $t\bar{t}$ production and decay. In the sample of 10 W+3 (or more) jets there are 7 events satisfying a more relaxed cut on the 4-th jet transverse energy, $E_T > 8$ GeV. Assuming that those events are of $t\bar{t}$ origin, a fit using SQUAW (an old bubble chamber fitting program) has been performed for the 7 W+4jets events, assigning the tagged jet as a b-jet in each case. Taking for the background distribution that obtained from similar fits to VECBOS W+jet events, the 7 tagged events have been found to agree best with a sum of $t\bar{t}$ and W+jets QCD background production, pointing to $M_t = 174 \pm 10^{+13}_{-12}$ GeV/c^2. The $t\bar{t}$ cross section for this value of M_t is $13.9^{+6.1}_{-4.8}$ pb.

The measured cross section value is larger than the theoretical prediction, which by itself is not a reason to be concerned, especially since the errors are very large. A statistical analysis of the CDF cross section measurement, CDF mass measurement and the theoretical cross section has found them compatible at an 11% confidence level. One should keep in mind, however, that the mass value was obtained by assuming that the 7 events are $t\bar{t}$ events, which may not be true. An unexpected result was found in the sample of Z+3 (or more) jets events. Two events were found to be SVX tagged, while the expected background is 0.31 events. No events were SLT tagged. Combining the SVX and SLT tagging techniques CDF observes 2 tagged Z+3jets (or more) events with the expected background of 0.64±0.06 events. This

M_t (GeV/c^2)	120	140	160	180
$\sigma(t\bar{t})$ (pb) (theory)	$38.9\pm^{10.8}_{5.2}$	$16.9\pm^{3.6}_{1.8}$	$8.2\pm^{1.4}_{0.8}$	$4.2\pm^{0.6}_{0.4}$
A_{top}	0.051 ± 0.007	0.067 ± 0.008	0.077 ± 0.008	0.081 ± 0.008
Expected # of events	39 ± 6	22 ± 3	12 ± 2	6.6 ± 0.7

Table 5: A_{top} is the acceptance for top events, including branching ratios, for the lepton+3 (or more) jets selection. The expected number of events is calculated with the central value of the theoretical expectation for $\sigma_{t\bar{t}}$. It includes a 3.6% normalization uncertainty on the luminosity of but no theoretical uncertainty.

Channel:	SVX	SLT	Dilepton
Expected # events $M_{top} = 120$ GeV/c^2	7.7 ± 2.5	6.3 ± 1.3	3.7 ± 0.6
Expected # events $M_{top} = 140$ GeV/c^2	4.8 ± 1.7	3.5 ± 0.7	2.2 ± 0.2
Expected # events $M_{top} = 160$ GeV/c^2	2.7 ± 0.9	1.9 ± 0.3	1.3 ± 0.1
Expected # events $M_{top} = 180$ GeV/c^2	1.4 ± 0.4	1.1 ± 0.2	0.68 ± 0.06
Expected Bkg.	2.3 ± 0.3	3.1 ± 0.3	$0.56^{+0.25}_{-0.13}$
Observed Events	6	7	2

Table 6: Numbers of $t\bar{t}$ events expected, assuming the theoretical production cross sections according to Ref.5, and the numbers of candidate events observed with expected backgrounds.

M_{top}	120 GeV/c^2	140 GeV/c^2	160 GeV/c^2	180 GeV/c^2
$\sigma^{SVX}_{t\bar{t}}$(pb)	$21.5^{+17.2}_{-11.3}$	$14.9^{+12.2}_{-7.9}$	$13.0^{+10.6}_{-6.9}$	$12.4^{+10.0}_{-6.5}$
$\sigma^{SLT}_{t\bar{t}}$(pb)	$28.9^{+18.5}_{-13.3}$	$22.7^{+14.3}_{-10.4}$	$20.4^{+12.7}_{-9.3}$	$18.8^{+11.7}_{-8.6}$
$\sigma^{DIL}_{t\bar{t}}$(pb)	$15.2^{+19.5}_{-12.2}$	$11.3^{+14.2}_{-9.0}$	$9.6^{+12.0}_{-7.6}$	$8.8^{+11.0}_{-7.0}$
$\sigma^{ALL}_{t\bar{t}}$(pb)	$22.7^{+10.0}_{-7.9}$	$16.8^{+7.4}_{-5.9}$	$14.7^{+6.5}_{-5.1}$	$13.7^{+6.0}_{-4.7}$

Table 7: $t\bar{t}$ production cross sections calculated from the individual analyses and from the combination of the three analyses.

may indicate an unknown source of boson+jets events, or another, yet unaccounted for, source of background.

2.6. Future plans

A precise measurement of M_t would allow more meaningful consistency checks of the Standard Model, using all available data (low energy, LEP, Tevatron...). This would provide an indirect measurement of the Higss mass. There is also a, long sought after, possibility that an inconsistency will appear, indicating new physics beyond the Standard Model. With more than 100 pb^{-1} expected to be available by next fall, the excess of events seen by CDF, if confirmed, will be very pronounced. The top quark mass could be measured with a precision of about 10 GeV/c^2. New analyses, for example more sophisticated kinematical techniques[7], should allow to study W+4jets events without tagging requirement, potentially increasing the top candidates sample by more than a factor of 2. D0, the second collider experiment at Tevatron, has published their result on top quark search recently. They observe a much smaller excess of events (if any), as you could learn at this conference[8]. There exist a possibility that an excess of events seen by CDF is a fluctuation, or that the events are not all due to $t\bar{t}$ production. The answer is in the new data.

3. Ackowledgements

I would like to thank the organizers of the Warsaw Workshop for their kind hospitality offered to me at Warsaw University during this very well organized conference. It was a particularly memorable experience for me, since it was my first longer visit to Warsaw University in more than 13 years.

4. References

1. Yu.A. Gelfand and E.P.Likhtman, JETP *Lett.* **13** (1973) 323.; D.V. Volkov and E.P. Akulov, *Phys. Lett.*46B (1973) 109; J. Wess and B. Zumino, *Nucl. Phys.*B70 (1974) 39.
2. for reviews see: H.P. Niles, *Phys. Rep.*110(1984) 1; H.E. Haber and G.L. Kane, *Phys. Rep.*117(1985) 75; S. Dawson, E. Eichten and C. Quigg, *Phys. Rev.*D31(1985) 1581.
3. J.L. Lopez, D.V. Nanopoulos and A. Zichichi, CERN-TH.7296/94.
4. F. Abe *et al*, submitted to Phys. Rev. **D**(1994).
5. E. Laenen, J. Smith and W.L. van Neerven, *Phys. Lett.* **321B** (1994) 254.
6. F.A. Berends, W.T. Giele, H. Kuijf and B. Tausk, *Nucl. Phys.* **B357**(1991) 32.
7. G. R. Goldstein, K. Sliwa and R. H. Dalitz, *Phys. Rev.*D47, (1993) 967; K. Kondo *et al.*, *J. Phys. Soc. Japn.*62, (1993) 1177;
8. A. White, D0 Collaboration, these proceedings.

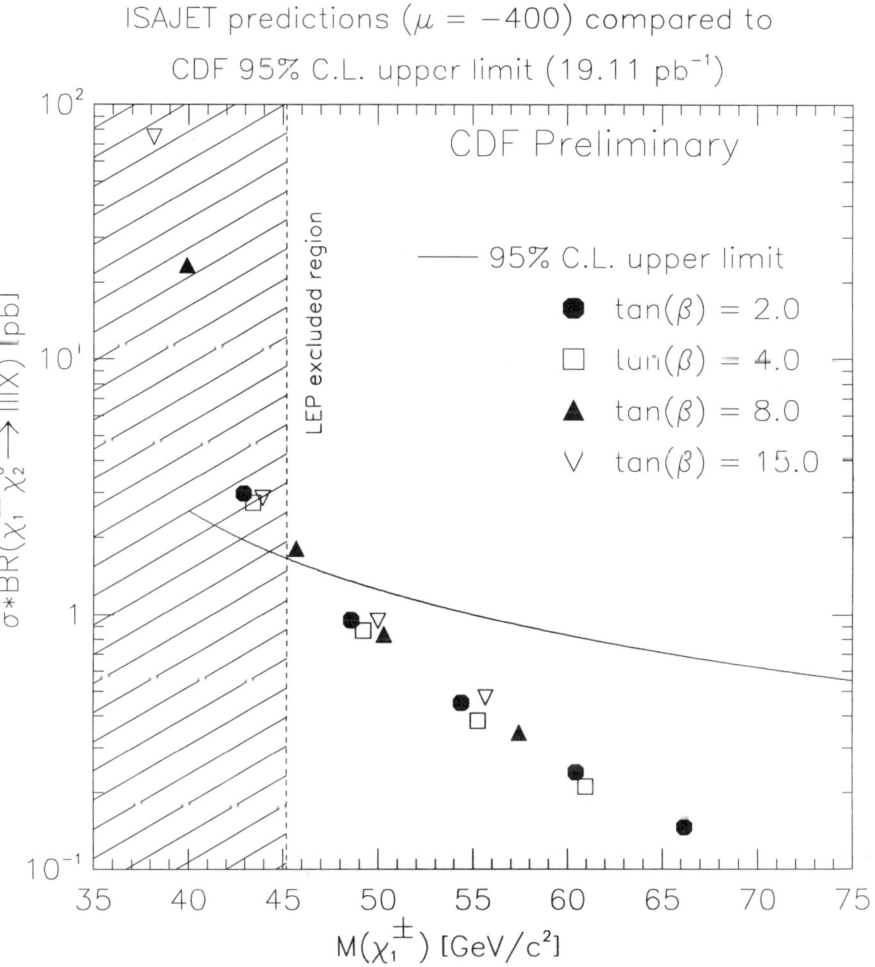
ISAJET predictions (μ = −400) compared to
CDF 95% C.L. upper limit (19.11 pb⁻¹)
CDF Preliminary
LEP excluded region
95% C.L. upper limit
tan(β) = 2.0
tan(β) = 4.0
tan(β) = 8.0
tan(β) = 15.0
σ*BR(χ₁±χ₂⁰ → lllX) [pb]
M(χ₁±) [GeV/c²]
10²
10¹
1
10⁻¹
35
40
45
50
55
60
65
70
75

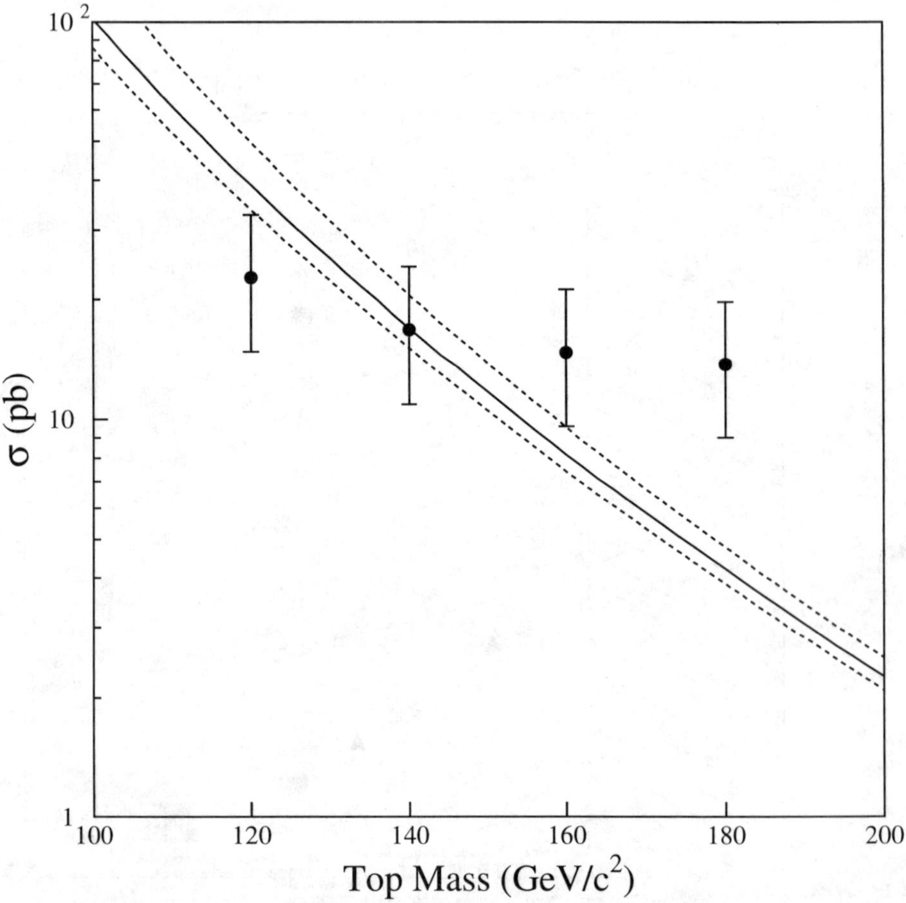
10 2
10
1
σ (pb)
100
120
140
160
180
200
Top Mass (GeV/c2)

Searches for Supersymmetry and Top in the D0 Experiment

Andrew P. White
Department of Physics, University of Texas at Arlington
Arlington, Texas 76019, USA

for

The D0 Collaboration

ABSTRACT

The statuses of the searches for evidence of supersymmetry and the top quark in the DØ experiment at Fermilab are presented. New limits are given on the masses of squarks and gluinos, and progress is reported in the search for charginos and neutralinos. Finally, results are presented on the search for the top quark in dilepton, lepton plus jet, and lepton plus jet plus tagging muon channels.

1. Introduction

Two of the most challenging areas of particle physics currently concern the search for evidence of supersymmetric particles, and the search for the top quark. Evidence for supersymmetric particles is needed to give credence to supersymmetry as a theoretically attractive way to stabilize the masses of the scalar sector and explain the origin of electroweak symmetry breaking. The top quark is required to complete the basic set of fermions in the Standard Model. There is considerable interest in the interconnection between the expected large mass of the top quark and the role it plays, through renormalization group equations, in driving negative the mass squared of one of the Higgs particles of the minimal supersymmetric standard model(MSSM). The negative mass squared is required to generate electroweak symmetry breaking through the usual form of the Higgs potential.

The supersymmetric particle searches described in this paper use the hermetic properties of the DØ detector to look for evidence of squarks and gluinos via their decay to channels with signatures of jets and missing transverse energy. The wide coverage of the DØ detector for charged leptons is used in the search for charginos and neutralinos via their decays to channels involving electrons and muons. Essentially all elements of the detector are used in the top search which involves jets, charged leptons, and missing transverse energy.

The approximately 15 pb^{-1} of data used were taken during Run Ia of the Tevatron collider in 1992-93.

This paper is organized as follows. First the essential elements of the DØ detector are described, then the results of the searches are presented, followed by conclusions.

2. The DØ Detector

The DØ detector, the trigger scheme, and the procedures for reconstructing jets, electrons, muons, and missing transverse energy have been fully described elsewhere[1 2].

We give here only a brief description of the elements of the detector. The three main subsystems of the detector are the central tracking system, the almost hermetic liquid argon calorimeter system, and the muon measurement system. The approximate ranges of coverage of these subsystems are

Central tracking	$\lvert\eta\rvert < 3.5$
Calorimetry	$\lvert\eta\rvert < 4.0$
Muon system	$\lvert\eta\rvert < 3.3$

The main components of the events studied here are charged leptons (electrons and muons), jets, and missing transverse energy ($\not\!\!E_T$). Muons were identified by their penetration through the calorimeter and muon toroid magnet iron, and by energy deposition, in a localized region of the calorimeter, consistent with the passage of a minimum ionizing particle. Various types of muon selection criteria were applied and the details are given in the appropriate analysis sections below.

Electrons were identified by their pattern of longitudinal and transverse energy deposition in the calorimeter. They were also required to have a matching track in the central tracking chambers. The electron energies were typically corrected by about 5% from a normalization of the masses of reconstructed Z bosons to the precise value from LEP.

Jets were reconstructed from the calorimeter data using a cone of radius 0.5 in η-ϕ space. Corrections were applied to the jet energies using a scale determined from a set of events in which a single jet was balanced by a photon. The size of the correction was typically about 25% depending on the jet transverse energy.

The missing transverse energy, $\not\!\!E_T$, was calculated as the negative of the vector sum of the individual calorimeter cells. A correction to $\not\!\!E_T$ for the momenta of any muons in an event was applied for some of the analyses described below.

3. Search for Squarks and Gluinos

For the masses of squarks and gluinos considered here it is essential to include the full range of cascade decays through intermediate neutralinos and charginos to final states that will contain many quarks and gluons. Since the lightest supersymmetric particle (LSP) does not interact, and thus produces $\not\!\!E_T$, and we wish to reduce backgrounds from W and Z production, we determined our basic experimental signature to be 1) significant $\not\!\!E_T$, 2) three or more jets, and 3) no charged leptons. Besides W and Z boson production there was expected background from mismeasured QCD multijet events.

Certain assumptions were made related to the MSSM for this analysis. All squarks (except the scalar top squark) were considered mass degenerate; the squarks were assumed to be lighter than the sleptons (which were not generated); finally, we chose the following MSSM parameter values to generate the decay tables from ISASUSY[3], the supersymmetry extended version of ISAJET:

$$\begin{aligned} \tan\beta &= 2.0 \\ \mu &= -250 \text{ GeV} \\ m(H+) &= 500 \text{ GeV} \end{aligned}$$

The trigger and first pass offline filter resulted in 9625 events for further analysis . The table below then gives the reductions in event numbers for each of the subsequent stages of analysis.

Trigger/offline filter	9625
Single interaction	3730
$\not\!\!E_T$ > 75 GeV	107
3 jets $E_T > 25$ GeV	47
Leading jet not in ICR	45
Jet-$\not\!\!E_T$ correlation	30
Reject charged leptons	25
Reject noise jets	17
Hand scan	14

Table 1. Squark/gluino analysis results

The single interaction requirement was made to avoid events from beam crossings with more than one event in which the $\not\!\!E_T$ was evaluated using the wrong vertex. From a study of the signal and background characteristics we were able to set a lower cut of 75 GeV on the $\not\!\!E_T$, and to requıre at least three jets in the passing events with a minimum of 25 GeV per jet. Other cuts were then made to supress known sources of background. The overlap, or "intercryostat" region (ICR) , between the central and end calorimeters systems, contains substantial dead material which can create an underestimate of the energy of a jet in this region. For this reason we did not allow events to pass that had their leading jet in this region. Also some events had erroneously high $\not\!\!E_T$ due to fluctuations of jet energy measurement in the calorimeter. However, for these events there tends to be an angular correlation between the jets and the $\not\!\!E_T$. We therefore rejected events with the $\not\!\!E_T$ nearly parallel or nearly opposite any of the three leading jets, and events where the fluctuations of the second leading jet could have masked the correlation with the first leading jet. Supression of W,Z boson related backgrounds was achieved by removing events with electrons having E_T > 20 GeV , or muons having p_T > 15 GeV . At this stage we also required all jets in the remaining events to have E_T of at least 15 GeV to supress spurious jets from noisy calorimeter cells. Finally a hand scan removed one event with a cosmic-ray muon, and two events where the wrong vertex had been reconstructed. There then remained a sample of 14 events for comparison with predictions of signals and backgrounds.

The background level expected from W and Z boson events was estimated using the VECBOS/ISAJET[4,5] event generator combination in conjuction with our GEANT[6] based simulation of the DØ detector. We expect a total of $16.7 \pm 1.7(^{+7.0}_{-6.6})$ events from this source.

The background from QCD multijet events was estimated from a fit to the $\not\!\!E_T$ distribution for a sample of events for which the trigger required one jet with $E_T >$ 20 GeV. The result was an expected background of 0.42 ± 0.37 events. The final background total was then $17.1 \pm 1.8(^{+7.0}_{-6.6})$ events which is consistent with the 14 events in the final candidate sample. We therefore see no excess of events above background that could be attributed to the production of squarks and gluinos.

On the basis of this result we can then set lower limits on the masses of squarks and gluinos in the m($\tilde{q}$) - m($\tilde{g}$) plane. Since the detection efficiency varies as a function of the squark and gluino masses, we determined this efficiency for a grid of points in the m($\tilde{q}$) - m($\tilde{g}$) plane and interpolated between the points for intermediate values.

We were then able to set 95% confidence level limits in this plane using the production cross-sections from ISASUSY. The results are shown in Fig. 1. For heavy squarks we set an asymptotic lower limit on m($\tilde{g}$) of 144 GeV/c^2, and a limit of 212 GeV/c^2 for the case of equal masses.

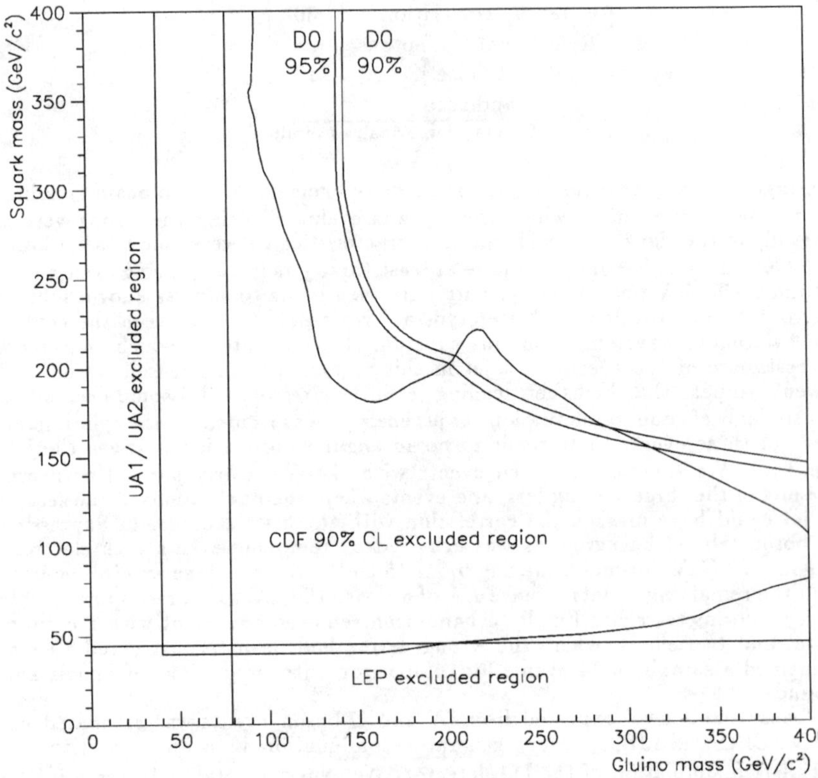

Fig. 1. Current squark and gluino mass limits. Shown are the D0 90% and 95% confidence level exclusion contours, together with the most recent CDF limit [7].

4. Search for Charginos and Neutralinos

An interesting, and complimentary, way to search for supersymmetry at the Tevatron is via the associated production of the lightest chargino($\tilde{W}_1$) and next to lightest neutralino($\tilde{Z}_2$)[8]. The interest stems from the expectation that the rate of associated production times decay branching ratio into final states with three charged leptons (electrons and/or muons) can be large for a range of MSSM parameters. In addition most events are expected to show little hadronic activity for a wide range of η. We have therefore searched for all combinations of three electrons and muons in the final state plus a limited amount of $\not\!\!E_T$ from unseen neutrino(s) and the LSP's. The practical difficulties of this search center around the relatively low p_T expected for the softest charged lepton (especially for low mass charginos and neutralinos), and the large number of potential background processes.

The triggers and offline selection criteria varied somewhat for the different electron muon final state combinations. We give here one example - for the $ee\mu$ channel. The triggers for this channel consisted of one electromagnetic object with p_T above 7 GeV or one muon with p_T above 5 GeV and $|\eta|$ below 2.4. Offline we required two electrons with p_T above 5 GeV , $|\eta|$ below 2.5 satisfying requirements on shower shape and isolation from other energy deposition in the calorimeter. Only one electron was allowed to have $|\eta|$ more than 1.7 to reduce background from converting photons. The muon was required to reconstruct with p_T above 5 GeV , $|\eta|$ less than 1.7, and to satisfy quality criteria: a "gold" quality muon had the expected energy deposition in the calorimeter for a minimum ionizing track, a matching central detector track, a minimum separation in η-ϕ space of 0.5 from any jet, and satisfied impact parameter criteria in both the bend ($<$ 15 cm.) and non-bend ($<$ 30 cm.) views.

At the time of writing this paper, the status of the analysis was that no candidate events had been found that survived all the selection criteria in approximately 12 pb^{-1} of data for the eee, $ee\mu$, and $e\mu\mu$ channels, and 9 pb^{-1} of data for the $\mu\mu\mu$ channel.

Many potential background processes have been studied. The main classes of backgrounds are Drell-Yan processes producing two charged leptons and either a photon or jet which can fake the third lepton, QCD processes in which there are heavy quarks decaying to charged leptons and/or jets faking leptons directly, and W and Z boson processes producing charged leptons in association with jet(s) that also fake lepton(s). The total background from all sources is currently estimated at approximately 2 events.

The final ingredient needed to set limits on $\tilde{W}_1$ $\tilde{Z}_2$ production and decay is the set of efficiencies for the four final states. These efficiencies are being estimated from a combination of data and Monte Carlo events. Events generated in ISASUSY are processed through a simulation of the trigger and reconstructed via the standard DØ offline reconstruction program D0RECO. Electron efficiencies are estimated by overlaying electrons from a very detailed simulation of the DØ calorimeter with real minimum bias data events. For the muons most of the elements of the fficiency calculation were taken from data samples of $Z \rightarrow \mu\mu$ and $J/\psi \rightarrow \mu\mu$.

The final results of this search for the Tevatron Run Ia data are expected to be published early in 1995.

5. Search for the Top Quark

The production of top/anti-top quark pairs, at the current center of mass energy of 1.8 TeV at the Tevatron, is expected to result from quark/anti-quark fusion. The top quark is expected to decay (assuming no light charged Higgs or other similar object) into a W boson and a b-quark. The subsequent decays of the W and b then lead to a number of possible final states. The W can decay to either jets or an electron or muon (plus neutrino), while the b quark gives further jet(s) and possibly a muon from semi-leptonic decay. We therefore divide our search into the following sub-categories:

- Dilepton - 2 high p_T central isolated leptons, $\not\!\!E_T$, and jets, accounting for approximately 5% of top decays and giving the cleanest signatures.

- Lepton plus jets - 1 high p_T central isolated lepton, $\not\!\!E_T$, and at least four jets, accounting for approximately 30% of top decays.

- Lepton plus jets plus tagging muon. As for the previous category, but also requiring a low p_T muon or a muon in a jet to enhance the signal by enriching the number of events containing b quarks.

Recently the DØ collaboration set a lower limit on the top mass of 131 GeV [2], and CDF announced[9] evidence for a top quark of mass $174 \pm 10^{+13}_{-12}$ GeV . Since then, with the knowledge that the top quark is heavy, we have re-optimized our search in order to reduce backgrounds, use the results of better background estimates, and we have extended the analysis to include the $\mu\mu$ and e+jets+μ tag channels.

5.1. Dilepton search

The basic requirements of this analysis are two isolated high p_T charged leptons (any combination of e and μ), a large $\not\!\!E_T$, and a minimum of two jets. This last requirement was previously one jet, but it was realized that a significant enhancement of signal to background can be achieved with respect to the WW and $Z \to \tau\tau$ processes by requiring at least two jets.

Electron candidates were identified as energy clusters in the electromagnetic calorimeters. Selection criteria for electrons consisted of a small energy leakage into the hadron calorimeter, correct longitudinal and transverse shower profiles, a track match with the inner chamber, isolation from other energy deposits, and correct $d\mathrm{E}/dx$ energy deposition in the tracking chambers. An $|\eta|$ cut of 2.5 was also applied. The electron selection efficiency varied from 45% to 80%.

Muons were identified as tracks in the muon drift chambers after successfully traversing the material in the calorimeter system and the muon toroid iron. Muon seletion criteria consisted of a minimum ionizing energy deposition in the calorimeter, a spatial match with the an inner chamber track, consistency with the interaction vertex, and isolation from other energy deposits in the calorimeter. A cut of $|\eta| < 1.7$ was applied to the muon candidates. The muon selection efficiency varied from 70% to 85%.

The dilepton analysis selection criteria are given in the table below.

	$e\mu$	ee	$\mu\mu$
Lepton p_T	$> 15/12$	$> 20/20$	$> 15/15$
$\not\!\!E_T$	> 10	> 25	-
No. of jets	$>= 2$	$>= 2$	$>= 2$
Jet E_T	> 15	> 15	> 15

Table 2. Dilepton event selection

Additional cuts were applied to remove events with a Z boson in the ee channel, and to reject cosmic ray muons in the $\mu\mu$ channel.

Application of the cuts and selection criteria resulted in one candidate event in the $e\mu$ channel, and no candidates in the ee or $\mu\mu$ channels. The physics backgrounds to the dilepton channels arise from heavy quark jets (b and c), WW and WZ pairs, and Z and continuum Drell-Yan production. Background events also resulted from jets being misidentified as electrons. The level of this background was estimated from data.

The results of the dilepton search are summarized in the table below:

	$e\mu$	ee	$\mu\mu$
Luminosity	13.5 ± 1.6	13.5 ± 1.6	9.8 ± 1.2
Data	1	0	0
Background	0.27 ± 0.09	0.16 ± 0.07	0.33 ± 0.06

Table 3. Dilepton event yields

These results will be combined with the results from the other channels at the end of this section.

5.2. Leptons plus Jets Search

The requirements for a candidate event in this search were one high p_T electron or muon, high jet multiplicity (defined in detail below), and large $\not{E}_T$. Although the combined electron and muon plus jets channels represent about 30% of the top decays, there are potentially large backgrounds from high multiplicity W boson plus jets events, and QCD events in which jet fluctuations can lead to fake electrons and/or spurious $\not{E}_T$. Two separate analysis paths were followed to reject the backgrounds: event shape criteria and soft muon tagging. The former analysis is discussed in this section, and the latter in the next section.

The electron plus jets selection consisted of a "tight electron", corrected $\not{E}_T$ > 25 GeV, at least four jets (R = 0.5) with E_T > 15 GeV, and no muon tag. The last requirement was imposed to maintain independence between this search and the tagging search. The "tight electron" criteria included shower shape cuts (transverse and longitudinal), matching between the shower and a track in the central tracking chambers, and a dE/$d$xcut to reduce background from photon conversions. Finally, electron candidates were required to have $|\eta| < 2.0$.

The muon plus jets selection criteria consisted of a "standard isolated muon", corrected E_T > 20 GeV, at least four jets (R = 0.5) with E_T > 15 GeV, and no muon tag. A 'standard isolated muon" was required to have p_T > 15 GeV, to pass a cosmic ray veto, to satisfy track quality cuts, to be confirmed in the calorimeter, to be separated from any jets with E_T > 15 GeV by $\delta R > 0.5$, and to have less than 5 GeV of energy in an annular cone $0.2 < \delta R < 0.4$ around the muon track. Additionally, only muons with $|\eta| < 1.7$ were accepted.

The signal and background levels present in the data were estimated by two independent methods. In the first method a "jet scaling" hypothesis was assumed for the W plus jets background: the addition of an additional jet was assumed to reduce the cross-section by a constant factor (depending only on the strong coupling constant and the minimum jet E_T). Then an exponential extrapolation can be made from one and two jets to four jets, and any excess observed in the high jet multiplicity data can be attributed to a top signal. The extrapolation was carried out after the subtraction of the QCD multijet background, which was estimated from the data.

In the second method cuts were made on the aplanarity of of jets in the laboratory frame, and on the scalar sum of the jet E_T 's. The two methods yielded very similar estimates of the backgrounds and errors. The results of this search are summarized below:

	$e+$ jets	$\mu+$ jets
Luminosity	13.5 ± 1.6	9.8 ± 1.2
Data	2	2
Background	1.2 ± 0.7	0.6 ± 0.5

Table 4. Lepton plus jets yields

5.3. Electron plus Jets plus Soft Muon Tag Search

Since each b quark from a top decay will decay semileptonically with a probability of 11% and since each charm quark from the b decays can also decay semileptonically with the same probability, then about 44% of top events will contain a soft muon associated with a b jet. With the cuts used in our analysis about 20% of the top events will have a detected soft muon.

The rate of tagging per jet was determined to be 0.5% from a study of an electron plus jets trigger sample where the electron was a "fake". Cross checks were made with other data samples and the 0.5% per jet tagging rate was found to be the same within errors. The soft muon tag requirements were either one muon with $p_T > 4$ GeV, or one non-isolated muon if the muon p_T was above 12 GeV . The details of the analysis before muon tagging were essentially the same as for the jet multiplicity analysis in the previous section, except that only three jets were required instead of four. After muon tagging two events passed all the cuts and the background was estimated to be 0.6 ± 0.2 events.

5.4. Combined Results of Top Searches

If all the results from the top searches described above are combined we observe 7 events with an estimated background of 3.2 ± 1.1 events. The probability that the background fluctuated to give the observed signal is 7.2%. If the observed excess of events over the estimated background is attributed to top production, the top cross section may be calculated if the top mass is specified. The results of such a calculation are plotted in Fig.2. as a function of the top quark mass. Also shown is the CDF measurement. The DØ result is consistent with a top quark mass above 141 GeV . It is also consistent with the CDF measurement, but does not demonstrate the existence of the top quark.

6. Conclusions

The DØ experiment has searched for evidence of supersymmetric squarks and gluinos, charginos and neutralinos, and for the existence of the top quark. In the case of squarks and gluinos we have significantly extended the previous CDF limits. The chargino and neutralino search is almost complete and will yield final results early in 1995. Further significant extensions of these searches are also expected with the greater than 100 pb^{-1} of data to be accumulated during the present Run Ib of the Tevatron.

The top search resulted in a small excess of events above background that could be attributed to top production. However, the significance of the excess is small and the additional data from Run Ib will be necessary to further investigate the existence of the top quark.

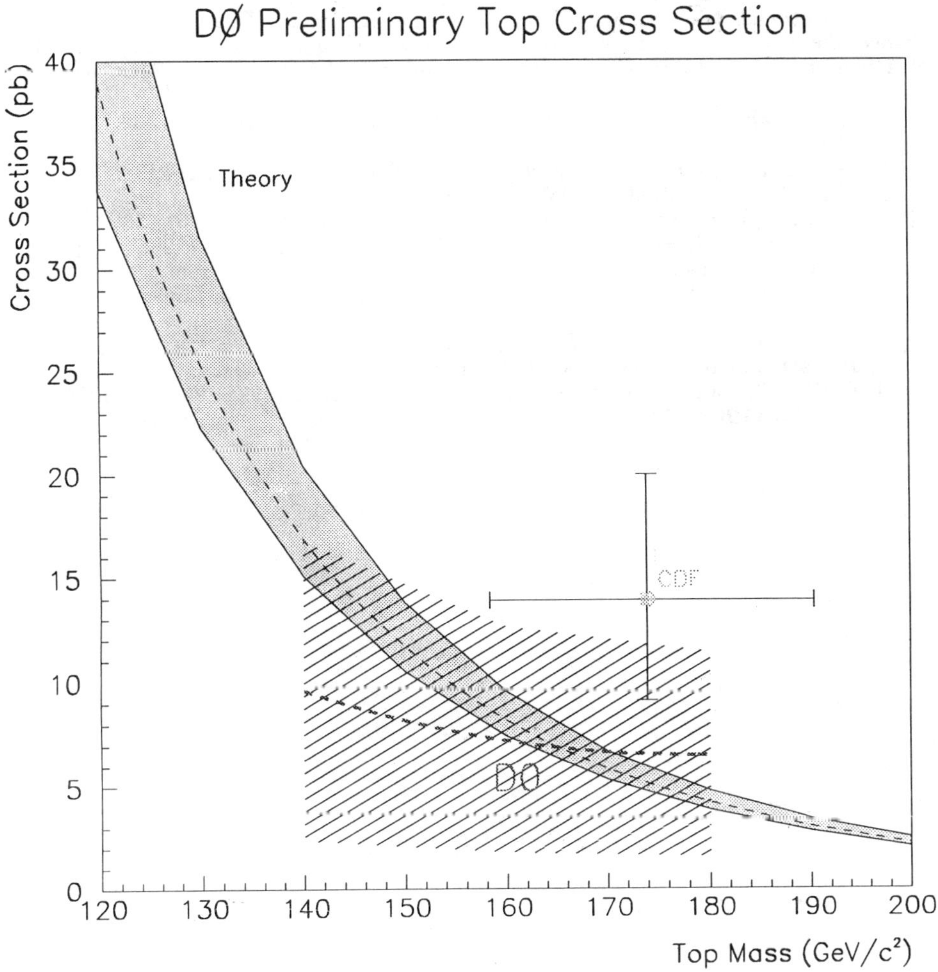

Fig. 2. Preliminary top quark mass limits.

7. Acknowledgements

We thank the Fermilab Accelerator, Computing, and Research Divisions, and the support staffs at the collaborating institutions for their contributions to the success of the experiment. We also acknowledge the support of the U.S. Department of Energy, the U.S. National Science Foundation, the Commissariat a L'Energie Atomique in France, the Ministry for Atomic Energy in Russia, CNPq in Brazil, the Department of Atomic Energy in India, Colciencias in Colombia, and CONACyT in Mexico.

1. DØ Collaboration, S. Abachi *et al.*, *Nucl. Instrum. Methods* **A338**, (1994) 185.
2. DØ Collaboration, S. Abachi *et al.*, *Phys. Rev. Lett.* **72**, (1994) 2138.
3. "The ISASUSY Monte Carlo" by H. Baer and X. Tata.
4. W. Giele, E. Glover, and D. Kosower, *Nucl. Phys.* **B403**, (1993) 633
5. F. Paige and S. Protopopescu, *BNL Report no. BNL 38034* (1986) (unpublished), release v.6.49.
6. F. Carminati *et al.*, *"GEANT Users Guide"* CERN Program Library, December 1991, (unpublished).
7. F. Abe *et al.*, *Phys. Rev. Lett.* **69** (1992).
8. P. Nath and R. Arnowitt, *Mod. Phys. Lett.* **A2**, (1987) 331
9. CDF Collaboration, F. Abe *et al.*, *Phys. Rev.* **D50** (1994) 2966, and *Phys. Rev.* **73 (1994) 225.**

SEARCHES FOR SUSY HIGGS BOSONS IN CMS

RITVA KINNUNEN
CMS COLLABORATION
Research Institute for High Energy Physics,
University of Helsinki, Siltavuorenpenger 20 C
Helsinki, Finland

ABSTRACT

Possibilities do detect the Higgs bosons of the Minimal Supersymmetric Standard Model (MSSM) in the CMS experiment, currently under design, are discussed.

1. Introduction

The main challenge of the of the experiments at LHC will be to discover the origin of the spontaneus symmetry breaking mechanism, either Higgs bosons or new phenomena in the weak boson scattering cross sections. The resent fits to all available precision electroweak data[1)], combined with the top quark mass from CDF[2)] lead to the lower bound of Higgs mass of 63.5 GeV. The upper bound has been calculated to be 1.1 TeV in ref.3[3)]. The CMS detector has been designed to allow investigations over this entire mass range, from the upper mass reach of LEPII at $m_H \simeq$ 80-90 GeV, up to $m_H \simeq 1$ TeV. There is a slight preference in favour of the region $m_H \simeq$ 80-400 GeV. Also the fits could be interpreted to prefer a SUSY scenario rather than the Standard Model (SM)[1)].

The Higgs sector of the minimal supersymmetric standard model (MSSM) contains one charged ($H^{\pm}$), two CP even (h, H) and one CP odd (A) neutral states. Higgs boson masses and couplings are normally expressed in terms of m_A and $tan\beta$. The decay channels of the MSSM neutral Higgs bosons are essentially the same as for the Standard Model Higgs, but the production rates may be significantly modified by MSSM couplings. There are regions of MSSM parameter space where extra decay channels like $A \to Zh$ or $H \to hh$ are not negligible and may be used for Higgs boson detection.

At lowest level the mass of the light scalar is bounded by $m_h < m_Z$. Radiative corrections modify the masses and couplings. At present, only leading (one-loop) corrections have been fully calculated and made available. Results on next-to-leading order corrections are not entirely consistent[4−6)]. In the following the 1-loop approximation is used to calculate masses, couplings, widths and branching ratios[7)]. The mass vs tanβ relations we use are shown in fig.1. The charged Higgs $H^{\pm}$ is always heavier than W and in most of the parameter space it is significantly more massive, with $H^{\pm}$, A, H^0 almost mass degenerate at large m_A. For the new CDF value of the top mass, $m_{top} = 174$ GeV[2]), LEPII may be put outside the reach of m_h and puts additional emphasis for the SUSY Higgs searches at LHC.

There are two main production processes for MSSM neutral Higgs bosons (ϕ): $gg \rightarrow \phi$ and $gg \rightarrow \phi b\bar{b}$. The second mechanism is important for large $tan\beta$ due to the enhanced $\phi b\bar{b}$ couplings. These associated production processes are interesting as b-tagging techniques open possibilities to enhance the signal to background ratio[8]).

Events are simulated in most cases with PYTHIA[9]) using the recent CTEQ2L[10]) structure functions and taking $m_{top} = 174\ GeV$. The new version produces correctly the cross sections obtained in the analytical calculations[11]). There is still a theoretical uncertainty of ~ 2 in the estimate of $gg \rightarrow \phi b\bar{b}$ process[7]).

A short introduction to the CMS detector is presented in sec. 2. Study of various SUSY Higgs searches in CMS are presented in secs. 3-7.

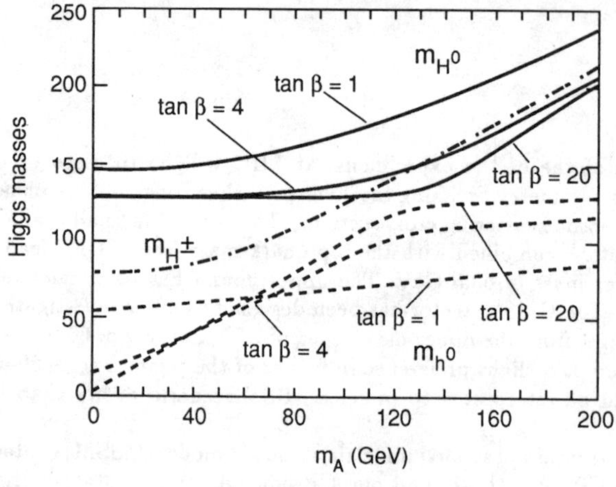

Fig. 1. Masses of the MSSM Higgses versus m_A for various $tan\beta$ values.

2. CMS detector

The CMS (Compact Muon Solenoid) detector is designed to measure precisely muons, electrons, photons, jets and missing energy at LHC[12,13]). A large superconducting solenoid magnet provides a high magnetic field of 4 Tesla and a compact design concept. The basic components of the detector are the central tracker, PbWO4 crystal electromagnetic calorimeter, hadronic calorimeter, the magnet and the muon chambers. A microvertex device is also included to measure secondary vertices and track impact parameters.

The rapidity coverage of the different detectors are shown in Fig.2. For of the tracker, the crystal calorimeter and the muon chambers it is up to $|\eta| = 2.4$. For jet and missing energy measurement the hadronic calorimeters are extented to $|\eta| = 4.6$. Muons are measured and identified with the muon chambers and with the tracker providing a momentum resolution $\Delta p/p < 2\%$ for $p_t < 100\ GeV$. Good intrinsic resolution and high granularity of the crystal calorimeter gives an energy resolution $< 1\%$ for electrons and photons with $E_t \geq 30$ GeV. The impact parameter resolution for b - and τ - tagging, determined by the performance of the microvertex detector and the tracker, is $\leq 30\mu m$ for $p_t > 10\ GeV$[14]. Details on the CMS detector and its expected performance can be found in refs. 12 and 13[12,13].

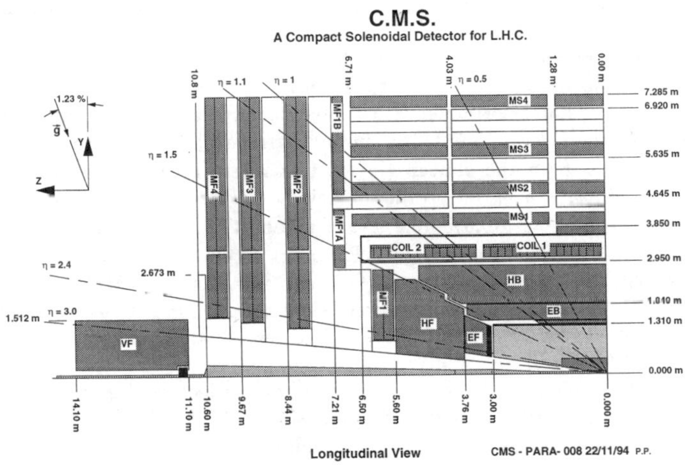

Fig. 2. CMS detector: the rapidity acceptances for different detector components.

3. $h, H \rightarrow \gamma\gamma$

In the MSSM the neutral Higgs remains extremely narrow in the kinematical regions for which the two photon decay has a useful branching ratio. Thus the mass resolution and geometrical acceptance of the electromagnetic calorimeter are of decisive importance. The background for $H \rightarrow \gamma\gamma$ signal may be devided into three categories: i) prompt di-photon production from the quark annihilation and gluon fusion diagrams, ii) prompt di-photon production due to bremsstrahlung from the outgoing quark line in the QCD Compton diagram, and iii) the background from jets, where the electromagnetic energy deposit results from the decay of neutral hadrons in a jet and from one jet plus one prompt photon processes.

The jet background can be reduced by imposing an isolation cut which also reduces the bremsstrahlung background[15]. Photon is defined isolated if there is no charged track or an electromagnetic shower with a transverse momentum greater than 2.5 GeV within a square region $\Delta = \pm 0.3$. With this isolation requirement the loss due to pile-up is at the acceptable level (12% at $10^{34} cm^{-2} s^{-1}$). The photons from the decays of π^0s are separated in calorimetry by a lateral distance of the order of 1 cm. The presence of two impacts can be separated using a lateral shower topology cut in the finally segmented barrel, and the silicon strips of the endcap preshower detector.

For the cuts, $p_t^{\gamma_1} > 40\ GeV$, $p_t^{\gamma_2} > 25\ GeV$ and $|\eta^\gamma| < 2.5$, the kinematic acceptance is 57% for $m_H = 110\ GeV$ and 63% for $m_H = 130\ GeV$. An efficiency of 64% is assumed for reconstruction of each photon. This is made up as follows: 19% loss due to conversions in the material before the last tracking element; 10% loss due to fiducial area cuts; 6% loss at high luminosity due to pileup in the isolation region; and 6% loss due to the π^0 rejection cut.

The background cross section for $m_{\gamma\gamma} = 130\ GeV$ after cuts is 134 fb/GeV, where 41 fb is due to $q\bar{q}$ fusion, 42 fb due to gg fusion and 50 fb due to isolated bremsstrahlung. The crystal calorimeter is assumed to have an energy resolution given by $\Delta E/E = 2\%/\sqrt{E} \oplus 0.5\% \oplus 0.2/E$ in the barrel ($|\eta| < 1.1$) and $\Delta E/E = 5\%/\sqrt{E} \oplus 0.5\% \oplus 0.2/E$ in the endcap ($|\eta| > 1.1$) where there is a preshower detector. At high luminosity, a barrel preshower detector covers $|\eta| < 1.1$, resulting in a resolution $\Delta E/E = 5\%/\sqrt{E} \oplus 0.5\% \oplus 0.2/E$ and an ability to measure the photon direction with resolution $\Delta\alpha = 40$ mrad$/\sqrt{E}$. The mass resolution for SM Higgs at $m_H = 130\ GeV$ is 600 MeV at low luminosity ($10^{33} cm^{-2} s^{-1}$) and 960 MeV at high luminosity ($10^{34} cm^{-2} s^{-1}$). The mass resolution is entirely dominated by detector resolution ($\Gamma_H = 5\ MeV$ for $m_H = 130\ GeV$). The results are therefore directly applicable to the MSSM Higgs.

Figure 3a shows the background substracted two-photon effective mass plot for a simulated single experiment with an integrated luminosity of $10^5 pb^{-1}$ for the SM Higgs at $m_H = 90$, 110, 130 and 150 GeV. The contours following the cross section times branching ratio required to give specified signal significances (N_S/N_B) as a function of mass are shown in Fig.3b. significancies have been calculated by counting events within a mass window of optimum width, which corresponds to the width containing about 75% of the signal. Figure 4 shows the 5σ discovery contour for $h \rightarrow \gamma\gamma$ and $H \rightarrow \gamma\gamma$ in MSSM in the ($tan\beta$, m_A) plane for $10^5 pb^{-1}$. More details can be found in refs. 13 and 15[13,15].

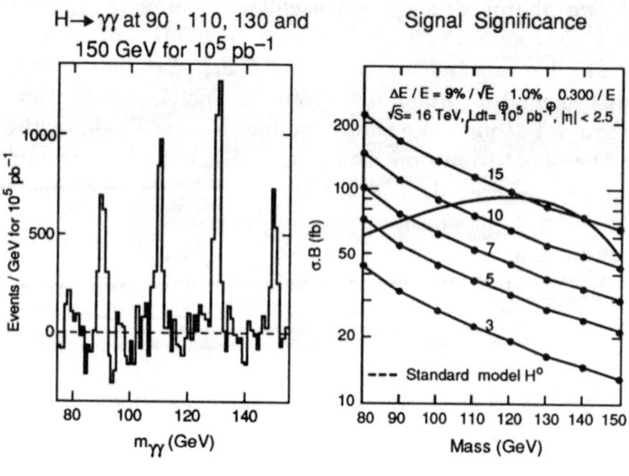

Fig. 3. a) Background substracted 2γ mass plot for $10^5 pb^{-1}$ with signals at m_H = 90, 110 and 150 GeV, in the crystal calorimeter. b) Signal significance contours for $10^5 pb^{-1}$ taken at high luminosity for various $tan\beta$ values.

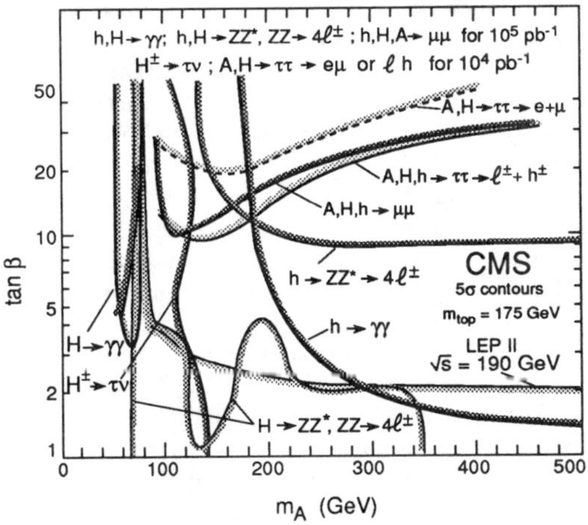

Fig. 4. Significance contours (5σ) in the MSSM parameter space for different Higgs boson decays accessible to CMS.

4. $H \rightarrow ZZ^*$, ZZ and $h \rightarrow ZZ^*$

The scalar Higgs bosons H and h couple to W and Z bosons allowing to search them in the 4-lepton final states from $H, h \rightarrow ZZ^*$ and ZZ. One-loop radiative corrections have a large effect on the upper bound of m_h and the $h \rightarrow ZZ^*$ channel is particularly sensitive for the changes in the top mass. For $m_{top} = 174$ GeV values as high as 128 GeV are possible making the $h \rightarrow ZZ^*$ channel detectable in CMS. Both H and h are narrow with widths well below 1 GeV. Thus a good lepton momentum resolution is mandatory. The geometrical and kinematical acceptances also play a crucial role in these channels[16].

The backgrouds in the low mass region, $m_H < 2m_Z$, for $h, H \rightarrow ZZ^* \rightarrow 4l^{\pm}$ are due to $t\bar{t}$, $Zb\bar{b}$ and ZZ^*. The ZZ^* background is irreducible and peaks charply near the ZZ threshold. The $Zb\bar{b}$ background cannot be reduced by a Z mass cut, but it can be suppressed by requiring lepton isolation. The $t\bar{t}$ background is reduced both by Z mass cut and by isolation cuts. For $m_H > 2m_Z$ the main background is non-resonant ZZ production.

In the simulation of the signal, both the gg fusion process and the WW/ZZ fusion processes were included[17,18]. The $Zb\bar{b}$ background was simulated using matrix element calculation. For the ZZ continuum background, the $q\bar{q} \rightarrow ZZ$ subprocess is simulated. The $gg \rightarrow ZZ$ contribution is included by multiplying the $q\bar{q} \rightarrow ZZ$ cross section by 1.33[19]. No k-factors are icluded. Effect of the pile-up is simulated by superimposing 15 QCD jet events with with low E_t threshold, whenever high luminosity of $\geq 10^{34} cm^{-2} s^{-1}$ is assumed.

High p_t electrons and muons radiate photons. For muons, only internal radiation is important. The radiation can lead to a significant loss of events in a chosen m_Z window. A dedicated photon bremsstrahlung program PHOTOS[20] is used to generate the the internal bremsstrahlung. At $m_H = 300\ GeV$, for instance, the fraction of Z's falling outside a $m_Z \pm 6\ GeV$ window, due to the internal bremsstrahlung, is 12% for $Z \rightarrow e^+e^-$ and 7% for $Z \rightarrow \mu^+\mu^-$[17,18,21]. About 60% of the radiated photons are within a cone $\Delta R \leq 0.1$ around the lepton direction. For electrons part of this loss can then be recovered by reconstructing the electron energy in the calorimeter.

To account for the detector effects, the muon momentum resolution as obtained from full simulation of muon tracks in the detector [22]. The results are parametrized as a function of p and η and used with Gaussian smearing. The reconstructed $Z \rightarrow \mu^+\mu^-$ has, in the Gaussian part of the peak, a resolution $\sigma_Z = 1.8\ GeV$. For electrons, full detector simulation is performed for the $H \rightarrow ZZ^*, ZZ$ events[23]. When traversing the pipe and the tracker, the electrons loose a significant amount of energy as a radiation. Electron energy is reconstructed in a 5x7 matrix of the crystal calorimeter cells, optimized to the shower size in the presence of the magnetic field. The mass resolution for $Z \rightarrow e^+e^-$ is $\sigma_Z = 2.3\ GeV$, and the reconstruction effciency is about 70%, in a $\pm 2\sigma_Z$ window. In the calculation of signal significances an additional reconstruction efficiency of 0.9 per lepton is included.

Selection cuts are the following: one electron with $p_t > 20\ GeV$, one with $p_t > 15\ GeV$ and the remaining two with $p_t > 10\ GeV$, all within $|\eta| < 2.5$. For muons the corresponding cuts are 20, 10 and 5 GeV and the rapidity range $|\eta| < 2.4$. For $m_H = 130\ GeV$, the kinematic and geometrical acceptance is 22% for the four-electron channel and 42% in the four-muon channel. At high masses, for $m_H = 600\ GeV$ for instance, these acceptances increase to 77 and 79%, respectively.

To select $H \rightarrow ZZ^*$ events one of the e^+e^- or $\mu^+\mu^-$ pairs and to select $H \rightarrow ZZ$ events two of these pairs are required to be on within $\pm 2\sigma$ of the Z mass. Due to internal and external bremsstrahlung and detector resolution this cut results in about 40% loss of events. At the low end of the mass range part of this loss is due to the events where both Z's are off-shell. A more sophisticated method, using the fine granularity of the crystal calorimeter, could allow recover a larger fraction of the radiative losses in the future. The m_Z cut reduces the $t\bar{t}$ background by a factor of 11 in the $Z \rightarrow \mu^+\mu^-$ channel and by a factor of 5 in the $Z \rightarrow e^+e^-$ channel. For $H \rightarrow ZZ^*$ a cut $m_{ll} > 12\ GeV$ is also required for the two softer leptons,

Figure 5a shows the reconstructed SM Higgs signals at 130, 150 and 170 GeV, for the sum of the 4e, 2e2μ and 4μ channels, for $10^5 pb^{-1}$. In this mass region the background is dominated by the $t\bar{t}$ production and is about 8 $events/GeV$. For $m_{4l} > 2m_Z$ the background from ZZ decreases fast with increasing mass.

The $t\bar{t}$ background bolow $2m_Z$ could be suppressed by requiring any three of the four leptons to be isolated in the tracker, demanding that there is no track with $p_t > 2.5\ GeV$ within the cone $\Delta R \leq 0.2$ around the lepton direction. Efficiency for the signal is 95% while the $t\bar{t}$ background is further reduced by a factor 5 and the $Zb\bar{b}$ background by a factor of 2. This type of isolation is not sensitive to the radiation effects and to the pile-up. Figure 5b shows the signal in the 4μ channel using isolation for $2 * 10^5 pb^{-1}$.

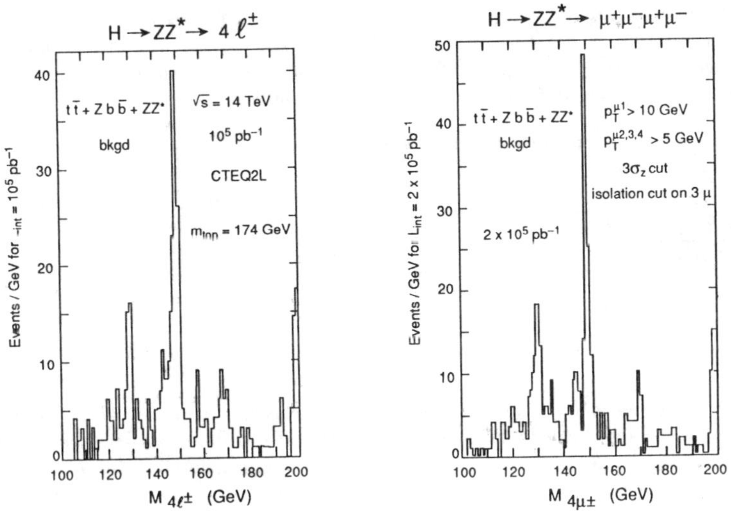

Fig. 5. a) Expected signals for $m_H = 130, 150$ and $170\ GeV$ in the $4l^{\pm}$ channel for $10^5 pb^{-1}$. b) Expected signals for $m_H = 130, 150$ and $170\ GeV$ in the four muon final state for $2 * 10^5 pb^{-1}$ including isolation cuts.

Some representative values of $\sigma * BR$ for $H \to ZZ/ZZ^* \to 4l^+$ and $h \to ZZ^* \to 4l^{\pm}$ in MSSM are given in Table 1. The domains of MSSM parameter space ($tan\beta$, m_A), where h and H can be searched in the ZZ^*/ZZ final states are disconnected: low $tan\beta$ region for H and high $tan\beta$ for h. The 5σ discovery limits at $10^5 pb^{-1}$ as a function of m_A and $tan\beta$ are shown in Fig. 4. We first discuss $H \to ZZ^*/ZZ$. Up to $m_A \sim 70\ GeV$ the $H \to hh$ and $H \to AA$ decay modes dominate suppressing the ZZ final states. Near its lower bound, $m_H \sim 128\ GeV$, the mass of the heavier scalar Higgs is almost independent on $tan\beta$. The couplings to ZZ^* are similar to the SM couplings and the H and the $H \to ZZ^*$ decay is observable for all $tan\beta$ values around

$m_A \sim 75\ GeV$. With increasing m_A the $H \rightarrow gg$ channel becomes the dominant decay mode, except for $tan\beta < 3$ and supresses the $H \rightarrow ZZ$ decay. The minimum around $m_A \sim 130\ GeV$ is due to the opening of the $H \rightarrow WW$ decay channel. Beyond the ZZ threshold the cross section decreases showing an enhancement at $m_A \sim 300\ GeV$ due to the top quark loo in the $gg \rightarrow H$ contribution. At higher m_A still, $H \rightarrow ZZ$ is closed by the opening of $H \rightarrow t\bar{t}$. Turning to $h \rightarrow ZZ^*$, the light scalar Higgs has essentially the SM couplings to ZZ^* for $m_A > 150\ GeV$. For $tan\beta > 10$ the mass is near its maximum value $m_h \sim 128\ GeV$ (depending on m_{top}), and the ZZ^* decay becomes observable, helped by the very small natural width, $\Gamma_h \sim 4\ MeV$. The $h \rightarrow ZZ^*$ discovery region is very sensitive to changes is m_h i.e. to the mass of the top quark. The contour in Fig.4 is for $m_{top} = 174\ GeV$ and m_h calculated with one-loop radiative corrections[7]. For $m_{top} = 190\ GeV$, the discovery region would be extented down to $tan\beta = 4$, whilst for $m_{top} = 150\ GeV$, it would vanish entirely.

Table 1. Cross sections times branching ratio (fb) for $H \rightarrow ZZ, ZZ^*$ and $h \rightarrow ZZ^*$ for different MSSM parameter values.

m_A	$80\ GeV$	$150\ GeV$	$200\ GeV$	$300\ GeV$
$\sigma * BR(H \rightarrow ZZ), tan\beta = 2$	2.2	1.6	1.8	1.2
$\sigma * BR(h \rightarrow ZZ^*), tan\beta = 10$	0.01	1.3	1.7	1.6

5. $h, H, A \rightarrow \tau\tau \rightarrow l^{\pm} + h^{\pm}$

The $\tau\tau$ channel is especially important for the pseudoscalar A which does not couple to ZZ- or significantly to the $\gamma\gamma$ - channel and is difficult to detect otherwise. The $\tau\tau$ final state can be searched for in a lepton+hadron final state $\tau_1 \rightarrow l\nu\nu$, $\tau_2 \rightarrow$hadron(s)+ ν_τ[24] or in e+μ final state, $\tau_1 \rightarrow e\nu\nu$ and $\tau_2 \rightarrow \mu\nu\nu$[25,26]. As there are always neutrinos in the final state, mass peaks are difficult to reconstruct and what is measured is an excess of $t\bar{t}$ production over backgrounds. The one lepton plus one hadron final states are discussed first. Irreducible backgrounds are due to Z, γ^*, $t\bar{t} \rightarrow \tau\tau + X$, $t\bar{t} \rightarrow \tau + X$ and reducible ones to events with one hard lepton and jets with a jet misidentified as a τ, $W + jets$, $Z + jets$, $t\bar{t}$ and $b\bar{b}$ production. Dominant backgrounds for $m_A < 300\ GeV$ are $W + jet$ and $t\bar{t}$ production and for low masses, $m_A \sim 100\ GeV$, the main backgrounds are $Z, \gamma^* \rightarrow \tau\tau$ and $b\bar{b}$. The study is made for for a luminosity of $10^{33} cm^{-2} s^{-1}$ i.e. without pile-up.

Events are required to have: i) one isolated lepton with $p_t > 15\ to\ 40\ GeV$ (depending on m_A within $|\eta| < 2$, ii) one τ jet candidate, and iii) no other significant jet activity. Lepton isolation is defined in a cone of $\Delta R = 0.3$ demanding that there is no other track with $p_t > 2.5\ GeV$ and no additional calorimeter cell with $E_t > 2.5\ GeV$. An overall lepton reconstruction efficiency of 0.9 is assumed. The τ's are selected asking for a collimated high E_t jet in the calorimeters in $|\eta| < 2.0$. The E_t cut varies with m_A: it is $50 < E_t < 12\ GeV$ for $m_A < 300\ GeV$ and $E_t > 60\ GeV$ for $m_A > 300\ GeV$. One high p_t charged track is required in the central tracker in a cone of $\Delta R = 0.1$ around the calorimeter jet axis. The p_t cut varies from $p_t > 25\ GeV$ for $m_A = 100\ GeV$ to $p_t > 40\ GeV$ for $m_A = 400\ GeV$. The τ jet is required to be isolated: applying the same criterion as for the lepton to an annulus of outer radius of $\Delta R = 0.4$ and inner radius $\Delta R = 0.1$. Ordinary jets are suppresed by a factor ~ 100 by this τ selection, whilst $\sim 40\%$ of τ's are retained. Additional jets with $E_t > 25\ GeV$ are vetoed in the region $|\eta| < 2.4$ to reduce the $t\bar{t}$ background. The small Q-value in the τ decays allowes the approxomative reconstruction of m_A, if A has transverse boost by taking the neutrinos parallel with the hadron and the lepton. The reconstructed mass distribution is shown in Fig.6 for $m_A = 300\ GeV$ and $tan\beta = 50$.

For the $\tau\tau \rightarrow e + \mu$ a pair of opposite sign isolated electrons and muons with $p_t > 20\ GeV$ within $|\eta| < 2$ is selected. There are large backgrounds from Z, γ^*, $t\bar{t}$, $b\bar{b}$ and W^+W^- production. The $t\bar{t}$ and W^+W^- backgrounds can be reduced to a level of $\sim 20\%$ of the Z, γ^* background by a calorimeter circularity cut and by a cut in the azimuthal angle between the electron and the muon. These backgrounds can be further reduced by tagging the τ's with an impact parameter measurement. The tagging efficiency for the signal is $\sim 40\%$ (depending on m_A), whilst $WW \rightarrow e + \mu$ contamination is reduced to $\sim 3\%$. The $Z, \gamma^* \rightarrow \tau\tau$ backround is still large but could be effectively reduced tagging the b-jets. The fraction of $Hb\bar{b}$, $hb\bar{b}$ and $Ab\bar{b}$ processes is $\sim 50\%$ of the total cross section at large tanβ. The usefulness of b-tagging in $e + \mu$ and the $l^\pm + h^\pm$ channels is still under study[25]. The 5σ discovery contours in the $(m_A, tan\beta)$ plot for the sum of $A, H, h \rightarrow \tau\tau \rightarrow l^\pm + h^\pm$ and of $A, H, h \rightarrow \tau\tau \rightarrow e + \mu$ are shown in Fig. 4 for $10^4 pb^{-1}$. The $\tau\tau$ decay modes allow allow to investigate a substantial region of parameter space. The discovery range of the $e + \mu$ channel can still be considerably extented at higher luminosity. As the signal is detected mainly through an excess of $\tau\tau$ events a good understunding of the backgrounds will be of paramount importance.

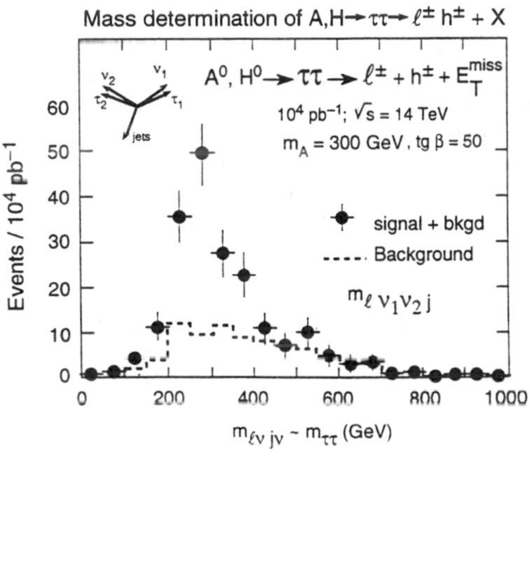

Fig. 6. Invariant mass distribution of the lepton-jet1- $\nu_1\nu_2$ system for selected events at $m_A = 300\ GeV$.

6. Charged Higgs in $t \to H^{\pm}b$, $H^{\pm} \to \tau\nu$

In the MSSM the top quark has two decay modes $t \to H^{\pm} + b$ and $t \to W + b$ provited $m_H < m_t - m_b$. Charged Higgs bosons can thus be produced through $t\bar{t}$ productionh $t \to H^{\pm}b$. The $t \to H^{\pm}b$ branching ratio is large at low and high $tan\beta$ values and has a minimum at $tan\beta \sim 6$. The $H^{\pm}$ has two main decay modes, $H^{\pm} \to c\bar{s}$ and $H^{\pm} \to \tau\nu_\tau$. The $H^{\pm} \to \tau\nu_\tau$ branching ratio is large, $\sim 98\%$, for $tan\beta > 2$ and only slightly dependent on $tan\beta$.

We search for a signal in $t\bar{t}$ events with one top decaying to a charged Higgs and the other decays leptonically giving a hard electron or muon[27,28]. The $t\bar{t}$ production cross section is 470 pb ($m_{top} = 174\ GeV$). We have investigated $H^{\pm} \to \tau\nu_\tau$ using the one-prong τ signature. Neutrinos in the final state prevent $H^{\pm}$ mass reconstruction. The existence of $H^{\pm}$ in the data can only be infered from an excess of τ production over $W \to \tau\nu$ production. The latter can be measured using $W \to \mu\nu$ production corrected for τ versus μ detection efficiencies. The backgrounds considered are of several types. The irreducible ones give real τ's from $t\bar{t}$ or $b\bar{b}$ production. The main reducible backgrounds are due $W + jets$ and $t\bar{t}$ and $b\bar{b}$ events, where one of the jets fakes a τ. Decay matrix elements calculated with explicit helicities[29,30] are implemented in PYTHIA and give the correct polarization for the τ's. The study is made for luminosities $10^{33}cm^{-2}s^{-1}$ and no pile-up effect has been simulated.

Events are selected requiring an isolated high p_t lepton, $p_t > 20\ GeV$ within $|\eta| < 2.4$. The isolation is defined as discussed in the previous section. In addition, one jet with $E_t > 40\ GeV$ within $|\eta| < 2.4$ must fulfill the τ selection criteria described in the previous section. The large $H^{\pm}$ mass and the polarization effects ensure better efficiency for signal (23%) than for the τ jets from $W \to \tau\nu$ (6.8%). To suppress further the dominant W+jet background we require that there is at least one tagged b-jet in the event. Muon tagging with $p_t^{\mu} > 7\ GeV$ has an efficiency $\sim 8\%$. Impact parameter information is used for more efficient tagging. The impact parameter resolution is $\sim 30\ \mu m$ for the tracks with $p_t > 10\ GeV$. The tagging efficiency for the signal events is $\sim 30\%$ and for the W+jet background $\sim 0.5\ \%$ with $E_t^{jet} > 30\ GeV$. Table 2 shows the number of events after kinematical cuts and after b-tagging. A contour in Fig.4 shows the $(m_A, tan\beta)$ domain that can be explored at 5σ significance with 10^4pb^{-1}.

Table 2. Number of events for 10^4pb^{-1} after standard cuts, b-tagging with micro-vertex and the requirement that there is no other jet with $E_t > 40\ GeV$ in addition to the t - and the b - jet.

	$H^{\pm} \to \tau\nu_\tau$	$W^{\pm} \to \tau\nu_\tau$	$t\bar{t}, t \to jet$	$W + jet$	$b\bar{b}, b \to jet$
$Cuts$	781	1573	1718	28600	1226
$b - tagging$	234	582	710	143	36

7. $h, H, A \rightarrow \mu\mu$

In the SM and in MSSM the branching ratio of $H \rightarrow \mu\mu$ is small $\sim 3 * 10^{-4}$. In the case of SM this channel is of little use due to the large Drell-Yan background (γ^*, $Z \rightarrow \mu\mu$). In MSSM however the cross sections are enhanced relative to SM for large $tan\beta$ and the widths of the MSSM Higgs bosons are narrow even for large $tan\beta$ and m_H. Thus the inclusive $\mu\mu$ channel is potentially interesting in a significant part of the parameter space.

The main background in the entire mass range $60\ GeV < m_H < 500\ GeV$ is Drell-Yan. Near the Z-peak internal bremsstrahlung is important and is taken in to account using PHOTOS[20]. The $t\bar{t}$ background is negligible for $|m_{\mu\mu} - m_Z| < 30\ GeV$. It is about 20 times smaller than DY for $m_{\mu\mu} < 150\ GeV$ and is important only for high $m_{\mu\mu}$ values. The WW background is ~ 4 times lower than the $t\bar{t}$ background.

Events are selected[31] by requiring i)two muons with $p_t^\mu > 10\ GeV$, within $|\eta| < 2.4$, ii) at most one jet with $E_t > 40\ GeV$ within $|\eta| < 2.4$ and iii) $E_t^{miss} < 50\ GeV$, to suppress $t\bar{t}$. For $m_A < 200\ GeV$ and $tan\beta > 25 - 30$ an integrated luminosity of $10^4 pb^{-1}$ is sufficient to obtain a signal at greater 7σ level. Figure 7 shows the signal for $10^5 pb^{-1}$. For large $tan\beta$ the h and A peaks remain unresolved when $m_A < 130\ GeV$ and similarly the A and H peaks cannot be separated for $m_A > 130\ GeV$. Only for the lowest accessible $tan\beta$ values $\sim$ 10-15, where the signal is less significant the mass peaks h and A or A and H become resolvable. Close to the Z - peak (115 GeV) the signal is difficult to observe, although statistically still significant, as it sits on the shoulder of a much larger Z - peak. B-tagging could be used to improve the signal to background ration, since in MSSM the $gg \rightarrow b\bar{b}H_{SUSY}$ is 30-50% of the total production rate for $m_A \sim 100\ GeV$ and 70-80% for $m_A \sim 300\ GeV$ for $tan\beta =$ 10-30. This is due to an enhanced $Hb\bar{b}$ couplings, proportional to $1/cos2\beta$. The 5σ contour for $H, h, A \rightarrow \mu\mu$ for an integrated luminosity of $10^5 pb^{-1}$ is shown in Fig. 4. It is close to the $H, h, A \rightarrow \tau\tau$ contour for $10^4 pb^{-1}$, but the $\mu\mu$ channel provides a much better signal identification and mass resolution.

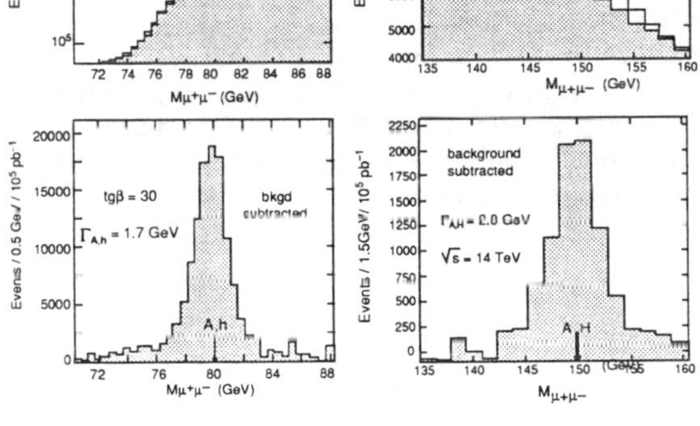

Fig. 7. Invariant mass distribution for inclusive $H_{SUSY} \rightarrow \mu\mu$ at $m_A = 80$ *and* $150\ GeV$ and $tan\beta = 30$ for $10^5 pb^{-1}$.

8. Conclusions

Figure 4 summarizes the situation concerning the regions of MSSM parameter space ($tan\beta$, m_A) that can be explored with the different channels investigated up to now in CMS. Further extentions can be expected for channels with a τ in the final states for luminosities $> 10^{33} cm^{-2} s^{-1}$. There are regions of parameter space that are explorable through more than one channel. This may eventually allow to clarify which among the various SUSY models is actually at play, if any. There remains however an inaccessible region for $100\ GeV < m_A < 190\ GeV$ and $3 < tan\beta < 10$ and a smaller one for $tan\beta \sim 2.5$ and $200\ GeV < m_A < 300\ GeV$. The only channel known which would allow to investigate these regions is $h, H \rightarrow b\bar{b}$ from Wh, Zh and $t\bar{t}$ final states[32]. The microvertex detectror of CMS might possibly allow this, the study is still under way.

9. Acknowledgements

I wish to thank T. Sjostrand and F. Zwirner for helpful discussions and my colleagues D. Bomestar, D. Denegri, I. Iashvili, A. Nikitenko, A. Starodumov and N. Stepanov for help and plesent collaboration.

10. References

1. J. Ellis, C.L. Fogli and E. Lisi, *The Top Quark and Higgs Boson Masses in the Standard Model and the MSSM*, CERN-TH, 7261/94.
2. CDF collaboration , F. Abe et al., *Phys. Rev.* **D50** (1994) 2966.
3. A. Blondel, *Precision Electroweak Physics at LEP*, CERN-PPE/94-133.
4. J. Kodaira, Y. Yasui and K. Sasaki, *The Mass of the Lightest Supersymmetric Higgs Boson beyond the Leading Logarithm Approximation*, HUPD-9316, YNUHEPTh-93-102 (1993).
5. R.Hempfling and A.H.Hoang, *Phys.Lett.* **B331** (1994) 99.
6. J.A.Casas et all, *The Lightest Higgs Boson Mass in the Minimal Supersymmetric Standard Model*, CERN-TH.7334/94 (1994).
7. Z.Kunszt and F.Zwirner, *Nucl.Phys.* **B385** (1992) 3;
8. D.Denegri, R.Kinnunen,A.Nikitenko, A.Starodumov, N.Stepanov, *Justifications for a microvertex detector in CMS*, CMS TN/94-258
9. H.-U. Bengtsson and T. Sjostrand, *Comp. Phys. Comm.* **46** (1987) 43.
10. J. Botts et al., *Phys. Lett.* **B304** (1993) 159.

11. T. Sjostrand, *private communication.*
12. CMS Collaboration, *Letter of Intent*, CERN/LHCC 93-3 , LHCC/I1, October 1992.
13. CMS Collaboration, *Technical Proposal*, CERN/LHCC 94-38, LHCC/P1, December 1994.
14. V. Karimaki, *Fast Tracker Response Simulation* CMS TN/94-275 (1994).
15. C. Seez, *H* $\rightarrow \gamma\gamma$*: An Update* , CMS TN/289 (1994
16. M. Dzelalija and R. Kinnunen, *Study of the Geometrical Acceptance Loss of Muons from H* $\rightarrow ZZ^* \rightarrow 4\mu$ *in the* $|\eta_{max}| \sim 2.5$ *Region*, CMS TN/94-214 (1994).
17. I. Iashvili, R. Kinnunen and A. Nikitenko, $H \rightarrow ZZ^* \rightarrow 4l^\pm$ *in CMS*, CMS TN in preparation.
18. D. Bomestar, D. Denegri and R. Kinnunen, A. Nikitenko, $H \rightarrow ZZ \rightarrow 4l^\pm$ *in CMS*, CMS TN in preparation.
19. E.W.N. Glover and J.J. van der Bij, *Nucl. Phys.* **B321** (1989) 561.
20. E. Barberio, B. van Eijk and Z. Was, *Comp. Phys. Comm.* **66** (1991) 115.
21. D. Denegri, R. Kinnunen and G. Roullet, *Investigation of* $H \rightarrow ZZ \rightarrow 4leptons$ *in CMS Including Radiative Effects*, CMS TN/93-101 (1993).
22. V.Karimaki, M. Pimia, G. Wrochna, *Optimization of Muon Momentum Measurement*, CMS TN/93-62 (1993).
23. A. Nikitenko, P. Verrechia and D. Bomestar, $H \rightarrow ZZ^* \rightarrow 4e^\pm$ *in CMS*, CMS TN in preparation.
24. R. Kinnunen, J. Tuominiemi, D.Denergri, *Search for the heavy neutral MSSM Higgs bosons with the CMS detector at the LHC*, CMS TN/93-98 (1993); TN/93-103 (1993).
25. R. Kinnunen, $H, h, A \rightarrow \tau\tau \rightarrow e + \mu$ *in CMS*, CMS TN in preparation.
26. C.J. Seez *Investigating SUSY with the CMS detector*, TN/93-84 (1993).
27. R. Kinnunen and J. Tuominiemi, *Search for Charge Higgs Boson with the CMS Detector at LHC*, CMS TN/92-48.
28. R. Kinnunen, D. Denegri and J. Tuominiemi, *Search for the Charged Higgs Boson from Top decays in CMS*, CMS TN/94-23 (1994).
29. D.P. Roy *Phys. Lett.* **B277** (1992) 183 and *Phys. Lett.* **B283** (1992) 403.
30. P.Aurenche, *private communication.*
31. N. Stepanov, *On the Possibility to Observe MSSM Neutral Higgses via the H* $\rightarrow \mu\mu$ *channel*" CMS TN/94-182 (1994).
32. C. Kao, *private communication*
33. J. Dai, J.F. Gunion and R. Vega, *Guaranted Detection of a Minimal Supersymmetric Model; Higgs Boson at Hadron supercolliders*, UCD-93-20, SLAC PUB-6277, (1993).

BOUNDS ON THE HIGGS MASS IN THE STANDARD MODEL AND MINIMAL SUPERSYMMETRIC STANDARD MODEL

MARIANO QUIROS *
TH Division, CERN, CH-1211
Geneva 23, Switzerland

ABSTRACT

We present bounds on the Higgs mass in the Standard Model and in the Minimal Supersymmetric Standard Model using the effective potential with next-to-leading logarithms resummed by the renormalization group equations, and physical (pole) masses for the top quark and Higgs boson. In the Standard Model we obtain lower bounds from stability requirements: they depend on the top mass and the cutoff scale. In the Minimal Supersymmetric Standard Model we obtain upper bounds which depend on the top mass and the scale of supersymmetry breaking. A Higgs mass measurement could discriminate, depending on the top mass, between the two models. Higgs discovery at LEP-200 can put an upper bound on the scale of new physics.

1. Introduction

In view of Higgs searches at future colliders (in particular at LEP-200) it is extremely important to compute theoretical bounds on the Higgs mass as accurately as possible. Indeed, the very measurement of the Higgs mass, apart from confirming our knowledge of renormalizable field theories, can shed some light on the particular model that Nature has chosen at the TeV scale.

The two most appealing models are: the Standard Model (SM), which is being confirmed at LEP with $\sim 1\%$ precision, and its minimal supersymmetric extension, the Minimal Supersymmetric Standard Model (MSSM), which is very well motivated theoretically (it helps in technically solving the gauge hierarchy problem) but has received no experimental support from present colliders (TEVATRON, LEP). On the contrary what present colliders are doing is to establish lower bounds on all supersymmetric parameters, and therefore sending the scale of supersymmetry breaking to higher values.

The direct confirmation or exclusion of the MSSM will probably have to wait for the LHC. In the meantime, while present and future accelerators will continue to rise the bounds on supersymmetric parameters, we would like to know whether the discovery of the Higgs boson and the measurement of its mass could, apart from confirming the SM, give information about the possibility that the SM is embedded in an extended electroweak theory: in particular the MSSM. On the one hand, if

*Work partly supported by CICYT under contract AEN94-0928, and by the European Union under contract No. CHRX-CT92-0004.

all supersymmetric parameters are at the TeV scale, the lightest Higgs boson of the MSSM should have the same couplings to ordinary matter than the SM Higgs. On the other hand LEP-200 will cover a range of Higgs masses up to $\sim M_Z$ and direct detection of supersymmetry will not happen if supersymmetric masses are not $\lesssim M_Z$. If this is the case, the best tool LEP-200 will have to uncover new physics is through the Higgs search and the measurement of its mass.

In this talk we will review the theoretical knowledge we have on the SM Higgs mass and on the MSSM lightest Higgs boson mass. In Section 2 we will present lower bounds on the SM Higgs mass from the requirement of stability of the effective potential, i.e. from the requirement that we do not live in a metastable minimum. We will see that it imposes strong lower bounds on the mass of the Higgs boson, M_H, which will be a function of the top mass M_t and of the SM cutoff Λ. In Section 3 we will review the upper bounds on the lightest Higgs boson mass M_H that will depend on various supersymmetric parameters and M_t. In Section 4 we will draw some conclusions concerning the possible detection of the Higgs at LEP-200.

2. Lower bounds in the Standard Model: stability bounds

The vacuum stability requirement in the SM imposes a severe lower bound on the mass of the Higgs boson M_H [1–5], which depends on the mass of the top quark M_t and on the cut-off Λ beyond which the SM is no longer valid. Roughly speaking, this is due to the fact that the top Yukawa coupling h_t drives the quartic coupling of the Higgs potential λ from its initial value at M_Z (which determines the Higgs mass) to negative values at large scales, thus destabilizing the standard electroweak vacuum.

In previous works, the stability bound was obtained from the tree level potential, improved by one-loop or two-loop renormalization group equations (RGE) for the β- and γ-functions of the running couplings, masses and the ϕ-field [2,3]. However, it has been shown that the one-loop corrections to the Higgs potential are important in order to fix the boundary conditions for the electroweak breaking and calculate the Higgs mass in a consistent and scale-independent way. As we will see, they are also significant to properly understand the whole structure of the potential. Typically we find that the lower bound on M_H is $\mathcal{O}(10\text{ GeV})$ lower than in previous estimates [3].

The renormalization group improved effective potential of the SM, V, can be written in the 't Hooft-Landau gauge and the $\overline{MS}$ scheme as [4]

$$V[\mu(t), \lambda_i(t); \phi(t)] \equiv V_0 + V_1 + \cdots \quad , \tag{1}$$

where $\lambda_i \equiv (g, g', \lambda, h_t, m^2)$ runs over all dimensionless and dimensionful couplings, and V_0, V_1 are respectively the tree level potential and the one loop correction, namely

$$V_0 = -\frac{1}{2}m^2(t)\phi^2(t) + \frac{1}{8}\lambda(t)\phi^4(t), \tag{2a}$$

$$V_1 = \sum_{i=1}^{5} \frac{n_i}{64\pi^2} M_i^4(\phi) \left[\log \frac{M_i^2(\phi)}{\mu^2(t)} - c_i\right] + \Omega(t), \tag{2b}$$

where $M_i^2(\phi,t) = \kappa_i\phi^2(t) - \kappa_i'$ ($1 \equiv W, 2 \equiv Z, 3 \equiv$ top, $4 \equiv$ Higgs and $5 \equiv$ Goldstones) are the tree-level expressions for the masses of the particles that enter in the one-loop radiative corrections, and $n_1 = 6$, $\kappa_1 = \frac{1}{4}g^2(t)$, $\kappa_1' = 0$, $c_1 = \frac{5}{6}$; $n_2 = 3$, $\kappa_2 = \frac{1}{4}[g^2(t) + g'^2(t)]$, $\kappa_2' = 0$, $c_2 = \frac{5}{6}$; $n_3 = -12$, $\kappa_3 = \frac{1}{2}h_t^2(t)$, $\kappa_3' = 0$, $c_3 = \frac{3}{2}$; $n_4 = 1$, $\kappa_4 = \frac{3}{2}\lambda(t)$, $\kappa_4' = m^2(t)$, $c_4 = \frac{3}{2}$; $n_5 = 3$, $\kappa_5 = \frac{1}{2}\lambda(t)$, $\kappa_5' = m^2(t)$, and $c_5 = \frac{3}{2}$. $\Omega(t) \equiv \Omega[\lambda_i(t), \mu(t)]$ is the one-loop contribution to the cosmological constant [4], which will turn out to be irrelevant in our calculation.

In the previous expressions the parameters $\lambda(t)$ and $m(t)$ are the SM quartic coupling and mass, whereas $g(t)$, $g'(t)$, $h_t(t)$ are the SU(2), U(1) and top Yukawa couplings respectively. All of them are running with the RGE. The running of the Higgs field is $\phi(t) = \xi(t)\phi_c$, ϕ_c being the classical field and $\xi(t) = \exp\{-\int_0^t \gamma(t')dt'\}$, where $\gamma(t)$ is the Higgs field anomalous dimension. Finally the scale $\mu(t)$ is related to the running parameter t by $\mu(t) = \mu e^t$, where μ is a fixed scale, we will take equal to the physical Z mass, M_Z.

It has been shown [6] that the L-loop effective potential improved by $(L+1)$-loop RGE resums all Lth-to-leading logarithm contributions. Consequently, we will consider all the β- and γ-functions of the previous parameters to two-loop order, so that our calculation will be valid up to next-to-leading logarithm approximation.

As has been pointed out [4], working with $\partial V/\partial\phi$ (and higher derivatives) rather than with V itself allows us to ignore the cosmological constant [†] term Ω. In fact, the structure of the potential can be well established once we have determined the values of ϕ, say ϕ_{ext}, in which V has extremals (maxima or minima). Thus we only need to evaluate $\partial V/\partial\phi$ and $\partial^2 V/\partial\phi^2$.

The structure of maxima and minima of V for large ϕ can be evaluated at a scale $\mu(t)$ within the region where V is scale-invariant. As was previously discussed [7], $\mu(t) = \phi(t)$ is always a correct choice (other choices, such that $\mu(t) = \phi(t)/2$, are equally valid and lead to essentially the same results). Then the extremal condition, neglecting the Higgs and Goldstone contributions, reads [8]

$$\phi_{ext}^2 = \frac{2m^2}{\tilde{\lambda}}, \tag{3}$$

$$\begin{aligned} \tilde{\lambda} &= \lambda - \frac{1}{16\pi^2}\left\{6h_t^4\left[\log\frac{h_t^2}{2} - 1\right] - \frac{3}{4}g^4\left[\log\frac{g^2}{4} - \frac{1}{3}\right]\right. \\ &- \left.\frac{3}{8}\left(g^2 + g'^2\right)^2\left[\log\frac{(g^2 + g'^2)}{4} - \frac{1}{3}\right]\right\}, \end{aligned} \tag{4}$$

[all quantities in (3,4) are evaluated at $\mu(t) = \phi_{ext}(t)$]. From (3) we see that, if V develops an extremal for large values of ϕ, this must occur for a value of ϕ such that

$$0 < \tilde{\lambda}[\mu(t) = \phi(t)] \ll 1. \tag{5}$$

[†] This holds even if we choose $\mu(t)$ to be a function of the ϕ-field since the scale-invariant properties of V allow the substitution to be performed either before or after taking the derivative [7] $\partial/\partial\phi$.

On the other hand, for large values of ϕ the second derivative of the potential (1) can be very accurately expressed as [8]

$$\left.\frac{\partial^2 V}{\partial\phi^2(t)}\right|_{\phi(t)=\phi_{ext}(t)} = \frac{1}{2}(\beta_\lambda - 4\gamma\lambda)\phi^2(t), \tag{6}$$

where β_λ is the one-loop β-function. Since near the extremum λ is very small, we see from (6) that depending on the sign of β_λ we will have a maximum or a minimum.

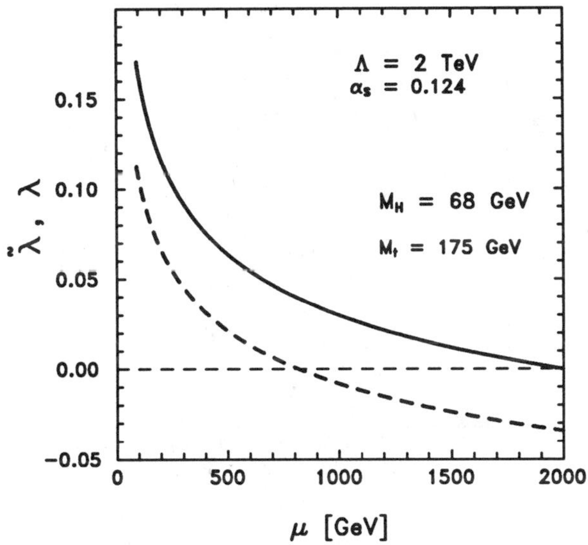

Figure 1: Plot of λ (dashed line) and $\tilde{\lambda}$ (solid line) as a function of the scale $\mu(t)$ for $M_t = 175$ GeV, $\Lambda = 2$ TeV, $M_H = 68$ GeV and $\alpha_S = 0.124$.

We have illustrated these features in Fig. 1 with a typical example. It represents the evolution of λ (dashed line) and $\tilde{\lambda}$ (solid line) with $\mu(t)$. It is worth noticing that they do not cross the horizontal axis at the same value of $\mu(t)$, but they differ by a relatively large amount. This is important since the point where the maximum of the potential is located, say ϕ_{MAX}, does correspond [8] to $\tilde{\lambda} \sim 0$ rather than $\lambda \sim 0$. This is so in Fig. 2, where the scalar potential, $V(\phi)$, has been represented for a typical choice of parameters. Notice that for values of ϕ very slightly higher than ϕ_{MAX}, the potential is negative and much deeper than the electroweak minimum. This is simply because for values of $\mu(t)$ just beyond $\mu_{MAX} = \phi_{MAX}$, the value of $\tilde{\lambda}$ becomes negative and the potential is dominated by the contribution $\frac{1}{8}\tilde{\lambda}\phi^4$. Consequently, a sensible criterion for a model to be safe is to require one of the two following conditions: ***a)*** The potential has no maximum; *b)* The maximum occurs for $\phi_M > \Lambda$. In the following we will assume $\Lambda \leq 10^{19}$ GeV. With this criterion [in particular condition (b)] we see that the model represented in Fig. 2 is acceptable for $\Lambda \leq 2.7 \times 10^{11}$ GeV. Beyond this scale, the stability of the vacuum requires the appearance of new physics. Note from this discussion that conditions (a), (b) are not equivalent to require $\lambda(\mu) > 0$ for $\mu(t) < \Lambda$, as is usually done. Instead, the significant parameter is $\tilde{\lambda}$ rather than λ.

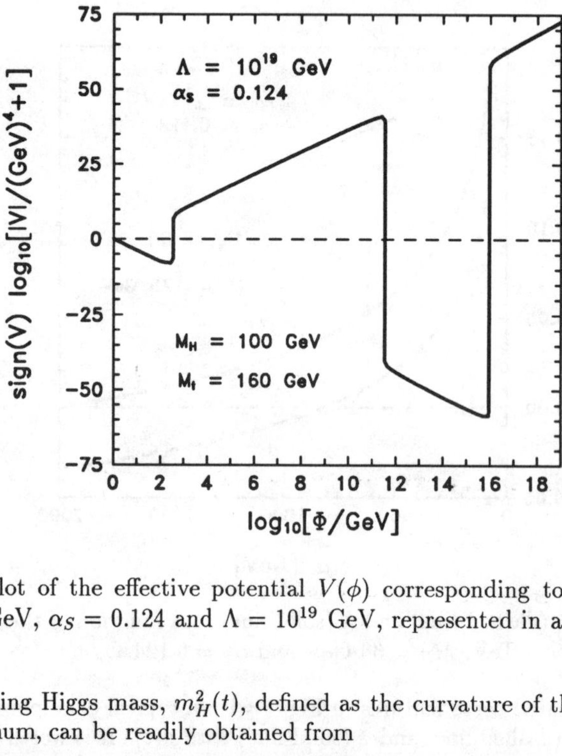

Figure 2: Plot of the effective potential $V(\phi)$ corresponding to $M_t = 160$ GeV, $M_H = 100$ GeV, $\alpha_S = 0.124$ and $\Lambda = 10^{19}$ GeV, represented in a convenient choice of units.

The running Higgs mass, $m_H^2(t)$, defined as the curvature of the scalar potential at the minimum, can be readily obtained from

$$m_H^2(t^*) = \left. \frac{\partial^2 V}{\partial \phi^2(t^*)} \right|_{\phi(t^*)=\langle\phi(t^*)\rangle}, \tag{7}$$

where t^* is the scale at which we define the electroweak minimum [7]. The scale invariance of the second derivative of the potential, $\frac{\partial}{\partial t}\left[\xi^2(t)\frac{\partial^2 V}{\partial \phi^2(t)}\right] = 0$, allows us to write $m_H^2(t)$ at any arbitrary scale

$$m_H^2(t) = m_H^2(t^*)\frac{\xi^2(t^*)}{\xi^2(t)}. \tag{8}$$

The physical (pole) Higgs mass, M_H^2, is then given by

$$M_H^2 = m_H^2(t) + \mathrm{Re}[\Pi(p^2 = M_H^2) - \Pi(p^2 = 0)], \tag{9}$$

where $\Pi(p^2)$ is the renormalized self-energy of the Higgs boson [7] (the t-dependence drops out from (9)).

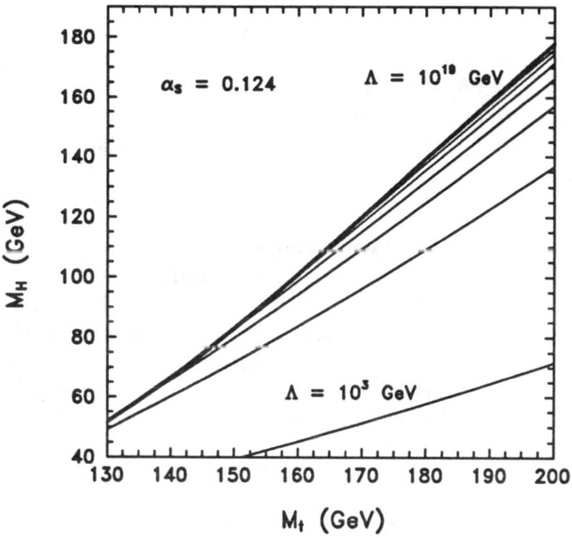

Figure 3: SM lower bound on M_H as a function of M_t for $\alpha_S(M_Z) = 0.124$ and different values of Λ in the range 10^3 GeV $\leq \Lambda \leq 10^{19}$ GeV. The values of Λ for consecutive curves differ by two orders of magnitude.

As has been stated above, the choice of μ^*, i.e. the scale at which we evaluate the minimum conditions, is not important for physical quantities, provided it is within a (quite wide) region around the optimal value [8]. The lack of flatness of M_H reflects the effect of all non-considered (higher-order) contributions in the calculation and, therefore, it is a measure of the total error in our estimate of M_H. We deduced [8] that the error is typically $\lesssim 3$ GeV, which is the uncertainty we should assign to our results. Had we performed the previous calculations just with the (RGE-improved) tree-level part of V in Eq. (1) [2,3], the Higgs mass would have a strong dependence on μ^*. Choosing [2,3] $\mu^* = M_Z$ results in an error in the estimate of M_H, whose precise value depends on the top mass and is typically of $\mathcal{O}(10\ GeV)$, showing the need of a more careful treatment of the problem, as the one exposed above [8].

Finally, let us note that in the previous equations the top Yukawa coupling $h_t(t)$ enters in several places. Therefore, the Higgs mass depends on the boundary condition chosen for $h_t(t)$, and thus on the top mass M_t. However the running top mass, defined as $m_t(t) = vh_t(t)$, does not coincide with the physical (pole) mass M_t. In the Landau gauge the relationship between the running m_t and the physical (pole) mass M_t is given by [9]

$$M_t = \left\{1 + \frac{4}{3}\frac{\alpha_S(M_t)}{\pi} + \left[16.11 - 1.04\sum_{i=1}^{5}\left(1 - \frac{M_i}{M_t}\right)\right]\left(\frac{\alpha_S(M_t)}{\pi}\right)^2\right\} m_t(M_t), \quad (10)$$

where M_i, $i = 1, \ldots, 5$ represent the masses of the five lighter quarks.

As has become clear from the previous discussion, the lower bound on M_H is a function of M_t and Λ. However, apart from the previously estimated error $\lesssim 3$ GeV in our calculation, there is an additional source of uncertainty coming from the value of α_s, which enters in several places in the previous calculation. The most recent estimate of α_s gives

$$\alpha_s = 0.124 \pm 0.006. \quad (11)$$

Using the central value of (11), we have represented in Fig. 3 the lower bound on M_H as a function of M_t for different values of Λ. The form of the curves is easily understandable from the previous discussion. In Fig. 4, we have fixed Λ at its maximum value, $\Lambda = 10^{19}$ GeV, and represented the lower bound on M_H for the central value of α_s in (11) (diagonal solid line) and the two extreme values (diagonal dashed lines).

If we use the recent evidence for the top quark production at CDF with a mass $M_t = 174 \pm 17$ GeV [10], we obtain the following lower bound on M_H:

$$M_H > 128 \pm 33\ GeV, \quad (12)$$

i.e. $M_H > 95$ GeV (1σ). If the Higgs is observed in the present or forthcoming accelerators with a mass below the bound of Eq. (12), this would be a clear signal of new physics beyond the SM.

Comparing these bounds with previous evaluations [3], we see that our values of M_H are lower by an amount increasing with M_t, going from ~ 5 GeV for $M_t \sim 130$ GeV to ~ 15 GeV for $M_t \sim 200$ GeV. As has been discussed previously the main reason of this difference is the way in which the Higgs mass was previously computed [3]. Accordingly, our results give more room to the Higgs mass in the framework of the Standard Model.

3. Upper bounds in the Minimal Supersymmetric Standard Model

The MSSM has an extended Higgs sector with two Higgs doublets with opposite hypercharges: H_1, responsible for the mass of the charged leptons and the down-type quarks, and H_2, which gives a mass to the up-type quarks. After the Higgs mechanism there remain three physical scalars, two CP-even and one CP-odd Higgs bosons. In particular, the lightest CP-even Higgs boson mass satisfies the tree-level bound

$$m_H^2 \leq M_Z^2 \cos^2 2\beta, \quad (13)$$

where $\tan\beta = v_2/v_1$ is the ratio of the Vacuum Expectation Values (VEVs) of the neutral components of the two Higgs fields H_2 and H_1. Relation (13) implies that $m_H^2 < M_Z^2$, for any value of $\tan\beta$, which, in turn, implies that it should be found at LEP-200 [11]. However, the tree-level relation (13) is spoiled by one-loop radiative corrections, which were computed by several groups using: the effective potential approach [12], diagrammatic methods [13] and renormalization group (RG) techniques [14]. All methods found excellent agreement with one another and large radiative corrections, mainly controlled by the top Yukawa coupling, which could make the lightest CP-even Higgs boson escape experimental detection at LEP-200. In particular, the RG approach (which will be followed in this talk) is based on the fact that supersymmetry decouples and, below the scale of supersymmetry breaking M_S, the effective theory is the SM, with some matching conditions at M_S. Assuming $M_Z^2 \ll M_S^2$ the tree-level bound (13) is saturated at the scale M_S and the effective SM at scales between M_Z and M_S contains the Higgs doublet $H = H_1 \cos\beta + i\sigma_2 H_2^* \sin\beta$, with a quartic coupling taking, at the scale M_S, the (tree-level) value of

$$\lambda = \frac{1}{4}(g^2 + g'^2)\cos^2 2\beta. \tag{14}$$

In these analyses [14] the Higgs mass was considered at the tree level, improved by one-loop renormalization group equations (RGE) in the γ- and β-functions, thus collecting all leading logarithm corrections.

Since the relative size of one-loop corrections to the Higgs mass is large (mainly for large top quark mass and/or small tree-level Higgs mass) it was compelling to analyse them at the two-loop level. A first step in that direction was given some time ago [15] where two-loop RGE-improved tree-level Higgs masses were considered. It was found that two-loop corrections were negative and small. The Higgs mass received all leading logarithm and part of the next-to-leading logarithm corrections [15]. As we have described in Section 2, for fully taking into account all next-to-leading logarithm corrections the one-loop effective potential (improved by two-loop RGE) is needed.

The tree-level quartic coupling (14) receives one-loop threshold contributions at the M_S scale. These are given by

$$\Delta\lambda = \frac{3h_t^4}{16\pi^2}\frac{X_t^2}{M_S^2}\left(2 - \frac{X_t^2}{6M_S^2}\right), \tag{15}$$

where h_t is the top Yukawa coupling in the SM and $X_t = A_t + \mu\cot\beta$ is the stop mixing.

The correction (15) has a maximum for $X_t^2 = 6M_S^2$. For that reason, in our numerical applications we will take $X_t^2 = 6M_S^2$, i.e. maximal threshold effect. Notice also that $X_t^2 = 6M_S^2$ is barely consistent with the bound from colour-conserving minimum [16], so that the case of maximal threshold really represents a particularly extreme situation. In addition to the previous effect, there appear effective higher-order operators ($D \geq 6$), which for $M_S \gtrsim 1$ TeV turn out to be negligible [7].

Upper bounds on the lightest Higss boson mass in the MSSM depend therefore on three supersymmetric parameters (besides M_t): M_S (from naturality reasons $M_S \lesssim$

1 TeV [17]), $\tan\beta$ and $X_t = A_t + \mu/\tan\beta$, which is responsible for the threshold correction to the Higgs quartic coupling. The larger the threshold correction and $\tan\beta$, the less stringent the supersymmetric bounds. Therefore, the most conservative situation takes place considering maximum threshold correction (which is achieved for $X_t^2 = 6M_S^2$) and $\tan\beta = \infty$. Likewise, the larger M_S, the less stringent the bounds; but, as mentioned above, it is not sensible to consider M_S much larger than 1 TeV. Consequently, to be on the safe side, we have represented in Fig. 4 the MSSM upper bounds (transverse solid and dashed lines), as recently obtained up to next-to-leading-log order [7], in the most conservative situation with $M_S = 1$ TeV.

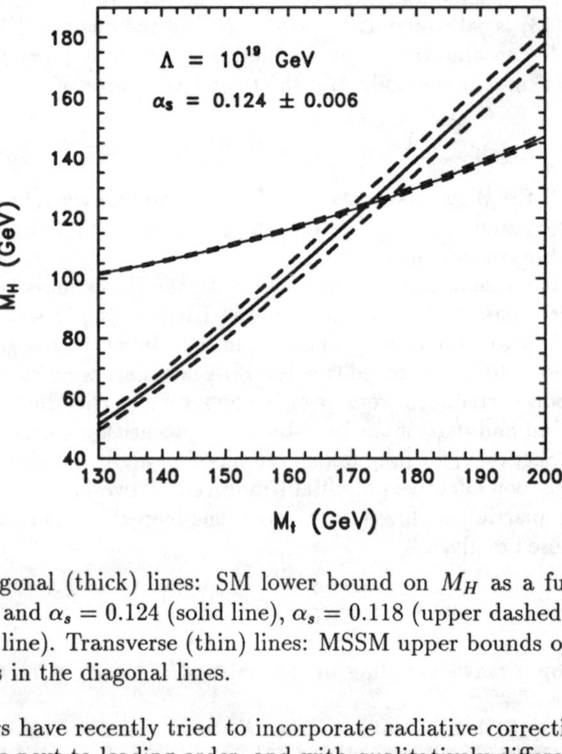

Figure 4: Diagonal (thick) lines: SM lower bound on M_H as a function of M_t for $\Lambda = 10^{19}$ GeV and $\alpha_s = 0.124$ (solid line), $\alpha_s = 0.118$ (upper dashed line), $\alpha_s = 0.130$ (lower dashed line). Transverse (thin) lines: MSSM upper bounds on M_H for $\Lambda_S = 1$ TeV and α_s as in the diagonal lines.

Two papers have recently tried to incorporate radiative corrections to the Higgs mass up to the next-to-leading order, and with qualitatively different results. Using the RG approach, positive and large next-to-leading corrections, with respect to the one-loop results, were found by Kodaira, Yasui and Sasaki (KYS) [18]. Using diagrammatic and effective potential methods in a particular MSSM, as well as various approximations, Hempfling and Hoang (HH) [19] found that two-loop corrections are also sizeable, but negative with respect to the one-loop result! Using the RG approach, we have found [7] that two-loop corrections are negative with respect to the one-loop result. We have traced back the origin of this disagreement with KYS [18] in their choice of the minimization scale. Furthermore KYS neglected various effects (as the contribution of gauge bosons to the one-loop effective potential, or the wave function renormalization of top quark and Higgs boson) and considered only the case with zero stop mixing. On the other hand, we have found that the abnormal size of the two-loop corrections obtained by HH [19] is a consequence of an excessively rough estimate of the one-loop result, but we are in agreement with their final two-loop result. In fact our two-loop results differ from those of HH [19] by less than 3%. Also our results show a large sensitivity of the Higgs mass to the stop mixing parameter.

Finally we would like to comment briefly on the generality of our results. As was already stated, we are assuming average squark masses $M_S^2 \gg M_Z^2$, and that all supersymmetric particle masses are $\gtrsim M_S$. If we relax the last assumption, i.e. if some supersymmetric particles were much lighter, the value of the quartic coupling at M_S (see Eq. (14)) would be slightly increased and, correspondingly, our bounds would be slightly relaxed. We have made an estimate of this effect. Assuming an extreme case where all gauginos, higgsinos and sleptons have masses $\sim M_Z$, we have found for $M_S = 1$ TeV and $\cos^2 2\beta = 1$ an increase of the Higgs mass $\sim 2\%$. For values of $\tan\beta$ close to 1 (as those appearing in infrared fixed point scenarios) the corresponding effect is negligible. On the other hand, our numerical results have been computed for $M_S = 1$ TeV. For values of $M_S < 1$ TeV the bounds on the lightest Higgs mass are lowered. Hence, in this sense, all our results can be considered as absolute upper bounds.

4. Conclusions

We want to conclude with comments on the two questions we raised at the beginning of this talk: **i)** Can the Higgs mass measurement disentangle between the SM and the MSSM? **ii)** Can eventually a Higgs mass measurement at LEP-200 help in putting upper bounds on the scale of new physics?

As for question **i)**, a quick glance at Fig. 4 shows that for $M_t = 173 \pm 4$ GeV, i.e. the crossing area of the SM and MSSM curves, the Higgs mass eventually measured will be compatible either with the pure SM or with the MSSM, but not with both at the same time. Accordingly, the experimental Higgs mass either will discard the MSSM or will be a clear signal of new physics beyond the SM compatible with the MSSM. For $M_t < 169$ GeV, the situation is analogous, but there is a wider range of Higgs masses (area within the two curves) compatible with both SM and MSSM. For $M_t > 177$ GeV, there is no region of Higgs masses simultaneously compatible with the SM and MSSM. On the contrary there is a range of M_H (within the two curves) that would discard both. This means that for $M_t > M_t^c = 177$ GeV a Higgs mass measurement could **always** discriminate between the SM and the MSSM. This statement has two caveats: one is that it is based upon assuming that the SM holds up to the Planck scale. Had we considered the validity of the SM up to a lower scale, we should have obtained a value $M_t^c > 177$ GeV, with a larger overlapping region where the Higgs mass measurement cannot discriminate between the SM and

MSSM. The second caveat is that we are requiring absolute stability for the effective potential: instead, the requirement of metastability of the electroweak minimum will decrease the lower bound and go along the same direction as the previous effect. A detailed calculation along these lines is at present being performed.

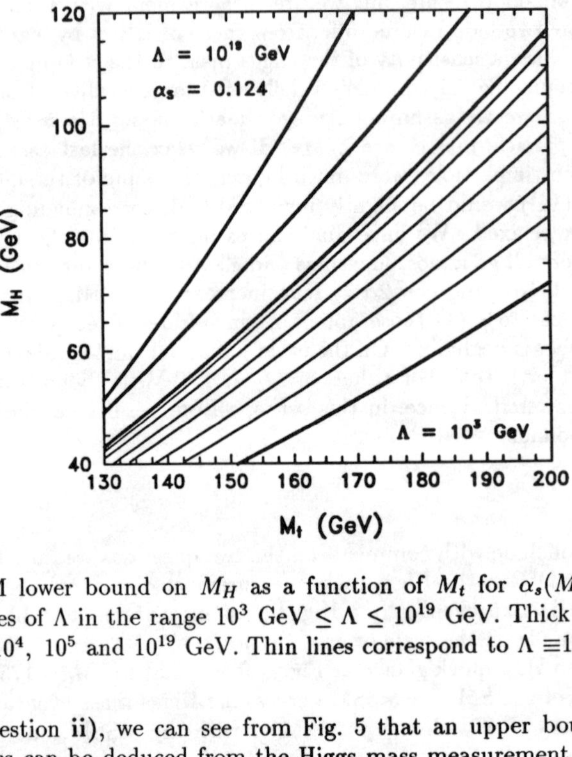

Figure 5: SM lower bound on M_H as a function of M_t for $\alpha_s(M_Z) = 0.124$ and different values of Λ in the range 10^3 GeV $\leq \Lambda \leq 10^{19}$ GeV. Thick lines correspond to $\Lambda = 10^3$, 10^4, 10^5 and 10^{19} GeV. Thin lines correspond to $\Lambda \equiv$1.5, 2, 3, 4, 6 and 8 TeV.

As for question **ii)**, we can see from Fig. 5 that an upper bound on the scale of new physics can be deduced from the Higgs mass measurement. For instance, if $M_H < M_Z$, and fixing $M_t \sim 175$ GeV, we obtain from Fig. 5 that $\Lambda \lesssim 10$ TeV. This bound crucially depends on the value of the top mass and will be softened by the second (metastability bound) effect just mentioned.

5. Acknowledgements

I want to express my deepest gratitude to my collaborators on the topics covered by this talk, Alberto Casas, José Ramón Espinosa and Antonio Riotto. I also acknowledge discussions with Marcela Carena, Gordy Kane, Santi Peris, Nir Polonsky, Stephan Pokorski, Carlos Wagner and Fabio Zwirner. I finally wish to thank Fabio Zwirner for a careful reading of the manuscript.

6. References

1. N. Cabibbo, L. Maiani, G. Parisi and R. Petronzio, *Nucl. Phys.* **B158** (1979) 295.
2. M. Lindner, *Z. Phys.* **C31** (1986) 295; M. Sher, *Phys. Rep.* **179** (1989) 273; M. Lindner, M. Sher and H.W. Zaglauer, *Phys. Lett.* **B228** (1989) 139.
3. M. Sher, *Phys. Lett.* **B317** (1993) 159; Addendum: [hep-ph/9404347].
4. C. Ford, D.R.T. Jones, P.W. Stephenson and M.B. Einhorn, *Nucl. Phys.* **B395** (1993) 17.
5. G. Altarelli and G. Isidori, *Phys. Lett.* **B337** (1994) 141.
6. B. Kastening, *Phys. Lett.* **B283** (1992) 287; M. Bando, T. Kugo, N. Maekawa and H. Nakano, *Phys. Lett.* **B301** (1993) 83; *Prog. Theor. Phys.* **90** (1993) 405.
7. J.A. Casas, J.R. Espinosa, M. Quirós and A. Riotto, preprint CERN-TH.7334/94 [hep-ph/9407389], to appear in *Nucl. Phys.* **B**.
8. J.A. Casas, J.R. Espinosa and M. Quirós, preprint IEM-FT-93/94 [hep-ph/9409458], to appear in *Phys. Lett.* **B**.
9. N. Gray, D.J. Broadhurst, W. Grafe and K. Schilcher, *Z. Phys.* **C48** (1990) 673.
10. F. Abe et al., CDF Collaboration, *Phys. Rev.* **D50** (1994) 2966 and *Phys. Rev. Lett.* **73** (1994) 225.
11. Z. Kunszt and W.J. Stirling, *Phys. Lett.* **B262** (1991) 54.
12. Y. Okada, M. Yamaguchi and T. Yanagida, *Prog. Theor. Phys.* **85** (1991) 1; J. Ellis, G. Ridolfi and F. Zwirner, *Phys. Lett.* **B257** (1991) 83 and **B262** (1991) 477; R. Barbieri and M. Frigeni, *Phys. Lett.* **B258** (1991) 395; D. Pierce, A. Papadopoulos and S. Johnson, *Phys. Rev. Lett.* **68** (1992) 3678.
13. H.E. Haber and R. Hempfling, *Phys. Rev. Lett.* **66** (1991) 1815; A. Yamada, *Phys. Lett.* **B263** (1991) 233; A. Brignole, *Phys. Lett.* **B281** (1992) 284; P.H. Chankowski, S. Pokorski and J. Rosiek, *Phys. Lett.* **B274** (1992) 191.
14. Y. Okada, M. Yamaguchi and T. Yanagida, *Phys. Lett.* **B262** (1991) 54; R. Barbieri, M. Frigeni and F. Caravaglios, *Phys. Lett.* **B258** (1991) 167; H.E. Haber and R. Hempfling, *Phys. Rev.* **D48** (1993) 4280.
15. J.R. Espinosa and M. Quirós, *Phys. Lett.* **B266** (1991) 389.
16. J.M. Frère, D.R.T. Jones and S. Raby, *Nucl. Phys.* **B222** (1983) 11; C. Kounnas, A.B. Lahanas, D.V. Nanopoulos and M. Quirós, *Nucl. Phys.* **B236** (1984) 438; J.-P. Derendinger and C.A. Savoy, *Nucl. Phys.* **B237** (1984) 307; J.A. Casas, A. Lleyda and C. Muñoz, in preparation.
17. R. Barbieri and G.F. Giudice, *Nucl. Phys.* **B306** (1988) 63; B. de Carlos and J.A. Casas, *Phys. Lett.* **B309** (1993) 320.
18. J. Kodaira, Y. Yasui and K. Sasaki, Hiroshima preprint HUPD-9316, YNU-HEPTh-93-102 (November 1993).
19. R. Hempfling and A.H. Hoang, *Phys. Lett.* **B331** (1994) 99.

Production of Sbottoms and Stops in e^+e^- Annihilation*

A. Bartl
Institut für Theoretische Physik, Universität Wien
A-1090 Vienna, Austria

H. Eberl, W. Majerotto
Institut für Hochenergiephysik
Österreichische Akademie der Wissenschaften
A-1050 Vienna, Austria

W. Porod
Institut für Theoretische Physik, Universität Wien
A-1090 Vienna, Austria

Abstract

This talk deals with the question how sbottoms and stops can be discovered in e^+e^- annihilation at $\sqrt{s} = 500$ GeV. We present numerical predictions within the Minimal Supersymmetric Standard Model for production cross sections and decay rates of these particles, and discuss their important signatures.

1 Introduction

Supersymmetry (SUSY) is one of the most attractive extensions of the Standard Model (SM) [1,2], because it provides a solution of the gauge hierarchy problem and allows us to inorporate in a natural way an elementary scalar Higgs boson. This also implies that the masses of SUSY particles can be expected to be not higher than 1 TeV, but may also be much smaller than this value. Therefore, the search for SUSY particles plays an essential rôle at present accelerators and will become even more important for future colliders.

*presented by A. Bartl

Theoretical arguments based mainly on renormalization group equations indicate that the scalar top quark, the SUSY partner of the top quark, may be the lightest squark [3–9]. It may even be the lightest charged SUSY particle [10–12]. On the other hand, if the parameter $tan\beta$ is large enough, i. e. if $tan\beta \gtrsim 10$, the scalar bottom quark could also be relatively light and thus turn out to be the lightest charged SUSY particle [13]. Both scenarios are possible, depending on the theoretical assumptions made. For example, SU(5) GUT's, together with the restrictions from proton decay, in general demand $tan\beta \lesssim 10$ [9]. Other scenarios, as for example SO(10) GUT's [14] or non-universal scalar mass parameters [15], in general prefer $tan\beta \gtrsim 10$.

As the stops and the sbottoms are charged and coloured particles, one can search for them at pp colliders and at e^+e^- colliders. Theoretical studies of stop production at the TEVATRON and at the LHC were recently carried out in [8,16,17]. Stop production in e^+e^- annihilation is treated in [11]. In this talk we shall present our results for the production of stops and sbottoms in e^+e^- annihilation at $\sqrt{s} = 500$ GeV. We shall also discuss the decay properties of these particles. If the splitting between the different mass eigenstates is large enough, new decay modes of these particles may be present which have to be taken into account when discussing their main signatures. We want to emphasize the aspects of left–right squark mixing, as they were not treated in the earlier studies of SUSY particle production at a linear e^+e^- collider with 500 GeV CMS energy [18].

We shall work in the Minimal Supersymmetric Standard Model (MSSM) [1,2], which offers a framework sufficiently general to describe the essential properties of stops and sbottoms, as well as of the SUSY particles involved in their decays. The MSSM contains the known particles plus the sleptons, $\tilde{\ell}^{\pm}$, $\tilde{\nu}_{\ell}$, squarks, $\tilde{q}$, gluinos $\tilde{g}$, two pairs of charginos, $\tilde{\chi}_i^{\pm}$, $i = 1, 2$, four neutralinos, $\tilde{\chi}_i^0$, $i = 1, \ldots, 4$, and five Higgs particles, h^0, H^0, A^0, $H^{\pm}$. The main parameters which determine the phenomenology of the SUSY particles are M, M', μ, $\tan\beta = v_2/v_1$, $M_{\tilde{Q}}$, $M_{\tilde{U}}$, $M_{\tilde{D}}$, A_t, A_b, and m_A. M and M' are the (soft breaking) $SU(2)$ and $U(1)$ gaugino masses, μ is the higgsino mass parameter, and v_1 and v_2 are the vacuum expectation values of the neutral members of the two Higgs doublets. $M_{\tilde{Q}}$, $M_{\tilde{U}}$, $M_{\tilde{D}}$, A_t, A_b, are soft–breaking parameters entering the squark mass matrices. m_A is the mass of the pseudoscalar Higgs boson A^0. We shall assume the GUT relations $M'/M = \frac{5}{3}\tan^2\Theta_W \approx 0.5$, and $m_{\tilde{g}}/M = \alpha_s/\alpha_2 \approx 3$, $m_{\tilde{g}}$ being the gluino mass.

Concerning the present experimental lower mass bounds for squarks, the CDF experiment at the TEVATRON [19] obtained $m_{\tilde{q}} \gtrsim 130$ GeV, for $m_{\tilde{q}} < m_{\tilde{g}} \lesssim 400$ GeV, taking into account cascade decays [20]. In the analysis of [19] it is assumed that all squark flavours are degenerate in mass. Because of t_L-t_R mixing, however, the mass of the stop could be much smaller, even below 100 GeV [21]. Moreover, if the mass of the LSP is larger than approximately 55 GeV, no bound on $m_{\tilde{q}}$ can be derived. In this case there is only the lower mass bound of 45 GeV for charged SUSY particles from LEP [22]. This bound does not apply for the stop if the mixing angle is in the intervall $0.85 \lesssim \Theta_{\tilde{t}} \lesssim 1.15$, because the stop coupling to the Z^0 vanishes for the special mixing angle $\Theta_{\tilde{t}} = 0.98$ [11]. The bound $m_{\tilde{t}_1} \geq 28$ GeV recently obtained at KEK is independent of L-R mixing [23]. Quite analogously, one can expect that the experimental situation concerning the sbottom system is also more complicated than for the squarks of the first and second generations, if $\tan\beta \gtrsim 10$. There are no experimental bounds for $\tan\beta$ yet, but for theoretical reasons one expects $1 \lesssim \tan\beta \lesssim m_t/m_b$.

In Section 2 we shortly review the basic facts about stop and sbottom mixing and present our numerical results for the production cross sections of $e^+ + e^- \to \tilde{t}_i + \bar{\tilde{t}}_j$ and $e^+ + e^- \to \tilde{b}_i + \bar{\tilde{b}}_j$ at $\sqrt{s} = 500$ GeV. In Section 3 we describe the decays of sbottoms and stops and their main signatures, while paying particular attention to squark mixing and to the Yukawa coupling terms. In Section 4 we give a short summary.

2 Cross Sections for Pair Production of Stops and Sbottoms

The mass terms in the Lagrangian of the stop and sbottom systems can be combined into two 2×2 mass mixing matrices. The diagonal elements are $M^2_{\tilde{t}_L}$, $M^2_{\tilde{t}_R}$, and $M^2_{\tilde{t}_L}$, $M^2_{\tilde{t}_R}$ for top and bottom squarks, respectively, with

$$M^2_{\tilde{t}_L} = M^2_{\tilde{Q}} + m^2_Z \cos 2\beta (\frac{1}{2} - \frac{2}{3}\sin^2\Theta_W) + m^2_t \tag{1}$$

$$M^2_{\tilde{t}_R} = M^2_{\tilde{U}} + \frac{2}{3} m^2_Z \cos 2\beta \sin^2\Theta_W + m^2_t \tag{2}$$

$$M^2_{\tilde{b}_L} = M^2_{\tilde{Q}} - m^2_Z \cos 2\beta (\frac{1}{2} - \frac{1}{3}\sin^2\Theta_W) + m^2_b \tag{3}$$

$$M^2_{\tilde{b}_R} = M^2_{\tilde{D}} - \frac{1}{3} m^2_Z \cos 2\beta \sin^2\Theta_W + m^2_b \tag{4}$$

The off-diagonal elements are $m_t(A_t - \mu\cot\beta)$ and $m_b(A_b - \mu\tan\beta)$ for top and bottom squarks, respectively. For scalar top quarks there can be an appreciable amount of $\tilde{t}_L$-$\tilde{t}_R$ mixing due to the large top quark mass. Similarly, for the scalar bottom quarks $\tilde{b}_L$-$\tilde{b}_R$ mixing is also non-negligible if $\tan\beta \gtrsim 10$. The mass eigenvalues for the top and bottom squarks are

$$M^2_{\tilde{t}_{1,2}} = \frac{1}{2}(m^2_{\tilde{t}_L} + m^2_{\tilde{t}_R} \mp \sqrt{(m^2_{\tilde{t}_L} - m^2_{\tilde{t}_R})^2 + 4m^2_t(A_t - \mu\cot\beta)^2} \tag{5}$$

$$M^2_{\tilde{b}_{1,2}} = \frac{1}{2}(m^2_{\tilde{b}_L} + m^2_{\tilde{b}_R} \mp \sqrt{(m^2_{\tilde{b}_L} - m^2_{\tilde{b}_R})^2 + 4m^2_b(A_b - \mu\tan\beta)^2} \tag{6}$$

where $\tilde{t}_1$ and $\tilde{b}_1$ denote the lighter eigenstates.

We show in Figs. 1a and b the stop and sbottom mass eigenvalues as a function of μ for $M_{\tilde{Q}} = M_{\tilde{U}} = M_{\tilde{D}} = A_t = A_b = 150$ GeV, $\tan\beta = 2$, and for $M_{\tilde{Q}} = M_{\tilde{U}} =$

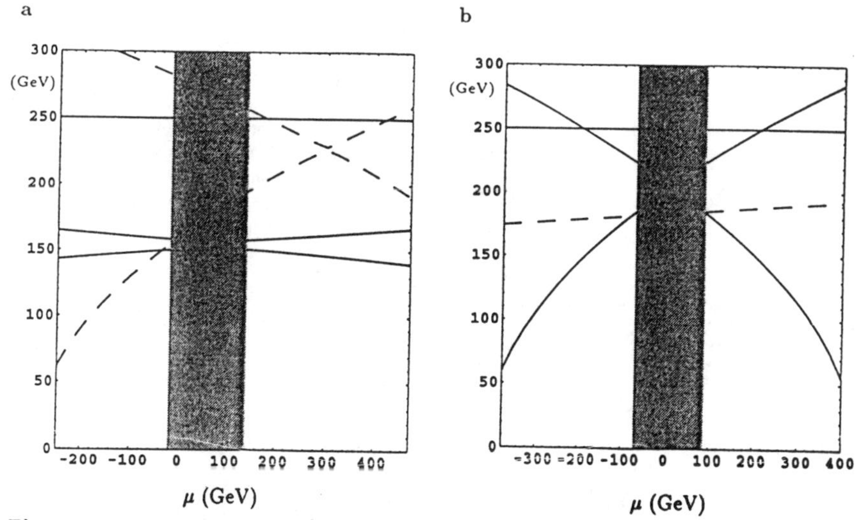

Fig.1: Mass eigenvalues of stops $\tilde{t}_{1,2}$ (– –) and sbottoms $\tilde{b}_{1,2}$ (—) as a function of μ, for m_b = 5 GeV, m_t = 175 GeV, M = 100 GeV, (a) $\tan\beta = 2$, $M_{\tilde{Q}} = M_{\tilde{U}} = M_{\tilde{D}} = A_t = A_b = 150$ GeV, (b) $\tan\beta = 20$, $M_{\tilde{Q}} = M_{\tilde{U}} = M_{\tilde{D}} = A_t = A_b = 200$ GeV. The grey area is covered by LEP.

$M_{\tilde{D}} = A_t = A_b = 200$ GeV, $\tan\beta = 20$. As can be seen, for $\tan\beta = 2$ (Fig. 1a) the $\tilde{t}_1$ is the lightest of the four states in a large region of μ. For $\tan\beta = 20$ (Fig. 1b) the $\tilde{b}_1$ is lighter then the $\tilde{t}_1$ and presumably also the lightest squark.

It is well-known that the cross section for $e^+ + e^- \rightarrow \tilde{t}_1 + \bar{\tilde{t}}_1$ has a peculiar dependence on the stop-mixing parameters, because the $Z^0\tilde{t}_1\bar{\tilde{t}}_1$ coupling vanishes for the mixing angle $\Theta_{\tilde{t}} = 0.98$. We expect a similar behaviour for the cross section of $e^+ + e^- \rightarrow \tilde{b}_1 + \bar{\tilde{b}}_1$, because the $Z^0\tilde{b}_1\bar{\tilde{b}}_1$ coupling vanishes at $\Theta_{\tilde{b}} = 1.17$. The cross sections have their minimum values at those values of the mixing angles, which are given by $\cos^2\Theta_{\tilde{t}} = \frac{4}{3}\sin^2\Theta_W \times F(s)$ and $\cos^2\Theta_{\tilde{b}} = \frac{2}{3}\sin^2\Theta_W \times F(s)$, respectively. The function $F(s) = 1 + 2\frac{s-m_Z^2}{s}\frac{\cos^2\Theta_W(1-2\sin^2\Theta_W)}{1-4\sin^2\Theta_W+8\sin^4\Theta_W} \approx 1 + 1.7\frac{s-m_Z^2}{s}$, takes into account the interference of the γ and Z^0-exchange contributions

We show in Fig. 2 the contour lines in the $M_{\tilde{Q}} - A_t$ plane for the total cross section of $e^+ + e^- \rightarrow \tilde{t}_1 + \bar{\tilde{t}}_1$, for $\sqrt{s} = 500$ GeV, taking $\tan\beta = 2$, $\mu = -200$ GeV, $M_{\tilde{U}} = M_{\tilde{Q}}$, $m_t = 175$ GeV. This cross section can go up to 120 fb. We plot only the parameter region $M_{\tilde{Q}} \leq 500$ GeV and $|A_t| \leq 1000$ GeV. As can be seen, the kinematically allowed parameter range may be larger than these values, but we expect further theoretical restrictions on these parameters, in particular $|A_t| \lesssim 3M_{\tilde{Q}}$, which we also show in Fig. 2. We also expect a lower bound on $M_{\tilde{Q}}$ coming from the lower bound on $M_{\tilde{b}_1}$. Since $M_{\tilde{b}_1}$ also depends on the parameters $M_{\tilde{D}}$ and A_b, this bound cannot be shown without making further assumptions. Assuming $M_{\tilde{D}} = M_{\tilde{Q}}$, $A_b = A_t$, and $m_b = 5$ GeV we

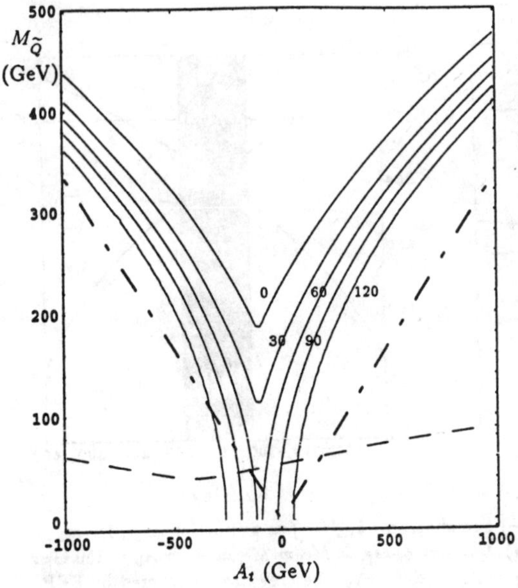

Fig.2: Contour lines for the cross section of $e^+e^- \to \tilde{t}_1\bar{\tilde{t}}_1$ (in fb) in the $M_{\tilde{Q}}$–A_t plane, for $\sqrt{s}$ = 500 GeV, m_t = 175 GeV, $\tan\beta$ = 2, μ = -200 GeV, $M_{\tilde{U}} = M_{\tilde{Q}}$. Also shown are the lines for $m_{\tilde{b}_1}$ = 45 GeV (– –, assuming $M_{\tilde{D}} = M_{\tilde{Q}}$) and $|A_t| = 3\ M_{\tilde{Q}}$(- -).

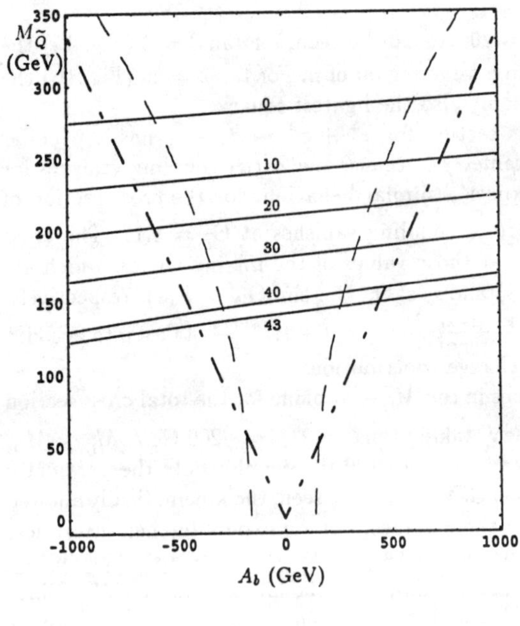

Fig.3: Contour lines for the cross section of $e^+e^- \to \tilde{b}_1\bar{\tilde{b}}_1$ (in fb) in the $M_{\tilde{Q}}$–A_b plane, for $\sqrt{s}$ = 500 GeV, m_b = 5 GeV, $\tan\beta$ = 20, μ = -200 GeV, $M_{\tilde{D}} = M_{\tilde{Q}}$. Also shown are the lines for $m_{\tilde{t}_1}$ = 45 GeV (– –, assuming $M_{\tilde{U}} = M_{\tilde{Q}}$) and $|A_b|$ = 3 $M_{\tilde{Q}}$(- -).

also plot the line $M_{\tilde{b}_1} = 45$ GeV in Fig. 2, indicating in this way the approximate position of the lower bound on $M_{\tilde{Q}}$.

Similarly, Fig. 3 is a contour plot of the total cross section of $e^+ + e^- \to \tilde{b}_1 + \bar{\tilde{b}}_1$ in the $M_{\tilde{Q}} - A_b$ plane for $\sqrt{s} = 500$ GeV, $\tan\beta = 20$, $\mu = -200$ GeV, $M_{\tilde{D}} = M_{\tilde{Q}}$, $m_b = 5$ GeV. The values of this section can go up to 40 fb. Here the theoretical constraint $|A_b| \lesssim 3M_{\tilde{Q}}$ is more restrictive than for stop production. In the present case it gives almost the same restrictions as the experimental lower bound on the stop mass, $M_{\tilde{t}_1} \gtrsim 45$ GeV. However, one must keep in mind that the particle masses and couplings also depend on the numerical values of $\tan\beta$ and μ (see Figs. 1). The situation will be different if other values for these parameters are taken.

In the calculations of the cross sections shown in Figs. 2 and 3 we have used the tree level formulae of [11,24]. QCD corrections and corrections due to initial state radiation and beamstrahlung are not included. Taking into account the QCD correction with the formulae given in [11] would lead to cross section values which are up to 40 percent higher. The corrections due to initial state radiation and beamstrahlung turn out to be of the order of 10 percent.

The cross sections for $e^+ + e^- \to \tilde{t}_1 + \bar{\tilde{t}}_1$ and $e^+ + e^- \to \tilde{b}_1 + \bar{\tilde{b}}_1$ show a characteristic dependence on the polarization of the initial beams. We have calculated the cross sections for polarized e^- beams. In the case of stop production we have obtained a polarization asymmetry of approximately 40 percent in a large part of the parameter region. For sbottom production the polarization asymmetry can reach values of 90 percent for most of the parameters considered. Similar to the case of production of sleptons or squarks of light flavours [25], experiments with polarized beams will be very helpful for the determination of the underlying model parameters.

At $\sqrt{s} = 500$ GeV also the higher mass eigenstates of stop and sbottom can be produced in a restricted region of the parameters. The parameter region of the process $e^+ + e^- \to \tilde{t}_1 + \bar{\tilde{t}}_2 +$ c.c. is approximately $M_{\tilde{Q}} \lesssim 250$ GeV, $|A_t| \lesssim 500$ GeV for $\tan\beta = 2$, $\mu = -200$ GeV. Its cross section can go up to 30 fb. The reaction $e^+ + e^- \to \tilde{t}_2 + \bar{\tilde{t}}_2$ is possible for $M_{\tilde{Q}} \lesssim 200$ GeV, $|A_t| \lesssim 250$ GeV if $\tan\beta = 2$, $\mu = -200$ GeV. Its cross section can be as high as 50 fb. Similarly, the parameter domains of the reactions $e^+ + e^- \to \tilde{b}_1 + \bar{\tilde{b}}_2 +$ c.c. and $e^+ + e^- \to \tilde{b}_2 + \bar{\tilde{b}}_2$ are approximately 130 GeV $\lesssim M_{\tilde{Q}} \lesssim 250$ GeV and 130 GeV $\lesssim M_{\tilde{Q}} \lesssim 200$ GeV, respectively, for $M_{\tilde{Q}} = M_{\tilde{D}}$, $\tan\beta = 20$, $\mu = -200$ GeV. Their cross sections can go up to 40 fb and 10 fb, respectively. At higher CMS energies these reactions will become even more important and will provide additional tests of the stop and sbottom systems.

3 Stop and Sbottom Decays

The squarks of the third generation can decay according to

$$\tilde{t}_i \to t + \tilde{\chi}^0_k, \qquad b + \tilde{\chi}^+_k \tag{7}$$

$$\tilde{b}_i \rightarrow b + \tilde{\chi}^0_k, \qquad t + \tilde{\chi}^-_k. \tag{8}$$

If these decays are kinematically allowed, they will be important. Due to the influence of the top and bottom Yukawa terms and because of L–R mixing the decay patterns of stops and sbottoms will be different from those of the squarks of the first two generations. Moreover, in case of strong L–R mixing the splitting between the two mass eigenstates may be so large that the following additional decay modes are also present [13]:

$$\tilde{t}_2 \rightarrow \tilde{t}_1 + Z^0(h^0, H^0, A^0), \qquad \tilde{b}_1 + W^+(H^+) \tag{9}$$

$$\tilde{b}_2 \rightarrow \tilde{b}_1 + Z^0(h^0, H^0, A^0), \qquad \tilde{t}_1 + W^-(H^-). \tag{10}$$

The transitions

$$\tilde{t}_1 \rightarrow \tilde{b}_1 + W^+(H^+), \tag{11}$$

or

$$\tilde{b}_1 \rightarrow \tilde{t}_1 + W^-(H^-), \tag{12}$$

can occur if the mass difference $|M_{\tilde{t}_1} - M_{\tilde{b}_1}|$ is large enough. If $M_{\tilde{t}_1} < m_{\tilde{\chi}^0_1} + m_t$ as well as $M_{\tilde{t}_1} < m_{\tilde{\chi}^\pm_1} + m_b$, then for the $\tilde{t}_1$ only the higher–order decay

$$\tilde{t}_1 \rightarrow c + \tilde{\chi}^0_1 \tag{13}$$

is possible [10].

We show in Fig. 4 a–d the branching ratios for the decays of $\tilde{b}_1$ and $\tilde{t}_1$ as a function of μ, for $\tan\beta = 2$, $M_{\tilde{Q}} = M_{\tilde{U}} = M_{\tilde{D}} = A_t = A_b = 150$ GeV, and for $\tan\beta = 20$, $M_{\tilde{Q}} = M_{\tilde{U}} = M_{\tilde{D}} = A_t = A_b = 200$, respectively, with the other parameters $M = 100$ GeV, $m_A = 150$ GeV, $m_t = 175$ GeV, $m_b = 5$ GeV. For the calculation of the various decay widths we have used the formulae of [13].

As can be seen in Figs. 4a and c, the main decay mode of the light stop is $\tilde{t}_1 \rightarrow b + \tilde{\chi}^+_1$. In a large region of μ its branching ratio is 100 percent. For $\mu \lesssim -150$ GeV and for the choice of the parameters as in Fig. 4a, only the decay $\tilde{t}_1 \rightarrow c + \tilde{\chi}^0_1$ is possible. For $|\mu| \lesssim 120$ GeV also $\tilde{t}_1 \rightarrow b + \tilde{\chi}^+_2$ is possible. The transition $\tilde{t}_1 \rightarrow \tilde{b}_1 + W^+$ can occur for $|\mu| \gtrsim 300$ GeV (Fig. 4c). Its branching ratio increases rapidly with $|\mu|$ and can exceed 50 percent. For $m_A = 150$ GeV, as assumed in Fig. 4, the transition $\tilde{t}_1 \rightarrow \tilde{b}_1 + H^+$ is kinematically forbidden. This decay would be possible for $|\mu| \gtrsim 300$ GeV, if $m_A \lesssim 120$ GeV, and the other parameters as in Fig. 4c.

The $\tilde{b}_1$ decays mainly into $b + \tilde{\chi}^0_1$ and $b + \tilde{\chi}^0_2$. The decays into charginos are kinematically forbidden in the whole parameter domain considered, because of the large top-quark mass. The branching ratio for $\tilde{b}_1 \rightarrow b + \tilde{\chi}^0_1$ is at least 50 percent, that for $\tilde{b}_1 \rightarrow b + \tilde{\chi}^0_2$ can go up to 50 percent. For $|\mu| \lesssim 120$ GeV also the decay $\tilde{b}_1 \rightarrow b + \tilde{\chi}^0_3$ is possible (Figs. 4b and d).

We have also calculated the branching ratios for the decays of $\tilde{b}_2$ and $\tilde{t}_2$. These particles can also be produced at $\sqrt{s} = 500$ GeV, although the available parameter

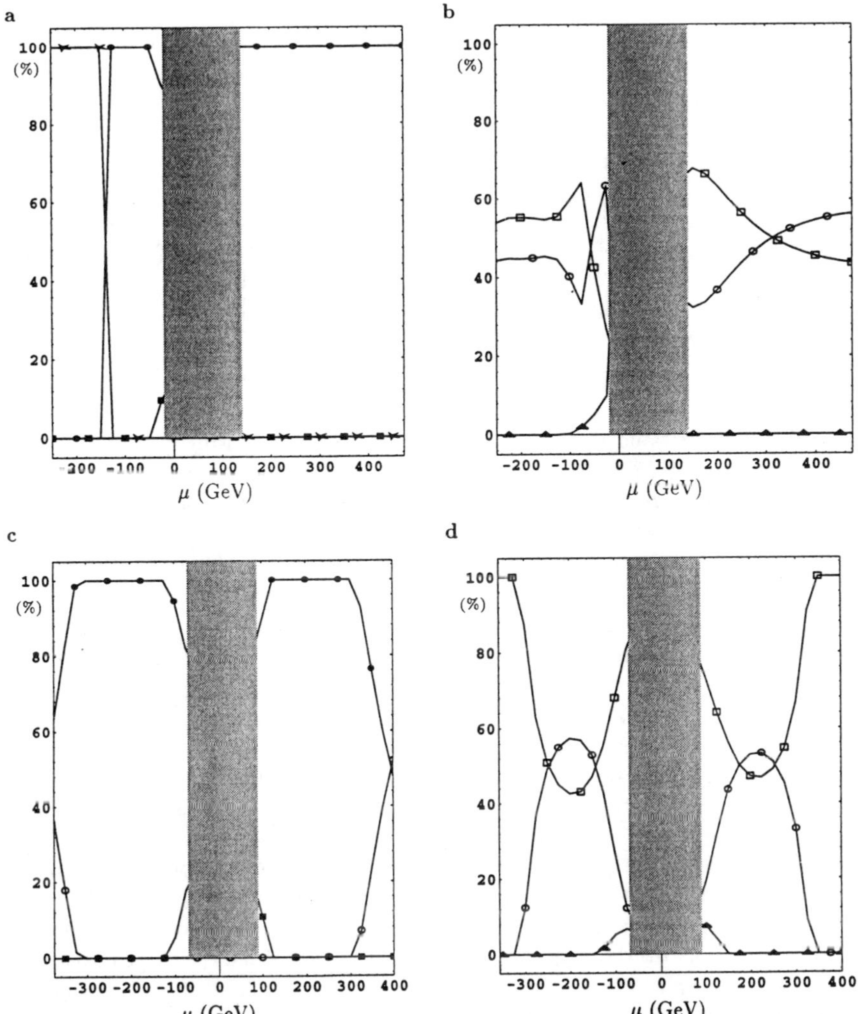

Fig.4: Branching ratios for $\tilde{t}_1$ and $\tilde{b}_1$ decays as a function of μ, for M = 100 GeV, m_t = 175 GeV, m_b = 5 GeV, **(a)**, **(b)** $\tan\beta$ = 2, $M_{\tilde{Q}} = M_{\tilde{U}} = M_{\tilde{D}} = A_t = A_b$ = 150 GeV, and **(c)**, **(d)** $\tan\beta$ = 20, $M_{\tilde{Q}} = M_{\tilde{U}} = M_{\tilde{D}} = A_t = A_b$ = 200 GeV. The curves in **(a)** and **(c)** correspond to the transitions of $\tilde{t}_1$ into • $\to b + \chi_1^{\pm}$, ■ $\to b + \chi_2^{\pm}$, ⋆ $\to c + \chi_1^0$, ∘ $\to W^+ + \tilde{b}_1$. The curves in **(b)** and **(d)** correspond to the transitions of $\tilde{b}_1$ into □ $\to b + \chi_1^0$, ∘ $\to b + \chi_2^0$, △ $\to b + \chi_3^0$. The grey area is covered by LEP.

domain is smaller. The branching ratios of their decays into $\tilde{\chi}_1^0$ are always less than 10 percent. The transitions into $\tilde{b}_1$ and $\tilde{t}_1$, eqs. (9) and (10), can have appreciable branching ratios. Obviously, the transitions into Higgs particles depend strongly on the parameter m_A, the mass of A^0. In these calculations the radiative corrections to the masses and couplings of the Higgs particles have to be included [26]. In our calculations we have used the formulae of [27].

From the preceeding discussion it follows that there is a high probability sbottoms and stops will decay into charginos or heavier neutralinos. These particles then have the two-body decays $\tilde{\chi}_i^\pm \to Z^0 + \tilde{\chi}_k^\pm, h^0(A^0 \text{ or } H^0) + \tilde{\chi}_k^\pm, W^\pm + \tilde{\chi}_k^0, H^\pm + \tilde{\chi}_k^0$ and $\tilde{\chi}_i^0 \to Z^0 + \tilde{\chi}_k^0, h^0(A^0 \text{ or } H^0) + \tilde{\chi}_k^0, W^\pm + \tilde{\chi}_k^\mp, H^\pm + \tilde{\chi}_k^\mp$ [18,24,28]. If these two-body decays are kinematically not allowed, then the charginos and neutralinos have leptonic or hadronic three-body decays, as for example $\tilde{\chi}_i^\pm \to \ell^\pm + \overset{(-)}{\nu} + \tilde{\chi}_k^0, q + \bar{q}\,' + \tilde{\chi}_k^0$, and $\tilde{\chi}_i^0 \to f + \bar{f} + \tilde{\chi}_k^0, f + \bar{f}' + \tilde{\chi}_k^\pm$, with f denoting a quark or lepton [21]. In this manner a decay chain may be developed which ends when the $\tilde{\chi}_1^0$ is reached.

The direct transitions into the LSP give rise to large missing p_T, which is the main signature for SUSY. Cascade decays play an important rôle in all SUSY particle decays, since they crucially influence the possible signatures. The cascade decays give rise to events with smaller missing p_T, but with leptons and/or jets in the final state. Qualitatively, cascade decays are possible if either $|\mu| \lesssim M_{\tilde{q}}$, or $M \lesssim M_{\tilde{q}}$. If $|\mu| \gtrsim M_{\tilde{q}}$ and $M_{\tilde{q}} \lesssim M \lesssim 2M_{\tilde{q}}$, then only transitions into the lightest neutralino are allowed.

It can be seen from Figs. 4b and d that the $\tilde{b}_1$ has a sufficiently large branching ratio for the decay into the LSP. Therefore, the most important signature for $\tilde{b}_1$ pair production is two b jets plus missing p_T. This signature will be applicable similar to the case of right squarks of light flavours. In order to distinguish the $\tilde{b}_1$ from the other squarks, however, b tagging will be crucial. Tagging of b jets will also be necessary for the reconstruction of the decays of the $\tilde{b}_1$ into $b + \tilde{\chi}_2^0$, as they will lead to final states containing b jets plus lepton pairs plus p_{miss}, or b jets plus light–quark jets plus p_{miss} etc.

Concerning the signatures of $\tilde{t}_1$ production, the situation is a bit more involved. If only the decay $\tilde{t}_1 \to c + \tilde{\chi}_1^0$ is allowed, the p_{miss} signature can be applied. However, in order to distinguish the $\tilde{t}_1$ from the other squarks, it would then be necessary to distinguish the c jets from the quark jets of other flavours. If the $\tilde{t}_1$ decays into $b + \tilde{\chi}_1^+$, then again b tagging will be important. For the reconstruction of the χ_1^+, techniques similar to those described in [29] may be applicable.

4 Conclusions

In this talk we discussed the production of stop and sbottom pairs in e^+e^- annihilation at $\sqrt{s} = 500$ GeV. We presented numerical predictions within the Minimal Supersymmetric Standard Model for the production cross sections and the decay rates and analyzed their SUSY parameter dependence. If $\tan\beta \gtrsim 10$, not only the top Yukawa terms must be included, but also the bottom Yukawa terms. The production cross sections as well as the decay rates of stops and sbottoms depend on the parameters of the mixing matrices in a characteristic way. If the differences between the eigenstates are large enough, transitions between these states by emmission of a vector boson or a Higgs boson can occur. The signature for sbottom pair production is b jets plus p_{miss}. An important signature for stops is b jets plus leptons and/or jets plus p_{miss}. Tagging of b jets is necessary to distinguish sbottoms and stops from other squarks.

Acknowledgements

A. B. is very grateful to Prof. S. Pokorski and the organizers for inviting him to this interesting workshop, and to the Institute of Theoretical Physics of the University of Warsaw for the hospitality extended to him.

References

[1] H.P. Nilles, Phys. Rep. 110 (1984); 1
A. B. Lahanas, D.V. Nanopoulos, Phys. Rep. 145 (1985) 1;
R. Barbieri, Riv. Nuov. Cim. 11 (1988) 1

[2] H.E. Haber, G.L. Kane, Phys. Rep. 117 (1985) 75

[3] J. Ellis, S. Rudaz, Phys. Lett. B128 (1983) 248
G. Altarelli, R. Rückl, Phys. Lett. B144 (1984) 126
I. Bigi, S. Rudaz, Phys. Lett. B153 (1985) 335

[4] M. Olechowski, S. Pokorski, Nucl. Phys. B404 (1993) 590; ibidem B426 (1994) 269

[5] S. Kelley, J. L. Lopez, D. V. Nanopoulos, H. Pois, K. Yuan, Phys. Rev. D47 (1993) 2461
J. L. Lopez, D. V. Nanopoulos, A. Zichichi, CERN-PPE/94-82

[6] G. L. Kane, C. Kolda, L. Roszkowski, J. D. Wells, Phys. Rev. D49 (1994) 6173

[7] V. Barger, M. S. Berger, P. Ohmann, Phys. Rev. D49 (1994) 4908

[8] H. Baer, M. Drees, C. Kao, M. Nojiri, X. Tata, Phys. Rev. D50 (1994) 2148

[9] R. Arnowitt, P. Nath, Phys. Rev. Lett. 69 (1992) 725; Phys. Rev. D49 (1994) 1479

[10] K. I. Hikasa, M. Kobayashi, Phys. Rev. D36 (1987) 724

[11] M. Drees, K. I. Hikasa, Phys. Lett. B252 (1990) 127

[12] M. Fukugita, H. Murayama, M. Yamaguchi, T. Yanagida, Phys. Rev. Lett. 72 (1994) 3009
T. Kon, T. Nonaka, preprint ITP-SU-94/02

[13] A. Bartl, W. Majerotto, W. Porod, Z. Phys. C64 (1994) 499

[14] S. Raby, these Proceedings

[15] M. Olechowski, these Proceedings

[16] H. Baer, J. Sender, X. Tata, Phys. Rev. D50 (1994) 4517

[17] W. de Boer, R. Ehret, D. I. Kazakov, Phys. Lett. B334 (1994) 220

[18] A. Bartl, W. Majerotto, B. Mößlacher, Proc. of the Workshop "e^+e^- Collisions at 500 GeV: The Physics Potential", Munich, Annecy, Hamburg, DESY 92-123B, p. 641 (P. M. Zerwas ed.); A. Bartl, W. Majerotto, ibidem, DESY 93-123C, p. 445

[19] CDF Collaboration, F. Abe et al., Phys. Rev. Lett. 69 (1992) 3439

[20] H. Baer, X. Tata, J. Woodside, Phys. Rev. D44 (1991) 207

[21] H. Baer, M. Drees, R. Godbole, J. F. Gunion, X. Tata, Phys. Rev. D 44 (1991) 725

[22] G. Coignet, Proc. of the 16th Int. Symposium on Lepton and photon Interactions, Ithaca, New York, 1993, p. 324, AIP Conference Proceedings 302 (P. Drell, D. Rubin, eds.)

[23] VENUS Collaboration, I. Shirai et al., Phys. Rev. Lett. 72 (1994) 3313

[24] A. Bartl, W. Majerotto, B. Mößlacher, Proc. of the Workshop "Phenomenological Aspects of Supersymmetry", München, 1991, Lecture Notes in Physics, Vol. 405, p. 119, Springer (W. Hollik, R. Rückl, J. Wess eds.)
A. Bartl, H. Fraas, W. Majerotto, Z. Phys. C34 (1987) 411

[25] JLC Group, "JLC-I", KEK-Report 92-16
R. Becker, C. Vander Velde, Proc. of the Workshop "e^+e^- Collisions at 500 GeV: The Physics Potential", Munich, Annecy, Hamburg, DESY 93-123C, p. 457 (P. M. Zerwas ed.)
L. J. Feng, D. E. Finnel, Phys. Rev. D49 (1994) 2369

[26] P. H. Chankowski, S. Pokorski, J. Rosiek, Phys. Lett. B274 (1992) 191; ibidem B281 (1992) 100; ibidem B286 (1992) 307; Nucl. Phys. B423 (1994) 437; ibidem B423 (1994) 497

[27] J. Ellis, G. Ridolfi, F. Zwirner, Phys. Lett. B262 (1991) 477
A. Brignole, J. Ellis, G. Ridolfi, F. Zwirner, Phys. Lett. B271 (1991) 123

[28] J. F. Gunion, H. E. Haber, Phys. Rev. D37 (1988) 2515
H. Baer, A. Bartl, D. Karatas, W. Majerotto, X. Tata, Int. J. Mod. Phys. A4 (1989) 4111

[29] J.-F. Grivaz, Proc. of the Workshop "e^+e^- Collisions at 500 GeV: The Physics Potential", Munich, Annecy, Hamburg, DESY 92-123B, p. 663 (P. M. Zerwas ed.)

REVEALING MINIMAL SUPERGRAVITY AT COLLIDING BEAM EXPERIMENTS

HOWARD BAER
Dep't of Physics, Florida State University
Tallahassee, FL, 32306 USA

ABSTRACT

I briefly review the theoretical framework which is thought to be the low energy limit of supergravity GUT or string models. I describe the current status of the event generator program ISAJET which links the above mentioned theoretical framework with experimental detection of π's, K's, μ's, etc., at pp, $p\bar{p}$ and e^+e^- colliders. Then, I review expectations for supersymmetry at LEP 180, the Tevatron and its various upgrades, and also the CERN LHC project.

1. Low Energy Effective Supergravity Framework

Incorporation of symmetries into particle physics has frequently led to the postulation of new particles. For instance, when Dirac incorporated Lorentz symmetry into quantum mechanics, it led to the prediction of an anti-matter partner particle for every known particle. Likewise, supersymmetry[1], as an extension of the Poincaré group of space-time symmetries, postulates a superpartner particle for every known particle state. When supersymmetry is promoted to a local symmetry (supergravity or SUGRA), one must necessarily introduce a graviton field into particle physics, and the Lagrangian is invariant under general co-ordinate transformations, reducing to Einsteinian gravity in the appropriate limit.

Specific motivations for weak-scale supersymmetry include *i*) cancellation of quadratic divergences associated with fundamental scalar (Higgs) fields, which is only natural if supersymmetry is broken somewhere around the Fermi scale ($(\sqrt{2}G_F)^{-\frac{1}{2}} = 248$ GeV), and *ii*) unification of gauge (and Yukawa) couplings, which works best for 100-1000 GeV super-particles.

The Minimal Supersymmetric Standard model (MSSM) is the resultant theory one obtains by supersymmetrizing the Standard Model (SM) with the least number of additional degrees of freedom or interactions introduced. Supersymmetry (SUSY) breaking is then included by introducing various soft SUSY breaking terms; without further constraints, this leads to an enormous proliferation of model parameters. At high energy scales of $\mathcal{O}(M_{pl})$ or $\mathcal{O}(M_{GUT})$, it is commonly postulated that supersymmetry is broken in some "hidden sector" of the theory, and that superparticles then gain weak scale masses via gravitational interactions. This leads to a common mass m_0 for all scalar particles, and a common mass $m_{1/2}$ for all fermionic superparticles. These parameters, along with the common trilinear and bilinear couplings A and B, are all expected to have values on the order of 100-1000 GeV. The resultant spectrum of sparticles expected at collider energies is then obtained via renormalization group equation (RGE) running of the various parameters from some ultra-high mass scale down to the weak scale. A surprising bonus of this procedure is that electroweak symmetry breaking results as a consequence of supersymmetry breaking at the ultra-high scale. In this case, the B parameter can be traded for $\tan\beta$, and the magnitude of the Higgsino mixing term μ is adjusted to give the correct value of M_Z. It is thus common in the literature to see the low energy super-particle spectrum parameterized in terms of the set of mixed weak and unification scale parameters ($m_0, m_{1/2}, A_0, \tan\beta, sign(\mu)$ and m_t)[2,3].

2. Event Simulation

The theoretical ideas of supersymmetry and supergravity can most directly be tested via the observation of super-particles at high energy collider experiments. The crucial link between theory on the one hand, and experimental detection of π's, e's, K's, μ's and γ's on the other hand, is the event generator program. Recently, the MSSM has been incorporated into the event generator ISAJET 7.13[4]. A minimal supergravity model with gauge coupling unification and radiative electroweak symmetry breaking option is also included, so that, given the above SUGRA parameter set, the RGE equations are iterated until a consistent low energy spectrum is found. Thus we hope future experimental analyses will have access to the popular constrained model parameter set.

To simulate sparticle production and decay at a hadron collider, the following steps are taken:

- input the parameter set ($m_0, m_{1/2}, A_0, \tan\beta, sign(\mu), m_t$), or a less constrained MSSM set,
- all sparticle and Higgs masses and couplings are computed,
- all sparticle, top and Higgs decay modes and branching fractions are calculated,
- all lowest order $2 \to 2$ sparticle production processes are calculated (if desired, subsets of the reactions can be selected),
- the hard scattering is convoluted with CTEQ2L PDF's[5],
- initial and final state QCD radiation is calculated with the parton shower model,
- particles and sparticles decay through their various cascades,
- quarks and gluons are hadronized, and mesons are further decayed,
- the underlying soft scattering of beam remnants is modelled,
- the resulting event and event history is generated for interface with detector simulations, or for direct analysis.

In addition, ISAJET 7.13 includes all lowest order $e^+e^- \to SUSY$[6] and Higgs particle production processes, so that cascade decays and minimal SUGRA can be simulated for electron machines as well. The e^+e^- generator does not at this time include initial state photon radiation, spin correlations between initial and final state sparticles (important only for chargino/neutralino production), or sparticle decay matrix elements beyond pure phase space.

In the near future, we hope to include in ISAJET the following:

- direct Higgs production mechanisms for hadron colliders,
- large $\tan\beta$ solution, (currently, ISAJET is only valid for $\tan\beta \lesssim 10$,
- further breaking of squark degeneracies.

Separate programs can be extracted from ISAJET 7.13 which generate just the SUGRA mass solution plus decay table (ISASUGRA), and also the sparticle decay table with MSSM input (ISASUSY). ISAJET is maintained in PATCHY format. The complete card image PAM file for ISAJET 7.13 can be copied across HEPNET, the high energy physics DECNET, from
`bnlcl6::$2$dua14:[isajet.isalibrary]isajet.car`. A Unix makefile `makefile.unix` and a VMS `isamake.com` are available in the same directory. The same files can be obtained by anonymous ftp from `bnlux1.bnl.gov:pub/isajet`.

3. Supersymmetry at LEP and LEP 180

Currently, there is no evidence for supersymmetry at LEP, although radiative correction analyses are completely consistent with supersymmetric theories, and precision measurement of α_s and $\sin^2\theta_W$ are consistent with SUSY-GUT grand unification of couplings. Direct searches for sparticles at LEP have lead[7] to mass limits of $m_{\widetilde{W}_1} > 47$ GeV, $m_{\tilde{\ell}} > 45$ GeV and $m_{\tilde{t}_1} > 45$ GeV (except for some unlikely regions of parameter space). Furthermore, precision measurements of Z-boson properties constrain $m_{\tilde{\nu}} > 43$ GeV (assuming degenerate sneutrinos), and lead to parameter-dependent constraints on $Z \to \widetilde{Z}_1\widetilde{Z}_1$ due to limits on the Z invisible width. In addition, search for anomalous Z decays put limits on $Z \to \widetilde{Z}_i\widetilde{Z}_j$ decay rates. Finally, search for SM Higgs bosons requires $m_H > 63$ GeV, which also constrains the Higgs sector of the MSSM.

In late 1995, CERN should begin ramping up towards LEP 180. LEP 180 should be able to search[8] thoroughly for charginos, sleptons and SM Higgs bosons up to $m_{\widetilde{W}_1} \sim 85 - 90$ GeV. In addition, some extra parameter space can be explored via the reaction $e^+e^- \to \widetilde{Z}_1\widetilde{Z}_2$. The approximately accessible regions[3] are shown in the m_0 vs. $m_{1/2}$ plane of Fig. 1, where we have taken $A_0 = 0$ and $\tan\beta = 2$ ($m_t = 170$ GeV).

4. Supersymmetry at the Tevatron

4.1 Multi-lepton events

In the past, most of the focus of CDF and D0 experiments has been on the search for gluino and squark production by searching for multi-jet plus $\not\!\!E_T$ events. In this case, the gluinos and squarks decay via hadronic cascades through the various super-partners, yielding jets plus $\not\!\!E_T$ from the final state $\widetilde{Z}_1$'s. Both experiments have presented recent bounds[9], which imply $m_{\tilde{g}} > 157$ GeV if $m_{\tilde{q}} >> m_{\tilde{g}}$, and $m_{\tilde{g}} > 218$ GeV if $m_{\tilde{q}} \sim m_{\tilde{g}}$.

In addition, one can also search for supersymmetry by searching for multi-jet+$\not\!\!E_T$

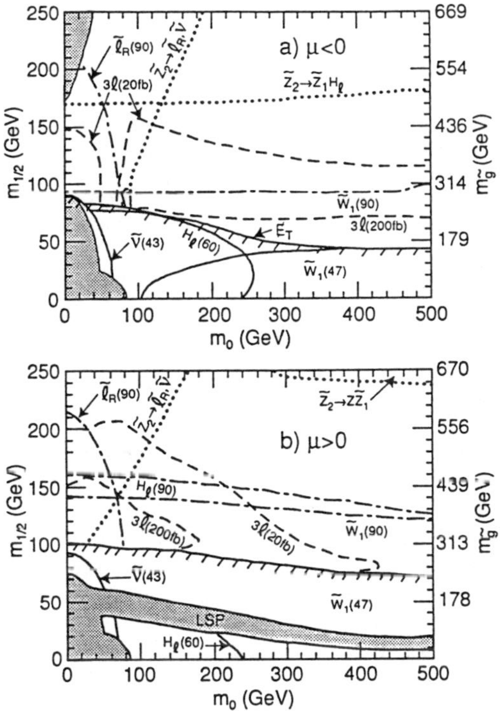

Figure 1: Excluded regions and the reach of LEP II and the Tevatron within the framework of the minimal supergravity model with radiative electroweak symmetry breaking. We show the results in the $m_0 - m_{1/2}$ plane for $A_0 = 0$, $\tan\beta = 2$ and $a)\mu < 0$ and $b)\mu > 0$. The shaded region is excluded for theoretical reasons discussed in the text, while the region below the hatched line is excluded by the experimental constraints from LEP and Tevatron. The area below the dashed-dotted lines labeled $\tilde{\ell}_R(90)$, $\widetilde{W}_1(90)$, $H_\ell(90)$ [$m_{H_\ell} < 90$ GeV over the whole plane exhibited in case (a)] will be probed at LEP II via direct searches for sleptons, charginos, and the lightest Higgs scalar, respectively, while the region below the dashed lines should be accessible via the isolated trilepton search for $\widetilde{W}_1\widetilde{Z}_2$ events a as discussed in the text. We terminate these dashed and dashed-dotted lines at the hatched boundary for reasons of clarity. Finally, the dotted lines denote the kinematic boundary of various two body $\widetilde{Z}_2$ decays that significantly alter the phenomenology.

Table: Discovery reach of various options of future hadron colliders.

Signal	Tevatron II 0.1 fb^{-1} 1.8 TeV	Main Injector 1 fb^{-1} 2 TeV	Tevatron* 10 fb^{-1} 2 TeV	DiTevatron 1 fb^{-1} 4 TeV	LHC 10 fb^{-1} 14 TeV
$\not{E}_T(\tilde{q} \gg \tilde{g})$	$\tilde{g}(210)/\tilde{g}(185)$	$\tilde{g}(270)/\tilde{g}(200)$	$\tilde{g}(340)/\tilde{g}(200)$	$\tilde{g}(450)/\tilde{g}(300)$	$\tilde{g}(1300)$
$l^\pm l^\pm(\tilde{q} \gg \tilde{g})$	$\tilde{g}(160)$	$\tilde{g}(210)$	$\tilde{g}(270)$	$\tilde{g}(320)$	
$\tilde{g}, \tilde{q} \to 3l\ (\tilde{q} \gg \tilde{g})$	$\tilde{g}(180)$	$\tilde{g}(260)$	$\tilde{g}(430)$	$\tilde{g}(320)$	
$\not{E}_T(\tilde{q} \sim \tilde{g})$	$\tilde{g}(300)/\tilde{g}(245)$	$\tilde{g}(350)/\tilde{g}(265)$	$\tilde{g}(400)/\tilde{g}(265)$	$\tilde{g}(580)/\tilde{g}(470)$	$\tilde{g}(2000)$
$l^\pm l^\pm(\tilde{q} \sim \tilde{g})$	$\tilde{g}(180-230)$	$\tilde{g}(320-325)$	$\tilde{g}(385-405)$	$\tilde{g}(460)$	$\tilde{g}(1000)$
$\tilde{g}, \tilde{q} \to 3l\ (\tilde{q} \sim \tilde{g})$	$\tilde{g}(240-290)$	$\tilde{g}(425-440)$	$\tilde{g}(550^*)$	$\tilde{g}(550^*)$	$>\sim \tilde{g}(1000)$
$\widetilde{W}_1 \widetilde{Z}_2 \to 3l$	$\tilde{g}(200)$	$\tilde{g}(250-450)$	$\tilde{g}(300-500)$	$\tilde{g}(250-470)$	$\tilde{g}(400\text{–}700)$
$\tilde{t}_1 \to c\widetilde{Z}_1$	$\tilde{t}_1(80\text{–}100)$	$\tilde{t}_1(120)$			
$\tilde{t}_1 \to b\widetilde{W}_1$	$\tilde{t}_1(80\text{–}100)$	$\tilde{t}_1(120)$			
$\Theta(\tilde{t}_1\tilde{t}_1^*) \to \gamma\gamma$	—	—	—	—	$\tilde{t}_1(250)$
$\tilde{l}\tilde{l}^*$	$\tilde{l}(50)$	$\tilde{l}(50)$	$\tilde{l}(50)$		$\tilde{l}(250\text{–}300)$

events which contain isolated leptons[10]. Here, it is important to generate all possible supersymmetric pair reactions, since frequently *e.g.* chargino/neutralino production can contribute significantly.

- The single isolated lepton topology is swamped by huge backgrounds mainly coming from direct W boson production, so is not a promising discovery channel.
- The dilepton sample can be divided into opposite-sign (OS) and same-sign (SS) isolated dileptons. The OS sample can have large backgrounds from $t\bar{t}$, WW and $Z \to \tau\bar{\tau}$ processes, although signal can be seen above background in many cases. The SS signal has much smaller backgrounds. Traditionally, the SS sample is heralded[11] as a signature for the Majorana nature of the gluino, where $p\bar{p} \to \tilde{g}\tilde{g}X$ occurs, with each gluino decaying via $\tilde{g} \to q\bar{q}'\widetilde{W}_1$, where the $\widetilde{W}_1$ are same-sign, and undergo further leptonic decay *e.g.* $\widetilde{W}_1 \to \ell\nu\widetilde{Z}_1$. In actual simulations, these events come from a wide variety of sources, including $\tilde{g}\tilde{q}$ production, various associated production mechanisms such as $\tilde{g}\widetilde{W}_1$, etc., and $\widetilde{W}_1\widetilde{Z}_2$ production where some of the final state leptons are lost due to cuts.
- Events containing three isolated leptons[12] can come again from several sources. One major source is from gluino and squark production, followed by cascade decays to $\widetilde{W}_1$ and $\widetilde{Z}_2$, where $\widetilde{W}_1 \to \ell\nu\widetilde{Z}_1$ and $\widetilde{Z}_2 \to \ell\bar{\ell}\widetilde{Z}_1$. These trilepton events are then expected to be accompanied by substantial jet activity. Another major source of such events is $p\bar{p} \to W^* \to \widetilde{W}_1\widetilde{Z}_2X \to 3\ell + \not{E}_T$ production, where the events are relatively free of hadronic activity.

We have evaluated the size of the various $\not{E}_T$ plus isolated multi-lepton signals after typical cuts, and evaluated backgrounds as well. The reach of the Tevatron[13] in these various channels is shown in Table 1. For small data samples on the order of 1-10 pb^{-1}, the $\not{E}_T$+jets channel offers the largest reach. But for larger data sets, $\sim 100 - 1000\ pb^{-1}$, the isolated trilepton sample offers the best opportunity for discovery of supersymmetry. In special cases, such as models with "no-scale" or "dilaton" boundary conditions, the reach can be even further[14].

4.2 Search for Top Squark

The lighter of the two super-partners of the top quark, the top squark ($\tilde{t}_1$), deserves special mention. In SUGRA theories, the diagonal masses $m^2_{\tilde{t}_L}$ and $m^2_{\tilde{t}_R}$ are driven to smaller values than the other squarks due to the contribution to the RGE's of the large top Yukawa coupling. Furthermore, large off-diagonal mass matrix terms can further reduce the $\tilde{t}_1$ mass. Hence, it is possible that $\tilde{t}_1$ may be accessible to collider searches, while the other sparticles are quite heavy. In fact, current collider limits on squark masses do not apply to $\tilde{t}_1$, and the best current bound is the above mentioned LEP limit.

Tevatron experiments can hope to better the LEP limit by searching for[15]:

- $p\bar{p} \to \tilde{t}_1\bar{\tilde{t}}_1X$ followed by $\tilde{t}_1 \to b\widetilde{W}_1$ decay, where the chargino decays either leptonically or hadronically. In this case, the visible final state particles are exactly as expected in the top-quark search, although the decay products are expected to be much softer, for the regions of sparticle masses accessible to Tevatron searches. By cutting away on "hard" event topologies, it seems $m_{\tilde{t}_1} \sim$ 100 GeV ought to be explorable.
- If the $\tilde{t}_1 \to b\widetilde{W}_1$ is kinematically closed, then $\tilde{t}_1$ is expected to decay via a flavor changing loop $\tilde{t}_1 \to c\widetilde{Z}_1$. In this case, the final state is $c\bar{c} + \not{E}_T$. Once again, it seems that suitable cuts ought to allow exploration of $m_{\tilde{t}_1} \sim 100$ GeV.

4.3 Multi-channel search for minimal supergravity

The question arises: can one combine the various search strategies within one framework? One needs a small enough parameter space that sparticle masses are correlated, but not too small so that the analyses is not sufficiently general. A well-motivated framework is the minimal supergravity model, where unification of gauge couplings is required, and radiative electroweak symmetry breaking takes place. The solution to this spectrum has been calculated in a number of places, and has also been recently incorporated into ISAJET. We see in Figure 1 the ranges[3] of m_0 vs. $m_{1/2}$ that can be explored by LEP 200, as well as by Tevatron trilepton searches given integrated luminosity of 100 pb^{-1} (200 fb contour) and 1000 pb^{-1} (20 fb contour). Also, one can correlate the projected reach in $\not{E}_T$ from Table 1 with this plot, since the corresponding gluino mass is listed on the right axis. Clearly, LEP 200 and Tevatron (Main Injector) are complementary machines.

5. Supersymmetry at the LHC

The LHC, operating at $\sqrt{s} = 14$ TeV, ought to be able to search for sparticles up to and slightly beyond the TeV scale. Polesello and the LHC Atlas group[16] have investigated the reach of LHC for gluinos and squarks via multi jet plus $\not{E}_T$ events, and find $m_{\tilde{g}} \sim 1300$ ought to be accessible if $m_{\tilde{g}} << m_{\tilde{q}}$, while $m_{\tilde{g}} \sim 2000$ GeV is accessible if $m_{\tilde{g}} \sim m_{\tilde{q}}$. Thus, if LHC doesn't find supersymmetry, it will be hard to believe low energy supersymmetry is realized in nature. If a signal in this channel is detected, then probably mass information can be gleaned by reconstructing hemispherical multi-jet invariant masses.

There exist as well the variety of multi-lepton topologies to be explored with LHC. These have been evaluated in Ref. [17], and further work is in progress. For large sparticle masses, the multileptons come from various cascade decay mechanisms, many involving cascade decay production of top quarks. These channels should provide solid confirmation, if not discovery, due to small backgrounds: again, the trilepton channel looks most promising.

Sleptons can be searched for at the LHC, and studies have shown that sleptons of mass $m_{\tilde{\ell}} \sim 250 - 300$ GeV ought to be accessible in the OS dilepton channel[18]. Stiff cuts must be made to remove SM and SUSY backgrounds. Mass measurement appears difficult.

Also, $pp \to \widetilde{W}_1\widetilde{Z}_2 \to 3\ell + \not{E}_T$ searches can be made[19]. In this channel, equivalent gluino masses up to $m_{\tilde{g}} \sim 550 - 650$ GeV can be explored. Beyond this range, the $\widetilde{Z}_2 \to \widetilde{Z}_1 H_\ell$ or $\widetilde{Z}_1 Z$ spoiler modes turn on, limiting exploration. If such a signal is detected, however, the dilepton invariant mass spectrum can be used to find the value of $m_{\widetilde{Z}_2} - m_{\widetilde{Z}_1}$.

Finally, at LHC, the Higgs bosons of the MSSM can be searched for, mainly via $\gamma\gamma$ and $ZZ \to 4\ell$ decays. If supersymmetric decay modes of the Higgs bosons are allowed[20], then the SM type signatures can be reduced, making detection more difficult. However, new detection modes such as $H_h, H_p \to \widetilde{Z}_2\widetilde{Z}_2 \to 4\ell + \not{E}_T$ can open up, and be detectable in some regions of parameter space.

6. Acknowledgements

I thank M. Bisset, C. H. Chen, J. Gunion, C. Kao, R. Munroe, F. Paige, H. Pois, J. Sender and X. Tata for collaborations on the various topics presented here. I also thank S. Pokorski, P. Nath and T. Taylor for their kind invitation to speak here. I thank them and the other conference organizers for organizing an excellent conference.

7. References

1. For reviews of SUSY, see H. P. Nilles, Phys. Rep. **110**, 1 (1984); H. Haber and G. Kane, Phys. Rep. **117**, 75 (1985); X. Tata, in *The Standard Model and Beyond*, p. 304, edited by J. E. Kim, World Scientific (1991); R. Arnowitt and P. Nath, *Lectures presented at the VII J. A. Swieca Summer School, Campos do Jordao, Brazil, 1993* CTP-TAMU-52/93.
2. Some recent analyses of supergravity mass patterns include, G. Ross and R. G. Roberts, Nucl. Phys. **B377**, 571 (1992); R. Arnowitt and P. Nath, Phys. Rev. Lett. **69**, 725 (1992); M. Drees and M. M. Nojiri, Nucl. Phys. **B369**, 54 (1993); S. Kelley *et. al.*, Nucl. Phys. **B398**, 3 (1993); M. Olechowski and S. Pokorski, Nucl. Phys. **B404**, 590 (1993); V. Barger, M. Berger and P. Ohmann, *ibid.* **49**, 4908 (1994); G. Kane, C. Kolda, L. Roszkowski and J. Wells, Phys. Rev. **D49**, 6173 (1994); D. J. Castaño, E. Piard and P. Ramond, Phys. Rev. **D49**, 4882 (1994); W. de Boer, R. Ehret and D. Kazakov,

Karlsruhe preprint, IEKP-KA/94-05 (1994); see also Ref. [3].
3. H. Baer, C. H. Chen, R. Munroe, F. Paige and X. Tata, FSU-HEP-940625 (1994) (Phys. Rev. **D**, in press).
4. F. Paige and S. Protopopescu, in *Supercollider Physics*, p. 41, ed. D. Soper (World Scientific, 1986); H. Baer, F. Paige, S. Protopopescu and X. Tata, in *Proceedings of the Workshop on Physics at Current Accelerators and Supercolliders*, ed. J. Hewett, A. White and D. Zeppenfeld, (Argonne National Laboratory, 1993).
5. J. Botts *et. al.* Phys. Lett. **B304**, 159 (1993).
6. H. Baer, A. Bartl, D. Karatas, W. Majerotto and X. Tata, Int. Jou. Mod. Phys. **A4**, 4111 (1989).
7. D. Decamp *et.al.* (ALEPH Collaboration), Phys. Lett. **B236**, 86 (1990); P. Abreu *et.al.* (DELPHI Collaboration), Phys. Lett. **B247**, 157 (1990); O. Adriani *et.al.* (L3 Collaboration), CERN-PPE-93-31 (1993); M. Akrawy *et.al.* (OPAL Collaboration), Phys. Lett **B240**, 261 (1990); for a review, see G. Giacomelli and P. Giacomelli, CERN-PPE/93-107 (1993).
8. See *e.g.* M. Chen, C. Dionisi, M. Martinez and X. Tata, Phys. Rep. **159**, 201 (1988).
9. M. Paterno, presented at the SUSY 94 Conference, Ann Arbor, MI and M. Paterno, Stony Brook Ph. D. thesis. See also, F. Abe *et. al.*, Phys. Rev. Lett. **60**, 3439 (1992) for an earlier analysis by the CDF Collaboration.
10. H. Baer, C. Kao and X. Tata, Phys. Rev. **D48**, R2978 (1993) and FSU-HEP-941001 (1994).
11. R. M. Barnett, J. Gunion and H. Haber, Phys. Lett. **315**, 349 (1993); H. Baer, X. Tata and J. Woodside, Phys. Rev. **D41**, 906 (1990).
12. P. Nath and R. Arnowitt, Mod. Phys. Letters **A2**, 331 (1987); R. Barbieri *et. al.*, Nucl. Phys. **B367**, 28 (1991); H. Baer and X. Tata, Phys. Rev. **D47**, 2739 (1993); H. Baer, C. Kao and X. Tata, Phys. Rev. **D48**, 5175 (1993); T. Kamon, J. Lopez, P. McIntyre and J. White, Phys. Rev. **D50**, 5676 (1994).
13. See contribution by H. Baer, H. Murayama and X. Tata, in *Proceedings of the 1994 DPF Study on Electroweak Symmetry Breaking and Beyond the Standard Model*, edited by T. Barklow, S. Dawson, H. Haber and J. Siegrist, (to be published).
14. H. Baer, J. Gunion, C. Kao and H. Pois, FSU-HEP-940501 (1994).
15. H. Baer, J. Sender and X. Tata, Phys. Rev. **D50**, 4517 (1994).
16. ATLAS Collaboration, Letter of Intent, CERN LHCC/92-4 (1992).
17. H. Baer, X. Tata and J. Woodside, Phys. Rev. **D45**, 142 (1992); H. Baer, M. Bisset, X. Tata and J. Woodside, Phys. Rev. **D46**, 303 (1992);
18. F. del Aguila and Ll. Ametller, Phys. Lett. **B261**, 326 (1991); H. Baer, C. H. Chen, F. Paige and X. Tata, Phys. Rev. **D49**, 3283 (1994).
19. H. Baer, C. H. Chen, F. Paige and X. Tata, Phys. Rev. **D50**, 4508 (1994).
20. H. Baer, M. Bisset, D. Dicus, C. Kao and X. Tata, Phys. Rev. **D47**, 1062 (1993); H. Baer, M. Bisset, C. Kao and X. Tata, Phys. Rev. **D50**, 316 (1994).

On Electromagnetic Duality in Locally Supersymmetric N=2 Yang–Mills Theory *

A. Ceresole[1], R. D'Auria[1], S. Ferrara[2] and A. Van Proeyen[3] †

[1] *Dipartimento di Fisica, Politecnico di Torino, Corso Duca Degli Abruzzi 24, 10129 Torino, Italy*
and
INFN, Sezione di Torino, Italy

[2] *CERN, 1211 Geneva 23, Switzerland*

[3] *Instituut voor Theoretische Fysica, K.U. Leuven Celestijnenlaan 200 D B–3001 Leuven, Belgium*

Abstract

We consider duality transformations in $N = 2$ Yang–Mills theory coupled to $N = 2$ supergravity, in a manifestly symplectic and coordinate covariant setting. We give the essential of the geometrical framework which allows one to discuss stringy classical and quantum monodromies, the form of the spectrum of BPS saturated states and the Picard–Fuchs identities encoded in the special geometry of $N = 2$ supergravity theories.

⋆ Supported in part by DOE grants DE-AC0381-ER50050 and DOE-AT03-88ER40384,Task E. and by EEC Science Program SC1*CI92-0789.

† Onderzoeksleider, NFWO, Belgium

1 Introduction

Recently, proposals for the quantum moduli space of $N = 2$ rigid Yang–Mills theories [1] have been given in terms of particular classes of genus r Riemann surfaces parametrized by r complex moduli[2], r being the rank for the gauge group G broken to $U(1)^r$ for generic values of the moduli. The effective action for such theories, with terms up to two derivatives, is described by $N = 2$ supersymmetric lagrangians of r abelian massless vector multiplets[3], whose dynamics is encoded in a holomorphic prepotential $F(X^A)$, function of the moduli coordinates X^A $(A = 1, \ldots, r)$. According to Seiberg and Witten [1] this effective theory has classical, perturbative and non perturbative duality symmetries which reflect on monodromy properties of certain holomorphic symplectic vectors $(X^A, F_A(X))$, eventually related to periods of holomorphic one–forms[1]

$$\omega = X^A \alpha_A + F_A \beta^A \ , \tag{1.1}$$

where α_A, β^A is a basis for the $2r$ homology cycles of a genus r Riemann surface. The Picard–Fuchs equations, satisfied by the holomorphic vector one–form $U_i = (\partial_i X^A, \partial_i F_A)$ $(i = 1, \ldots, r)$ can be regarded as differential identities for "rigid special geometry"[4]. To attach a particular algebraic curve to "rigid special geometry" is therefore equivalent to exactly compute the holomorphic data U_i and therefore to exactly reconstruct the effective action for the self interaction of the r massless gauge multiplets once the massive states, both perturbative and non perturbative, have been integrated out. Indeed it is a virtue of $N = 2$ supersymmetry that all the couplings in the effective Lagrangian, including 4–fermion terms, can be computed purely in terms of the holomorphic data. Quite remarkably the quantum monodromies dictate the monopole and dyon spectrum of the effective theory [1,2]which turns out to be "dual" to non–perturbative instanton effects [5] in the original G–invariant microscopic theory[6]. In this paper we consider several issues in order to extend the approach pursued in the rigid case to the more challenging case of coupling a $N = 2$ Yang–Mills theory to gravity. In particular we shall include in the $N = 2$ supergravity theory a dilaton–axion vector multiplet which is an essential ingredient to describe effective $N = 2$ theories which come from the low energy limit of $N = 2$ heterotic string theories in four dimensions[7]. Another ingredient is the extension of the "classical monodromies" to $N = 2$ local supersymmetry. For rigid theories the classical metric is essentially the Cartan matrix of the group G and the classical monodromies are related to the Weyl group of the Cartan subalgebra of G[2]. For $N = 2$ supergravity theories coming from $N = 2$ heterotic strings, the classical metric of the moduli space of the pure gauge sector is based on the homogeneous space $O(2,r)/O(2)\times O(r)$ [3][7][8][9] and the classical monodromies are related to the T–duality group $O(2,r;\mathbb{Z})$ which in particular is an invariance of the massive charged states[10]. This state of affair is quite analogous to the analysis performed by Sen and Schwarz[11] for the $N=4$ heterotic string compactifications, in which case an exact quantum duality symmetry $SL(2,\mathbb{Z})\times O(6,r;\mathbb{Z})$ was conjectured [11–15] and a resulting spectrum for BPS states with both electric and magnetic states was proposed. In the $N=4$ theory the $SL(2,\mathbb{Z})\times O(6,r;\mathbb{Z})$ symmetry, using general arguments [16,17], has a natural embedding in $Sp(2(6+r);\mathbb{Z})$, acting on the $6+r$ vector self–dual field strengths $\mathcal{F}^{+A}_{\mu\nu}$ and their "dual" defined through $G^{\mu\nu}_{+A}\equiv -i\frac{\delta\mathcal{L}}{\delta\mathcal{F}^{+A}_{\mu\nu}}$. In generic $N=2$ theories, because of quantum corrections [18,19], we do not expect such factorized $S-T$ duality to occur anymore[4]. Indeed this can be argued with a pure supersymmetry argument, related to the fact that once the classical moduli space $O(2,r)/O(2)\times O(r)$ is deformed by quantum corrections, then the factorized structure with the dilaton degrees of freedom is lost and a non trivial moduli space, mixing the S and T degrees of freedom should emerge. This result is in fact a consequence of a theorem on "special geometry" [20,21] which asserts that the only factorized special manifolds are the $\frac{SU(1,1)}{U(1)}\times\frac{O(2,r)}{O(2)\times O(r)}$ series, which precisely describe the "classical moduli space" of $S-T$ moduli. Because of the coupling to gravity, the symplectic structure and identification of periods, coming from special geometry, is also remarkably different from rigid special geometry. Indeed the interpretation of (X^Λ,F_Λ), $\Lambda=0,1,\ldots,r+1$ as periods of algebraic curves is no longer appropriate to genus r Riemann surfaces, as it can be seen from the Picard–Fuchs [22,23] equations and from the form of the metric $g_{i\bar{\jmath}}=-\partial_i\partial_{\bar{\jmath}}\log i(\overline{F}_A X^A-\overline{X}^A F_A)$ of the moduli space [24–28,23,29] . In fact special geometry is known to be appropriate to a particular class of complex manifolds (Calabi–Yau manifolds or their mirrors) and to describe the deformations of the complex structure [26][17]. It is therefore tempting to argue that the quantum moduli space including $S-T$ duality and its monodromies is related to 3–manifolds (or their mirrors) with $h_{(2,1)}=r+1$.

The paper is organized as follows: In chapter 2 we give a resumé of rigid theories, also discussing duality for the fermionic sector and the physical significance of monodromies and geometrical data, such as the holomorphic tensor C_{ijk}, related to the gaugino anomalous magnetic moment. In chapter 3 we describe in detail the coupling to gravity, the extension of duality to the fermionic sector and the existence of symplectic bases which do not admit a prepotential function F, as it occurs in certain formulations of $N=2$ supergravities coming from $N=2$ heterotic strings.

In chapter 4 we discuss classical and quantum duality symmetries and give generic formulae for the spectrum of the BPS states and the "semiclassical formulae" when the non perturbative spectrum is computed in terms of the "classical periods". The explicit expression for the $r = 1, 2$ cases are given as examples. The paper ends with some concluding remarks.

An expanded version of these results is contained in [30].

2 Resumé of rigid special geometry

2.1 Basics

$N = 2$ supersymmetric gauge theory on a group G broken to $U(1)^r$, with $r = rank\ G$, corresponds to a particular case of the most general $N = 1$ coupling of r chiral multiplets (X^A, χ^A) to r $N = 1$ abelian vector multiplets $(\mathcal{A}^A_\mu, \lambda^A)$ in which the Kähler potential K and the holomorphic kinetic term function $f_{AB}(X^A)$ are given by

$$\begin{aligned} K &= i(\overline{F}_A X^A - F_A \overline{X}^A) \quad , \quad (F_A = \partial_A F) \\ f_{AB} &= \partial_A \partial_B F \equiv F_{AB} \end{aligned} \tag{2.1}$$

in terms of the single prepotential $F(X)$[3]. One can show that the Kähler geometry is constrained because the Riemann tensor satisfies the identity [24,4]

$$R_{A\overline{B}C\overline{D}} = -\partial_A \partial_C \partial_P F \ \partial_{\overline{B}} \partial_{\overline{D}} \partial_{\overline{Q}} \overline{F} \ g^{P\overline{Q}} \ , \tag{2.2}$$

with

$$g_{P\overline{Q}} = \partial_P \partial_{\overline{Q}} K = 2 \ \mathrm{Im} \ \partial_P \partial_Q F \ . \tag{2.3}$$

The bosonic lagrangian has the form

$$\begin{aligned} \mathcal{L} &= g_{A\overline{B}} \partial_\mu X^A \partial_\mu \overline{X}^B + \ (g_{A\overline{B}} \lambda^{IA} \sigma^\mu \mathcal{D}_\mu \overline{\lambda}^{\overline{B}}_I + \mathrm{h.c.}) \\ &+ \ \mathrm{Im} \ (F_{AB} \mathcal{F}^{-A}_{\mu\nu} \mathcal{F}^{-B}_{\mu\nu}) + \mathcal{L}_{\mathrm{Pauli}} + \mathcal{L}_{4-\mathrm{fermi}} \ , \end{aligned} \tag{2.4}$$

where $A, B, \ldots$ run on the adjoint representation of the gauge group G, $I = 1, 2$ and $\mathcal{F}^{+A}_{\mu\nu} = \mathcal{F}^A_{\mu\nu} - \frac{i}{2} \epsilon_{\mu\nu\rho\sigma} \mathcal{F}^{\rho\sigma}$ $(\mathcal{F}^{-A}_{\mu\nu} = \overline{\mathcal{F}^{+A}_{\mu\nu}})$. As we shall see, also $\mathcal{L}_{\mathrm{Pauli}}$ and $\mathcal{L}_{4-\mathrm{Fermi}}$ contain the function F and its derivatives up to the fourth.

The previous formulation, derived from tensor calculus, is incomplete because it is not coordinate covariant. It is written in a particular coordinate system ("special coordinates") which is not uniquely selected. In fact, eq.(2.1) is left invariant under particular coordinate changes of the $X^A \to \widetilde{X}^A$ with some new function $\widetilde{F}(\widetilde{X})$ described by

$$\begin{aligned} \widetilde{X}^A(X) &= A^A_{\ B} X^B + B^{AB} F_B(X) + P^A \\ \widetilde{F}_A(\widetilde{X}^A(X)) &= C_{AB} X^B + D_A^{\ B} F_B(X) + Q_A \ , \end{aligned} \tag{2.5}$$

where $\begin{pmatrix} A & B \\ C & D \end{pmatrix}$ is an $Sp(2r, \mathrm{R})$ matrix

$$A^T C - C^T A = 0 \ , \quad B^T D - D^T B = 0 \ , \quad A^T D - C^T B = \mathbb{1} \ , \tag{2.6}$$

and P^A, Q_A can be complex constants which from now on will be set to zero.

It can be shown that[3]

$$\widetilde{F}_A = \frac{\partial \widetilde{F}}{\partial \widetilde{X}^A} \ , \tag{2.7}$$

provided the mapping $X^A \to \widetilde{X}^A$ is invertible.

It is well known that the equations of motion and the Bianchi identities [3][16][17]

$$\begin{aligned} \partial^\mu \mathrm{Im}\ \mathcal{F}^{-A}_{\mu\nu} &= 0 \quad \text{Bianchi identities} \\ \partial_\mu \mathrm{Im}\ G^{\mu\nu}_{-A} &= 0 \quad \text{Equations of motion} \end{aligned} \tag{2.8}$$

transform covariantly under (2.5) (with $P^A = Q_A = 0$), so that $(\mathcal{F}^{-A}_{\mu\nu}, G^{\mu\nu}_{-A})$ is a symplectic vector. Here, $G^{\mu\nu}_{-A} \equiv i\frac{\delta \mathcal{L}}{\delta \mathcal{F}^{-A}_{\mu\nu}} = \overline{\mathcal{N}}_{AB} \mathcal{F}^{-B}_{\mu\nu}$ + fermionic terms, where we have set $F_{AB} = \overline{\mathcal{N}}_{AB}$ in order to unify the notations to the gravitational case[3]. The transformations (2.5) leave invariant the whole lagrangian but the vector kinetic term. Indeed, neglecting for the moment fermion terms (see section 2.2) and setting for simplicity $\mathcal{F}^{-A}_{\mu\nu} = \mathcal{F}^A$ and $G^{\mu\nu}_{-A} = G_A$ the vector kinetic lagrangian transforms as follows

$$\begin{aligned} \mathrm{Im}\ \mathcal{F}^A \overline{\mathcal{N}}_{AB} \mathcal{F}^B \to\ & \mathrm{Im}\ \widetilde{\mathcal{F}}^A \widetilde{G}_A = \\ &= \mathrm{Im}\ (\mathcal{F}^A G_A + 2\mathcal{F}^A (C^T B)_A^{\ B} G_B + \\ &+ \mathcal{F}^A (C^T A)_{AB} \mathcal{F}^B + G_A (D^T B)^{AB} G_B) \ . \end{aligned} \tag{2.9}$$

If $C = B = 0$ the lagrangian is invariant. If $C \neq 0, B = 0$ it is invariant up to a four–divergence. In presence of a topologically non–trivial $\mathcal{F}^{-A}_{\mu\nu}$ background,

$(C^TA)_{AB}\int \text{Im}\, \mathcal{F}^{-A}_{\mu\nu}\mathcal{F}^{-B}_{\mu\nu} \neq 0$, one sees that in the quantum theory duality transformations must be integral valued in $Sp(2r,\mathbb{Z})$[1] and transformations with $B=0$ will be called perturbative duality transformations.

If $B \neq 0$ the lagrangian is not invariant. As it is well known, then the duality transformation is only a symmetry of the equations of motion and not of the lagrangian.

Since $\widetilde{G}^{\mu\nu}_{-A} = \widetilde{\overline{\mathcal{N}}}_{AB}\widetilde{\mathcal{F}}^{-B}_{\mu\nu}$ one also has

$$\widetilde{\mathcal{N}} = (C + D\mathcal{N})(A + B\mathcal{N})^{-1} \ . \tag{2.10}$$

Note that $B \neq 0$ means that the coupling constant $\widetilde{\mathcal{N}}$ is inverted and transformations with $B \neq 0$ will be called quantum non perturbative duality symmetries.

The perturbative duality rotations are of the form

$$\begin{pmatrix} A & 0 \\ C & (A^T)^{-1} \end{pmatrix} \quad , \; A \subset GL(r) \quad , \; A^TC \;\; \text{symmetric} \ . \tag{2.11}$$

In rigid supersymmetry the tree level symmetries are of the form $\begin{pmatrix} A & 0 \\ 0 & (A^T)^{-1} \end{pmatrix}$ while the quantum perturbative monodromy introduces a $C \neq 0$.

If the original unbroken gauge group is $G = SU(r+1)$, then $A \in$ Weyl group and A^TC is the Cartan matrix $< \alpha_i|\alpha_j >$ of $SU(r+1)$[2].

Special coordinates do not give a coordinate free description of the effective action. A coordinate free description is obtained by introducing a holomorphic symplectic bundle $V = (X^A(z), F_A(z))$ and holomorphic $(1,0)$ forms on the Kähler manifold[4,1]

$$U_i \equiv \partial_i V = (\partial_i X^A, \partial_i F_A) \qquad \text{with } i = 1, \dots, r \ . \tag{2.12}$$

In rigid special geometry the U_i satisfy the constraints[4]

$$\begin{aligned} \mathcal{D}_i U_j &= iC_{ijk}g^{k\bar{l}}\overline{U}_{\bar{l}} \\ \partial_i \overline{U}_{\bar{j}} &= 0 \ . \end{aligned} \tag{2.13}$$

From these one may derive the metric and the tensor C_{ijk}

$$\begin{aligned} g_{i\bar{j}} &= \partial_i\partial_{\bar{j}}K = i(\partial_{\bar{j}}\overline{F}_A\partial_i X^A - \partial_{\bar{j}}\overline{X}^A\partial_i F_A) \\ &= i\partial_i X^A\partial_{\bar{j}}\overline{X}^B(\mathcal{N}_{AB} - \overline{\mathcal{N}}_{AB}) \ , \end{aligned} \tag{2.14}$$

$$
\begin{aligned}
C_{ikp} &= \partial_i X^A \mathcal{D}_k \partial_p F_A - \partial_i F_A \mathcal{D}_k \partial_p X^A \\
&= \partial_i X^B (\partial_k \partial_p F_B - \partial_k \partial_p X^A \overline{\mathcal{N}}_{AB}) \ ,
\end{aligned}
\tag{2.15}
$$

where we have used

$$
\partial_{\bar{i}} \overline{F}_A = \mathcal{N}_{AB} \partial_{\bar{i}} \overline{X}^B \ . \tag{2.16}
$$

The integrability conditions on (2.13) yields

$$
R_{i\bar{j}k\bar{l}} = -C_{ikp} \overline{C}_{\bar{j}\bar{l}\bar{p}} g^{p\bar{p}} \ . \tag{2.17}
$$

The Bianchi identities of (2.17) also imply that C_{ijk} is a holomorphic completely symmetric tensor obeying $\mathcal{D}_{[i} C_{j]kl} = 0$.

Note that from (2.15) it also follows

$$
C_{ijk} = \partial_i X^A \partial_j X^B \partial_k X^C \partial_A \partial_B \partial_C F \ , \tag{2.18}
$$

which in special coordinates reduces to

$$
C_{ABC} = \partial_A \partial_B \partial_C F \ . \tag{2.19}
$$

2.2 Fermions

The coordinate free description of fermions is given by $SU(2)$ doublets $(\lambda^{iI}, \lambda^{\bar{i}}_I)$ where upper and lower $SU(2)$ indices I mean positive and negative chiralities respectively[3][27][24]. As such the spinors are symplectic invariant and contravariant vector fields. The antiselfdual field strength $\mathcal{F}^{-A}_{\alpha\beta}$ and positive chiralities spinors are in the same $N=2$ multiplet, which is, in two component spinor notation,*

$$
(X^A, \partial_i X^A \lambda^{iI}_\alpha, \mathcal{F}^{-A}_{\alpha\beta}) \ , \tag{2.20}
$$

with $\alpha, \beta \in SL(2,\mathbb{C})$.

It is easy to see that a Pauli term which is both coordinate and symplectic invariant up to four fermion terms is uniquely fixed to be[27]

$$
\mathcal{L}_{\text{Pauli}}(\lambda) = a \ \text{Im} \ [(\partial_{\bar{i}} \overline{X}^A G^{b,\alpha\beta}_{-A} - \partial_{\bar{i}} \overline{F}_A \mathcal{F}^{-A\alpha\beta}) g^{\bar{i}j} C_{jkp} \lambda^{kI}_\alpha \lambda^{pJ}_\beta \epsilon_{IJ}] \ , \tag{2.21}
$$

* $\mathcal{F}^{-A}_{\alpha\beta}$ is $\sigma^{\mu\nu}_{\alpha\beta} \mathcal{F}^{-A}_{\mu\nu}$.

where $G^b_{-A} \equiv \frac{i\partial \mathcal{L}_{\rm gauge}}{\partial \mathcal{F}^{-A}}$ and a is a real constant fixed by supersymmetry. The full field strength then contains fermionic corrections, as due to (2.21)

$$G^{\alpha\beta}_{-A} \equiv \frac{i\partial \mathcal{L}}{\partial \mathcal{F}^{-A}_{\alpha\beta}} = \overline{\mathcal{N}}_{AB}\mathcal{F}^{-B\alpha\beta} + \frac{1}{2}a\partial_{\bar{i}}\overline{X}^{B}(\overline{\mathcal{N}}_{BA} - \mathcal{N}_{BA})g^{\bar{i}j}C_{jkp}\lambda^{\alpha kI}\lambda^{\beta pJ}\epsilon_{IJ}\,. \quad (2.22)$$

Replacing G^b_{-A} by G_{-A} in $\mathcal{L}_{\rm Pauli}$ does not yet lead to a full duality invariant action.

According to a general argument given in ref [16], duality of the full action (2.4) determines those four fermion terms which are not invariant by themselves. In fact, it is sufficient to demand that (2.4) be duality invariant at the point $\frac{\partial \mathcal{L}}{\partial \mathcal{A}^A} = 0$ where $\mathcal{A}^A$ is the gauge potential. Dropping $\mathcal{L}_{\rm scalar}$ and $\mathcal{L}_{\rm fermions}$ (the first line of (2.4)), which are already invariant, we consider only $\widehat{\mathcal{L}} = \mathcal{L}_{\rm gauge} + \mathcal{L}_{\rm Pauli} + \mathcal{L}_{4-{\rm fermi}}$. One finds

$$\widehat{\mathcal{L}}({\rm Im}\ \partial^{\mu}G^{-}_{\mu\nu A} = 0) = \frac{1}{2}\mathcal{L}_{\rm Pauli} + \mathcal{L}_{4-{\rm fermi}} \quad (2.23)$$

and the coefficient of the non–invariant 4–fermi term is fixed by replacing here G^b_{-A} with G_{-A} as given by (2.22). This yields

$$\begin{aligned}\mathcal{L}_{4-{\rm fermi}} =& \frac{a^2}{4}\partial_{\bar{i}}\overline{X}^{A}\partial_{\bar{l}}\overline{X}^{B}(\overline{\mathcal{N}}_{AB} - \mathcal{N}_{AB})g^{\bar{i}j}g^{\bar{l}q}C_{jkp}C_{qrs}\lambda^{k\alpha I}\lambda^{p\beta J}\epsilon_{IJ}\lambda^{rK}_{\alpha}\lambda^{sL}_{\beta}\epsilon_{KL} + {\rm h.c.}\\ &+ \ \text{invariant terms}\ ,\end{aligned} \quad (2.24)$$

in agreement with Cremmer et al. [31].

In special coordinates, setting $\lambda^{i1}_{\alpha} = \chi^{i}_{\alpha}$, $\lambda^{i2}_{\alpha} = \lambda^{i}_{\alpha}$ and using eq.(2.21), the Pauli term reduces to

$$\mathcal{L}_{\rm Pauli} = a\ {\rm Re}\ \partial_A\partial_B\partial_C F(\chi^A_{\alpha}\lambda^B_{\beta} - \lambda^A_{\alpha}\chi^B_{\beta})\mathcal{F}^{-C\alpha\beta}\ , \quad (2.25)$$

in agreement with the standard $N = 1$ supersymmetric action with $f_{AB} = F_{AB}$ [31]. We see from (2.21) and (2.25) that in rigid supersymmetry the physical meaning of C_{ijk} is that of an anomalous magnetic moment. Note that C_{ijk} vanishes at tree–level and it is $\sim \frac{1}{<X>}$ at one loop-level as it must be [1,18]. It is obviously singular at $< X >= 0$. In the $SU(2)$ theory, because of non–perturbative corrections[1], one expects such terms to behave instead as $\frac{c_0}{\Lambda}$ where c_0 is a dimensionless number. The vanishing at tree-level of both Pauli terms and the corresponding four fermions terms is consistent with renormalizability arguments.

The other fermionic terms which are already duality invariant read

$$\lambda^{iI}_{\alpha}\lambda^{kJ}_{\beta}\epsilon^{\alpha\beta}\overline{\lambda}^{\overline{j}}_{\dot{\alpha}I}\overline{\lambda}^{\overline{l}}_{\dot{\beta}J}\epsilon^{\dot{\alpha}\dot{\beta}}R_{\overline{i}\overline{j}k\overline{l}} \tag{2.26}$$

and

$$\mathcal{D}_i C_{jlm}\lambda^{iI}_{\alpha}\lambda^{jK}_{\beta}\epsilon^{\alpha\beta}\lambda^{lJ}_{\gamma}\lambda^{mL}_{\delta}\epsilon^{\gamma\delta}\epsilon_{IJ}\epsilon_{KL} \ . \tag{2.27}$$

Note that, because of eq. (2.17), all couplings in the lagrangian are expressed through the tensors C_{ijk}.

From a tensor calculus point of view all quartic terms but the last come from the equations of motion of the Y^i_{IJ} auxiliary field triplet[3].

2.3 Positivity and monodromies

Let us consider a submanifold $\mathcal{M}_r$ of the moduli space of a Riemann surface of genus r such that its tangent space is isomorphic to the Hodge bundle. In particular the dimension of $\mathcal{M}_r$ is equal to the genus r of the Riemann surface $\mathcal{C}_r$ * . In this case, decomposing an abelian differential in terms of the $2r$ harmonic forms dual to the canonical basis of cycles, we have

$$\begin{aligned} &\omega = X^A(z^i)\alpha_A + F_A(z^i)\beta^A \qquad A, i = 1,\ldots,r \\ &\int \alpha_A \wedge \beta^B = \delta_A^{\ B} \quad , \int \alpha_A \wedge \alpha_B = \int \beta_A \wedge \beta_B = 0 \ , \end{aligned} \tag{2.28}$$

where z^i are coordinates on the moduli space submanifold, and

$$\partial_i \omega = \partial_i X^A \alpha_A + \partial_i F_A \beta^A \ . \tag{2.29}$$

Then the metric, given by the norm

$$g_{\overline{i}j} = i\int \partial_i\omega \wedge \partial_{\overline{j}}\overline{\omega} = i\partial_i\partial_{\overline{j}}\int \omega\wedge\overline{\omega} \tag{2.30}$$

is manifestly positive. Using eqs. (2.28), (2.29) we find

$\star$ We are aware of the fact that to find an intrinsic characterization of such an algebraic locus is far from obvious. We thank D. Dubrovin, P. Fré and Reina for clarifying discussions on this point.

$$
\begin{aligned}
g_{i\bar{j}} &= i\partial_i\partial_{\bar{j}}(\overline{F}_A X^A - \overline{X}^A F_A) = \\
&= i(\partial_i X^A \partial_{\bar{j}}\overline{F}_A - \partial_i F^A \partial_{\bar{j}}\overline{X}^A) = \\
&= i\partial_i X^A \partial_{\bar{j}}\overline{X}^B (\mathcal{N}_{AB} - \overline{\mathcal{N}}_{AB}) \ ,
\end{aligned}
\tag{2.31}
$$

which coincides with the metric of $N = 2$ rigid special geometry [1,4].

Formula (2.29) implies by supersymmetry a similar expansion for the full multiplet (2.20). For the upper component $\mathcal{F}^{-A}_{\mu\nu}$ we get a self dual three form

$$
w = \mathcal{F}^A \alpha_A + G_A \beta^A \tag{2.32}
$$

on $\mathrm{R}_4 \times \mathcal{C}_r$ when (2.8) hold. We observe that in six dimensions an abelian vector multiplet is dual to a tensor multiplet containing a self-dual three form. This remarkable coincidence actually suggests a physical picture for the characterization of this subclass $\mathcal{C}_r$ of Riemann surfaces. Namely, they should appear in the compactification on $\mathrm{R}_4 \times \mathcal{C}_r$ of $N = 1$ six–dimensional theory of a self interacting tensor multiplet.

As shown in ref. [4], the Picard–Fuchs equations for $\mathcal{C}_r$ have a general form dictated by the differential constraints of rigid special geometry. A general proposal for $\mathcal{C}_r$ has been given in [2] and can be used to write down the Picard–Fuchs equations for the periods and to determine their monodromies. Such proposal can be checked by comparing the explicit form of the Picard–Fuchs equations with their general form given by rigid special geometry.

In the one parameter case ($G = SU(2)$), where $\mathcal{C}_1$ is given by the elliptic curve of ref. [1], the special geometry equations reduce to one ordinary second order equation

$$
\left(\frac{d}{dz} + \widehat{\Gamma}\right)C^{-1}\left(\frac{d}{dz} - \widehat{\Gamma}\right)U = 0 \tag{2.33}
$$

where $\widehat{\Gamma} = \frac{d}{dz}\log e$, $e = \frac{dX}{dz}$ and C is the 3–tensor appearing in (2.13). This agrees with the Picard–Fuchs equations derived from $\mathcal{C}_1$. The general solution of this equation is[4]

$$
U = \left(e, e\frac{d^2F}{dX^2}\right) \quad , \tag{2.34}
$$

with $\tau = \frac{\partial^2 F}{\partial X^2}$ being the uniformizing variable for which the differential equation reduces to $\frac{d^2}{d\tau^2}(\ ) = 0$.

3 Coupling to gravity

The coupling to gravity modifies the constraints of rigid special geometry because of the introduction of a $U(1)$ connection due to the $U(1)$ Kähler –Hodge structure of moduli space. For n vector multiplets one introduces $2(n+1)$ covariantly holomorphic sections [24–26,23,27]

$$V = (L^\Lambda, M_\Lambda) \quad (\Lambda = 0, \dots, n) \, , \tag{3.1}$$

where 0 is the graviphoton index.

The new differential constraints of special geometry are

$$\begin{aligned} U_i &\equiv (\mathcal{D}_i L^\Lambda, \mathcal{D}_i M_\Lambda) = (f_i^\Lambda, h_{i\Lambda}) \\ \mathcal{D}_i U_j &= i C_{ijk} g^{k\bar{l}} \overline{U}_{\bar{l}} \\ \mathcal{D}_i \overline{U}_{\bar{j}} &= g_{i\bar{j}} \overline{V} \\ \mathcal{D}_i \overline{V} &= 0 \, , \end{aligned} \tag{3.2}$$

where now $\mathcal{D}_i$ is the covariant derivative with respect to the usual Levi-Civita connection and the Kähler connection $\partial_i K$. That is, under $K \to K + f + \overline{f}$ a generic field ψ^i which under $U(1)$ transforms as $\psi^i \to e^{-(\frac{p}{2} f + \frac{\overline{p}}{2} \overline{f})} \psi^i$ has the following covariant derivative

$$\mathcal{D}_i \psi^j = \partial_i \psi^j + \Gamma^j_{ik} \psi^k + \frac{p}{2} \partial_i K \psi^j \, , \tag{3.3}$$

and analogously for $\mathcal{D}_{\bar{i}}$ with $p \to \overline{p}$. The (L^Λ, M_Λ) have been given conventionally weights $p = -\overline{p} = 1$.

Since L^Λ, M_Λ are covariantly holomorphic, it is convenient to introduce holomorphic sections $X^\Lambda = e^{-K/2} L^\Lambda$, $F_\Lambda = e^{-K/2} M_\Lambda$.

The Kähler potential is fixed by the condition[3][24]

$$i(\overline{L}^\Lambda M_\Lambda - L^\Lambda \overline{M}_\Lambda) = 1 \tag{3.4}$$

to be

$$K = -\log i(\overline{X}^\Lambda F_\Lambda - X^\Lambda \overline{F}_\Lambda) \, . \tag{3.5}$$

As it is well known[3][32], the differential constraints (3.2) can in general be solved in terms of a holomorphic function homogeneous of degree two $F(X)$. However, as we will see in the sequel, there exist particular symplectic sections for which such prepotential F does not exist. In particular this is the case appearing in the effective theory of the $N = 2$ heterotic string. For this reason it is convenient to have the fundamental formulas of special geometry written in a way independent of the existence of F.

First of all we note that quite generally we may write

$$h_{i\Lambda} = \overline{\mathcal{N}}_{\Lambda\Sigma} f_i^{\Sigma} \ . \tag{3.6}$$

Indeed, when F exists, $\mathcal{N}_{\Lambda\Sigma}$ has the form[3][27]

$$\mathcal{N}_{\Lambda\Sigma} = \overline{F}_{\Lambda\Sigma} + 2i \frac{(\text{Im } F_{\Lambda\Gamma})(\text{Im } F_{\Sigma\Pi}) L^{\Gamma} L^{\Pi}}{(\text{Im } F_{\Xi\Omega}) \ L^{\Xi} L^{\Omega}} \ , \tag{3.7}$$

which turns out to be the coupling matrix appearing in the kinetic term of the vector fields. However, as we show below, (3.6) is symplectic covariant and therefore it always holds even in some specific coordinate system in which F does not exist.

In the same way as in the rigid case, from eqs. (3.2) we find

$$g_{i\bar{\jmath}} = i(f_i^{\Lambda} \overline{h}_{\bar{\jmath}\Lambda} - h_{i\Lambda} \overline{f}_{\bar{\jmath}}^{\Lambda}) = i(\mathcal{N}_{\Lambda\Sigma} - \overline{\mathcal{N}}_{\Lambda\Sigma}) f_i^{\Lambda} f_{\bar{\jmath}}^{\Sigma} \tag{3.8}$$

$$C_{ijk} = f_i^{\Lambda} \mathcal{D}_j h_{k\Lambda} - h_{i\Lambda} \mathcal{D}_j f_k^{\Lambda} = f_i^{\Lambda} \partial_j \overline{\mathcal{N}}_{\Lambda\Sigma} f_k^{\Sigma} \ , \tag{3.9}$$

which are symplectic invariant. (Note that $\mathcal{N}_{\Lambda\Sigma}$ has zero Kähler weight).

Furthermore, the integrability conditions (3.2) give[3][24][25][26][23][27]

$$R_{i\bar{\jmath}l\bar{k}} = g_{i\bar{\jmath}} g_{l\bar{k}} + g_{i\bar{k}} g_{l\bar{\jmath}} - C_{ilp} C_{\bar{\jmath}\bar{k}\bar{p}} g^{p\bar{p}} \ , \tag{3.10}$$

replacing eq. (2.6).

Here C_{ilp} is a covariantly holomorphic tensor of weight $p = -\overline{p} = 2$,

$$\mathcal{D}_{\bar{l}} C_{ijk} = \partial_{\bar{l}} C_{ijk} - \partial_{\bar{l}} K C_{ijk} = 0 \ , \tag{3.11}$$

which implies $\partial_{\bar{l}} W_{ijk} = 0$ with $C_{ijk} = e^K W_{ijk}$.

Duality transformations are now in $Sp(2n+2, \mathbb{Z})$ and act on X^{Λ}, F_{Λ} as in the rigid case. The symplectic action on $(L^{\Lambda}, M_{\Lambda})$ (or $X^{\Lambda}, F_{\Lambda})$) is

$$\begin{pmatrix} L \\ M \end{pmatrix}' = \begin{pmatrix} A & B \\ C & D \end{pmatrix} \begin{pmatrix} L \\ M \end{pmatrix} = \mathcal{S} \begin{pmatrix} L \\ M \end{pmatrix} \quad \mathcal{S} \in Sp(2n+2, \mathbb{Z}) \ . \tag{3.12}$$

Then it follows, because of eq. (3.2) and (3.6),

$$\begin{pmatrix} f_i^\Lambda \\ h_{i\Lambda} \end{pmatrix}' = \begin{pmatrix} A & B\overline{\mathcal{N}} \\ C & D\overline{\mathcal{N}} \end{pmatrix} \begin{pmatrix} f_i^\Lambda \\ f_i^\Lambda \end{pmatrix} , \tag{3.13}$$

which also implies

$$\widetilde{\mathcal{N}} = (C + D\mathcal{N})(A + B\mathcal{N})^{-1} . \tag{3.14}$$

These two transformations laws imply the covariance of (3.6).

The symplectic action on $\mathcal{F}^{+\Lambda}_{\mu\nu}, G^{\mu\nu}_{+\Lambda}$ is the same as on (L^Λ, M_Λ), so eq. (2.8)is unchanged. Therefore the discussion of the previous section on perturbative and non perturbative duality transformations in the rigid case remains unchanged when gravity is turned on.

When the sections (X^Λ, F_Λ) are chosen in such a way that a function F exists*, from (3.4) and the degree two homogeneity of F it follows that [24][27]

$$\mathrm{Im}\ F_{\Lambda\Sigma} L^\Lambda \overline{f}^\Sigma_{\overline{\imath}} = 0 , \tag{3.15}$$

so that eq. (3.6) becomes $h_{i\Lambda} = F_{\Lambda\Sigma} f^{\Sigma i}$. Furthermore from (3.9) and (3.15) it also follows

$$e^{K/2} C_{ijk} = f_i^\Lambda f_j^\Gamma f_k^\Sigma F_{\Lambda\Gamma\Sigma} . \tag{3.16}$$

By the same token, we have

$$\begin{pmatrix} f_i^\Lambda \\ h_{i\Lambda} \end{pmatrix}' = \begin{pmatrix} A & B\mathcal{F} \\ C & D\mathcal{F} \end{pmatrix} \begin{pmatrix} f_i^\Lambda \\ f_i^\Lambda \end{pmatrix} , \tag{3.17}$$

where $\mathcal{F} = F_{\Lambda\Sigma}$. Note that in these cases

$$\begin{aligned} 2\widetilde{F}(\widetilde{X}) = \widetilde{F}_\Lambda \widetilde{X}^\Lambda = \\ 2F + 2X^\Lambda (C^T B)_\Lambda^\Sigma F_\Sigma + X^\Lambda (C^T A)_{\Lambda\Sigma} X^\Sigma + F_\Lambda (D^T B)^{\Lambda\Sigma} F_\Sigma . \end{aligned} \tag{3.18}$$

Note also that the homogeneity of F implies

$$\widetilde{X} = (A + B\mathcal{F})X , \tag{3.19}$$

$\star$ A resumé of the duality transformations for this case, including the supergravity corrections has been given in appendix C of [32].

where $\mathcal{F} = F_{\Lambda\Sigma}$ and

$$\widetilde{F} = (C + D\mathcal{F})X \ . \tag{3.20}$$

Special coordinates in supergravity are defined by $t^\Lambda = X^\Lambda / X^0$ since we now have a set of $n+1$ homogeneous coordinates. If we assume that $\mathcal{D}_i(\frac{X^\Lambda}{X^0})$ is an invertible matrix, then we may choose a frame for which $\partial_i(\frac{X^\Lambda}{X^0}) = \delta_i^\Lambda$. This is possible only if X^Λ are unconstrained variables and so $F_\Lambda = F_\Lambda(X)$, which implies $F_\Lambda = \partial_\Lambda F(X)$ with F homogeneous of degree 2.

We now discuss the possible non-existence of $F(X)$. If we start with some special coordinates $X^\Lambda, F_\Lambda(X)$, it is possible that in the new basis the $\widetilde{X}^\Lambda$ are not good special coordinates in the sense that the mapping $X \to \widetilde{X}$ is not invertible. This happens whenever the $(n+1) \times (n+1)$ matrix $A + B\mathcal{F}$ is not invertible (its determinant vanishes). This does not mean that $\widetilde{X}, \widetilde{F}$ are not good symplectic sections since the symplectic matrix $\mathcal{S} = \begin{pmatrix} A & B \\ C & D \end{pmatrix}$ is always invertible. It simply means that $\widetilde{F}_\Lambda \neq \widetilde{F}_\Lambda(\widetilde{X})$ and therefore a prepotential $\widetilde{F}(\widetilde{X})$ does not exist. However our formulation of special geometry never explicitly used the fact that F_Λ be a functional of the X's and indeed the quantities (X^Λ, F_Λ), $(f_i^\Lambda, h_{i\Lambda})$, $\mathcal{N}_{\Lambda\Sigma}$ and $C_{ijk}, g_{i\bar{j}}$ are well defined for any choice of the symplectic sections (X^A, F_A) since they are symplectic invariant or covariant. For example, to compute the "gauge coupling" $\widetilde{\mathcal{N}}$ in such a basis $(\widetilde{X}^\Lambda, \widetilde{F}_\Lambda)$ one uses the formula

$$\widetilde{\mathcal{N}}(\widetilde{X}, \widetilde{F}) = (C + D\mathcal{N}(X))(A + B\mathcal{N}(X))^{-1} \ , \tag{3.21}$$

and expresses the $X - X(\widetilde{X}, \widetilde{F})$ by using the fact that the symplectic mapping can be inverted. All other quantities can be computed in this way.

We will see the relevance of this observation in the sequel, while discussing low energy effective action of $N = 2$ heterotic string. A simple example is the following. Consider $F = iX^0X^1$, leading to

$$\mathcal{N} = \begin{pmatrix} i\frac{X^1}{X^0} & 0 \\ 0 & i\frac{X^0}{X^1} \end{pmatrix} \ . \tag{3.22}$$

This appears in the $N = 2$ reduction of pure $N = 4$ supergravity in the so-called $SO(4)$ formulation [33]. Consider now the symplectic mapping defined by

$$A = D = \begin{pmatrix} 1 & 0 \\ 0 & 0 \end{pmatrix} \ ; \qquad C = -B = \begin{pmatrix} 0 & 0 \\ 0 & 1 \end{pmatrix} \ . \tag{3.23}$$

Then the transformation is

$$\begin{aligned} \widetilde{X}^0 &= X^0 \qquad \widetilde{X}^1 = -F_1 \\ \widetilde{F}_0 &= F_0 \qquad \widetilde{F}_1 = X^1 \;. \end{aligned} \tag{3.24}$$

Using in the first line $F_1 = iX^0$ would lead to a non–invertible mapping $X \to \widetilde{X}$, and using (3.18) would lead to $\widetilde{F} = 0$. One observes also that $A + B\mathcal{F}$ is non–invertible. However, $A + B\mathcal{N}$ is invertible, and one obtains $\widetilde{N} = iX^1(X^0)^{-1}1\!\!1 = i\widetilde{F}_1(\widetilde{X}^0)^{-1}1\!\!1$. This form appears in the $N = 2$ reduction of the $SU(4)$ formulation of pure $N = 4$ supergravity [34]. These two forms of the $N = 2$ reduced action and the duality transformation have been studied in [35] to relate electric and magnetic charges of black holes.

As far as the fermions are concerned, the gravitational $N = 2$ multiplet is now

$$(L^\Lambda, f_i^\Lambda \lambda^{iI}, \mathcal{F}^{-\Lambda}_{\alpha\beta}) \,, \tag{3.25}$$

and, correspondingly, the gaugino Pauli terms have the form

$$a \text{ Im } [(\mathcal{D}_{\bar{\imath}}\overline{L}^\Lambda G^{\alpha\beta}_{-\Lambda} - \mathcal{D}_{\bar{\imath}}\overline{M}_\Lambda \mathcal{F}^{-\Lambda\alpha\beta}) g^{\bar{\imath}j} C_{jkl} \lambda^{kI}_\alpha \lambda^{lJ}_\beta \epsilon_{IJ}] \,, \tag{3.26}$$

quite analogous to eq. (2.21).

The Pauli terms for gravitino currents which are manifestly duality invariant are easy to find. They are [3][31][27]

$$\begin{aligned} &a_1 \text{Im } [(\mathcal{F}^{-\Lambda}_{\mu\nu} M_\Lambda - G_{\mu\nu-\Lambda} L^\Lambda) \overline{\psi}^I_\rho \psi^J_\sigma \epsilon_{IJ} \epsilon^{\mu\nu\rho\sigma}] + \\ &+ a_2 \text{Im } [(\mathcal{F}^{-\Lambda}_{\mu\nu} \mathcal{D}_{\bar{\imath}} \overline{M}_\Lambda - G_{\mu\nu-\Lambda} \mathcal{D}_{\bar{\imath}} \overline{L}^\Lambda) \overline{\lambda}^{\bar{\imath}}_I \gamma_\rho \psi_{\sigma J} \epsilon^{IJ} \epsilon^{\mu\nu\rho\sigma}] \,. \end{aligned} \tag{3.27}$$

Again, as before, they generate unique quartic terms by requiring duality invariance of the action, on the equations of motion of the vector fields. Of course many of these terms are absent in $N = 1$[31] theories because of the absence of the second gravitino. This is one of the differences between rigid supersymmetry and local supersymmetry. What happens is that in $N = 2$ supergravity, one introduces an extra $(\frac{3}{2}, 1)$ multiplet, with respect to the $N = 1$ case. This has the effect of having extra auxiliary fields in the supergravity multiplet[36]

$$\mathcal{V}^I_{J\mu} \;, \quad A_\mu \;, \quad T^-_{\mu\nu} \;, \quad D \tag{3.28}$$

other than the matter auxiliary field of the vector multiplet Y^{iIJ} (traceless, real, symmetric in IJ), $i,j = 1,2$, *i.e.* a real $SU(2)$ triplet. The meaning of the auxiliary fields is straightforward. The Y's correspond to the three auxiliary fields of a $N = 1$ vector multiplet and a chiral multiplet. The D auxiliary field gives the equation (3.4) (i.e (3.5)), $T^-_{\mu\nu}$ is the graviphoton (symplectic invariant) combination of the gauge fields $T^-_{\mu\nu} = L^\Lambda G^{\mu\nu}_{-\Lambda} - M_\Lambda \mathcal{F}^{-\Lambda}_{\mu\nu}$, and $\mathcal{V}^I_{J\mu}$, A_μ are the composite $SU(2)$ and $U(1)$ connections of the quaternionic manifold and Kähler–Hodge manifold respectively. Note that comparison between $N = 1$ and $N = 2$ theories shows that the spinors χ^i of the scalar multiplet and λ^Σ of the vector multiplet of the $N = 1$ theory are related to the doublet λ^{iI} of the $N = 2$ theory by

$$\chi^i = \lambda^{i1} \quad , \lambda^\Sigma = f_i^\Sigma \lambda^{i2} \tag{3.29}$$

3.1 The three–form cohomology

We recall that special geometry in $N = 2$ supergravity, unlike rigid special geometry, is suitable for three form cohomology for Calabi–Yau manifolds. Let's define a holomorphic three–form [25,26]

$$\Omega = X^\Lambda \alpha_\Lambda + F_\Lambda \beta^\Lambda \tag{3.30}$$

where $\alpha_\Lambda, \beta^\Lambda$ is a $2n + 2$ dimensional cohomology basis dual to the $2n + 2$ homology cycles ($n = h_{21}$). Ω is a holomorphic section of a line bundle. Then it follows that if one defines

$$e^{-K} = i \int \Omega \wedge \overline{\Omega} > 0 \tag{3.31}$$

then

$$g_{i\bar{\jmath}} = \frac{-i \int \mathcal{D}_i\Omega \wedge \mathcal{D}_{\bar{\jmath}}\overline{\Omega}}{i \int \Omega \wedge \overline{\Omega}} = \partial_i \partial_{\bar{\jmath}} \log i \int \Omega \wedge \overline{\Omega} > 0 \, . \tag{3.32}$$

The $(2n + 2)$ three–forms $\mathcal{D}_i\Omega, \mathcal{D}_i\overline{\Omega}, \Omega, \overline{\Omega}$ with the cohomology basis (α_A, β^A) correspond to the decomposition

$$H^3(\mathrm{R}) = H^{(2,1)}(\mathbb{C}) + H^{(1,2)}(\mathbb{C}) + H^{(3,0)}(\mathbb{C}) + H^{(0,3)}(\mathbb{C}) \, . \tag{3.33}$$

Note that since $\Omega = (X^\Lambda, F_\Lambda)$, then $\mathcal{D}_i\Omega = (\mathcal{D}_i X^\Lambda, \mathcal{D}_i F_\Lambda)$, with $f_i^\Lambda = e^{\frac{K}{2}} \mathcal{D}_i X^\Lambda, h_{i\Lambda} = e^{\frac{K}{2}} \mathcal{D}_i F_\Lambda$. The relations

$$\int \Omega \wedge \Omega = \int \Omega \wedge \mathcal{D}_i\overline{\Omega} = 0 \tag{3.34}$$

are obvious since $\mathcal{D}_i\Omega = \partial_i\Omega - \frac{1}{(\Omega,\overline{\Omega})}(\partial_i\Omega,\overline{\Omega})\Omega$. However the relation

$$\int \Omega \wedge \mathcal{D}_i\Omega = 0 \, , \tag{3.35}$$

which is suitable for three–form cohomology, implies

$$\int \Omega \wedge \partial_i\Omega = 0 \, , \tag{3.36}$$

i.e.

$$\partial_i X^\Lambda F_\Lambda - \partial_i F_\Lambda X^\Lambda = 0 \tag{3.37}$$

for any choice of the symplectic section. Eq. (3.37) is equivalent to

$$X^\Lambda \mathcal{D}_i F_\Lambda - \mathcal{D}_i X^\Lambda F_\Lambda = 0 \, . \tag{3.38}$$

From $\mathcal{D}_i F_\Lambda = \overline{\mathcal{N}}_{\Lambda\Sigma}\mathcal{D}_i X^\Sigma$, applying $\mathcal{D}_{\bar{\jmath}}$ to both sides we also find

$$\mathcal{D}_{\bar{\jmath}}\mathcal{D}_i F_\Lambda = \partial_{\bar{\jmath}}\overline{\mathcal{N}}_{\Lambda\Sigma}\mathcal{D}_i X^\Sigma + \overline{\mathcal{N}}_{\Lambda\Sigma}\mathcal{D}_{\bar{\jmath}}\mathcal{D}_i X^\Sigma \, , \tag{3.39}$$

which implies, using the third line of (3.2),

$$F_\Lambda = \overline{\mathcal{N}}_{\Lambda\Sigma} X^\Sigma + \frac{1}{n} g^{i\bar{\jmath}}\partial_{\bar{\jmath}}\overline{\mathcal{N}}_{\Lambda\Sigma}\mathcal{D}_i X^\Sigma \, . \tag{3.40}$$

4 Duality symmetries

Duality transformations in generic $N = 2$ supergravity theories are a different choice of the symplectic representative (X^Λ, F_Λ) of the underlying special geometry. If the fields $\mathcal{F}^{+\Lambda}_{\mu\nu}, G^{+\Lambda}_{\mu\nu}$ have no electric or magnetic sources these dualities are simply a different equivalent choice of sections (X^Λ, F_Λ) since they are defined up to a symplectic transformation[3][17]. However if the gauge fields are coupled to (abelian) sources then duality transformations map theories into different theories with a duality transformed source. Since the matrix $\mathcal{N}_{\Lambda\Sigma}$ plays the role of a coupling constant it is clear that in perturbation theory the only possible duality transformations are those with $B = 0$ and of a lower triangular block form

$$\mathcal{S} = \begin{pmatrix} A & 0 \\ C & A^{T-1} \end{pmatrix} \, . \tag{4.1}$$

Under such change, the action changes in a total derivative which, up to fermion terms, is

$$\mathcal{L}'(A,C) = \mathcal{L} + \mathrm{Im}\ \mathcal{F}^{-\Lambda}(CA)_{\Lambda\Sigma}\mathcal{F}^{-\Sigma}\ . \tag{4.2}$$

So the lagrangian is invariant up to a surface term. A duality transformation is a symmetry if

$$\widetilde{\mathcal{N}}(\widetilde{X},\widetilde{F}) = \mathcal{N}(\widetilde{X},\widetilde{F}) = (C + D\mathcal{N}(X,F))(A + B\mathcal{N}(X,F))^{-1}\ . \tag{4.3}$$

If $F_\Lambda = F_\Lambda(X)$ this implies

$$\widetilde{F}(\widetilde{X}) = F(\widetilde{X})\ . \tag{4.4}$$

Then using (3.18) we should have [3][37]

$$\begin{aligned} 2F[(A+B\mathcal{F})X] = 2F &+ 2X^\Lambda(C^TB)_\Lambda{}^\Sigma F_\Sigma \\ &+ X^\Lambda(C^TA)_{\Lambda\Sigma}X^\Sigma + F_\Lambda(D^TB)^{\Lambda\Sigma}F_\Sigma\ , \end{aligned} \tag{4.5}$$

which is a functional relation for F given A, B, C, D. Note that because of (3.18) it may happen that $\ddot{F}(\ddot{X}) = 0$. This is so when $\frac{\partial X^\Lambda}{\partial \widetilde{X}^\Sigma}$ is not an invertible matrix.

4.1 Heterotic $N=2$ superstring theories

In $N=2$ heterotic string theories, as the one obtained by the fermionic construction or by compactification on $T_2 \times K_3$, one often encounters classical moduli spaces which are locally of the form[38][39][18][40][9][41]

$$\frac{O(2,n_v)}{O(2)\times O(n_v)} \times \frac{O(4,n_h)}{O(4)\times O(n_h)}\ , \tag{4.6}$$

where n_v and n_h are respectively the number of the moduli in vector and hypermultiplets. If there are no charged massless hypermultiplets with respect to the gauge group $U(1)^r$, with $r = n_v$, we may avoid holomorphic anomalies [42][43][44][45] and the situation for this theory may be similar to the rigid Yang–Mills theory coupled to supergravity with an additional dilaton axion multiplet. According to the previous discussion, all perturbative duality symmetries are those for which the previous formula holds for a subgroup of lower triangular matrices

$$\begin{pmatrix} A & 0 \\ C & A^{T-1} \end{pmatrix} \tag{4.7}$$

with $A^T C$ symmetric.

The $(r+2)\times(r+2)$ block A contains the target space T duality and C contains the Peccei–Quinn axion symmetry [11] (for the definition of S in the $N=2$ context, see below)

$$S \to S+1\ . \tag{4.8}$$

These are the tree level stringy symmetries of the massive states with $M=|Z|$ where Z is the central charge of the $N=2$ supersymmetry algebra. If the number of T–moduli is r then the duality symmetries are in $Sp(2r+4;\mathbb{Z})$.

An important point is that we would like to make the tree level (string) symmetry manifest. This means that the gauge fields

$$\mathcal{A}^\Lambda_\mu = (G_\mu, B_\mu, \mathcal{A}^A_\mu)\ \ A=2,\ldots,r+1 \tag{4.9}$$

(G_μ is the graviphoton and the B_μ is the vector of the dilaton–axion multiplet) should transform in the $2+r$ dimensional (vector) representation of the target space duality symmetry

$$\mathcal{A}' = A\mathcal{A}\ ; \qquad A^T\eta A=\eta\ ; \qquad \eta_{\Lambda\Sigma} = Diag(1,1,-1,-1,\ldots)\ , \tag{4.10}$$

with $A\in O(2,r;\mathbb{Z})$. Under the axion Peccei–Quinn symmetry $S\to S+1$

$$\mathcal{A}^{\Lambda\prime} = \mathcal{A}^\Lambda\ \ ,\ \ G_{\Lambda\mu\nu} \to G_{\Lambda\mu\nu} + \eta_{\Lambda\Sigma}F^\Sigma_{\mu\nu}\ , \tag{4.11}$$

where

$$\mathcal{N}_{\Lambda\Sigma}(S+1) = \mathcal{N}_{\Lambda\Sigma}(S) + \eta_{\Lambda\Sigma}\ . \tag{4.12}$$

This formulation is directly obtained by $N=2$ reduction of the standard form of the $N=4$ supergravity action[11][7] with a moduli space of the type $O(6,r)/O(6)\times O(r)/\Gamma$ and duality group $\Gamma = O(6;r;\mathbb{Z})$. However to get this in a standard $N=2$ supergravity form, one must introduce $2+r$ symplectic sections (X^Λ, F_Λ) $(\Lambda=0,1,\ldots,r+1)$ for which $O(2,r)$ is block diagonal and the $S\to S+1$ shift is lower triangular. This formulation can be obtained by making a symplectic rotation, with $\mathcal{S}$ given by

$$\mathcal{S} = \frac{1}{\sqrt{2}}\begin{pmatrix} \mathbb{1} & -\mathbb{1} \\ \mathbb{1} & \mathbb{1}\end{pmatrix}\ , \tag{4.13}$$

from a representation in which only $O(2)\times O(r)$ is block diagonal [46], namely

$$
\begin{aligned}
SO(2,r): &\quad \begin{pmatrix} A & 0 \\ 0 & \eta A \eta \end{pmatrix} = \mathcal{S} A_1 \mathcal{S}^{-1} \\
S \to S+1: &\quad \begin{pmatrix} \mathbb{1} & 0 \\ \eta & \mathbb{1} \end{pmatrix} = \mathcal{S} A_2 \mathcal{S}^{-1} ,
\end{aligned}
\tag{4.14}
$$

where A_1, A_2 are the matrices given in ref. [46]. Since $O(2,r)$ is block diagonal and $S \to S+1$ is lower triangular, it follows that the new sections $(\widehat{X}^\Lambda, \widehat{F}_\Lambda)$ are $O(2,r)$ vectors with the property $\widehat{X}^\Lambda \eta_{\Lambda\Sigma} \widehat{X}^\Sigma = \widehat{F}_\Lambda \eta^{\Lambda\Sigma} \widehat{F}_\Sigma = 0$. Under $S \to S+1$

$$
\begin{aligned}
\widehat{X}^\Lambda &\to \widehat{X}^\Lambda \\
\widehat{F}_\Lambda &\to \widehat{F}_\Lambda + \eta_{\Lambda\Sigma} \widehat{X}^\Sigma ,
\end{aligned}
\tag{4.15}
$$

which induces a Kähler transformation on the Kähler potential. It follows that $\widehat{F}_\Lambda = S \eta_{\Lambda\Sigma} \widehat{X}^\Sigma$ and from eq. (3.18) we find $\widehat{F}(\widehat{X}) = 0$. Note that this is precisely the case for which $\widehat{F}_\Lambda = \widehat{F}_\Lambda(\widehat{X}^\Lambda)$ does not hold. In the same basis the (non–perturbative) inversion $S \to -\frac{1}{S}$ is given by the symplectic matrix $\begin{pmatrix} 0 & \eta \\ -\eta & 0 \end{pmatrix}$. This element, together with the one corresponding to $S \to S+1$ generates an $Sl(2,\mathbb{Z})$ commuting with the $O(2r,\mathbb{Z})$ in $Sp(2r+4,\mathbb{Z})$. The inversion is actually the only symmetry generator with $B \neq 0$. This transformation will be a symmetry of the spectrum of electrically and magnetically charged states discussed in chapter 5 .

The new sections are given explicitly by eqs. (3.19),(3.20),

$$
\begin{aligned}
\widehat{X}^\Lambda &= \frac{1}{\sqrt{2}} (\delta_{\Lambda\Sigma} - F_{\Lambda\Sigma}) X^\Sigma \\
\widehat{F}_\Lambda &= \frac{1}{\sqrt{2}} (\delta_{\Lambda\Sigma} + F_{\Lambda\Sigma}) X^\Sigma ,
\end{aligned}
\tag{4.16}
$$

where the function

$$
F = -\sqrt{X_i^2}\sqrt{X_\alpha^2} \quad i = 0,1; \quad \alpha = 2, \ldots, r+1 \tag{4.17}
$$

was obtained in ref. [46]. The Kähler potential is

$$
K = -\log i(\widehat{X}^\Lambda \widehat{\overline{F}}_\Lambda - \widehat{\overline{X}}^\Lambda \widehat{F}_\Lambda) = -\log i(\overline{S} - S) - \log \widehat{X}^\Lambda \eta_{\Lambda\Sigma} \widehat{\overline{X}}_\Sigma . \tag{4.18}
$$

The holomorphic sections $\widehat{X}^\Lambda$ can be written as follows [7]

$$\widehat{X}^\Lambda = (\frac{1}{2}(1+y_\alpha^2), \frac{i}{2}(1-y_\alpha^2), y^\alpha) , \tag{4.19}$$

where the y^α are coordinates of the $O(2,r)/O(2)\times O(r)$ manifold. If one introduces the objects

$$\Phi^\Lambda = \frac{\widehat{X}^\Lambda}{\sqrt{\widehat{X}^\Sigma \eta_{\Sigma\Pi} \overline{\widehat{X}}^\Pi}} , \tag{4.20}$$

then the kinetic matrix $\widehat{\mathcal{N}}_{\Lambda\Sigma}$ is given by [7][9][11]

$$\widehat{\mathcal{N}}_{\Lambda\Sigma}(\widehat{X}) = (S-\overline{S})(\Phi_\Lambda \overline{\Phi}_\Sigma + \overline{\Phi}_\Lambda \Phi_\Sigma) + \overline{S}\eta_{\Lambda\Sigma} , \tag{4.21}$$

where $\Phi_\Lambda = \eta_{\Lambda\Sigma}\Phi^\Sigma$, and we will also further raise or lower indices with η. This can be obtained from the formulae

$$\widehat{\mathcal{N}}(\widehat{X},\widehat{F}) = (\mathbb{1} + \mathcal{N}(X))(\mathbb{1} - \mathcal{N}(X))^{-1} \tag{4.22}$$

by writing on the right hand side $X^\Lambda = X^\Lambda(\widehat{X},\widehat{F})$. Formula (4.21) is precisely what is obtained from $N=4$ supergravity. Because of target space duality we expect that also the $\widehat{X}^\Lambda, \widehat{F}_\Lambda$ become, because of one loop corrections, a lower triangular representation of $Sp(2r+4,\mathbb{Z})$

$$\begin{pmatrix} \widehat{X}^\Lambda \\ \widehat{F}_\Lambda \end{pmatrix} \to \begin{pmatrix} A & 0 \\ A^{T-1}C & A^{T-1} \end{pmatrix} \begin{pmatrix} X^\Lambda \\ F_\Lambda \end{pmatrix} , \tag{4.23}$$

where the matrix C comes from the monodromy of the one–loop term [1,2].

5 On monodromies in string effective field theories

We have just seen that the tree-level values of the symplectic sections $(X^\Lambda(z), F_\Lambda(z))$ are given by

$$X^\Lambda \equiv X^\Lambda_{\text{tree}} \quad , \; F_\Lambda = S\eta_{\Lambda\Sigma} X^\Sigma_{\text{tree}} . \tag{5.1}$$

The target space duality group $O(2,r;\mathbb{Z})$ acts non–trivially on them

$$\Gamma_{\text{cl}} : \quad \begin{pmatrix} X^\Lambda \\ F_\Lambda \end{pmatrix}_{\text{tree}} \to \begin{pmatrix} A & 0 \\ 0 & \eta A \eta \end{pmatrix} \begin{pmatrix} X^\Lambda \\ F_\Lambda \end{pmatrix}_{\text{tree}} , \tag{5.2}$$

generalizing the action of the Weyl group of the rigid case [2].

At the one loop level, one expects that F_Λ^{tree} is changed to [45]

$$F_\Lambda^{\text{tree}} \to SX^\Sigma \eta_{\Lambda\Sigma} + f_\Lambda(X) \tag{5.3}$$

where $f_\Lambda(X)$ is a modular covariant structure.

The associated perturbative monodromy can be obtained assuming, according to ref. [1], that the rigid perturbative monodromy does not affect the gravitational sector X^0, X^1, F_0, F_1. Thus the perturbative lower triangular monodromy matrix is $\Gamma_{\text{cl}} T$, where[1][2]

$$T = \begin{pmatrix} \mathbb{1} & 0 \\ C & \mathbb{1} \end{pmatrix} \tag{5.4}$$

and C is an $(r+2)\times(r+2)$ symmetric matrix with non–vanishing entries on the $r \times r$ block

$$C = \begin{pmatrix} 0 & 0 & \dots & 0 \\ 0 & 0 & \dots & 0 \\ 0 & 0 & & \\ \vdots & \vdots & C_{ij} & \\ 0 & 0 & & \end{pmatrix} \qquad i,j = 1,\dots,r\,. \tag{5.5}$$

Indeed, we may think of decomposing $Sp(4+2r)$ into $Sp(4) \times Sp(2r)$ and simply assume that the rigid monodromy $\Gamma_r \in Sp(2r)$ commute with the gravitational $Sp(4)$ sector. This argument should at least apply when the vectors of the Cartan subalgebra of the enhanced gauge symmetry belong to the compact $O(r)$ in $O(2,r)$.

In string theory, the classical stringy moduli space corresponds to the broken phase $U(1)^r$ of several gauge groups with the same rank. For instance, for $r = 2$, $O(2,2;\mathbb{Z})$ interpolates between $SU(2)\times U(1)$, $SU(2)\times SU(2)$ and $SU(3)$[47]. In the $N = 4$ theory the $O(6;22)$ moduli space corresponds to broken phases of several gauge groups of rank 22 such as, $U(1)^6 \times E_8 \times E_8$ or $SO(32)\times U(1)^6$ or $SO(44)$ which are not subgroups one of the other [38].

It is obvious that generically this means that the one loop β–function term [18][19] should have non–trivial monodromies at the points where some higher symmetry is restored. For instance, for $r = 2$ we may expect non trivial monodromies around $t = u$ ($SU(2)\times U(1)$ symmetry restored) and $t = u = i$, $t = u = e^{i\pi/3}$ ($SU(2)\times SU(2)$ or $SU(3)$ symmetry restored) , t, u being the parameters defined below.

This means that in supergravity theories derived from strings, because of target space T–duality, the enhanced symmetry points are richer than in the rigid case. Since different enhancement points are consequence of $O(2, r; \mathbb{Z})$ duality, we expect that a modular invariant treatment of quantum monodromies will automatically ensure non trivial monodromy at the enhanced symmetry points.

In the sequel we shall discuss in some more detail the classical and perturbative monodromies in the $r = 1$ case ($O(2,1;\mathbb{Z})$) and the classical monodromies for $r = 2$ ($O(2,2;\mathbb{Z})$).

Consider the tree level prepotential F in the so–called cubic form [3] for $\frac{SU(1,1)}{U(1)} \times \frac{O(2,1)}{O(2)}$:

$$F = \frac{1}{2}(X^0)^2 st^2 , \tag{5.6}$$

where $S = \frac{X^1}{X^0}$ is the dilaton coordinate and $t = \frac{X^2}{X^0}$ is the single modulus of the classical target space duality. We parametrize the $O(2,1;\mathbb{Z})$ vector as follows

$$\begin{aligned} X^0 &= \frac{1}{2}(1-t^2) \\ X^1 &= -t \qquad\qquad\qquad (X^0)^2 + (X^1)^2 - (X^2)^2 = 0 \\ X^2 &= -\frac{1}{2}(1+t^2) \end{aligned} \tag{5.7}$$

The symplectic transformation relating (X^Λ, F_Λ), $(\Lambda = 0,1,2)$ to the $(\widehat{X}^\Lambda, \widehat{F}_\Lambda)$ where $O(2,1)$ is linearly realized is easily found to be

$$\begin{pmatrix} \widehat{X}^\Lambda \\ \widehat{F}_\Lambda \end{pmatrix} = \begin{pmatrix} P & -2R \\ R & 2P \end{pmatrix} \begin{pmatrix} X^\Lambda \\ F_\Lambda \end{pmatrix} , \tag{5.8}$$

where

$$P = \begin{pmatrix} \frac{1}{2} & 0 & 0 \\ 0 & 0 & -1 \\ -\frac{1}{2} & 0 & 0 \end{pmatrix} \quad ; \quad R = \begin{pmatrix} 0 & \frac{1}{2} & 0 \\ 0 & 0 & 0 \\ 0 & \frac{1}{2} & 0 \end{pmatrix} . \tag{5.9}$$

Let us now implement the t-modulus $Sl(2,\mathbb{Z})$ transformations $t \to -\frac{1}{t}$, $t \to t+n$ (note that while $t \to -\frac{1}{t}$ corresponds to the $SU(2)$ Weyl transformation of the rigid theory, $t \to t+n$ has no counterpart in the rigid case, being of stringy nature). Using the parametrization (5.7) we find

$$\begin{aligned} t \to -\frac{1}{t} : &\quad \begin{pmatrix} -1 & 0 & 0 \\ 0 & -1 & 0 \\ 0 & 0 & 1 \end{pmatrix} \equiv -\eta \in O(2,1;\mathbb{Z}) \\ t \to t+n : &\quad \begin{pmatrix} 1-\frac{n^2}{2} & n & \frac{n^2}{2} \\ -n & 1 & n \\ -\frac{n^2}{2} & n & 1+\frac{n^2}{2} \end{pmatrix} \equiv V(n) \in O(2,1;\mathbb{Z}) . \end{aligned} \tag{5.10}$$

Note that (5.10) implies $n \in 2\mathbb{Z}$, i.e. the subgroup $\Gamma_{(0)}(2)$ of $SL(2,\mathbb{Z})$. Actually this gives a projective representation in the subgroup in $O(2,1;\mathbb{Z})$ of the matrices congruent to the identity *mod* 2.

It follows that $\Gamma_{\rm cl}$ is generated by (Γ_1, Γ_2) where

$$\begin{aligned} \Gamma_1 &= \begin{pmatrix} -\eta & 0 \\ 0 & -\eta \end{pmatrix} \in Sp(6,\mathbb{Z}) \\ \Gamma_2 &= \begin{pmatrix} V(2) & 0 \\ 0 & \eta V(2)\eta \end{pmatrix} \in Sp(6,\mathbb{Z}) . \end{aligned} \tag{5.11}$$

On the other hand it is possible to go to a stringy basis with a new metric $X_0^2 + X_1^2 - X_2^2 = 2\widetilde{X}_1^2 + XY$ such that $SL(2,\mathbb{Z})$ is integral valued in $O(2,1;\mathbb{Z})$.

The $O(2,1;\mathbb{Z})$ generators corresponding to translation and inversion are respectively given by:

$$\begin{pmatrix} 1 & -2n & 0 \\ 0 & 1 & 0 \\ n & -n^2 & 1 \end{pmatrix} \; ; \qquad \begin{pmatrix} -1 & 0 & 0 \\ 0 & 0 & -1 \\ 0 & -1 & 0 \end{pmatrix} . \tag{5.12}$$

To make contact with the rigid theory it is convenient to define the inversion generator in $O(2,1;\mathbb{Z})$ with the opposite sign with respect to the previous definition.

Let us now examine the perturbative monodromy matrices T [1]. If we assume as before that the $t \to -\frac{1}{t}$ pertaining to the rigid theory does not affect the gravitational sector (X^0, X^1, F_0, F_1), then we have

$$T = \begin{pmatrix} \eta & 0 \\ C & \eta \end{pmatrix} \quad , \quad C = \begin{pmatrix} 0 & 0 & 0 \\ 0 & 0 & 0 \\ 0 & 0 & 2 \end{pmatrix} \tag{5.13}$$

corresponding to the embedding of the $Sp(2,\mathbb{Z})$ rigid transformations acting on the rigid section (X^2, F_2) in $Sp(6,\mathbb{Z})$. Furthermore, considering the transformation of the $\mathcal{N}_{\Lambda\Sigma}$ matrix and setting $D = A = \eta$, $B = 0$ we find

$$\widehat{\mathcal{N}}_{22} = -2 + \mathcal{N}_{22} \tag{5.14}$$

for all other entries $\widehat{\mathcal{N}}_{\Lambda\Sigma} = \mathcal{N}^{\Lambda\Sigma}$. This is exactly the rigid result[1]. However conjugating the T matrix with Γ_2 one gets

$$C_{\Lambda\Sigma} = \begin{pmatrix} 8 & -8 & -12 \\ -8 & 8 & 12 \\ -12 & 12 & 18 \end{pmatrix} \tag{5.15}$$

which shows that $O(2,1;\mathbb{Z})$ introduces non–trivial perturbative monodromies for all couplings. The other perturbative lower diagonal monodromy is the dilaton shift (4.14) which commutes with $O(2,1;\mathbb{Z})$.

Analogous considerations hold for $O(2,n;\mathbb{Z})$, $n>1$. We limit ourselves to write down the generators of $\Gamma_{\rm cl}$ for the $O(2,2;\mathbb{Z})$ case. We use the parametrization of $O(2,2)/O(2)\times O(2)$ given by

$$\begin{aligned} X^0 &= \frac{1}{2}(1-tu) \\ X^1 &= -\frac{1}{2}(t+u) \\ X^2 &= -\frac{1}{2}(1+tu) \\ X^3 &= \frac{1}{2}(t-u) \qquad (X^0)^2+(X^1)^2-(X^2)^2-(X^3)^2=0\,, \end{aligned} \tag{5.16}$$

where t,u are the moduli appearing in the F function $F=stu$. In the same way as for the $r=1$ case it is easy to find the symplectic transformations relating the sections of the cubic parametrization to the X^Λ defined in (5.16). They are given by

$$\begin{pmatrix} X \\ F \end{pmatrix} \to \begin{pmatrix} A & B \\ -B & A \end{pmatrix} \begin{pmatrix} X \\ F \end{pmatrix}, \tag{5.17}$$

with

$$\begin{gathered} X=(X^0,X^1,X^2,X^3)^T \quad , \quad F=(F_0,F_1,F_2,F_3)^T \\ A=\frac{1}{\sqrt{2}}\begin{pmatrix} 1 & 0 & 0 & 0 \\ 0 & 0 & -1 & -1 \\ -1 & 0 & 0 & 0 \\ 0 & 0 & 1 & -1 \end{pmatrix} \quad , \quad B=\frac{1}{\sqrt{2}}\begin{pmatrix} 0 & -1 & 0 & 0 \\ 0 & 0 & 0 & 0 \\ 0 & -1 & 0 & 0 \\ 0 & 0 & 0 & 0 \end{pmatrix}. \end{gathered} \tag{5.18}$$

It is convenient to use the string basis where the metric η takes the form[11]

$$\eta=\begin{pmatrix} 0 & \mathbb{1}_{2\times 2} \\ \mathbb{1}_{2\times 2} & 0 \end{pmatrix}, \tag{5.19}$$

corresponding to the basis $\frac{1}{\sqrt{2}}(X^0\mp X^2)$, $\frac{1}{\sqrt{2}}(X^1\mp X^3)$. Then one finds the following

$O(2,2;\mathbb{Z})$ representation

$$\begin{aligned}
&ut \to +\frac{1}{ut} : \begin{pmatrix} 0 & -\mathbb{1} \\ -\mathbb{1} & 0 \end{pmatrix} = \gamma_{ut} \\
&t \to -\frac{1}{t} : \begin{pmatrix} \epsilon & 0 \\ 0 & \epsilon \end{pmatrix} = \gamma_t \\
&u \to -\frac{1}{u} : \begin{pmatrix} 0 & \epsilon \\ \epsilon & 0 \end{pmatrix} = \gamma_u \\
&t \to t+n : \begin{pmatrix} \mathbf{N}^t(-n) & 0 \\ 0 & \mathbf{N}(n) \end{pmatrix} = \gamma_n \\
&t \to u : \begin{pmatrix} a & b \\ b & a \end{pmatrix} = \gamma \; ; \qquad a = \begin{pmatrix} 1 & 0 \\ 0 & 0 \end{pmatrix} \; ; \qquad b = \begin{pmatrix} 0 & 0 \\ 0 & 1 \end{pmatrix}
\end{aligned} \tag{5.20}$$

where $\epsilon = \begin{pmatrix} 0 & -1 \\ 1 & 0 \end{pmatrix}$ and $\mathbf{N}(n) = \begin{pmatrix} 1 & n \\ 0 & 1 \end{pmatrix}$.

$\Gamma_{\rm cl}$ is then generated by the matrices:

$$\Gamma_{ut} = \begin{pmatrix} \gamma_{ut} & 0 \\ 0 & \gamma_{ut} \end{pmatrix} \; ; \; \Gamma_t = \begin{pmatrix} \gamma_t & 0 \\ 0 & \gamma_t \end{pmatrix} \; ; \; \Gamma_u = \begin{pmatrix} \gamma_u & 0 \\ 0 & \gamma_u \end{pmatrix} \; ; \Gamma_n = \begin{pmatrix} \gamma_n & 0 \\ 0 & \gamma^T_{-n} \end{pmatrix} . \tag{5.21}$$

We note that the points $t = u$; $t = u = i$; $t = u = e^{\frac{\pi i}{3}}$ are enhanced symmetry points corresponding to $SU(2) \times U(1)$, $SU(2) \times SU(2)$, and $SU(3)$ respectively [47]. Therefore we expect non–trivial quantum monodromies at these points according to the previous discussion.

5.1 The BPS mass formula

The classical and one loop monodromies are of course reflected in symmetries of the electrically charged massive states belonging to $O(2,n;\mathbb{Z})$ lorentzian lattice[38]. The BPS mass formula [48] in the gravitational case is

$$M = |Z| = |n^{(e)}_\Lambda L^\Lambda - n^\Lambda_{(m)} M_\Lambda| = e^{K/2} |n^{(e)}_\Lambda X^\Lambda - n^\Lambda_{(m)} F_\Lambda| \, . \tag{5.22}$$

Note that the central charge Z has definite $U(1)$ weight

$$Z \to e^{(f-f)/2} Z \, , \tag{5.23}$$

while the mass M is Kähler invariant. The symplectic invariance of M also implies that $(n^\Lambda_{(m)}, n^{(e)}_\Lambda)$ transforms as (X^Λ, F_Λ)

$$\begin{pmatrix} n^\Lambda_{(m)} \\ n^{(e)}_\Lambda \end{pmatrix} \to \begin{pmatrix} A & B \\ C & D \end{pmatrix} \begin{pmatrix} n^\Lambda_{(m)} \\ n^{(e)}_\Lambda \end{pmatrix} , \tag{5.24}$$

where according to our previous discussion the perturbative symmetries have $B = 0$. Note that $n^\Lambda_{(m)}, n^{(e)}_\Lambda$ must satisfy a lattice condition. We consider here the particular orbit:

$$n^\Lambda_{(m)} n^\Sigma_{(m)} \eta_{\Lambda\Sigma} = n^{(e)}_\Lambda n^{(e)}_\Sigma \eta^{\Lambda\Sigma} = n^\Lambda_{(m)} n^{(e)}_\Lambda = 0 . \tag{5.25}$$

In the tree level approximation we may write

$$M = |(n^{(e)}_\Lambda - n^\Sigma_{(m)} \eta_{\Lambda\Sigma} S) X^\Lambda | e^{K/2} \tag{5.26}$$

which is invariant under the tree level symmetry $S \to S + 1$, but also under the non–perturbative inversion $S \to -\frac{1}{S}$[34][12][11][13][14] taking into account that

$$K = -\log i(\overline{S} - S) - \log \frac{X^\Lambda \overline{X}^\Sigma}{M^2_{Pl}} \eta_{\Lambda\Sigma} \tag{5.27}$$

Formula (5.26) is therefore invariant under the $S - T$ duality symmetry $Sl(2, \mathbb{Z}) \times O(2, r; \mathbb{Z}) \subset Sp(2r + 4; \mathbb{Z})$.

The electric mass spectrum can be written as

$$M^2 = |Z|^2 = M^2_{Pl} \frac{1}{2i(\overline{S} - S)} \mathcal{Q}^{\Lambda\Sigma} n^{(e)}_\Lambda n^{(e)}_\Sigma , \tag{5.28}$$

where $i(\overline{S} - S) = \frac{8\pi}{g^2} > 0$ and $\mathcal{Q}^{\Lambda\Sigma} = \Phi^\Lambda \overline{\Phi}^\Sigma + \overline{\Phi}^\Lambda \Phi^\Sigma$. Formula (5.28) has exactly the same form as the analogous one obtained in $N = 4$ (see ref [11]). When also magnetic charges are present then

$$M^2 = \frac{1}{4} M^2_{Pl} (n_m, n_e)(\mathcal{M}\mathcal{Q} + \mathcal{L}\widehat{\mathcal{Q}}) \begin{pmatrix} n_m \\ n_e \end{pmatrix} , \tag{5.29}$$

where $\mathcal{M} = \frac{1}{\text{Im } S} \begin{pmatrix} S\overline{S} & -\text{Re } S \\ -\text{Re } S & 1 \end{pmatrix}$, $\mathcal{L} = \begin{pmatrix} 0 & -1 \\ 1 & 0 \end{pmatrix}$ and $\widehat{\mathcal{Q}} = i \left(\overline{\Phi}^\Lambda \Phi^\Sigma - \Phi^\Lambda \overline{\Phi}^\Sigma \right)$. The last term actually vanishes if in solving the constraints (5.25) we set $n^{(e)}_\Lambda = m_1 n_\Lambda$, $n^\Lambda_{(m)} = m_2 n_\Sigma \eta^{\Lambda\Sigma}$, with $n^\Lambda n_\Lambda = 0$.

In the $O(2,1;\mathbb{Z})$ case the constraint $n_0^2 + n_1^2 - n_2^2 = 0$ is solved by setting

$$\begin{aligned} n_1 &= m r_1 r_2 \\ n_0 - n_2 &= -m r_2^2 \\ n_0 + n_2 &= m r_1^2 \ , \end{aligned} \tag{5.30}$$

where $m, r_i \in \mathbb{Z}$ and r_1, r_2 are relatively prime integers. If $t \to t + n$, n even, then if m is odd, r_1, r_2 are odd. Using the parametrization (5.7) we find

$$n_\Lambda X^\Lambda = \frac{M_{Pl}}{2} m (r_1 - r_2 t)^2 \ , \tag{5.31}$$

and therefore the electric charged states have mass

$$M_{el}^2 = -\frac{M_{Pl}^2}{2} m^2 \frac{(r_1 - r_2 t)^2 (r_1 - r_2 \bar{t})^2}{(t - \bar{t})^2 i(\overline{S} - S)} \ , \tag{5.32}$$

which at $t = i$ $(X^2 = 0)$ $(SU(2)$ restored) becomes

$$M_{el}^2(t = i) = \frac{M_{Pl}^2}{8} \frac{m^2}{i(\overline{S} - S)} (r_1^2 + r_2^2)^2 \ . \tag{5.33}$$

Similarly, in the $O(2,2;\mathbb{Z})$ case the constraint $n_0^2 + n_1^2 - n_2^2 - n_3^2 = 0$ is solved by

$$\begin{aligned} n_0 - n_2 &= r_1 r_3 \\ n_1 - n_3 &= r_1 r_4 \\ n_0 + n_2 &- r_2 r_4 \\ n_1 + n_3 &= -r_2 r_3 \ , \end{aligned} \tag{5.34}$$

where $r_i \in \mathbb{Z}$ and (r_3, r_4) relatively prime. From the parametrizations (5.16) we find

$$n_\Lambda X^\Lambda = \frac{M_{Pl}}{2} (r_2 - r_1 u)(r_4 + r_3 t) \ , \tag{5.35}$$

and the corresponding electric mass is

$$M_{el}^2 = -\frac{1}{8} M_{pl}^2 \frac{V^T C(t, \bar{t}) V U^T C(u, \bar{u}) U}{i(\overline{S} - S)(\bar{t} - t)(\bar{u} - u)} \ , \tag{5.36}$$

where

$$V = \begin{pmatrix} -r_3 \\ r_4 \end{pmatrix} \quad U = \begin{pmatrix} r_1 \\ r_2 \end{pmatrix} \quad C = \begin{pmatrix} 2\bar{t}t & -(t + \bar{t}) \\ -(t + \bar{t}) & 2 \end{pmatrix} \ . \tag{5.37}$$

The three enhancement points (5.36) become

$$\begin{aligned} t=u: \ & M_{el}^2 = -\frac{1}{8}M_{Pl}^2\frac{V^T C(t,\bar{t})VU^T C(t,\bar{t})U}{i(\overline{S}-S)(\bar{t}-t)^2} \\ t=u=i: \ & \frac{1}{8}M_{Pl}^2\frac{V^T VU^T U}{i(\overline{S}-S)} \\ t=u=e^{i\pi/3}: \ & M_{el}^2 = \frac{1}{24}M_{Pl}^2\frac{V^T C(e^{i\pi/3})VU^T C(e^{i\pi/3})U}{i(\overline{S}-S)} \ . \end{aligned} \tag{5.38}$$

The real symmetric matrix $C_{ij}(t,\bar{t})$ is essentially the σ–model metric G_{ij}[10], and it reduces to the Cartan matrix of $SU(2)\times SU(2)$,$SU(3)$, at the enhanced symmetry points $t=i$, $t=e^{i\pi/3}$. Analogous considerations can be drawn for the classical spectrum of magnetic and dyon charges using the general formula (5.29) . Obviously the quantum spectrum is found by substituting $F_{\Lambda\mathrm{tree}} \equiv S\eta_{\Lambda\Sigma}X^\Sigma \to F_{\Lambda\mathrm{tree}}+$quantum corrections.

6 Conclusions

In this paper we have formulated electromagnetic duality transformations in generic $D=4$, $N=2$ supergravities theories in a form suitable to investigate non–perturbative phaenomena. Our formulation is manifestly duality covariant for the full Lagrangian, including fermionic terms, which unlike the rigid case, cannot be retrieved from the $N=1$ formulation, nor from the $N=2$ tensor calculus approach. Particular attention has been given to classical T-duality symmetries which actually occur in string compactifications and whose linear action on the gauge potential fields do not allow for the existence of a prepotential function F for the $N=2$ special geometry. As examples we described the "classical" electric and monopole spectrum for T–duality symmetries of the type $O(2,r;\mathbb{Z})$, with particular details for the $r=1,2$ cases, by using the $N=2$ formalism.

For "classical" monodromies this spectrum is of course related to the spectrum of $N=4$ theories studied by Sen and Schwarz [11]. Possible extensions of duality symmetries to type II strings have been conjectured by Hull and Townsend [49] and also discussed in [2]. In the present context of $N=2$ heterotic strings the corresponding type II theories, having $N=2$ space–time supersymmetry would correspond to $(2,2)$ superconformal field theories, i.e. quantum Calabi–Yau manifolds.

Due to the non–compact symmetries the BPS saturated states with non–vanishing central charges have a spectrum quite different from the rigid case. Indeed in rigid theories the "classical" central charge $Z_{(cl)}$ vanishes at the enhanced symmetry points where the original gauge group is restored since there is no dimensional scale other than the Higgs v.e.v.. On the contrary, in the supergravity theory the BPS spectrum at these particular points corresponds in general to electrically and magnetically charged states with Planckian mass (black holes, gravitational monopoles and dyons) [50][11][51][52][53][54]. The only charged states which become massless at the enhanced symmetric point are those with $\eta^{\Lambda\Sigma} n^{(e)}_{\Lambda} n^{(e)}_{\Sigma} < 0$.

We also discussed perturbative monodromies and their possible relations with the rigid case. Non perturbative duality symmetries are more difficult to guess, but it is tempting to conjecture that a quantum monodromy consistent with positivity of the metric and special geometry may be originated by a 3-dimensional Calabi-Yau manifold or its mirror image. If this is the case this manifold should embed in some sense the class of Riemann surfaces studied[1][2] in connection with the moduli space of $N = 2$ rigid supersymmetric Yang Mills theories.

Acknowledgements

We would like to acknowledge useful discussions with the following colleagues: L. Alvarez-Gaumé, I. Antoniadis, M. Bianchi, P. Fré, E. Kiritsis, C. Kounnas, W. Lerche, K. Narain, M. Porrati, T. Regge, G. Rossi and T. Taylor.

References

[1] N. Seiberg and E. Witten, Nucl. Phys. B426 (1994) 19, hep-th/9407087; Nucl. Phys. B431 (1994) 484, hep-th/9408099.

[2] A. Klemm, W. Lerche, S. Yankielowicz and S. Theisen, *Simple Singularities and $N = 2$ Supersymmetric Yang-Mills Theory*, preprint CERN-TH.7495/94, LMU-TPW 94/16, hep-th/9411048 and *On the monodromies of N=2 supersymmetric Yang-Mills theory*, contribution to these Proceedings, preprint CERN-TH-7538-94, hep-th/9412158; P. C. Argyres and A. E. Faraggi, *The Vacuum Structure and Spectrum of $N = 2$ Supersymmetric $SU(n)$ Gauge Theory*, preprint IASSNS-HEP-94/94, hep-th/9411057

[3] B. de Wit and A. Van Proeyen, Nucl. Phys. B245 (1984) 89; E. Cremmer, C. Kounnas, A. Van Proeyen, J. P. Derendinger, S. Ferrara, B. De Wit and L. Girardello, Nucl. Phys. B250 (1985) 385; B. de Wit, P. G. Lauwers and A. Van Proeyen, Nucl. Phys. B255 (1985) 569; S. Cecotti, S. Ferrara and L. Girardello, Int. J. Mod. Phys. A4 (1989) 2475.

[4] A. Ceresole, R. D'Auria and S. Ferrara, Phys. Lett. 339B (1994) 71, hep-th/9408036 .

[5] D. Amati, K. Konishi, Y. Meurice, G. C. Rossi and G. Veneziano, Phys. Rep. 162 (1988) 169.

[6] E. Witten, *Monopoles and Four–Manifolds*, preprint IASSNS-HEP-94-96, hep-th/9411102.

[7] S. Ferrara, L. Girardello, C. Kounnas and M. Porrati, Phys. Lett. B192 (1987) 368.

[8] N. Seiberg, Nucl. Phys. B303 (1988) 206.

[9] S. Ferrara and M. Porrati, Phys. Lett. B216 (1989) 289.

[10] For a review on target space duality see A. Giveon, M. Porrati and E. Rabinovici, Phys. Rep. 244 (1994) 77, hep-th/9401139.

[11] A. Sen, Nucl. Phys. B388 (1992) 457 and Phys. Lett. 303B (1993) 22; A. Sen and J. H. Schwarz, Nucl. Phys. B411 (1994) 35; Phys. Lett. 312B (1993) 105.

[12] M. J. Duff and R. R. Khuri, Nucl. Phys. B411 (1994) 473.

[13] J. Gauntlett and J. A. Harvey, *S–Duality and the Spectrum of Magnetic Monopoles in Heterotic String Theory* , preprint EFI-94-36, hep-th/9407111.

[14] L. Girardello, A. Giveon. M. Porrati and A. Zaffaroni, Phys. Lett. B334 (1994) 331, hep-th/9406128.

[15] J. H. Schwarz, *Evidence for Non–perturbative String Symmetries* , preprint CALT-68-1965, hep-th/9411178.

[16] S. Ferrara, J. Scherk and B. Zumino, Nucl. Phys. B121 (1977) 393.

[17] M. K. Gaillard and B. Zumino, Nucl. Phys. B193 (1981) 221.

[18] P. Di Vecchia, R. Musto, F. Nicodemi and R. Pettorino, Nucl. Phys. B252 (1985) 635.

[19] N. Seiberg, Phys. Lett. 206B (1988) 75.

[20] S. Ferrara and A. Van Proeyen, Class. Quantum Grav. 6 (1989) 124.

[21] E. Cremmer and A. Van Proeyen, Class. Quantum Grav. 2 (1985) 445.

[22] S. Ferrara and J. Louis, Phys. Lett. 278B (1992) 240.

[23] A. Ceresole, R. D'Auria, S. Ferrara, W. Lerche and J. Louis, Int. J. Mod. Phys. A8 (1993) 79

[24] L. Castellani, R. D'Auria and S. Ferrara, Phys. Lett. 241B (1990) 57; Class. Quantum Grav. 1 (1990) 317.

[25] S. Ferrara and A. Strominger, in *Strings '89* , eds. R. Arnowitt, R. Bryan, M.J. Duff, D.V. Nanopoulos and C.N. Pope(World Scientific, 1989), p. 245; A. Strominger, Comm. Math. Phys. 133 (1990) 163.

[26] P. Candelas and X. de la Ossa, Nucl. Phys. B355 (1991) 455; P. Candelas, X. de la Ossa, P. S. Green and L. Parkes, Phys. Lett. 258B (1991) 118; Nucl. Phys. B359 (1991) 21.

[27] R. D'Auria, S. Ferrara and P. Fré, Nucl. Phys. B359 (1991) 705.

[28] A. C. Cadavid and S. Ferrara, Phys. Lett. 267B (1991) 193; W. Lerche, D. Smit and N. Warner, Nucl. Phys. B372 (1992) 87.

[29] A. Ceresole, R. D'Auria and T. Regge, Nucl. Phys. B414 (1994) 517; for a recent review, see A. Ceresole, R. D'Auria, S. Ferrara, W. Lerche, J. Louis and T. Regge, *Picard–Fuchs Equations, Special Geometry and Target Space Duality* ,

preprint CERN-TH.7055/93, POLFIS-TH.09/93, to appear on *Essays on Mirror Manifolds* , vol. II, S. T. Yau editor, International Press (1994).

[30] A. Ceresole, R. D'Auria, S. Ferrara and A. Van Proeyen, *Duality transformations in supersymmetric Yang–Mills theories coupled to supergravity* , preprint CERN-TH 7547/94, to appear.

[31] E. Cremmer, S. Ferrara, L. Girardello and A. Van Proeyen, Nucl. Phys. B212 (1983) 413.

[32] B. de Wit, F. Vanderseypen and A. Van Proeyen, Nucl. Phys. B400 (1993) 463.

[33] A. Das, Phys. Rev. D15 (1977) 2805; E. Cremmer, J. Scherk and S. Ferrara, Phys. Lett. 68B (1977) 234; E. Cremmer and J. Scherk, Nucl. Phys. B127 (1977) 259.

[34] E. Cremmer, J. Scherk and S. Ferrara, Phys. Lett. 74B (1978) 61.

[35] R. Kallosh, A. Linde, T. Ortín, A. Peet and A. Van Proeyen, Phys. Rev. D46 (1992) 5278.

[36] E. S. Fradkin, M. A. Vasiliev, Nuovo Cim. Lett. **25** (1979) 79; B. de Wit and J. W. van Holten, Nucl. Phys. B155 (1979) 530; B. de Wit, J. W. van Holten and A. Van Proeyen, Nucl. Phys. B167 (1980) 186.

[37] M. Villasante, Phys. Rev. D45 (1992) 831.

[38] K. Narain, Phys. Lett. 169B (1986) 41; K. Narain, M. Sarmadi and E. Witten, Nucl. Phys. B279 (1987) 369.

[39] A. Shapere and F. Wilczek, Nucl. Phys. B320 (1989) 669.

[40] I. Antoniadis, C. Bachas and C. Kounnas, Nucl. Phys. B289 (1987) 87; H. Kawai, D. C. Lewellen and S. H. Tye, Nucl. Phys. B288 (1987)1.

[41] S. Ferrara, C. Kounnas, D. Lust and F. Zwirner, Nucl. Phys. B365 (1991) 431.

[42] L. J. Dixon, V. S. Kaplunovsky and J. Louis, Nucl. Phys. B355 (1990) 27.

[43] G. L. Cardoso and B. A. Ovrut, Nucl. Phys. B369 (1992) 351.

[44] J. P. Derendinger, S. Ferrara, C. Kounnas and F. Zwirner, Nucl. Phys. B372 (1992) 145.

[45] I. Antoniadis, E. Gava, K. S. Narain and T. R. Taylor, Nucl. Phys. B413 (1994) 162.

[46] P. Fré and P. Soriani, Nucl. Phys. B371 (1992) 659.

[47] R. Dijkgraaf, E. Verlinde, H. Verlinde, Comm. Math. Phys. 115 (1988) 649.

[48] M. K. Prasad and C. M. Sommerfield, Phys. Rev. Lett. 35 (1975) 760; E. B. Bogomol'nyi, Sov. J. Nucl. Phys. 24 (1976) 449; E. Witten and D. Olive, Phys. Lett. *78B* (1978) 97; C. Montonen and D. Olive, Phys. Lett. *72B* (1977) 117; P. Goddard, J. Nuyts and D. Olive, Nucl. Phys. *B125* (1977) 1.

[49] C. M. Hull and P. K. Townsend, *Unity of Superstring Dualities* , preprint QMW 94-30, hep-th/9410167.

[50] G. Gibbons and C. Hull, Phys. Lett. *B109* (1982) 190; G. Gibbons and N. Manton, Nucl. Phys. *B274* (1986) 183.

[51] A. Sen, Int. J. Mod. Phys. *A9* (1994) 3707, hep-th/9402002 and references there in.

[52] J. Harvey and J. Liu, Phys. Lett. *B268* (1991) 40.

[53] R.R. Khuri, Phys. Lett. *B259* (1991) 201; Phys. Lett. *B294* (1992) 325.

[54] R. Kallosh and T. Ortin, Phys. Rev. *D48* (1993) 742, hep-th/9302109.

Low-Energy Kähler Potentials In Supersymmetric Gauge Theories*
With (Almost) Flat Directions

Erich Poppitz†
Enrico Fermi Institute, University of Chicago
5640 S. Ellis Avenue, Chicago, IL 60637 USA

and

Lisa Randall‡
Center for Theoretical Physics, Laboratory for Nuclear Science and Department of Physics
Massachusetts Institute of Technology
Cambridge, MA 02139 USA

ABSTRACT

We derive the supersymmetric low-energy effective theory of the D-flat directions of a supersymmetric gauge theory. The Kähler potential of Affleck, Dine and Seiberg is derived by applying holomorphic constraints which manifestly maintain supersymmetry. We also present a simple procedure for calculating all derivatives of the Kähler potential at points on the flat direction manifold. Together with knowledge of the superpotential, these are sufficient for a complete determination of the spectrum and the interactions of the light degrees of freedom. We illustrate the method on the example of a chiral abelian model, and comment on its application to more complicated calculable models with dynamical supersymmetry breaking.

1. Introduction

When considering Lagrangians with many different mass scales, it is often useful to integrate out the heavy degrees of freedom and derive an effective low-energy theory of the light fields. In supersymmetric theories, it is useful to follow this procedure while manifestly preserving supersymmetry.

Supersymmetric gauge theories often exhibit directions in the scalar field space, along which the scalar potential identically vanishes, the so-called "flat" directions.[1] When studying the low-energy dynamics of the theory at a given point on the flat direction manifold it is useful to consider an effective Lagrangian where the fields are constrained to these flat directions.[2] However, the vanishing of the D term is a nonholomorphic constraint which generally cannot be solved in terms of chiral superfields. It is therefore not clear a priori how to construct an effective theory of the flat directions which manifestly maintains supersymmetry.

---

*Talk presented by Lisa Randall based on paper published in *Physics Letters B* **B336** (1994) 402–408.
†Robert R. McCormick Fellow. Supported in part by DOE contract DE-FGP2-90ER40560.
‡NSF Young Investigator Award, Alfred P. Sloan Foundation Fellowship, DOE Outstanding Junior Investigator Award. This work is supported in part by funds provided by the U.S. Department of Energy (D.O.E.) under cooperative agreement #DF-FC02-94ER40818, by NSF grant PHY89-04035.

The method for finding the Kähler potential of the low-energy sigma model used by Affleck, Dine and Seiberg (ADS)[2] is based on the procedure of using the flat direction equations to project the full theory Kähler potential without gauge fields onto the light gauge invariant chiral superfields. For simple gauge groups and matter representations, such as SU(2) one-flavor SQCD, it leads to unambiguous results for the Kähler potential. However, even in the case of the simplest model with dynamical supersymmetry breaking, the SU(3)×SU(2) model,[2] the Kähler potential is determined by the solution of a cubic equation.[2] The choice of the correct solution can only be made by examining the positivity of the Kähler metric for each of the roots of this equation.[3]

In this paper, we address both of these ambiguities. We derive the ADS effective Lagrangian by applying a *holomorphic* constraint and explicitly integrating out the heavy vector fields. We show how this procedure is equivalent, via a *nonholomorphic* field redefinition, to the ADS procedure.

We also show how to compute all derivatives of the Kähler potential at the flat direction without solving the complicated equations for the vanishing of the D terms. Moreover, our procedure yields a manifestly positive definite Kähler metric, as we show in Section 2. It is very general and can be applied to calculable models of dynamical supersymmetry breaking, with virtually unknown ground state properties.

2. Gauge Invariant Description of the Light Degrees of Freedom

In this section we derive the gauge invariant theory of the light degrees of freedom along a flat direction with completely broken gauge symmetry. It can be generalized to the case where there is some unbroken nonabelian gauge group, so long as we are interested in the effective theory at scales below the scale where all particles carrying charge under the unbroken group acquire a dynamical mass. The matter part of the classical Lagrangian of a general supersymmetric gauge theory with gauge group G (of dimension $N-n$) and no classical superpotential has the form of a D-term:

$$L_D = Q^\dagger e^{V^a T^a} Q \,, \tag{1}$$

where Q_i are chiral matter superfields. Hereafter i runs over both the gauge index and the different representations, $i = 1, \ldots, N$, where N is the number of chiral matter superfields. V^a, $a = 1, \ldots, N-n$, denote the vector superfields, corresponding to the various factors in G. Under a supergauge transformation the matter and gauge superfields transform as:[4]

$$\begin{aligned} Q &\to e^{-i\Omega^a T^a} Q \\ e^{V^a T^a} &\to e^{-i\Omega^{a\dagger} T^a} e^{V^b T^b} e^{i\Omega^a T^a} \,, \end{aligned} \tag{2}$$

where the parameters of the transformation Ω^a are chiral superfields. The scalar potential of this theory has classically flat directions along which it identically vanishes. They are given by the solutions of

$$Q^{\dagger i} T_i^{aj} Q_j = 0 \,, \tag{3}$$

where Q_i now means the scalar components of the corresponding chiral superfields and a sum over the different representations is again implicit. If we expand the theory around a solution of Eq. (3) sufficiently far from the origin the theory is weakly coupled and can be analyzed perturbatively.[1] We consider the theory in the vicinity of a such solution of Eq. (3), which completely breaks the gauge symmetry[2] (up to possible abelian factors). The number of broken generators of the gauge group is $N-n$. Then $N-n$ of the N chiral superfields Q_i are massive and n chiral superfield are massless (in the absence of superpotential).

Below the scale of the gauge boson masses the heavy gauge fields and their superpartners can be integrated out. We assume that the theory of the light degrees of freedom can be given a gauge invariant description,[2] where the light supermultiplets are represented by a set of gauge invariant chiral superfields X^A, $A=1,\ldots,n$, which are independent polynomials in the matter fields Q_i:

$$X^A = X^A(Q)\,. \tag{4}$$

The dynamics of the theory below the scale of the gauge boson masses is described by a supersymmetric Kähler sigma model with coordinates spanned by the light chiral superfields Eq. (4).[2] The derivation of the Kähler potential of this sigma model is the focus of this letter.

The derivation is nontrivial, because if one follows the referenced procedure[2] and applies the nonholomorphic constraint of Eq. (3) it is not clear why the resulting Lagrangian should be supersymmetric. Explicitly, for particular models one observes that the flat directions equations cannot be promoted to chiral superfields. Furthermore, because the constraint equations are real, there is an insufficient number of constraining equations to determine the light chiral fields (see discussion below).

Another difficulty with the ADS procedure is more a matter of practice. To find the Kähler potential of the effective theory, one needs to solve for it along the flat directions, which can be very complicated. Even for the simplest model of dynamical supersymmetry breaking based on SU(3) $\times$ SU(2), one needs to solve a cubic polynomial equation and only the correct root gives a positive definite kinetic energy for the light degrees of freedom.

In the rest of this section, we will show how to derive the ADS potential. We will use holomorphic constraints when separating the light from the heavy degrees of freedom. We will show using our procedure how one can compute derivatives of the Kähler potential (which are all that are required for finding the spectrum and interactions) simply, without explicitly solving for the full form of the low energy Kähler potential.

In order to separate the light and heavy degrees of freedom in a gauge invariant way, it is necessary to make a field redefinition

$$\begin{aligned} Q &\rightarrow e^{-iG^aT^a}Q(X) \\ e^{V^aT^a} &\rightarrow e^{-iG^{c\dagger}T^c}e^{V^aT^a}e^{iG^bT^b}\,, \end{aligned} \tag{5}$$

where G^a are Goldstone chiral superfields, transforming as

$$e^{-iG^aT^a} \to e^{-i\Omega^aT^a}e^{-iG^bT^b} \tag{6}$$

under supersymmetric gauge transformations. The vector superfield $\mathcal{V}^a$ is gauge invariant, as follows from Eqs. (3), (5), and (6). On-shell it describes a massive vector supermultiplet.[4]

We are interested in making the field redefinition Eq. (5) locally, around a point $Q_i = v_i$ on the flat direction manifold ("moduli space") where the gauge symmetry is completely broken. Hence the $N - n$ vectors (N-dimensional) $T_i^{aj}v_j$ are all nonvanishing and linearly independent. We assume that one can find $X^A(Q)$ such that all derivatives $X_i^A(Q) \equiv \partial X^A(Q)/\partial Q_i$ are nonvanishing at this point. The functional independence of $X^A(Q)$ then assures that the n vectors (N-dimensional) $X_i^A(v)$ are linearly independent. Gauge invariance of $X^A(Q)$ implies that $X_j^A(v)T_i^{aj}v_j = 0$, for any A, a. Taken together with the linear independence, this implies that the N vectors $X_i^{A*}(v^*) \equiv \partial X^{A*}(Q^*)/\partial Q^{*i}|_{Q^*=v^*}$ and $T_i^{a\ j}v_j$ form a complete basis in the complex space spanned by Q_i. The gauge invariant fields X^A have the expansion around $Q_i = v_i$

$$X^A = X^A(v_i) + x^A \,. \tag{7}$$

Then the expansion of $Q(X)$ around points on the flat direction manifold is

$$Q_i(X) = v_i + q_i(x) \,. \tag{8}$$

The most general form of $q_i(x)$ is

$$q_i(x) = X_i^{A*}(v)\lambda_A(x) + T_i^{aj}v_j\chi_a(x) \,. \tag{9}$$

An important application of this formalism is to theories which possess flat directions only in a certain limit, the so-called almost flat directions. Examples of such models are massive supersymmetric QCD (with masses much smaller than the strong coupling scale of the theory),[1] the SU(2)×SU(3),[2] and the calculable SU(5)[5] models of dynamical supersymmetry breaking. For our purposes, the common property of these models is that the superpotential can be considered as a perturbation so long as the scale of the vacuum expectation values along the flat directions is larger than the strong coupling scale of the theory. The superpotential is a gauge invariant holomorphic function[6] of the chiral superfields Q_i, $W = W(X(Q))$. After the field redefinition Eq. (5), by gauge invariance of W, the resulting superpotential is independent of the Goldstone superfields:

$$W = W(X(Q(X))) = W(X) \,. \tag{10}$$

Here we required that $Q(X)$ obey

$$X(Q(X)) = X \,. \tag{11}$$

Notice that requiring Eq. (11) allows a nonholomorphic $Q(X^\dagger, X)$ of a special form

$$Q(X^\dagger, X) = e^{B^a(X^\dagger,X)T^a}Q(X) \,, \tag{12}$$

with $\mathcal{B}^a(X^\dagger, X)$ an arbitrary complex function. Since $X(Q)$ is invariant under the complex extension of the gauge group, the nonholomorphic factor disappears from Eq. (11).

With the most general form of $Q_i(X)$ given in terms of the N functions $q_i(x)$ Eq. (9), the n holomorphic functions $X(Q)$ can only be inverted after imposing $N-n$ holomorphic constraints. The field redefinition Eq. (5) amounts to introducing new coordinates on the space of the chiral superfields spanned by Q_i:

$$\{Q_i\} \to \{X^A(Q), S^a(Q)\}\,, \tag{13}$$

the only condition on $S^a(Q)$ being the nondegeneracy of the mapping Eq. (13):

$$\det\left(\frac{\partial X^A(Q)}{\partial Q_i}, \frac{\partial S^a(Q)}{\partial Q_i}\right) \neq 0\,. \tag{14}$$

In the vicinity of $Q_i = v_i$ the additional $N-n$ coordinates S^a are

$$S^a = v^{*j}\, T^{ai}_j\, Q_i\,, \tag{15}$$

since $\det(X^A_i(v), v^{*j}\, T^{ai}_j) \neq 0$. Now the holomorphic functions $X(Q)$, obeying Eq. (11) can be inverted by imposing $N-n$ holomorphic constraints. We choose the constraints

$$S^a = v^{*j}\, T^{ai}_j\, Q_i = 0\,, \tag{16}$$

since then the fields $Q(X)$ Eq. (17) are flat to linear order, $Q^*\, T^a\, Q = 0 + O(x^2)$ (this follows from gauge invariance of $X^A(Q)$).

Notice that this equation can be interpreted as the fact that the projection of the light field on the Goldstone boson direction vanishes. Then the fields Q_i Eqs. (8), (9) obeying the constraint from Eq. (16) have the expansion

$$Q_i(X) = v_i + X^{A*}_i \lambda_A(x)\,. \tag{17}$$

The n functions $\lambda_A(x)$ are determined by inverting Eq. (11)

$$x^A = X^A(v_i + X^{B*}_i \lambda_B(x)) - X^A(v_i)\,, \tag{18}$$

in terms of a Taylor series in x^A:

$$\lambda_A(x) = (X^C_i X^{*D}_i)^{-1}_{AB} x^B + O(x^2)\,. \tag{19}$$

Expressing the Lagrangian Eq. (1) in terms of the new fields $Q(X)$ and $\mathcal{V}$, and expanding in powers of the heavy field $\mathcal{V}$, we obtain

$$\begin{aligned} L_D &= Q(X)^\dagger\, e^{\mathcal{V}^a T^a} Q(X) \\ &= Q(X)^\dagger\, Q(X) + \mathcal{V}^a Q(X)^\dagger\, T^a Q(X) + \ldots\,, \end{aligned} \tag{20}$$

where dots denote higher powers of the massive vector superfield.* Below the scale of the mass of the gauge fields, the heavy vector multiplet can be integrated out. Since we are only interested in the leading term of the low-energy expansion, we can neglect the kinetic term of the gauge field. The zero-momentum tree graphs are easily computed by perturbatively solving the equation of motion for $\mathcal{V}$ that follow from Eq. (20). Then the low-energy Kähler potential $K(X^\dagger, X)$ is given by Eq. (20) with the field $\mathcal{V}_c$ substituted by the solution to its classical equation of motion, i.e.

$$K = Q^\dagger (X) e^{\mathcal{V}} Q(X)|_{\mathcal{V}=\mathcal{V}_c} , \tag{21}$$

with $\mathcal{V}_c$ determined by

$$\left.\frac{dK}{d\mathcal{V}}\right|_{\mathcal{V}=\mathcal{V}_c} = 0 . \tag{22}$$

If the functions $Q(X)$ obey the flat direction equations (3) there are no additional tree-level contributions to the low-energy Kähler potential, as follows from Eq. (20).† In general, however, the holomorphic functions Eqs. (17), (18) do not obey the flat directions equations. This is the case, e.g. in the SU(3)×SU(2) model with dynamical supersymmetry breaking[3] and in the simple abelian chiral model considered in the next section. Then there are zero-momentum tree-level graphs due to the heavy vector supermultiplet, which contribute to the low-energy Kähler potential.

Since the holomorphic functions $Q(X)$ are flat up to quadratic order in the expansion around the flat direction, the leading contribution of the tree graphs is fourth order in x^A. So when computing derivatives of fourth order or higher one needs to incorporate the contribution to the potential from the nonzero vector field. However, to compute the second and third derivatives, it is sufficient to invert Q in terms of X. The Kähler metric at the minimum does not receive tree-level corrections from the heavy fields and is given by the manifestly positive definite expression

$$K_{A^* B} = (X_i^C X_i^{*D})^{-1}_{AB} . \tag{23}$$

Similarly, using Eqs. (17), (18) and (21), one can derive expressions involving three derivatives. However, because the Q fields we define are not D-flat, to calculate fourth or higher order derivatives of the Kähler potential at the minimum, one needs to incorporate the additional contribution from integrating out the vector field.

However, there does exist a nonholomorphic redefinition of the fields such that $\tilde{Q}_i(X^\dagger, X)$ obey the (nonholomorphic) flat direction equations Eq. (3). The Kähler potential Eq. (21) can be written as

$$K = Q^\dagger(X) e^{\mathcal{V}_c/2} e^{\mathcal{V}'} e^{\mathcal{V}_c/2} Q(X)\Big|_{\mathcal{V}'=0} \tag{24}$$

*The gauge invariant field $\mathcal{V}$, defined by Eq. (5), includes the Goldstone fields and their superpartners and therefore there are infinitely many terms in the expansion of $e^{\mathcal{V}}$.

†In the one-flavor SU(2) SQCD finding such a redefinition is simple. Let Q and $\bar{Q}$ be the two doublets of chiral matter transforming in the 2 and 2* representations. For dimensional reasons $Q(X) = \sqrt{X}\eta$ and $\bar{Q}(X) = \sqrt{X}\xi$, with η and ξ being two constant spinors with unit norm, obeying $\eta^\dagger T^a \eta - \xi T^a \xi^\dagger = 0$. Then the Kähler potential $K(X^\dagger, X)$ becomes $2\sqrt{X^\dagger X}$, which coincides with the potential derived in the reference.[2]

Defining the new fields

$$\tilde{Q}(X^\dagger, X) \equiv e^{\mathcal{V}_c(X^\dagger,X)/2} Q(X)\,, \tag{25}$$

the Kähler potential Eq. (24) becomes

$$\tilde{K} = \tilde{Q}^\dagger(X^\dagger, X) e^{\mathcal{V}'} \tilde{Q}(X^\dagger, X)\Big|_{\mathcal{V}'=0}\,. \tag{26}$$

Now, from Eq. (26) it follows that

$$\frac{d\tilde{K}}{d\mathcal{V}'^a}\Big|_{\mathcal{V}'=0} = \tilde{Q}^\dagger(X^\dagger, X) T^a \tilde{Q}(X^\dagger, X) = \frac{dK}{d\mathcal{V}}\Big|_{\mathcal{V}=\mathcal{V}_c} = 0\,, \tag{27}$$

where the second equality follows from the redefinition Eq. (25) and the last holds by virtue of Eq. (22). Equation (27) shows that the nonholomorphic functions Eq. (25) $\tilde{Q}(X^\dagger, X)$ obey the flat direction equations. These fields correspond to those of Affleck, Dine, and Seiberg,[1] and differ from the fields Q Eq. (17) at quadratic order in x^A.

Note that the D-flat conditions only give $N-n$ real constraints whereas we need $N-n$ complex ones, hence there is an insufficient number of constraints. In the ADS construction the $N-n$ additional real constraints correspond to fixing the gauge.

The important fact is that even though the fields $\tilde{Q}$ are not holomorphic, the superpotential constructed from these fields is nonetheless manifestly supersymmetric. This is because the low-energy superpotential is still a holomorphic function of the light superfields, since by gauge invariance, the nonholomorphic factor $e^{\mathcal{V}_c(X^\dagger,X)/2}$ in the redefinition drops out of the superpotential. This justifies the ADS construction.

Therefore, when deriving the effective theory, one is faced with several possibilities. One can apply a holomorphic constraint to restrict oneself to the light degrees of freedom. It is easy with this procedure to derive all derivatives of the Kähler potential[‡] When applying this procedure to the derivatives of fourth or higher order, one needs to include the contribution from explicitly integrating out the heavy vector fields by solving their equations of motion.

Alternatively, one can use nonholomorphic fields which exactly satisfy the D flat equations. This procedure is justified by the fact that the nonholomorphic field contribution will cancel out from gauge invariant superpotential terms. Here, the classical vector field is zero, so there are no additional contributions to the Kähler potential for the light fields. The leading order derivative terms are the same as those calculated with the previous procedure. The higher order terms can be computed perturbatively as well; these are the same as those derived from the Lagrangian with a holomorphic constraint, so long as the classical vector field contribution is incorporated.

[‡]Alternatively, one can explicitly solve the holomorphic constraint equation, similar to the procedure of ADS (as in the one-flavor SU(2) example above).

We should also note that if there is some unbroken nonabelian gauge group, our procedure can be applied, as long as we are interested in the effective theory below the dynamically generated mass scale of the fields carrying gauge charge with respect to the unbroken gauge group. So long as we can find $X^A(Q)$ such that $T_j^{\alpha i} X_i^A(v) = 0$ holds for the unbroken generators T^α, it is easy to see that the fields Eq. (17) obey the unbroken group D-flat equations and are decoupled from the (strongly interacting) gauge field of the unbroken group.

3. The Low-Energy Kähler Potential: an Abelian Example

In this section we illustrate our method on a simple abelian chiral model. The advantage is that we can solve the vector field equations of motion to all orders and explicitly demonstrate the equivalence of our procedure with that of ADS via a nonholomorphic field redefinition.

Consider a chiral U(1) supersymmetric gauge theory with three chiral matter superfields: S_1, S_2 of charge 1, and T, of charge -2. The Lagrangian of the model is

$$L_D = S_1^\dagger e^V S_1 + S_2^\dagger e^V S_2 + T^\dagger e^{-2V} T \,. \tag{28}$$

Along the flat directions, given by

$$S_1^\dagger S_1 + S_2^\dagger S_2 - 2T^\dagger T = 0 \,, \tag{29}$$

the gauge symmetry is broken. There are two independent gauge invariant chiral superfields in this theory

$$X_1 = S_1^2 T \,, \qquad X_2 = S_2^2 T \,. \tag{30}$$

Performing the field redefinition Eq. (5), with $Q_i(X)$ determined by Eqs. (17), (18), the low-energy Kähler potential is given by Eq. (28), where the corresponding functions $S_1 = S_1(X_1, X_2)$, etc., are substituted, and V denotes the heavy vector multiplet. Substituting the solution to the equation of motion for V,

$$e^V = \left(\frac{2T^\dagger T}{S_1^\dagger S_1 + S_2^\dagger S_2} \right)^{1/3} , \tag{31}$$

into Eq. (28), the low-energy Kähler potential becomes

$$K(X^\dagger, X) = 3 \left(\frac{\sqrt{(S_1^2 T)^\dagger \, S_1^2 T} + \sqrt{(S_2^2 T)^\dagger \, S_2^2 T}}{2} \right)^{2/3} . \tag{32}$$

Finally, recalling that $S_1(X_1, X_2)$, etc., obey Eq. (11), e.g. $S_1^2(X_1, X_2) T(X_1, X_2) = X_1$, we find the expression for the Kähler potential of the light degrees of freedom

$$K(X^\dagger, X) = 3 \left(\frac{\sqrt{X_1^\dagger X_1} + \sqrt{X_2^\dagger X_2}}{2} \right)^{2/3} , \tag{33}$$

This coincides with the Kähler potential obtained by the referenced method.[2]

Notice that the vector field contribution only affected fourth and higher order derivatives, since its expansion in light fields begins at second order, as follows from the fact that the fields Eq. (17) are D flat to linear order. In more complicated examples, where one cannot explicitly solve the vector field equation of motion to all orders, one can nonetheless perturbatively derive the vector field contribution. Explicitly, the scalar contribution corresponds to integrating out the auxiliary complex scalar component (which vanishes in the Wess-Zumino gauge; it is denoted by $M + iN$in the reference[4]) of the vector field.

4. Conclusion

In this letter we developed a procedure for finding the Kähler potential of the light degrees of freedom in supersymmetric theories, where the gauge symmetry is completely broken along a flat direction of the D-term scalar potential. The resulting Kähler potential is determined as a power series expansion around the given point of the flat direction. The Kähler metric is manifestly positive definite.

The method is quite general and can be applied to any calculable model of dynamical supersymmetry breaking. It is satisfying that one can derive the low energy theory without exactly solving the flat directions equations in terms of the gauge invariant superfields, particularly in the case of more complicated models. This might prove useful when deriving the physics of specific models. Of particular interest is the SU(5) model with two generations,[5] since little is known about its ground state or broken symmetries.

5. Acknowledgements

We are grateful to Jon Bagger, Ken Intriligator and Joe Polchinski for useful discussions, and the ITP at Santa Barbara for hospitality during the workshop *Weak Interactions '94*.

6. References

1. I. Affleck, M. Dine and N. Seiberg, *Nucl. Phys.* **B241** (1984) 493.
2. I. Affleck, M. Dine and N. Seiberg, *Nucl. Phys.* **B256** (1985) 557.
3. J. Bagger, E. Poppitz and L. Randall, *The R Axion From Dynamical Supersymmetry Breaking*, preprint JHU-TIPAC-940007, MIT-CTP-2309, NSF-ITP-94-48; *Nucl. Phys* **B**, in press.
4. J. Wess and J. Bagger, *Supersymmetry and Supergravity*, (Princeton University Press, Princeton, NJ, 1992).
5. I. Affleck, M. Dine and N. Seiberg, *Phys. Rev. Lett.* **52** (1984) 1677.
6. M.A. Shifman and A.I. Vainshtein, *Nucl. Phys.* **B277** (1986) 456, *Nucl. Phys.* **B359** (1991) 571; N. Seiberg, *Phys. Lett.* **B318** (1993) 469; M. Dine and Y. Shirman, *Some Explorations in Holomorphy*, preprint SCIPP-94/11.

ON THE MONODROMIES OF N=2 SUPERSYMMETRIC YANG-MILLS THEORY

A. Klemm, W. Lerche, S. Yankielowicz [a, b]

CERN, Geneva, Switzerland

and

S. Theisen[b]

Sektion Physik, University Munich, Germany

Abstract

We review the generalization of the work of Seiberg and Witten on $N=2$ supersymmetric $SU(2)$ Yang-Mills theory to $SU(n)$ gauge groups. The quantum moduli spaces of the effective low energy theory parametrize a special family of hyperelliptic genus $n-1$ Riemann surfaces. We discuss the massless spectrum and the monodromies.

1. Introduction

In two recent papers [1,2], Seiberg and Witten have investigated $N=2$ supersymmetric $SU(2)$ gauge theories and solved for their exact nonperturbative low energy effective action. Some of their considerations have recently been extended to $SU(n)$ in [3,4]. We like to review here our work and present a global describtion of the monodromies of $SU(3)$, previewing some results to appear in a more complete account [5].

For arbitrary gauge group G, $N=2$ supersymmetric gauge theories without matter hypermultiplets are characterized by having flat directions for the Higgs vacuum expectation values, along which the gauge group is generically broken to the Cartan subalgebra. Thus, the effective theories contain $r=\text{rank}(G)$ abelian $N=2$ vector supermultiplets, which can be decomposed into r $N=1$ chiral multiplets A^i plus r

a On leave of absence from the School of Physics, Raymond and Beverly Sackler Faculty of Exact Sciences, Tel-Aviv University.

b Work supported in part by the US Israel Binational Science Foundation and GIF - the German-Israeli Foundation for Scientific Research.

$N\!=\!1$ $U(1)$ vector multiplets W^i_α. The $N\!=\!2$ supersymmetry implies that the effective theory up to two derivatives depends only on a single holomorphic prepotential $\mathcal{F}(A)$. More precisely, the effective lagrangian in $N\!=\!1$ superspace is

$$\mathcal{L} \;=\; \frac{1}{4\pi}\mathrm{Im}\Big[\int d^4\theta\,\Big(\sum \frac{\partial\mathcal{F}(A)}{\partial A^i}\bar{A}^i\Big) + \int d^2\theta\,\frac{1}{2}\Big(\sum \frac{\partial^2\mathcal{F}(A)}{\partial A^i\partial A^j}W^i_\alpha W^j_\alpha\Big)\Big]\,. \tag{1}$$

The holomorphic function $\mathcal{F}$ determines the quantum moduli space and, in particular, its metric. This space has singularities at subspaces of complex codimension one, where additional fields become massless. At these regions the effective action description breaks down. A crucial insight is that the electric and magnetic quantum numbers of the fields that become massless at a given singularity are determined by the left eigenvectors (with eigenvalues equal to $+1$) of the monodromy matrix associated with that singularity.

For $G = SU(2)$ considered in [1,2], besides the point at $u = \infty$ there are singularities at $u = \pm\Lambda^2$, where Λ is the dynamically generated scale of the theory, and $u = \frac{1}{2}\langle a^2\rangle$, where $a \equiv A|_{\theta=0}$. On the other hand, $u = 0$ is not singular in the exact quantum theory, which means that, in contrast to the classical theory, no massless non-abelian gauge bosons arise here (nor at any other point in moduli space). The singularities at $u = \pm\Lambda^2$ correspond to a massless monopole and a massless dyon, respectively. The parameter region near $u = \infty$ describes the semiclassical regime and is governed by the perturbative beta function with only one-loop contributions [6,7]. It gives rise to a non-trivial monodromy as well (arising from the logarithm in the effective coupling constant), but there are no massless states associated with it. Although in refs.[1,2] it appears as an assumption, we believe that it can be proven that only two extra singularities besides infinity can exist for $G = SU(2)$. On physical grounds this would follow from the fact that to the best of our understanding of gauge theories only one mass scale Λ is generated. Having more singularities would probably imply that there are more independent scales in the problem. What is easy to prove is that if indeed there are only two singularities at finite points, they correspond up to conjugation to monopole and dyon states with (magnetic;electric) charges $(1;-4n)$ and $(1;-2-4n)$ $(n \in \mathbb{Z})$. The proof follows by looking for solutions to the set of diophantine equations $M_{(g_1;q_1)}\,M_{(g_2;q_2)} = M_\infty$, where

$$M_{(g;q)} \;=\; \begin{pmatrix} 1-gq & -q^2 \\ g^2 & 1+gq \end{pmatrix} \tag{2}$$

is the monodromy matrix for a massless dyon with electric charge q and magnetic charge g and $M_\infty = \begin{pmatrix} -1 & -4 \\ 0 & -1 \end{pmatrix}$. We also have considered the situation of four singular points and could exclude solutions to the corresponding diophantine equations, $\prod_{i=1}^4 M_{(g_i;q_i)} = M_\infty$ for $g_i, q_i \leq 10$.

The singularity structure and knowledge of the monodromies allow to completely determine the holomorphic prepotential $\mathcal{F}$. The monodromy group is $\Gamma_0(4) \subset SL(2,\mathbb{Z})$ consisting of all unimodular integral matrices $\left(\begin{smallmatrix} a & b \\ c & d \end{smallmatrix}\right)$ with $b = 0 \bmod 4$. The matrices act on the vector $(a_D; a)^t$, where a_D is the magnetic dual of a, with $a_D \equiv \frac{\partial F(a)}{\partial a}$. The quantum moduli space is the u-plane punctured at $\pm\Lambda^2$ and ∞ and can be thought of as $\mathbb{H}/\Gamma_0(4)$ ($\mathbb{H}$ is the upper half-plane).

The basic idea [1] in solving for the effective theory is to consider a family of holomorphic curves whose monodromy group is $\Gamma_0(4)$ and which can be represented as follows:

$$y^2 = (x^2 - u)^2 - \Lambda^4 \,. \tag{3}$$

By transforming to Weierstrass form, this curve can be shown to be equivalent to the curve given in [2], which, in contrast to the curve given in [1], is the form appropriate for generalization to $SU(n), n > 2$. The curves (3) represent a double cover of the x-plane with the four branch points at $\pm\sqrt{u \pm \Lambda^2}$, and describe a genus one Riemann surface. That is, the quantum moduli space of the $SU(2)$ super Yang-Mills theory coincides with the moduli space of a particular torus; this torus becomes singular when two branch points in (3) coincide. The derivatives of the electric and magnetic coordinates $(a_D; a)^t$ with respect to u are given by the periods of the holomorphic one-form $\frac{dx}{y}$ with respect to a symplectic homology basis. Their ratio, the modular parameter τ, is positive definite and well-defined in the u-plane, and equals the metric of the moduli space. Integrating the periods yields $a(u), a_D(u)$ and integrating a_D finally determines the prepotential $\mathcal{F}(a)$.

We want to indicate next how these ideas of Seiberg and Witten generalize to gauge groups $G = SU(n)$. To be specific, we will mainly consider $G = SU(3)$, but the generalization to higher n is straightforward.

2. Semi-classical Regime

We will denote the gauge invariant order parameters (Casimirs) of $SU(n)$ by

$$u_k = \frac{1}{k}\mathrm{Tr}\langle\phi^k\rangle\,, \quad k = 2,\ldots,n \tag{4}$$

where we can always take the scalar superfield component to be $\phi = \sum_{k=1}^{n-1} a_k H_k$ with $H_k = E_{k,k} - E_{k+1,k+1}$, $(E_{k,l})_{i,j} = \delta_{ik}\delta_{jl}$. The anomaly free global $\mathbb{Z}_{2n}$ subgroup of $U(1)_{\mathcal{R}}$ acts as $u_k \to e^{i\pi k/n}u_k$. For $SU(3)$ this means that classically $u \equiv u_2 = {a_1}^2 + {a_2}^2 - a_1 a_2$, $v \equiv u_3 = a_1 a_2(a_1 - a_2)$ with $\mathbb{Z}_6$ action $u \to e^{2\pi i/3}u$, $v \to -v$. For generic eigenvalues of ϕ, the $SU(3)$ gauge symmetry is broken to $U(1) \times U(1)$, whereas if any two eigenvalues are equal, the unbroken symmetry is $SU(2) \times U(1)$. These classical symmetry properties are encoded in the following, gauge and globally $\mathbb{Z}_6$ invariant discriminant:

$$\Delta_0 \;=\; 4u^3 - 27v^2 \;=\; (a_1 + a_2)^2(2a_1 - a_2)^2(a_1 - 2a_2)^2 \quad . \tag{5}$$

The lines $\Delta_0 = 0$ in (u, v) space correspond to unbroken $SU(2)\times U(1)$ and have a cusp singularity at the origin, where the $SU(3)$ symmetry is restored. As we will see, in the full quantum theory this cusp is resolved, $\Delta_0 \to \Delta_\Lambda = 4u^3 - 27v^2 + O(\Lambda^6)$, which, in particular, prohibits a phase with massless non-abelian gluons. Other singularities will however appear, signalling the appearance of massless monopoles and dyons in the spectrum.

The prepotential $\mathcal{F}$ in the semi-classical, perturbative regime can easily be computed with the result

$$\mathcal{F}_{\substack{\text{semi-}\\ \text{class}}} \;=\; \frac{i}{4\pi}\sum_{i<j}^{3}(e_i - e_j)^2 \log[(e_i - e_j)^2/\Lambda^2]\;. \tag{6}$$

Here, e_i denote the roots of the equation

$$W_{A_2}(x, u, v) \;\equiv\; x^3 - u\,x - v \;=\; 0\;, \tag{7}$$

whose bifurcation set is given by the discriminant Δ_0 in (5), i.e.

$$\begin{aligned} e_1 - e_2 &= a_1 + a_2 \\ e_1 - e_3 &= 2a_1 - a_2 \\ e_2 - e_3 &= a_1 - 2a_2. \end{aligned} \tag{8}$$

The Casimirs u, v are gauge invariant and, in particular, invariant under the Weyl group $\mathcal{W}$ of $SU(3)$. This group is generated by any two of the transformations

$$\begin{aligned} r_1: \quad (a_1, a_2) &\rightarrow (a_2 - a_1, a_2) \\ r_2: \quad (a_1, a_2) &\rightarrow (a_1, a_1 - a_2) \\ r_3: \quad (a_1, a_2) &\rightarrow (-a_2, -a_1) \,. \end{aligned} \tag{9}$$

Due to the multi-valuedness of the inverse map $(u, v) \rightarrow (a_1, a_2)$, closed paths in (u, v) space will, in general, not close in (a_1, a_2) space, but will close up to Weyl transformations. Such a monodromy will be non-trivial if a given path encircles a singularity in (u, v) space — in our case, the singularities will be at "infinity" and along the lines where the discriminant vanishes.

It is indeed well-known [8] that the monodromy group of the simple singularity of type A_2 (7) is given by the Weyl group of $SU(3)$, and acts as Galois group on the e_i (and analogously for $W_{A_{n-1}}$ related to $SU(n)$). This will be the starting point for our generalization.

What we are interested in is however not just the monodromy acting on (a_1, a_2), but the monodromy acting on $(a_{D1}, a_{D2}; a_1, a_2)^t$, where

$$a_{Di} \equiv \mathcal{F}_i = \frac{\partial}{\partial a_i} \mathcal{F}(a_1, a_2) \,. \tag{10}$$

Performing the Weyl reflection r_1 on $(a_1, a_2)^t$, we easily find

$$\begin{pmatrix} \mathcal{F}_1 \\ \mathcal{F}_2 \end{pmatrix} \rightarrow \begin{pmatrix} -1 & 0 \\ 1 & 1 \end{pmatrix} \begin{pmatrix} \mathcal{F}_1 \\ \mathcal{F}_2 \end{pmatrix} + N \begin{pmatrix} 6 & -3 \\ -3 & -3 \end{pmatrix} \begin{pmatrix} a_1 \\ a_2 \end{pmatrix} \,. \tag{11}$$

The second term, i.e., the "quantum shift", arises from the logarithms and is not determined by the finite, "classical" Weyl transformation acting on the a_i, but rather depends on the particular path in (u, v) space. (We considered in (11) only paths for which all three logarithms in (6) contribute with the same sign; other paths do exist where the logarithms contribute differently and yield a quantum shift matrix different from that in (11) [5]). For example, for the closed loop given by $(u(a_i(t)), v(a_i(t))$ for $t \in [0, 1]$, where $a_1(t) = e^{i\pi t} a_1 + \frac{1}{2}(1 - e^{i\pi t}) a_2$, $a_2(t) = a_2$ we find $N = 1$. Therefore, the matrix representation of r_1 acting on $(a_{D1}, a_{D2}; a_1, a_2)^t$ is: †

$$r_1 = \begin{pmatrix} -1 & 0 & 6 & -3 \\ 1 & 1 & -3 & -3 \\ 0 & 0 & -1 & 1 \\ 0 & 0 & 0 & 1 \end{pmatrix} \equiv r_1^{\text{class}} T^{-3} \,, \tag{12}$$

† In [3] we represented these matrices for rescaled a_i. However, the present normalization is more appropriate if one wants the other monodromies to be given by integral matrices.

where $r_1^{\rm class}$ is the "classical" Weyl reflection (given by the block diagonal part of r_1), and T the "quantum monodromy"

$$T = \begin{pmatrix} 1\!\!1 & C \\ 0 & 1\!\!1 \end{pmatrix}, \qquad \text{where } C = \begin{pmatrix} 2 & -1 \\ -1 & 2 \end{pmatrix} \tag{13}$$

is the Cartan matrix of $SU(3)$. The other Weyl reflections are given analogously by $r_i = r_i^{\rm class}\, T^{-3}$. The r_i are related to each other by conjugation, and, in particular, rotate into each other via the Coxeter element, $r_{\rm cox}^{\rm class} = r_1^{\rm class} r_2^{\rm class}$.

3. The Curves for $SU(n)$

Let us now generalize the $SU(2)$ curve to a sequence of curves $\mathcal{C}_n$ whose moduli spaces are supposed to coincide with the quantum moduli space of effective low energy $N = 2$ supersymmetric $SU(n)$ Yang-Mills theories.

We first list the requirements we impose on the curves $\mathcal{C}_n$: (i) We seek surfaces with $2(n-1)^2$ periods (corresponding to $(\partial_{u_j} a_{Di}; \partial_{u_j} a_i)$), whose period matrices $\Omega_{ij} = \frac{\partial a_{Di}}{\partial a_j}$ are positive definite. (ii) For $\Lambda \to 0$ the classical situation must be recovered. That is, the discriminant of $\mathcal{C}_n$ should have, for $\Lambda = 0$, a factor of Δ_0 (cf. eq.(5)). This means that for $\Lambda = 0$ the curves should have the form $y^m = \mathcal{C}_n(x) \equiv W_{A_{n-1}} \times (\ldots)$ for some m. (iii) The curves must behave properly under the cyclic global transformations acting on the Casimirs u_k; in other words, there should be a natural dependence on Casimirs, for all groups. Finally, from [2] we know that Λ should appear in $\mathcal{C}(x)$ with a power that corresponds to the charge violation of the one-instanton process, which is $2n$ for $SU(n)$.

Taking these requirements together leads us to consider the following genus $g = n-1$ hyperelliptic curves for $SU(n)$:

$$y^2 = \mathcal{C}_n(x) \equiv \Big(W_{A_{n-1}}(x, u_i)\Big)^2 - \Lambda^{2n}, \tag{14}$$

where

$$W_{A_{n-1}}(x, u_i) = x^n - \sum_{i=2}^{n} u_i\, x^{n-i}. \tag{15}$$

are the A-type simple singularities related to $SU(n)$. The square of W reflects having both electric and magnetic degrees of freedom. Since there is a general relationship [8] between Arnold's simple A-D-E singularities, perturbations by Casimirs, monodromy and Weyl groups, we conjecture that (14) describes surfaces for the other simply laced gauge groups G as well. One simply replaces $W_{A_{n-1}}(x, u_i)$ by the corresponding D- or E-type singularity, and Λ^{2n} by Λ^{2h}, where h is the corresponding Coxeter number. For the following it will be useful to write

$$\begin{aligned} \mathcal{C}_n(x) &= \Big(W_{A_{n-1}}(x, u_i) + \Lambda^n\Big)\Big(W_{A_{n-1}}(x, u_i) - \Lambda^n\Big) \\ &= \prod_{i=1}^{n} (x - e_i^+)(x - e_i^-) \,. \end{aligned} \tag{16}$$

Critical surfaces occur whenever two roots of $\mathcal{C}_n(x)$ coincide, that is, whenever the discriminant $\Delta_\Lambda = \prod_{i<j}(e_i^\pm - e_j^\pm)^2$ vanishes. We expect this to happen when monopoles or dyons, whose quantum numbers are determined by the corresponding monodromy matrices, become massless. For example, for $G = SU(3)$ the quantum discriminant is:

$$\Delta_\Lambda = \Lambda^{18} \Delta_\Lambda^+ \Delta_\Lambda^- \,, \qquad \Delta_\Lambda^\pm = 4u^3 - 27(v \pm \Lambda^3)^2. \tag{17}$$

By construction, the hyperelliptic curves (14) are represented by branched covers over the x-plane. More precisely, we have n $\mathbb{Z}_2$ cuts, each linking a pair of roots e_i^+ and e_i^-, $i = 1, \ldots, n$. As an example, we present the picture for $G = SU(3)$ in Fig.1. In the classical theory, where $\Lambda \to 0$, the branch cuts shrink to n doubly degenerate points: $e_i^- \to e_i^+ \equiv e_i$. These points, given for $SU(3)$ in eq. (8), correspond to the weights of the n-dimensional fundamental representation (the picture represents a deformed projection of the weights onto the unique Coxeter eigenspace with $\mathbb{Z}_n$ action). This means that the branched x-plane transforms naturally under the finite "classical" Weyl group that permutes the points. This finite Weyl group is just the usual monodromy group of the A_{n-1} singularity alluded to earlier. In the quantum theory, where the degenerate points are resolved into branch cuts, there are in addition possibilities for "quantum monodromy", which involves braiding of the cuts.

4. Monodromies for G=SU(3)

The semi-classical monodromy discussed so far was determined from properties relying on asymptotic freedom of the $SU(3)$ theory. We now turn to the monodromies M_λ around singularities related to massless dyons of charges $\lambda \equiv (\vec{g}, \vec{q})^t \equiv (g_1, g_2; q_1, q_2)^t$ (the charge vectors will be given in a basis generated by the two fundamental weights of $SU(3)$). Analogous to [1], the charge vectors are left eigenvectors of the corresponding monodromy matrices with eigenvalue equal to one, i.e. $\lambda^t M_\lambda = \lambda^t$.

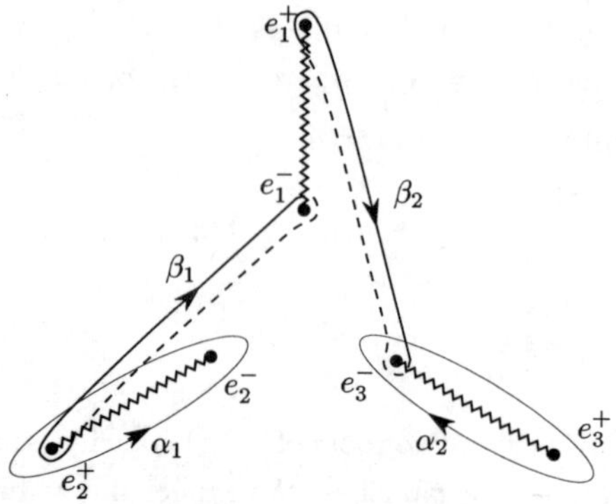

Fig.1: *Branched x-plane with cuts linking pairs of roots of $\mathcal{C}_3 = 0$. We depicted a choice of basis for the homology cycles that is adapted to the cyclic $\mathbb{Z}_3$ symmetry generated by the classical Coxeter element. Note that the locations of the cuts are $\mathbb{Z}_3$-symmetric only for $u = 0$.*

Obviously, the monodromies of any two dyons whose charge vectors are related via $\lambda_2 = N^t\lambda_1$, $N \in Sp(4, \mathbb{Z})$, are conjugated: $M_{\lambda_2} = N^{-1}M_{\lambda_1}N$. If we require the monodromies to depend only on the charges of the corresponding states, the matrices are determined to take the following form, which is the generalization of (2):

$$M_{(\vec{g};\vec{q})} = \begin{pmatrix} \mathbb{1} - \vec{q}\otimes\vec{g} & -\vec{q}\otimes\vec{q} \\ \vec{g}\otimes\vec{g} & \mathbb{1} + \vec{g}\otimes\vec{q} \end{pmatrix} \tag{18}$$

This form of the matrix is obviously the same for all $SU(n)$. Here we have fixed some freedom with hindsight to our results below. Note that for purely magnetically charged monopoles, where $q_i = 0$, one has: $a_{Di} \to a_{Di}$, and this reflects the fact that the a_{Di} are good local coordinates in the vicinity of the singular loci where the monopoles become massless.

We now indicate how to determine the monodromies M_λ of the curves $\mathcal{C}_n$. Fixing a base point u_0 in the moduli space $\mathcal{M}_n$ of $\mathcal{C}_n$, there is a homomorphism of the fundamental group $\pi_1(\mathcal{M}_n, p)$ into $Sp(2n\text{–}2; \mathbb{Z})$ whose image is the monodromy group. Note that the singular locus of $\mathcal{C}_n$, $\Delta_\Lambda = 0$, does itself have singularities. For $n = 3$ there are cusps at $u = 0$, $v = \pm\Lambda^3$ ($4\delta u^3 - 27\delta v^2 = 0$) and nodes at $u^3 = \frac{27}{4}\Lambda^6$, $v = 0$ ($\rho\,\delta u^2 - \delta v^2 = 0$, $\rho^3 = \Lambda^6/4$), as discussed in [4]. We have depicted in Fig.2 the moduli space for real v.

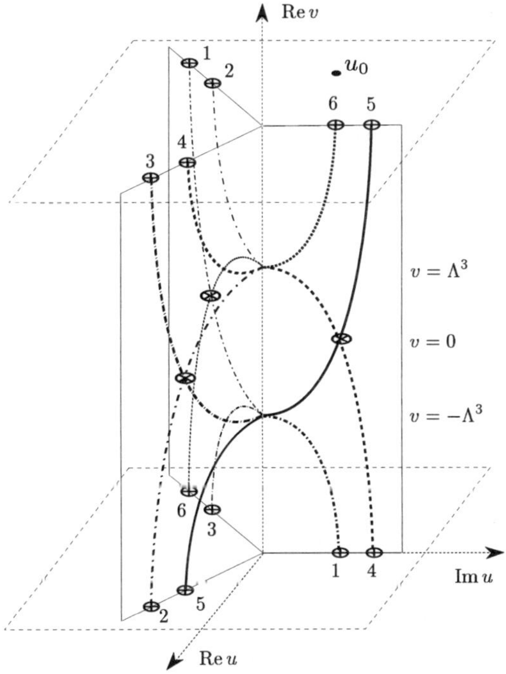

Fig.2: *$SU(3)$ quantum moduli space for real v. The six lines are the singular loci where $\Delta_\Lambda = 0$ and where certain dyons become massless. At each of the three crossing points at $v = 0$, two mutually local dyons become simultaneously massless, and the theory is semi-classical in the corresponding dual variables. The various markings of the lines indicate how the association with particular monodromy matrices changes when moving through the cusps.*

The monodromy matrices reflect the action of braiding and permuting the cuts on the vector $(a_{Di}; a_i)^t$. This action is expressed in terms of the action on the homology cycles via

$$a_{Di} = \oint_{\beta_i} \lambda \,, \qquad a_i = \oint_{\alpha_i} \lambda \,, \tag{19}$$

where α_i, β_j is some symplectic basis of $H_1(\mathcal{C}_n; \mathbb{Z})$: $\langle \alpha_i, \beta_j \rangle = -\langle \beta_j, \alpha_i \rangle = \delta_{ij}$, $\langle \alpha_i, \alpha_j \rangle = \langle \beta_i, \beta_j \rangle = 0$, $i, j = 1, \ldots, g$ and λ is some suitable chosen meromorphic differential. From the theory of Riemann surfaces it is clear that the monodromy group is contained in $Sp(2g,\mathbb{Z}) = Sp(2n-2,\mathbb{Z})$. For $G = SU(3)$, we have depicted our choice of homology basis in Fig.1.

The holomorphic differentials on a genus $g = n-1$ hyperelliptic curve are $\omega_{n-i} = x^i \frac{dx}{y}$, $i = 0,\ldots,g-1$. The $g \times 2g$ period matrix is $(A,B) = A(\mathbf{1},\Omega)$ with $A_{ij} = \int_{\alpha_j}\omega_i$ and $B_{ij} = \int_{\beta_j}\omega_i$, which are related to (a_{Di}, a_i) as follows:

$$A_{ij} = \frac{\partial a^j(u)}{\partial u_i}, \quad B_{ij} = \frac{\partial a_D^j(u)}{\partial u_i}\,. \tag{20}$$

Note that due to (20), the second Riemann bilinear relation, $\mathrm{Im}\,\Omega > 0$, ensures the positivity of the metric

$$(ds)^2 = \mathrm{Im}\,\frac{\partial\mathcal{F}}{\partial a_i \partial a_j} da_i d\bar{a}_j = \mathrm{Im}\sum_{i=1}^{g} da_D^i d\bar{a}_i\,.$$

This generalizes the situation at genus one [1]. Note also that (20) represents a non-trivial integrability condition for the periods, $\partial_i A_{jk} = \partial_j A_{ik}$, $\partial_i B_{ik} = \partial_j B_{ik}$, which holds for our parametrization of the curve (14). The periods, which can be obtained as solutions of the Picard-Fuchs equations [5], can thus be integrated to yield $a_{D\,i}(u,v)$, $a_i(u,v)$ and by further integration to yield $\mathcal{F}(u,v)$. Alternatively one can choose the meromorphic differential λ in (19) s.t. $\omega_i \sim \partial_i\lambda$ up to exact forms, e.g. $\lambda \sim (3x^3 - ux)\frac{dx}{y}$ as in [4], and try to integrate (19).

Now, if ν denotes the vanishing cycle for the braiding of the branch points induced by a loop γ in moduli space, the action on the homology cycle $\delta \in H_1(C_n;\mathbb{Z})$ can be simply obtained via the Picard-Lefshetz formula [8]

$$S_\nu\delta = \delta + \langle\delta,\nu\rangle\nu$$

We find that when decomposing the given vanishing cycle as $\nu = \sum_{i=1,2}(q_i\alpha_i + g_i\beta_i)$, the action on the homology basis $(\beta_1,\beta_2,\alpha_1,\alpha_2)^t$ is given by a monodromy matrix M_λ precisely of the form (18), with labels just given by the expansion coefficients: $M_\lambda \equiv M_{(g_1,g_2;q_1,q_2)}$. That is, we can read off the electric and magnetic quantum numbers of a given massless dyon just by looking at the picture of the corresponding vanishing cycle !

More specifically, we studied the monodromies of $G = SU(3)$ by fixing a base point u_0 and by carefully tracing the effects of loops in moduli space on the motions of the branch points in the x-plane. With reference to results by Zariski and van Kampen (cf. [9] and references therein) it suffices to study loops in a generic complex line through the base point. Across cusps and nodes the monodromies are related through the "van Kampen relations". In Fig.3 the six marked points are the intersections of the complex u-plane at $\mathrm{Re}(v) = \mathrm{const} > \Lambda^3$, $\mathrm{Im}(v) = 0$ with the singular set $\Delta_\Lambda = 0$. The monodromies around the six loops γ_i can be characterized by the corresponding vanishing cycles, which are depicted in Fig.4. According to what we said above, the quantum numbers of the dyons that become massless at the encircled singular lines in moduli space can be directly obtained from Fig.4, by comparing the vanishing cycles with the basis cycles in Fig.1. The corresponding monodromy matrices (18) then turn out to be:

$$\begin{aligned} M_{\lambda_1} &= M_{(1,1;-1,0)}, & M_{\lambda_2} &= M_{(1,1;0,1)}, \\ M_{\lambda_3} &= M_{(1,0;-1,1)}, & M_{\lambda_4} &= M_{(1,0;1,0)}, \\ M_{\lambda_5} &= M_{(0,1;0,-1)}, & M_{\lambda_6} &= M_{(0,1;-1,1)} . \end{aligned} \tag{21}$$

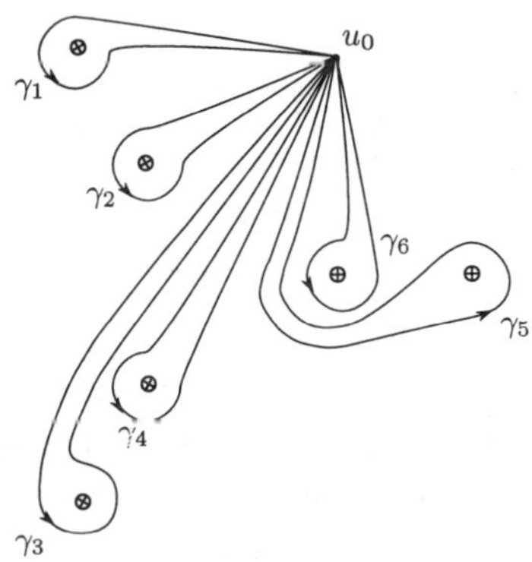

Fig.3: *Loops γ_i in the u-plane at $\mathrm{Re}(v) = const > \Lambda^3$, $\mathrm{Im}(v) = 0$ (cf., Fig.2). u_0 denotes the base point. A loop encircling all the points yields essentially the monodromy matrix r_1 given in (12).*

We remark that the charge vectors of each pair of lines which cross in the nodes at $u = 0$ satisfy $\lambda_i^t \, \Sigma \, \lambda_j = 0$, where $\Sigma = \begin{pmatrix} 0 & 1 \\ -1 & 0 \end{pmatrix}$ is the symplectic metric [4]. This is a necessary condition for two types of dyons to condense simultaneously [10]. (It follows from (18) that if two charge vectors are symplectically orthogonal, the corresponding

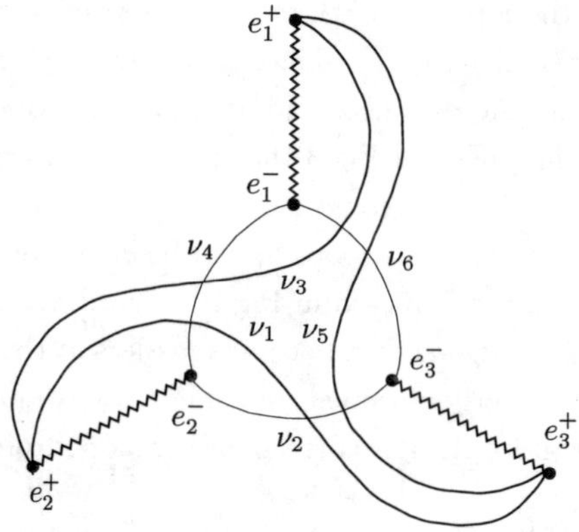

Fig.4: *Vanishing cycles ν_i associated with the loops γ_i in Fig.3. The dyon charges can be directly inferred from this picture, by expanding the cycles in terms of the homology basis given in Fig.1 (e.g. $\nu_6 = \beta_2 - \alpha_1 + \alpha_2 \Rightarrow \lambda_6 = (0,1;-1,1)$). We have depicted here only the paths on the upper sheet, and not the return paths on the lower sheet.*

monodromy matrices commute.) More precisely, the charge labels $[\lambda_i, \lambda_j]$ of the condensing pairs are $\{[\lambda_1,\lambda_6],\,[\lambda_2,\lambda_3],\,[\lambda_4,\lambda_5]\}$. Note that by appropriately changing the homology basis, one can always conjugate the charge vectors of any given one such pair of dyons to purely magnetic charges, ie., to $(1,0;0,0)$ and $(0,1;0,0)$. At the nodes both dual $U(1)$'s are weakly coupled and one can verify that the monodromies are consistent with the beta functions of the effective dual theory that contains two monopole hypermultiplets.

One type of monodromy at infinity is given by a loop encircling all the singular points in Fig.3, that is, by the product of the matrices in (21). It turns out to be precisely the monodromy (12) deduced from the semi-classical effective action (6), up to a change of basis:

$$M_\infty \equiv \prod_{i=1}^{6} M_{\lambda_i} \;=\; S^{-1}\,(r_1)^{-1} S,$$

with $S = \begin{pmatrix} 1 & s \\ 0 & 1 \end{pmatrix}$, $s \equiv \begin{pmatrix} -2 & 0 \\ 0 & 2 \end{pmatrix}$. (In this basis, the other matrices r_2, r_3 can be obtained as well, by starting from a different base point.) We take this as a non-trivial physical consistency check.

One can also consider monodromies for paths lying in other complex lines in the moduli space, like in the line $u = const.$ as discussed in [4]. This plane is however not in generic position but is special, in that it does not cut through all six, but only through four singular lines. (The corresponding monodromy matrices form a subset of the matrices in (21)). The total monodromy obtained by encircling the four points in the v-plane is different from the monodromy in the u-plane discussed above, however it is closely related to it: its classical (block-diagonal) part is just given by the Coxeter element $r_{\rm cox}^{\rm class} \equiv r_1^{\rm class} r_2^{\rm class}$ of the Weyl group of $SU(3)$. The Coxeter element is of order three, and its $\mathbb{Z}_3$ action corresponds to the cyclic rotation of the cuts in Fig.1. The complete monodromy in the v-plane, which includes the quantum shift, is the same as the semi-classical monodromy obtained in [4], up to change of basis. It can be obtained from the effective action (6) as well, by chosing an appropriate path.

This Coxeter monodromy is closely related to a well-known fact in singularity theory [8]. Here one considers perturbations $W = W_0(u_k = 0) + \Lambda$, where the parameter circles around the origin: $\Lambda = e^{2\pi i t}$, $t \in [0, 1]$. This induces what is called, ironically, "classical monodromy", and for the A-D-E simple singularities, this classical monodromy is precisely given by the Coxeter element of the corresponding A-D-E type Weyl group. For our curves (14), such loops in the Λ-plane indeed reproduce the above-mentioned Coxeter monodromy in the v-plane, if we set $u, v = 0$. On the other hand, if $|\Lambda| < |v|$, these loops do not induce the cyclic rotation of the cuts in Fig.1, but only induce simultaneous braiding of the cuts. This corresponds to pure "quantum monodromy", given by the shift matrix T in (13) [3].

Summarizing, the classical piece of the total monodromy in the u-plane is given by any one of the Weyl group generators, say $r_1^{\rm class}$, whereas for the v-plane one obtains the Coxeter element, $r_1^{\rm class} r_2^{\rm class}$. We expect for $SU(n)$ that when looping around infinity in a parameter plane u_k related to a Casimir of degree k, the classical part of the monodromy is given by a Weyl group conjugacy class of the corresponding order. In particular, encircling the top Casimir parameter plane u_n will induce the Coxeter monodromy, which has order n.

References

[1] N. Seiberg and E. Witten, Nucl. Phys. <u>*B*426</u> (1994) 19, `hep-th/9407087`.

[2] N. Seiberg and E. Witten, Nucl. Phys. <u>*B*431</u> (1994) 484, `hep-th/9408099`.

[3] A. Klemm, W. Lerche, S. Theisen and S. Yankielowicz, *Simple Singularities and N=2 Supersymmetric Yang-Mills Theory*, preprint CERN-TH.7495/94, LMU-TPW 94/16; `hep-th/9411048`

[4] P. Argyres and A. Faraggi, *The Vacuum Structure and Spectrum of N=2 Supersymmetric SU(n) Gauge Theory*, preprint IASSNS-HEP-94/94; `hep-th/9411057`

[5] A. Klemm, W. Lerche, S. Theisen and S. Yankielowicz, in preparation.

[6] V. Novikov, M. Schifman, A. Vainstein, M Voloshin and V. Zakharov, Nucl. Phys. <u>*B*229</u> (1983) 394; V. Novikov, M. Schifman, A. Vainstein and V. Zakharov, Nucl. Phys. <u>*B*229</u> (1983) 381, 407; M. Schifman, A. Vainstein and V. Zakharov, Phys. Lett. <u>166</u> (1986) 329.

[7] N. Seiberg, Phys. Lett. <u>206</u> (1988) 75.

[8] See e.g., V. Arnold, A. Gusein-Zade and A. Varchenko, *Singularities of Differentiable Maps I, II*, Birkhäuser 1985.

[9] A. Ceresole, R. D'Auria and T. Regge, Nucl. Phys. <u>*B*414</u> (1994) 517; see also: A. Ceresole, R. D'Auria, S. Ferrara, W. Lerche, J. Louis and T. Regge, *Picard-Fuchs Equations, Special Geometry and Target Space Duality*, preprint CERN-TH.7055/93, POLFIS-TH.09/93, to be published in "Essays on Mirror Symmetry, Vol. 2", B. Green and S.-T. Yau, eds.

[10] G. 't Hooft, Nucl. Phys. <u>*B*190</u> (1981) 455.

SOFT SUPERSYMMETRY–BREAKING TERMS AND THE μ PROBLEM

C. MUÑOZ
Departamento de Física Teórica C-XI, Universidad Autónoma de Madrid, Cantoblanco Madrid, E-28049, Spain

ABSTRACT

We analyze the soft supersymmetry–breaking terms in Supergravity and Superstring. Several solutions proposed to solve the μ problem are also discussed. In particular, the B soft term is given for the different solutions.

1. Introduction

Despite the absence of experimental verification, relevant theoretical arguments can be given in favour of Supersymmetry[1] (SUSY): i) It is a new symmetry which relates bosons and fermions. The importance of this is twofold: we know from the past that symmetries are crucial in particle physics and, it implies a new kind of unification, between particles of different spin. The latter implies that the Higgs is no longer a mysterious particle as it stands in the Standard Model (SM): the only fundamental scalar particle that should exist. Now, the Supersymmetric Standard Model (SSM) is naturally full of fundamental scalars (squarks, sleptons and Higgses) related through SUSY with their fermionic partners (quarks, leptons and Higgsinos). ii) The local version of SUSY leads to a partial unification of the SM with gravity, Supergravity[1,2] (SUGRA). iii) String Theory[3] needs to be SUSY (Superstring) in order to avoid tachyons in the spectrum and besides, after compactification of extra dimensions leads to an effective SUGRA. iv) Whereas the quadratic and quartic terms of the Higgs potential which are necessary in order to break the electroweak symmetry have to be postulated "ad hoc" in the case of the SM, they appear in a natural way in the context of the SSM. v) The joining of the three gauge coupling constants of the SM at a single unification scale which only taking into account SUSY agrees with the experimental results. But still, the most important argument in favour of SUSY, is that vi) it solves the so–called gauge hierarchy problem.

If we believe that the SM should be embedded within a more fundamental theory including gravity with a characteristic scale of order the Planck mass, $M_P \equiv \frac{G_N^{-1/2}}{\sqrt{8\pi}} \simeq 2.4 \times 10^{18} GeV$, then we are faced with the hierarchy problem. There is no symmetry protecting the masses of the scalar particles against quadratic divergences in perturbation theory. Therefore they will be proportional to the huge cut off scale $\sim M_P$. The Higgs particle is included in the SM because of its good properties: can have vacuum expectation value (vev), without breaking Lorent invariance, inducing the spontaneous breaking of the electroweak symmetry at the same time that generates the fermion masses through Yukawa couplings. But, of course, all these properties are due to the fact that the Higgs is a scalar particle. As mentioned above this leads to a huge mass for it and as a consequence for the W and Z gauge bosons. This problem of naturalness, to stabilize $M_W \ll M_P$ against quantum corrections, is solved in SUSY since now the scalar mass and the mass of its superpartner, the fermion, are related. As a consequence, only a logarithmic divergence in the scalar mass is left. In diagrammatic language, the dangerous diagrams of SM particles are cancelled with new ones which are present due to the existence of the additional partners and couplings.

Of course, SUSY cannot be an exact symmetry of nature and should be broken since a SUSY particle with the same mass than its SM partner has never been detected (e.g. there is no selectron with mass $\sim 0.5 MeV$). One may introduce terms in the Lagrangian which explicitly break SUSY but, they should not induce quadratic divergences in order not to spoil the SUSY solution to the gauge hierarchy problem. This restricted type of terms exist and they are called soft SUSY–breaking terms. The simplest SSM is the so–called Minimal Supersymmetric Standard Model (MSSM), where the matter consists of three generations of quark and lepton superfields plus two Higgs doublets* H_1 and H_2 of opposite hypercharge, and the gauge sector consists of $SU(3) \times SU(2)_L \times U(1)_Y$ vector superfields. Assuming certain universality of soft terms these can be parametrized by only four parameters: a universal gaugino mass (where the gauginos are the superpartners of the gauge bosons) M, a universal scalar mass m, a universal trilinear scalar parameter A and an extra bilinear scalar parameter B. These soft terms are very important since they determine the SUSY spectrum, like gaugino, squark and slepton masses, and contribute to the Higgs potential generating the radiative breakdown of the electroweak symmetry. Although in principle they are constrained since they must reproduce the experimental results ($M_W \simeq 80 GeV$), the correct numbers can be obtained for wide ranges of the above parameters. In this sense the predictivity of the theory is limited.

The previous mechanism for breaking SUSY explicitly looks arbitrary but it turns out to be the most natural one when global SUSY is promoted to local, i.e. in SUGRA. When SUGRA is spontaneously broken in a "hidden sector" the soft SUSY-breaking terms are generated. The process is the following: Singlet scalar fields under the observable gauge group ("hidden sector" fields) acquire vevs giving rise to spontaneous breaking of SUGRA. The goldstino, which is a combination of the fermionic partners of the above fields, is swallowed by the gravitino (the spin 3/2 superpartner of the graviton in $N = 1$ SUGRA) which becomes massive. This is the so–called super–Higgs effect. It is completely analogous to the usual Higgs mechanism. Then, the hidden fields, which only have gravitational interactions with the observable sector, decouple from the low–energy theory and the only signals they produce are the soft terms. These are characterized by the gravitino mass ($m_{3/2}$) scale and therefore in order not to introduce a new problem of naturalness, $m_{3/2}$ should be of the electroweak scale order (recall that the soft terms contribute to the Higgs masses). An interesting non–perturbative source of SUSY–breaking, capable of generating this large mass

*SUSY demands the presence of two Higgs doublets unlike the SM where only one is needed.

hierarchy, is gaugino condensation in some hidden sector gauge group[4]. Below M_P (i.e. in the so–called flat limit where $M_P \to \infty$ but $m_{3/2}$ is kept fixed), one is left with a global supersymmetric Lagrangian plus the soft SUSY–breaking terms. In summary, the MSSM is just an effective low–energy theory derived from SUGRA when this is spontaneously broken.

Two main arguments can be used in order to criticize the previous SUSY–breaking mechanism: First of all, SUGRA is a non–renormalizable theory and then it is not clear that we are doing a consistent analysis. Notice however that the effective Lagrangian below the Planck scale is renormalizable and we are interested only in this region. The general idea below this approach is that we are considering the SUGRA Lagrangian as an effective phenomenological Lagrangian which comes from a bigger structure, renormalizable or even finite (Superstring?). The situation is similar to the one of the old Fermi Theory. Second, the existence of the "hidden sector" has to be postulated "ad hoc". However, we will see below that in the context of Superstring it appears in a natural way.

The full SUGRA Lagrangian is specified in terms of two functions which depend on the hidden and observable scalars of the theory: the real analytic gauge–invariant Kähler function G which is a combination of two functions K and W, and the analytic gauge kinetic function f. K is the Kähler potential whose second derivative determines the kinetic terms for the fields in the chiral supermultiplets and is thus important for obtaining the proper normalization of the fields. W is the complete analytic superpotential which is related with the Yukawa couplings (which eventually determine the fermion masses) and also includes possibly non–perturbative effects. Finally, f determines the kinetic terms for the fields in the vector supermultiplets, and in particular the gauge coupling constant $Re f_a = 1/g_a^2$. The subindex a is associated with the different gauge groups of the theory $G = \prod_a G_a$. For example, in the case of the SM coupled to SUGRA it is associated with $SU(3)$,$SU(2)_L$, $U(1)_Y$. Then, once we know these functions the soft SUSY–breaking terms are calculable. Unfortunately for the predictivity of the theory, G and f are arbitrary and therefore the soft terms become supergravity model dependent. All the above mentioned problems can be solved in Superstring.

As in the case of SUSY, relevant theoretical arguments can be given in favour of Superstring: i) It is possible to obtain models resembling the SSM at low energy[5]. Of course, this is crucial in order to connect Superstring with the observable world. ii) It is the only hope to answer fundamental questions that in the context of the SM, SSM or Grand Unified Theories (GUT) cannot even be posed: why the gauge and Yukawa couplings should have a particular value?. First, the gauge coupling constant is dynamical because it arises as the vev of a gauge singlet field S called the dilaton, $\langle Re S \rangle = 1/g^2$. This result can be understood taking into account that in Superstring $f_a = S$ (at tree level). It is worth noticing here that the gauge coupling constants are unified even in the absence of a GUT. Thus GUT gauge groups, as e.g. $SU(5)$ or $SO(10)$, are not mandatory in order to have unification in the context of Superstring. Second, the Yukawa couplings, which determine the quark and lepton masses, can be explicitly calculated[6] and they turn out to be also dynamical. They depend in general on other gauge singlet fields T_m called the moduli whose vev determine the size and shape of the compactified space. E.g. for the overall modulus $\langle ReT \rangle \sim R^2$. Of course, experimental data demand $ReS \sim 2$ and $ReT \sim 1$ (in M_P units). Therefore the initial questions translate as how are the vev determined. This will be discussed below. Finally, the most outstanding virtue of Superstring is that iii) It is the only finite theory which can unify all the known interactions including gravity.

In particular, the Heterotic Superstring after compactification of six extra dimensions (on some compact manifolds) leads to a $N = 1$ effective SUGRA. Now, K and f *are in principle calculable* from Superstring scattering amplitudes. Besides, whereas in SUGRA (non–Superstring) models we do not have the slightest idea of what fields could be involved in SUSY–breaking, four–dimensional *Superstring automatically have natural candidates for that job*: the dilaton S and the moduli T_m. These gauge singlet fields are generically present in four–dimensional Heterotic Superstring since S is related with the gravitational sector of the theory and T_m are related with the extra dimensions. While other extra fields could also play a role in specific models, the dilaton and moduli constitute in some way the *minimal possible SUSY–breaking sector in Superstring*.[†]

Concerning the superpotential W, the situation is more involved. It is known that the process of SUSY–breaking in Superstring has to have a non–perturbative origin since SUSY is preserved order by order in perturbation theory (the scalar potential $V(S,T)$ is flat) and hence S and T_m are undetermined at this level. On the other hand, very little is known about non–perturbative effects in Superstring, particularly in the four–dimensional case. It is true that gaugino condensation gives rise to an effective W which breaks SUSY at the same time that S and T_m acquire reasonable vev[7] (as the ones explained in point ii)), and determines explicitly the values of the soft SUSY–breaking terms[8]. However one should keep in mind that this analysis requires the assumption that the dominant non–perturbative effects in Superstring are the field theory ones. This is because gaugino condensation is not a pure "stringy" mechanism. Thus a pessimist would say that Superstring does not look particularly promising in trying to get information about the SUSY–breaking sector of the theory.

However, since K and f are known, and the degrees of freedom involved in the process of SUSY–breaking have been identified (the "hidden fields" S and T_m), the effect of SUSY–breaking can be parametrized by the vev of those fields. The purpose of the present paper is to review this approach for addressing the problem, trying to provide a theory of soft terms which could enable us to interpret the (future) experimental results on supersymmetric spectra.

The structure of the paper is as follows. In sect.2 we discuss some general formulae for the computation of soft terms in SUGRA. We analyze how the existence of "hidden sector" dependent masses and couplings modify the usual soft terms taking a simple case as a guideline example. Finally, we also discuss the so–called μ problem and various solutions proposed to generate a μ term in the low–energy theory. The B soft

---

[†]Starting with this minimal sector one can also study the possible role on SUSY–breaking of other extra fields (see the example discussed in sect.8 of ref.[9]).

term is given for the different solutions. In sect.3 we apply the formulae obtained in the previous sect. to the computation of soft terms in Superstring[‡]. It turns out to be specially useful to introduce a "goldstino angle" whose value tells us where the dominant source of SUSY–breaking resides. All formulae for soft parameters take on particularly simple forms when written in terms of this variable. We also allow for a non–vanishing vacuum energy and arbitrary complex phases in the relevant vevs. Allowing for these turns out to be important for some relevant issues concerning soft terms. We also discuss the more model–dependent B soft term for the solutions to the μ problem which were analyzed in the previous sect. This kind of solutions are naturally present in Superstring.

2. Soft terms from SUGRA and the μ problem

Soft masses, trilinear and bilinear terms arise from the expansion of the SUGRA[§] scalar potential

$$V = e^G \left[G_\alpha \, (G^{-1})^\alpha_\beta \, G^\beta - 3 \right] \tag{1}$$

and soft gaugino masses for the canonically normalized gaugino fields can be obtained from the fermionic part of the SUGRA Lagrangian

$$M_a = \frac{1}{2} \, (Re f_a)^{-1} \, e^{G/2} \, f^\alpha_a \, (G^{-1})^\beta_\alpha \, G_\beta \tag{2}$$

where the real gauge–invariant Kähler function G is given by

$$G(z_\alpha, z^*_\alpha) = K(z_\alpha, z^*_\alpha) + \log |W(z_\alpha)|^2 \tag{3}$$

and we use standard SUGRA conventions on derivatives (e.g. $G_\alpha = \frac{\partial G}{\partial z^*_\alpha}$, $G^\alpha = \frac{\partial G}{\partial z_\alpha}$). K and W are given in general by the form

$$\begin{aligned} K &= K_0(h_l, h^*_l) + K_{ij}\phi_i\phi^*_j + (Z_{ij}\phi_i\phi_j + h.c.) + ... & (4)\\ W &= W_0(h_l) + \mu_{ij}\phi_i\phi_j + Y_{ijk}\phi_i\phi_j\phi_k + ... & (5) \end{aligned}$$

where we assume two different types of scalar fields $z_\alpha = h_l, \phi_i$. ϕ_i correspond to the observable sector (they include the SSM fields) and h_l correspond to a hidden sector. The latter are responsible for the SUSY breaking when some of them acquire large ($\gg M_W$) vevs. The ellipsis indicates terms of higher order in ϕ_i, ϕ^*_i and *the quantities* $\mu_{ij}, Y_{ijk}, K_{ij}$ *and* Z_{ij} *are in general* h_l, h^*_l *dependent.*

‡For an extended list of references see ref.[9]

§We use from now on the standard SUGRA mass units where $M_P = 1$.

Then the form of the effective soft Lagrangian obtained from eqs.(1,2) is

$$\mathcal{L}_{soft} \;=\; -\frac{1}{2}\sum_a M_a \bar{\widehat{\lambda}}_a \widehat{\lambda}_a - \sum_i m_i^2 |\widehat{\phi}_i|^2 - (A_{ijk}\widehat{Y}_{ijk}\widehat{\phi}_i\widehat{\phi}_j\widehat{\phi}_k + B_{ij}\widehat{\mu}_{ij}\widehat{\phi}_i\widehat{\phi}_j + h.c.) \quad (6)$$

The mass parameter $\widehat{\mu}$ is related with Z and μ terms of eqs.(4) and (5) respectively and will be discussed below in the context of the so–called μ problem. We recall that the passage to the effective low–energy theory involves a number of rescalings. In particular, in the previous equation

$$\begin{aligned} \widehat{\phi}_i &= K_i^{1/2}\phi_i \\ \widehat{\lambda}_a &= (Ref_a)^{1/2}\lambda_a \\ \widehat{Y}_{ijk} &= Y_{ijk}\,\frac{W_0^*}{|W_0|}\,e^{K_0/2}\,(K_iK_jK_k)^{-1/2} \end{aligned} \quad (7)$$

where $\widehat{\phi}_i, \widehat{\lambda}_a$ are the canonically normalized fields.¶

In the case of the MSSM the Kähler potential (to first order in the observable fields ϕ_i) and superpotential have the form (see eqs.(4),(5))

$$K \;=\; K_0(h_l, h_l^*) + \sum_i K_i\phi_i\phi_i^* + (ZH_1H_2 + h.c.) \quad (8)$$

$$W \;=\; W_0(h_l) + \mu H_1H_2 + \sum_{generations} (Y_uQ_LH_2u_R + Y_dQ_LH_1d_R + Y_eL_LH_1e_R) \quad (9)$$

where now $i = Q_L, u_R, d_R, L_L, e_R, H_1, H_2$. These equations include the usual Yukawa couplings ($Y_{ijk} = Y_u, Y_d, Y_e$, in a self–explanatory notation) and *we have allowed a possible mass μ and coupling Z for the Higgses* (recall that they have opossite hypercharges), where μ and Z are in principle free parameters.

For the sake of definiteness let us consider the simple case of canonical kinetic terms for hidden and observable fields (i.e. $K_0 = \sum_l h_l h_l^*$ and $K_i = 1$ in eq.(8)), $Z = 0$ and constant f_a. Then it is straightforward to compute the form of the soft terms

$$M_a \;=\; 0 \quad (10)$$

$$m_i^2 \;=\; m_{3/2}^2 + V_0 \quad (11)$$

$$A_{ijk} \;=\; m_{3/2}\sum_l G_{0h_l}\left(h_l^* + \frac{Y_{ijk}^{h_l}}{Y_{ijk}}\right) \quad (12)$$

$$B_\mu \;=\; m_{3/2}\sum_l G_{0h_l}\left(h_l^* + \frac{\mu^{h_l}}{\mu}\right) - m_{3/2} \quad (13)$$

¶For phenomenological reasons related to the absence of flavour changing neutral currents (FCNC) in the effective low–energy theory (see sect.5 of ref.[9] for a discussion of this point) from now on we will assume a diagonal form for the part of the Kähler potential associated with matter fields, $K_{ij} = K_i\delta_j^i$ in eq.(4).

where $m_{3/2}^2 = e^{G_0}$ and $V_0 = e^{G_0}(\sum_l |G_{0h_l}|^2 - 3)$ is the vev of the scalar potential (i.e. the cosmological constant). $M_a = 0$ is a consecuence of the assumption that f_a is a constant as can be seen from eq.(2). We call B_μ the B-term since it is associated only (recall that we are analyzing the case $Z = 0$) with the μ term of eq.(9). It is worth noticing here that only if Y_{ijk} and μ are h_l independent, the well known result for the B–term

$$B_\mu = A - m_{3/2} \tag{14}$$

is recovered (where $A = A_{ijk}$). One can, of course, think of many possible SUGRA models (with different K and f) leading to SUSY–breaking in a "hidden sector" of the theory leading in turn to different results for the soft terms

Now, let us discuss the μ problem and the different solutions proposed in the literature. From eqs.(6,11,13) and taking into account SUSY D–terms the relevant Higgs scalar potential along the neutral direction for the electroweak breaking is readily obtained

$$V(H_1, H_2) = \frac{1}{8}(g^2 + g'^2)(|H_1|^2 - |H_2|^2)^2 + m_1^2|H_1|^2 + m_2^2|H_2|^2 - m_3^2(H_1H_2 + h.c.) \tag{15}$$

with

$$\begin{aligned} m_{1,2}^2 &= m_{3/2}^2 + V_0 + |\widehat{\mu}|^2 \\ m_3^2 &= -B_\mu \widehat{\mu} \\ \widehat{\mu} &= \mu\, e^{K_0/2} \frac{W_0^*}{|W_0|} \end{aligned} \tag{16}$$

where the Higgsino mass $\widehat{\mu}$ gives a SUSY contribution to the Higgs scalar masses. This is the SUSY version of the usual Higgs potential in the SM. In order for the potential to be bounded from below, the condition $m_1^2 + m_2^2 - 2|m_3|^2 > 0$ must be impose all over the energy range $[M_W, M_P]$. This in particular implies $\langle H_{1,2} \rangle = 0$ at the Planck scale. Below the Planck scale, one has to consider the radiative corrections to the scalar potential. Then the boundary conditions of eq.(16) are substantially modified in such a way that the determinant of the Higgs mass–squared matrix becomes negative, triggering $\langle H_{1,2} \rangle \neq 0$ and $SU(2) \times U(1)_Y$ symmetry breaking.

For this scheme to work, the presence of the second term in eq.(9) is crucial. If $\mu = 0$, then the form of the renormalization group equations (RGEs) implies that such a term is not generated at any Q scale since $\mu(Q) \propto \mu$. The same occurs for μ_3, i.e. $\mu_3(Q) \propto \mu$. Then, the minimum of the potential of eq.(15) occurs for $H_1 = 0$ and, therefore, d–type quarks and e–type leptons remain massless. Besides, the superpotential of eq.(9) with $\mu = 0$ possesses a spontaneously broken Peccei-Quinn symmetry[10] leading to the appearance of an unacceptable Weinberg–Wilczek axion[11].

Once it is accepted that the presence of the μ–term in the superpotential is essential, there arises an inmediate question: Is there any dynamical reason why μ should be small, of the order of the electroweak scale? Note that, to this respect, the μ-term is different from the soft SUSY-breaking terms, which are characterized by the small scale $m_{3/2}$ once we assume correct SUSY breaking. In principle the natural scale of μ would be M_P, but this would re-introduce the problem of naturalness since the Higgs scalars get a contribution μ^2 to their squared mass (see eq.(16)). Thus, any complete explanation of the electroweak breaking scale must justify the origin of μ. This is the so-called μ-problem. This problem has been considered by several authors and different possible solutions have been proposed[12,13,14,15,16]. We will mention here two interesting scenarios in order to solve the problem.

In ref.[14] was pointed out that the presence of a non-renormalizable term in the superpotential

$$\lambda W_0 H_1 H_2 \tag{17}$$

characterized by the coupling λ, which mixes the observable sector with the hidden sector, yields dynamically a μ parameter when W_0 acquires a vev

$$\mu = \lambda W_0 \tag{18}$$

The fact that μ is of the electroweak scale order is a consequence of our assumption of a correct SUSY-breaking scale $m_{3/2} = e^{K_0/2}|W_0| = O(M_W)$. Now, with this solution to the μ problem, eq.(13) gives

$$B_\mu = m_{3/2}\left(2 + \frac{V_0}{m_{3/2}^2} + \sum_l G_{0h_l}\frac{\lambda^{h_l}}{\lambda}\right) \tag{19}$$

and $\widehat{\mu} = m_{3/2}\lambda$ in eq.(16). For a vanishing cosmological constant, $V_0 = 0$, and λ independent of h_l a very simple result is obtained

$$B_\mu = 2m_{3/2} \tag{20}$$

The value of A is still given by eq.(12), but the relation $B_\mu = A - m_{3/2}$ is no longer true even in the case of Y_{ijk} independent of h_l.

In refs.[13,14] was shown that if the coupling Z is present in the Kähler potential (see eq.(8)), an effective low-energy B-term is naturally generated of the order of magnitude of the gravitino mass. We will call it B_Z. We can compute its form in our guideline example with canonical kinetic terms. The result is

$$\begin{aligned} B_Z &= \frac{m_{3/2}}{X}\left(2 + \frac{V_0}{m_{3/2}^2} + \sum_l [G_{0h_l}\frac{Z^{h_l}}{Z} - G_0^{h_l}\frac{Z_{h_l}}{Z} - |G_0{}^{h_l}|^2\frac{Z^{h_l}_{h_l}}{Z}]\right) \\ X &= 1 - \sum_l G_0^{h_l}\frac{Z_{h_l}}{Z} \end{aligned} \tag{21}$$

and now the Higgsino mass $\widehat{\mu}$ in eq.(16) is modified to $\widehat{\mu} = m_{3/2}XZ$. Notice that the dependence on Z and λ in eqs.(21,19) is the same for the case of an analytic Z, i.e. Z independent of h_l^*. This is not a surprising result since both mechanisms for solving the μ problem are equivalent in this particular case[14]. Indeed, in the case now considered, the SUGRA theory is equivalent to the one with a Kähler potential K (without the terms $ZH_1H_2 + h.c.$ of eq.(8) and a superpotential $We^{ZH_1H_2}$, since the function $G = K + \log|W|^2$ is the same for both. After expanding the exponential, the superpotential will have a contribution $ZW_0H_1H_2$, i.e. a term of the type of eq.(17). Of course, in the simple case $V_0 = 0$ and Z independent of h_l, B_Z coincides with the result of eq.(20).

$$B_Z = 2m_{3/2} \tag{22}$$

Notice that it is conceivable that both mechanisms could be present simultaneously. In that case the general expresions for the B–term and Higgsino mass are easily obtained

$$\begin{aligned} B &= \frac{1}{\widehat{\mu}}(B_\mu m_{3/2}\lambda + B_Z m_{3/2} X Z) \\ \widehat{\mu} &= m_{3/2}\lambda + m_{3/2} X Z \end{aligned} \tag{23}$$

where B_μ and B_Z, X are given in eqs.(19) and (21) respectively. The solutions discussed here in order to solve the μ problem are naturally present in Superstring.

Then, let us know analyze the soft terms in Superstring models where K, f and the "hidden sector" are known.

3. Soft terms from Superstring

Following the Introduction and the notation of Sect.2, the "hidden" and observable sectors are given by $h_l = S, T_m$ and ϕ_i respectively. In order to proceed we need to make some simplifying assumptions. We will comment below what changes are to be expected if the assumptions are relaxed. Amongst the moduli T_m we will concentrate on the overall modulus T whose classical value gives the size of the manifold. Apart from simplicity, this modulus is the only one which is always necessarily present in any (0,2) (but left-right symmetric) 4-D strings. We believe that studying the one modulus case is enough to get a feeling of the most important physics of soft terms We will disregard for the moment any mixing between the S and T fields kinetic terms. In fact this is strictly correct in all 4-D strings at tree level. However, it is known that this type of mixing may arise at one loop level in some cases. On the other hand, these are loop effects which should be small and in fact can be easily incorporated in the analysis in some simple cases (orbifolds) as shown below.

Under the above conditions, the functions which appear in K and W eqs.(8,9) have the following general expressions:

$$K_0 = -log(S+S^*) + K_0(T,T^*) \; ; \; K_i = K_i(T,T^*) \; ; \; Z = Z(T,T^*) \tag{24}$$

$$W_0 = W_0(S,T) \; ; \; \mu = \mu(S,T) \; ; \; Y_{ijk} = Y_{ijk}(T) \tag{25}$$

The tree-level expression for f_a for any four-dimensional string, as mentioned in the Introduction, is well known, $f_a = k_a S$, where k_a is the Kac-Moody level of the gauge factor. Normally (level one case) one takes $k_3 = k_2 = \frac{3}{5}k_1 = 1$. Since a possible T dependence may appear at one–loop, then in general

$$f_a \ = \ f_a(S,T) \tag{26}$$

where we assume that other possible chiral fields do not contribute to SUSY–breaking. Finally, the cosmological constant is (see eq.(1))

$$V_0 \ = \ G^S_S|F^S|^2 \ + \ G^T_T|F^T|^2 \ - \ 3e^G \tag{27}$$

Of course, the first two terms in the right hand side of eq.(27) represent the contributions of the S and T auxiliary fields, $F^S = e^{G/2}(G^S_S)^{-1}G^S$ and $F^T = e^{G/2}(G^T_T)^{-1}G^T$.

As we will show below, it is important to know what field, either S or T, plays the predominant role in the process of SUSY-breaking. This will have relevant consequences in determining the pattern of soft terms, and therefore the spectrum of physical particles. That is why it is very useful to define an angle θ in the following way (consistently with eq.(27)):

$$\begin{aligned}(G^S_S)^{1/2}\ F^S \ &= \ \sqrt{3}Cm_{3/2}\ e^{i\alpha_S} sin\theta \\ (G^T_T)^{1/2}\ F^T \ &= \ \sqrt{3}Cm_{3/2}\ e^{i\alpha_T} cos\theta\end{aligned} \tag{28}$$

where α_S, α_T are the phases of F^S and F^T, and the constant C is defined as follows:

$$C^2 \ = \ 1 \ + \ \frac{V_0}{3m^2_{3/2}} \tag{29}$$

If the cosmological constant V_0 is assumed to vanish, one has $C = 1$, but we prefer for the moment to leave it undetermined. Of course, the way one deals with the cosmological constant problem is important.

Notice that, with the above assumptions, the goldstino field which is swallowed by the gravitino in the process of supersymmetry breaking is proportional to

$$\tilde{\eta} \ = \ sin\theta\ \tilde{S} \ + \ cos\theta\ \tilde{T} \tag{30}$$

where $\tilde{S}$ and $\tilde{T}$ are the canonically normalized fermionic partners of the scalar fields S and T (we have reabsorbed here the phases by redefinitions of the fermions $\tilde{S}, \tilde{T}$). Thus the angle defined above may be appropriately termed *goldstino angle* and has a clear physical interpretation as a mixing angle.

Then it is straightforward to compute the general form of the soft terms

$$M_a = \frac{C\sqrt{3}}{2Ref_a} m_{3/2}(k_a 2ReS\ e^{-i\alpha_S} sin\theta \ + \ f^T_a(G^T_T)^{-1/2}\ e^{-i\alpha_T} cos\theta) \tag{31}$$

$$
\begin{aligned}
m_i^2 &= 2m_{3/2}^2\,(C^2-1) \;+\; m_{3/2}^2 C^2 (1+N_i(T,T^*) cos^2\theta) \\
N_i(T,T^*) &= \frac{3}{K_{0T}^{\;T}}\Big(\frac{K_{iT}K_i^{\;T}}{K_i^{\;2}} \;-\; \frac{K_{iT}^{\;T}}{K_i}\Big) \;=\; -3\frac{(log K_i)_T^T}{K_{0T}^{\;T}} \qquad (32) \\
A_{ijk} &= -\sqrt{3} m_{3/2} C (e^{-i\alpha_S} sin\theta \;+\; e^{-i\alpha_T}\omega_{ijk}(T,T^*) cos\theta) \\
\omega_{ijk} &= (K_{0T}^{\;T})^{-1/2}\Big(\sum_{l=i,j,k}\frac{K_l^T}{K_l} \;-\; K_0^T \;-\; \frac{Y_{ijk}^T}{Y_{ijk}}\Big) \qquad (33)
\end{aligned}
$$

The formula for the B–term can be found in ref.[9]. In order to obtain more concrete expressions for the soft terms one has to compute the functions $N_i(T,T^*)$, $\omega_{ijk}(T)$ and $f_a(T)$. In order to evaluate these functions one needs a minimum of information about the Kähler potential K, the structure of Yukawa couplings $Y_{ijk}(T)$ and the one-loop threshold corrections $f_a(T)$. This type of information is only known for some classes of four-dimensional strings which deserve special attention: large T limit of Calabi-Yau compactifications and orbifold compactifications. This analysis can be found in ref.[9]. Likewise in ref.[17], the analysis of the different moduli case can be found.

4. References

1. For a review, see: H.P. Nilles, *Phys. Rep.* **110** (1984) 1, and references therein.
2. For a historical review, see: S. Ferrara, *Dirac Lecture delivered at ICTP, Trieste (1994)*, CERN–TH.7285/94, and references therein.
3. For a review, see: M. Green, J. Schwarz and E. Witten, *Superstring Theory* (Cambridge University Press,1987), and references therein.
4. For a review, see: H.P. Nilles, *Int. J. Mod. Phys.* **A5** (1990) 4199, and references therein.
5. For a review, see: J.A. Casas and C. Muñoz, *Nucl. Phys. (Proc. Suppl.)* **B16** (1990) 624, and references therein.
6. See e.g.: J.A. Casas, F. Gómez and C. Muñoz, *Phys. Lett.* **B292** (1992) 42, and references therein.
7. See e.g.: B. de Carlos, J.A. Casas and C. Muñoz, *Nucl. Phys.* **B399** (1993) 623, and references therein.
8. B. de Carlos, J.A. Casas and C. Muñoz, *Phys. Lett.* **B299** (1993) 234.
9. A. Brignole, L.E. Ibáñez and C. Muñoz, *Nucl. Phys.* **B422** (1994) 125.
10. R. Peccei and H. Quinn, *Phys. Rev. Lett.* **38** (1977) 1440.
11. S. Weinberg, *Phys. Rev. Lett.* **40** (1978) 223; F. Wilczeck, *Phys. Rev. Lett.* **40** (1978) 229
12. J.E. Kim and H.P. Nilles, *Phys. Lett.* **B138** (1984) 150, **B263** (1991) 79; E.J. Chun, J.E. Kim and H.P. Nilles, *Nucl. Phys.* **B370** (1992) 105.
13. G.F. Giudice and A. Masiero, *Phys. Lett.* **B206** (1988) 480.
14. J.A. Casas and C. Muñoz, *Phys. Lett.* **B306** (1993) 288.
15. G.F. Giudice and E. Roulet, *Phys. Lett.* **B315** (1993) 107.

16. I. Antoniadis, C. Muñoz and M. Quirós *Nucl. Phys.* **B397** (1993) 515; S. Ferrara, C. Kounnas, M. Porrati, F. Zwirner *Nucl. Phys.* **B318** (1989) 75.
17. A. Brignole, L.E. Ibáñez and C. Muñoz, unpublished; A. Brignole, L.E. Ibáñez, C. Muñoz and C. Scheich, in preparation.

NJL APPROACH TO GAUGINO CONDENSATION

C.A. SAVOY
Service de Physique Théorique, CE-Saclay
F-91191 Gif-sur-Yvette Cedex, FRANCE

ABSTRACT

Gaugino condensation in the hidden sector of supergravity models is approached à la Nambu-Jona-Lasinio by minimization of a one-loop scalar potential. Auxiliary superfields whose *v.e.v.'s* give the scales of gaugino condensation and supersymmetry breaking are introduced through a Laplace transform. The case of several group factors in the hidden gauge sector is also solved.

In this talk I would like to summarize some recent work in collaboration with R. Peschanski (Saclay preprint SPhT-T/94-04), on the question of supersymmetry breaking through gaugino condensation.

In supergravity models, the spontaneous breaking of local supersymmetry, or super-Higgs mechanism may generate soft supersymmetry breaking terms that allow to fulfill such phenomenlogical requirements as scalar masses and spontaneous gauge symmetry breaking at scales of $O(M_{Z^0})$. However, the super-Higgs mechanism implies the existence of a supergravity breaking scale, intermediate between M_{Planck} (or M_c, the grand-unification scale) and M_{Z^0}. This intermediate breaking scale (of order $10^{-11} - 10^{-13}$GeV) is difficult to implement in a natural way, except for the mechanism of *gaugino condensation*[1] in a "hidden" sector gravitationally coupled to the "observable" sector of physical matter and gauge fields. One interest of this mechanism is its compatibility[2] with supergravity models coming from superstring unification, where the absence of tree-level intermediate scales and the existence of large hidden gauge sectors follows from the heteretotic string constructions.

Gaugino condensation is expected to occur from the renormalization group flow of the coupling constant from the unification scale downwards. Indeed, with an appropriate large gauge group in the hidden sector, and a gauge coupling as the one obtained by the observed approximate unification of the Standard Model couplings, one expects condensates to form at the suitable scale. Gaugino condensation has been derived for globally sypersymmetric theories[3] within both the effective potential and the instanton approaches. Thus, a global chiral symmetry is found to be spontaneously broken. However, it does not lead to supersymmetry breaking unless one includes a dilaton-axion supermultiplet coupled to the gauge sector, the so called S-fields[4]. Interestingly enough, this pattern is indeed suggested in string theory.

Basically, only effective potential approaches have been extended to the supergravity models, in order to justify an effective S-dependent superpotential leading to the super-Higgs mechanism. Since this problem of supersymmetry breaking is already discussed in the review talk of Tom Taylor in these Proceedings, let me just mention a series of more or less technical difficulties in the formulation of gaugino condensation in supergravity models. First, the presence of "modulis" (so-callet T-fields) in superstring compactification is quite relevant in this respect. While often responsible for the "no-scale" structure of the models, their properties, such as T-duality, gives problems to implement supersymmetry breaking by gaugino condensation[5].

Another problem is the proper mechanism of supersymmetry breaking involving the S-field. The original proposal along this direction within the superstring framework has been challenged because of T-duality requirements[5],. A mechanism based on the existence of two or more condensates has been put forward[6]. However, it is not clear how well the different scales introduced in this case can be determined. Moreover, it has been argued that supersymmetry would be restored[7] at the minimum of the scalar potential.

These obstacles encountered in the previous studies are our main motivation to adapt the Nambu-Jona-Lasinio approach[8] to supersymmetry breaking by gaugino condensation in the framework of supergravity models as already suggested in Ref.[9]. In the most appropriate version of this approach one introduces an auxiliary composite field $\sum_a (\lambda^a \lambda^a)$ (λ^a are the gaugino fields). The one-loop effective potential is then evaluated and provides the non-trivial dependence on the composite field. However, this effective potential is quadratically divergent and defined by the introduction of a cut-off. Then, the minimization of the one-loop potential gives a non-trivial vacuum expectation value to the composite field, in the form of a "gap equation", for couplings above some critical value defined in terms of the cut-off. In this way, the fermion chiral symmetry is broken and so are other symmetries acting non-trivially on the fermion (i.e., gaugino) bilinear. Therefore the one-loop quantum correction is implementing the necessary dynamical dependence on the composite fields. Moreover, kinetic terms are also generated at one loop, supporting the physical interpretation that the composite field become propagating below the critical scale defined by the cut-off. All this can be made precise in the large-N limit, where N is the number of fermions (gauginos).

The supersymmetric version of the NJL is somewhat more complex but it has been widely discussed in the literature[10]. In this context, one considers "super-composite" chiral fields $Q^c Q$ and soft supersymmetry breaking terms are added by hand. Instead, gaugino condensates are defined as the first components of bilinears, $(W_a^\alpha W_\alpha^\alpha)_{\theta=0}$, where W_α is the spinor chiral field that contain the gauge strength among its components. One aims to induce the dynamical breaking of supersymmetry from non-vanishing last components $(W_a^\alpha W_\alpha^a)_{\theta^2}$ in the wake of the symmetry breaking by the first components. A further complication is the fact that there could be several orders of magnitude between the unification scale where the theory is settled and the condensation scale under investigation. Since condensation is presumably induced as a non-perturbative phenomenon by the increasing of the gauge coupling as the energy scale decreases, this scale dependence has to be accounted for in the NJL effective theory.

As stated above, the relevant framework in this study the supergravity enriched by a dilaton-dilatino-axion sector and some modulus superfields suggested by superstring theory. The superfields contain auxiliary fields (F components) that control supersymmetry breaking. Therefore we choose to keep these auxiliary components in the calculation of the one-loop effective potential and eliminate them at the end from the overall (cut-off dependent) potential. Finally, let us take $M_{\text{Planck}} = 1$ as our unity for energy scales unless contrarily stated.

Let us assume a hidden gauge sector with a semisimple gauge group $G = \bigotimes_{i=1}^{n} G_i$ and dimension $N_G = \sum_{i=1}^{n} N_i$ where N_i is the dimension (i.e., the number of gaugino) of each one of the n factors G_i in G. Let us neglect all matter fields, all gauge superfields, and any additional component of the observable sector that are irrelevant for the discussion hereafter. Instead the dilaton chiral superfield S play a major rôle in supersymmetry breaking and it must be retained together with the essential part of the moduli fields. At the level of our approach it is enough to introduce only one chiral superfield T, the overall modulus field. Correspondly, we set the relevant part of the Kahler potential to the usual tree-level expression:

$$K(S,T) = -\ln\left(S + S^t\right) - 3\ln\left(T + T^t\right) \tag{1}$$

This suits our purposes here, since we do not seek for the vacuum expectation values of the first components of S and T, which we assume to be correctly fixed by a thorough treatment of duality symmetries in the framework of the underlying superstring theory. The supersymmetric Yang-Mills lagrangian is

$$\left[\frac{S}{4}\text{Tr } W^{\alpha}W_{\alpha}\right]_{\theta^2} + \text{h.c.} \tag{2}$$

where W^{α} are the chiral (Majorana) spinor superfield whose first components are the gaugino fields $\lambda^{\alpha}_{a_i}$ $(a_i = 1, ..., N_i;\ i = 1, ..., n)$ in the hidden sector. The v.e.v. of the dilaton component of S sets the gauge coupling at the compactification scale M_c as follows:

$$S + S^{*} = 2g^{-2}\left(M_c\right) \tag{3}$$

$$M_c = \frac{2M_{\text{Planck}}}{\left(S + S^{*}\right)^{1/2}\left(T + T^{*}\right)^{1/2}}$$

For simplicity we are assuming a unique gauge coupling $g^2\left(M_c\right)$ for all the factors G_i. This amounts to all the levels of their Kac-Moody algebra being equal to 1, but the generalization to $k_i \neq 1$ is obvious. However, those couplings will be splitted by the RG running. At one-loop they obey the equations:

$$\frac{1}{g_i^2(\Lambda)} = \frac{1}{g^2(M_c)} + 4\nu_i \ln\frac{\Lambda^3}{M_c^2}$$

$$B_i = \frac{C_2\left(G_i\right)}{32\pi^2} \tag{4}$$

where $C_2\left(G_i\right)$ is the Casimir eigenvalue of $adj\left(G_i\right)$.

In order to describe gaugino condensation and to determine $\left\langle\lambda^{\alpha}_{a_i}\lambda_{\alpha a_i}\right\rangle \equiv \left\langle\lambda\lambda\right\rangle_i$ we perform a supersymmetric Laplace transformation by introducing the auxiliary chiral superfields U_i and, correspondingly, the conjugate chiral superfields X_i and by adding to the lagrangian the following expression

$$\sum_{i=1}^{n} X_i \left(U_i - \mathrm{Tr}_i W^\alpha W_\alpha\right) \tag{5}$$

The v.e.v.'s $\langle U_i \rangle$ set the condensation scale for each sector G_i of the hidden gauge sector.

On quite general grounds, the NJL approach to spontaneous symmetry breaking amounts to a minimization of the effective scalar potential for the auxiliary scalars associated to the fermion condensates. The effective potential calculated at one-loop depends on a cut-off scale Λ. The minimum equation is the so-called "gap equation" that fixes the symmetry breaking scale. The couplings entering this effective potential should be calculated at scales of $O\left(\langle U_i \rangle\right)$. The gauge couplings explicitly appear in the masses of gaugino $M_i(\langle U \rangle) \propto g_i^2$, that circulate around the loop. Equation (4) suggests to replace (2) by

$$\left\{ \left[\frac{S}{4} + \beta_i \ln \frac{U_i}{M_c^3} \right] \mathrm{Tr}_i W^\alpha W_\alpha \right\}_{\theta^2} + \text{h.c.} \tag{2'}$$

(more on the justification of (2') later).

The supergravity theory defined by the lagrangian given by (1), (2') and (5) can be now contemplated in two forms, that are equivalent as a matter of fact:

a) From (2') and (5) one derives the superpotential

$$W = \sum_i X_i U_i \tag{6}$$

and the following expression for the field dependent gaugino masses

$$M_i = \frac{\left(\hat{X}_i\right)_{\theta^2}}{\hat{X}_i + \hat{X}_i^*} = \frac{\hat{F}_X^i}{\hat{X}_i + \hat{X}_i^*}$$

$$\hat{X}_i \equiv \frac{S}{4} + X_i + \beta_i \ln \frac{U_i}{M_c^3} \tag{7}$$

b) Because (5) defines n Laplace transformations, one can replace $\mathrm{Tr}_i W^\alpha W_\alpha$ by U_i in (2'). In this form, that turns out to be handier than a), the superpotential becomes:

$$W = \sum_i U_i \left(\frac{S}{4} + X_i + \beta_i \ln \frac{U_i}{M_c^3} \right) \tag{8}$$

and the gaugino masses reduce to

$$M_i = \frac{F_X^i}{X_i + X_i^*} \tag{9}$$

where F_X^i are the auxiliary fields in the chiral multiplets X_i.

Let us formulate the NJL approach in the more convenient form b). The results in a) can be checked to be identical with the replacements $F_X^i \longrightarrow \hat{F}_X^i$, $X_i \longrightarrow \hat{X}_i$: this is just a check of the self-consistence of the Laplace transformation.

The scalar potential at one-loop, $V = V_0 + V_1$, is obtained from (1), (8) and (9) as follows:

(i) V_0 is given by the classical supergravity lagrangian[11] :

$$V_0 = -\alpha \left|m_{3/2}\right|^2 - \frac{\left|F^S\right|^2}{S+S^*} - \left[m_{3/2} \sum_{I=S,U_i,X_i} F^I \frac{\partial}{\partial I}(K + \ln W) + \text{h.c.}\right] \tag{10}$$

where $m_{3/2}$ is the gravitino mass,

$$m_{3/2} = \frac{W}{(S+S^*)^{1/2}(T+T^*)^{3/2}} \tag{11}$$

The factor α represents the result of the minimization with respect to the auxiliary field F^T of the expression:

$$3\left|m_{3/2}\right|^2 + \frac{3\left|F_T\right|^2}{\left|T+T^*\right|^2} - \left[m_{3/2} F_T \frac{\partial}{\partial T}(K + \ln W) + \text{h.c.}\right]$$

Thus, with K given by (1) and a T-independent W, one gets $\alpha = 0$. However, we prefer to keep a parameter $\alpha \sim O(1)$ to account for the necessary corrections to the over-simplified treatment of the S and T sectors herein.

(ii) The one-loop contribution V_1 to the effective potential is obtained by the addition of the contribution of all the physical particles[12] :

$$V_1 = -2\sum_i N_i\ H\left(\left|M_i\right|^2, \Lambda\right) - \sum_A \nu_A H\left(K_A \left|m_{3/2}\right|^2, \Lambda\right) \tag{12}$$

where H is defined by the integral

$$H\left(m^2, \Lambda\right) = \int^\Lambda \frac{d^4 p}{(2\pi)^4} \ln\left(p^2 + m^2\right) \tag{13}$$

and m is the mass of the state as a function of the scalar fields. The cut-off Λ defines the physical region for the composite auxiliary fields, and the sums in (12) include all propagating fields such that $m < \Lambda$. The gaugino masses M_i are given by (9). In the second term ν_A are multiplicities of states whose mass is $\sqrt{K_A}\left|m_{3/2}\right|$. The factors K_A are assumed to be independent of the fields X_i, U_i. Thus, the gravitino sector contributes to V_1 with the well-known[13] term,

$$2H\left(4m_{3/2}^2\right) - 4H\left(m_{3/2}^2\right) \tag{14}$$

The NJL is better justified in the large-N limit. For this reason, the gaugino contribution in the one-loop potential is instrumental while other particle contributions seem circumstancial. However we shall keep (12) as it is for completeness and for some further discussion.

The essence of the present approach is the non-trivial dependence of V_1 on the auxiliary fields F_X^i ($\hat{F}_X^i$ in form a)) in the gaugino masses. Thus, these Lagrange multipliers acquire a dynamical rôle and contribute to the vacuum energy. This is the effective potential reflex of the NJL transmutation of the auxiliary fields into dynamical composite fields below some critical scale characterized by the cut-off Λ.

The next step is the minimization of $V = V_0 + V_1$ with respect to the fields. In default of a thorough understanding of duality symmetries we dismiss the minimum conditions with respect to the fields S (dilaton) and T (modulus). The extrema of $V = V_0 + V_1$ are defined by the vanishing of its gradient along the various field directions. Thus, we eliminate F_S and F_U^i and work in terms of the variables

$$x_i = \frac{X_i + X_i^*}{2\beta_i} \qquad M_i = \frac{F_X^i}{2\beta_i x_i} \tag{15}$$

The scalar potential at the extrema can now be written as,

$$\begin{aligned} V = {} & 2\sum_i N_i \left(2\left|M_i\right|^2 H'\left(\left|M_i^2\right|, \Lambda\right) - H\left(\left|M_i^2\right|, \Lambda\right)\right) \\ & + \sum_A \nu_A \left(K_A \left|m_{3/2}\right|^2 H'\left(K_A \left|m_{3/2}\right|^2, \Lambda\right) - H\left(K_A \left|m_{3/2}\right|^2, \Lambda\right)\right) \\ & - \frac{2}{(S+S^*)^{1/2}(T+T^*)^{3/2}} \sum_i \beta_i M_i U_i (1 - x_i) \end{aligned} \tag{16}$$

and the minimum conditions are:

$$\begin{aligned} M_i &= \frac{-\beta_i x_i \hat{U}_i}{N_i H'(M_i^2)} \qquad (\textit{gap equation}) \\ \hat{U}_i &= \frac{M_c^3 e^{-1-\frac{S}{4\beta_i}-x_i}}{(S+S^*)^{1/2}(T+T^*)^{3/2}} \qquad (\textit{condensation scale}) \\ 2M_i(x_i - 1) &= \sum_j \beta_j \hat{U}_j \left[\left(1 + \frac{S+S^*}{4\beta_j}\right)\left(1 + \frac{S+S^*}{4\beta_i}\right)\right] + (\alpha + \varepsilon) m_{3/2} \\ \varepsilon &= \sum_A \nu_A K_A H'\left(K_A \left|m_{3/2}^2\right|\right) \\ m_{3/2} &= -\sum_i \beta_i \hat{U}_i \end{aligned} \tag{17}$$

A manifest problem in the above relations concerns the dependence on M_i, thus on the F_X^i auxiliary fields. The usual couplings of chiral multiplets to supergravity only produce lagrangians quadratic in F's. Instead, F_X^i's appear in the potential (12) within the cut-off integral,

$$H(Z,\Lambda) = \frac{\Lambda^4}{64\pi^2}\left[\frac{Z}{\Lambda^2} + \ln\left(1+\frac{Z}{\Lambda^2}\right) - \frac{Z^2}{\Lambda^4}\ln\left(1+\frac{\Lambda^2}{Z}\right)\right] \tag{18}$$

This dependence could arise the question of the supersymmetric consistency of the NJL approach. At one-loop level the two options for eliminating the F -components, namely before or after the calculation of the one-loop theory should produce identical results to lowest order in $\hbar$ (the loop-expansion parameter). The paradox should be solved by a complete calculation of the one-loop action (I thank M.K. Gaillard for a discussion on this problem).

In the absence of a definite answer to that, we proceed on more pragmatic grounds by retaining only the dominant terms in the functions H and H' in (16)-(17)

$$\begin{gathered} H(Z,\Lambda) \sim \rho^2 Z \\ H'(Z,\Lambda) \sim \rho^2 \qquad \rho^2 \equiv \frac{\Lambda^2}{32\pi^2} \\ (Z \ll \Lambda^2) \end{gathered} \tag{19}$$

This approximation that we shall check for consistency at the end by requiring $M_i \ll \Lambda$, has also another advantage. It results in a quadratic lagrangian in the F_X^i, such that the minimization with respect to the F_X^i amounts to an integration on these auxiliary fields.

An interesting property of the scalar potential (16) is the vanishing of the term $O\left(m_{3/2}^2\Lambda^2\right)$ in the contribution of the gravitino sector as well as the dilaton-dilatino-axion one to the one-loop potential. This cancellation also occurs for the matter sector in the limit $F^T \to 0$. It is due to the combination $H\left(m^2\right) - m^2 H'\left(m^2\right)$ that appear in the scalar potential after the minimization. However it could be spoiled beyond the approximation in which the scalar potential (11), (12) has been derived, in particular, if modular invariance is implemented and S and T mixing is brought in.

Let us first analyse the simplest case with a simple group G, i.e., $n = 1$. In this case the above relations reduce to the simpler expressions,

$$\begin{aligned} m_{3/2} &= -\beta\hat{U} \\ &= \frac{-\beta M_c^3 e^{-1-\frac{S+S^*}{8\beta}-x}}{(S+S^*)^{1/2}(T+T^*)^{3/2}} \end{aligned}$$

$$\frac{2}{N_G\rho^2}x(1-x) = \left(1+\frac{S+S^*}{4\beta}\right)^2 - (\alpha+\varepsilon)$$

$$M_{1/2}^* = x\frac{m_{3/2}}{N_G\rho^2}$$

$$V_{\min} = \frac{2xm_{3/2}^2}{N_G\rho^2} M_{\text{Planck}}^2 \tag{20}$$

where the value of the potential at the minimum could be an artefact of our approximations. Let us examine (20) under the assumption that the value of $\langle S + S^* \rangle = 2g^{-2}(M_c)$ is correctly fixed by physical conditions beyond the scope in the present work. Then $\langle S + S^* \rangle / 4\beta = (4\pi)^2/g^2 C_2(G) \simeq 32\pi^2/C_2(G)$, so that

$$\frac{S + S^*}{4\beta} \gg 1 \, , \qquad (\alpha + \varepsilon) \sim O(1) \tag{21}$$

and from (20) one derives

$$x(1-x) \simeq \frac{N_G}{(C_2(G))^2} \frac{(2\pi)^2}{g^4(M_c)} \frac{\Lambda^2}{M_{\text{Planck}}^2} \tag{22}$$

This relation implies $0 < x < 1$, as well as,

$$\Lambda^2 < \left(\frac{g^2}{4\pi}\right)^2 \frac{C_2^2}{N_G} M_{\text{Planck}}^2 \tag{23}$$

which is perfectly consistent with a cut-off one or two orders of magnitude below the Planck mass. The other result, namely, that x is finite (and $x < 1$) fixes the value of the condensate at the expected scale

$$U = \mathrm{e}^{-1-x} M_c^3 \mathrm{e}^{-\frac{1}{4\beta g^2}} \tag{24}$$

This is the scale of chiral symmetry spontaneous breaking. The supersymmetry breaking scale is $\left(m_{3/2} M_{\text{Planck}}^2\right)$ which by (20) is indeed non-zero for $\langle U \rangle \neq 0$. Moreover, the gaugino mass is related to the gravitino mass as follows,

$$M \simeq \frac{128\pi^4}{C_2^2(G) g^4 (1-x)} m_{3/2} \gg m_{3/2} \tag{25}$$

Note that the conditions $M, m_{3/2} < \Lambda$, necessary for the consistency of the NJL picture leads to an effective *lower bound* for the cut-off, namely

$$\frac{\Lambda^2}{M_{\text{Planck}}^2} > \left(\frac{g^2}{4\pi}\right)^2 \left(\frac{C_2^2}{N_G}\right)^{2/3} \left(\frac{M_{\text{Planck}}}{T + T^*}\right)^2 \frac{\mathrm{e}^{-2}}{(4\pi^2)^{1/3}} \tag{26}$$

where we have successively used the expressions (3), (20), (22), (24).

Let us compare the results obtained here with the original calculation of gaugino condensation in the framework of global supersymmetry in[3]. These authors write an effective superpotential for the composite field U in the form,

$$W_{VY} = \left(\frac{1}{4g^2} + \beta \ln \frac{U}{\mu^3}\right) U \tag{27}$$

The first term replaces the gauge multiplet kinetic term. The second term reproduces the anomaly in the conformal supercurrent, that ?? dilation, superconformal and chiral symmetry currents, which would be generated by the gaugino loops. The superpotential (27) is combined with some kinetic term defined by a Kahler potential $K_V\left(U,U^{\dagger}\right)$ to determine an effective theory at energies below the confinement scale. The gauginos are confined and assumed to be faithfully replaced at low energies by the U-condensate. Minimization ot the scalar potential,

$$V_V = \left|F^U\right|^2$$

$$F^U = \left[\frac{\partial}{\partial U}\frac{\partial}{\partial U^*}K\left(U,U^*\right)\right]^{-1/2}\beta\ \ln\frac{U}{\langle U\rangle}$$

$$\langle U\rangle = \mu^3 e^{-\frac{1}{2\beta g^2}-1} \tag{28}$$

so that $V_V = 0$ at the minimum $U = \langle U\rangle$, where supersymmetry remains unbroken while chiral symmetry is spontaneously broken. The scale μ is the cut-off of the effective theory, which should be related to the confinement energy. The results of[3] are reproduced in the NJL approach if one assumes $\mu \simeq M_c$ and set $X = 0$, so that there is no loop-corrections to the scalar potential. In order to allow for supersymmetry breaking one has to replace the gauge coupling by the dynamical superfield S in W_{VY}.

This effective theory formulation of gaugino condensation is formally reproduced in our approach by integration on the auxiliary superfields X_i in (5) and then with drawing the gaugino fields. This reproduces the superpotential W_{VY}.

One can also consider the globally supersymmetric version of the NJL approach considered here. The results are similar to (20) with $F^S = \left(S+S^*\right)^2 U$ and the scalar potential,

$$V = \frac{\left|F^S\right|^2}{\left(S+S^*\right)^2}\left(1+\rho^2\right) + 2N_G\rho^2 M^2$$

$$x(1-x) = \frac{N\rho^2\left(S+S^*\right)^2}{\beta^2}$$

$$M = \frac{m_S}{1-x}\frac{\left(S+S^*\right)}{2\beta} \tag{29}$$

where $m_S = 2F^S/\left(S+S^*\right)$ is the dilatino mass. Again, with g^{-2} replaced back for S, the minimum of V is for $M=0$ and no supersymmetry breaking.

Let us turn back for a while to the form a) of the theory, in order to compare our approach to that in Ref.[9]. As already stated, it is easily checked that a) results in the same relations as b) with $\hat{X}_i$ and $\hat{F}^i_X$ replaced for X_i and F^i_X, respectively. In Ref.[9], the U-dependences of the superpotential and the gauge kinetic term are derived by requiring $U = W^\alpha W_\alpha$ to be a solution of the equations of motion. As a matter of fact this leads to an ambiguity which is resolved a posteriori by fixing the U coupling to $W^\alpha W_\alpha$ to be as in (2$'$) to reproduce the running of the gauge coupling.

However, instead of the Lagrange multipliers X_i in the superpotential (5), an arbitrary constant appears in Ref.[9] which is then roughly estimated by physical arguments. Since the auxiliary fields X_i, F^i_X are absent there are less minimum conditions and the parameter corresponding to $X = \hat{X}(X = 0)$ is not bounded and has to be fixed by hand. In summary, the main difference with[9], is that in our case the equivalence between U and $(W^\alpha W_\alpha)$ is imposed as a constraint by a Laplace transformation while in[9] it is only a on-shell relation, so that some ambiguities remain.

Moreover, only the case $n = 1$ was considered in[9], while we allow for $n > 1$ by introducing X_i $(i = 1, ..., n)$. With our methods, the coefficients can then be calculated once the values of $SandT$ are given. This amounts to the solution of simple algebraic relations. As already remarked, the case of more than one gauge factors is the relevant one. I refer to our paper for further details and for the explicit calculations in this more general situation.

REFERENCES

[1] H.P. Nilles, Phys. Lett. **B 115** (1982) 193.

[2] M. Dine, R. Rohim, N. Seiberg and E. Witten, Phys. Lett. **B 156** (1985) 55.
J.P. Deredinger, Lit. Ibañez and H.P. Nilles, Phys. Lett. **B 155** (1985) 65; Nucl. Phys. **B 267** (1986) 365.

[3] C. Veneziano and S. Yankielowicz, Phys. Lett. **B 113** (1982) 231.

[4] J. Affleck, M. Dine and N. Serberg, Nucl. Phys. **B 241** (1984) 493; **B 256** (1985) 557.
T.R. Taylor, Phys. Lett. **B 164** (1985) 43.

[5] K. Kikkawa and M. Yamasaki, Phys. Lett. **B 149** (1984) 357.
N. Sakai and I. Senda, Prog. Theo. Phys. **75**(1986) 692.

[6] N.V. Krasnikov, Phys. Lett. **B 193** (1987) 37.

[7] T.R. Taylor, Phys. Lett. **B 252** (1990) 59.

[8] Y. Nambu and G. Jona-Lasinio, Phys. Rev. **122** (1961) 231.

[9] A. de la Macorra and G.G. Ross, Nucl. Phys. **B 404** (1993) 321.

[10] W. Buchmüller and S.T. Love, Nucl. Phys. **B 204** (1982) 213.
W. Buchmüller and U. Ellwanger, Nucl. Phys. **B 245** (1984) 237.

[11] E. Cremner, S. Ferrara, L. Girardello and A. van Procyen, Nucl. Phys. **B 212** (1983) 413.

[12] S. Coleman and E. Weinberg, Phys. Rev. **D 7** (1973) 1883.

[13] R. Barbieri and S. Cecotti, Z. Phys. **C 17** (1983) 183.

SUPERSYMMETRIC AND NON-EXTREME $U(1)_M \times U(1)_E$ BLACK HOLES IN KALUZA-KLEIN THEORY

MIRJAM CVETIČ[*] AND DONAM YOUM[†]

Physics Department
University of Pennsylvania, Philadelphia PA 19104-6396

ABSTRACT

We present recent results for supersymmetric four-dimensional Abelian black holes in Kaluza-Klein theory. Such solutions, which saturate the corresponding Bogomol'nyi bound, can be obtained if and only if the Abelian isometry group of the internal space is broken down to the $U(1)_M \times U(1)_E$ group, *i.e.*, supersymmetric Kaluza-Klein black holes are dyonic ones with the magnetic charge P and the electric charge Q necessarily associated with different $U(1)$ factors. We also present the corresponding non-extreme $U(1)_M \times U(1)_E$ solutions, which have the singularity hidden behind the horizon, finite Hawking temperature T and non-zero entropy. In the extreme (supersymmetric) limit the singularity becomes null, the temperature saturates the upper bound $T = 1/\sqrt{4\pi|PQ|}$ and the entropy is zero.

1. Introduction

Solitons, defined as spatially localized time-independent solutions of classical field equations, are vacuum configurations in a sense that they saturate the Bogomol'nyi bounds on their energies. These bounds are characterized by topological charges which are defined by the boundary conditions on fields.

When embedded in supersymmetric theories, solitons form supermultiplets of states with masses given by ADM masses of configurations and central charges identified as their topological charges. Namely, such configurations correspond to bosonic configurations which are invariant under (constrained) supersymmetry transformations, *i.e.*, they satisfy the corresponding Killing spinor equations. In addition, one can show that in this case the corresponding Bogomol'nyi bound for their ADM masses are saturated.

⋆ E-mail address: cvetic@cvetic.hep.upenn.edu
† E-mail address: youm@cvetic.hep.upenn.edu

Over the past decades a number of solitonic configurations, *e.g.*, black holes[1,2,3,4,5] as well as domain walls[6,7], have been embedded into supersymmetric theories. It turns out that in most cases the Bogomol'nyi bound for the black hole ADM mass coincides with the corresponding inequality relation which characterize (non-) extreme black holes. Thereby, the terminologies extreme and supersymmetric are often used interchangeably.

Dilaton(s), *i.e.*, scalar field(s) without self-interactions, arise naturally in basic theories that unify gravity with other interactions, including certain supergravity theories, effective theory from superstrings as well as Kaluza-Klein theories. In general, the dilaton couples to the gauge field kinetic energy, *i.e.*, $e^{2\alpha\varphi}F_{\mu\nu}F^{\mu\nu}$, as well as to the matter potential, *i.e.*, $e^{2\varphi/\alpha}V(T)$. It is therefore of interest to address dilatonic topological configurations in such theories. Dilatonic black holes[8] as well as dilatonic domain walls[7,9] have been subjects of intense studies. Within $N = 1$ supergravity with a linear supermultiplet (whose coupling is parameterized by the parameter α) it has been observed[7] that the space-time structure and thermal properties of extreme magnetically charged dilatonic black holes and extreme dilatonic domain wall solutions exhibit complementary features; the global space-time of the extreme black holes with the dilaton coupling α is the same as the one of the extreme domain walls with the dilaton coupling $1/\alpha$. Interestingly, the theory with $\alpha = 1$, *i.e.*, an effective theory from strings, corresponds to the case where both extreme dilatonic black holes as well as domain walls are void of naked singularities.

While supersymmetric embeddings of domain wall solutions with arbitrary couplings α have been completed in Ref.7, supersymmetric embeddings of black hole solutions are known only for special values of dilaton-Maxwell couplings α, *i.e.*, $\alpha = 0$[1], 1[2], $\sqrt{3}$[3]. Therefore, it is of interest to address supersymmetric embedding(s) of black hole solutions in theories where the coupling α is different from $0, 1$ and $\sqrt{3}$. Kaluza-Klein (KK) theories, *i.e.*, $(4+n)$-dimensional ($(4 + n)$-d) gravity theories compactified down to 4-d, may serve as such an example. For example, if the internal radii of the compactified dimensions are fixed, the dilaton coupling is of the type $\alpha = \sqrt{\frac{n+2}{n}}$. Such configurations may also be of interest since for $n > 1$ additional scalar fields associated with compactified degrees of freedom could play a role in determining the space-time structure of the configuration. In addition, such configurations may shed light on the nature of non-perturbative 4-d vacua of $N \geq 2$ supergravity theories, including those corresponding to non-trivial vacuum configurations associated with the KK modes of certain $N \geq 2$ string vacua.

We would like to summarize recent results[10] for the supersymmetric embedding of

Abelian 4-d black hole solutions in $(4+n)$-d KK theory. Surprisingly, it turns out that supersymmetric embeddings of the corresponding 4-d Abelian (the isometry group of the compactified internal space is Abelian) configurations exist if and only if the isometry group of the internal space is broken down to the $U(1)_E \times U(1)_M$ group, *i.e.*, such configurations correspond to dyonic black holes with the magnetic P and the electric Q charges necessarily associated with different $U(1)$ factors. We further present solutions for the corresponding non-extreme $U(1)_E \times U(1)_M$ black holes.

In Chapter 2, we discuss the Abelian supersymmetric black holes in KK theory. Non-extreme $U(1)_M \times U(1)_E$ solutions are derived in Chapter 3. Global space-time structure and thermal properties of this class of solutions are discussed in Chapter 4. Conclusions are given in Chapter 5.

2. Supersymmetric embedding of black hole configurations

The starting point of supersymmetric embeddings of 4-d KK black holes is the "minimal" ($N=1$ or $N=2$) supergravity theories in $(4+n)$-d. $N=1$ extensions of $(4+n)$-d gravity are possible in $D=4,8,9,11$, where Majorana spinors exist[11]. In other dimensions, one has to introduce $N=2$ extended supergravity with the pair of spinors satisfying alternative reality conditions, *e.g.*, $SU(2)$ or $USp(2)$ (pseudo-) Majorana conditions. Supersymmetric extensions in $(4+n)$-d, in general, require an introduction of additional bosonic fields. However, since we are interested in non-trivial configurations associated with the $(4+n)$-d pure gravity, only, we shall set all the additional bosonic fields to zero.

In this case the bosonic part of the Lagrangian density is of the form:

$$\mathcal{L} = -\frac{1}{2\kappa^2}\sqrt{-g^{(4+n)}}\mathcal{R}^{(4+n)} \quad , \tag{2.1}$$

where $\mathcal{R}^{(4+n)}$ is the Ricci scalar defined in terms of a $(4+n)$-dimensional metric $g^{(4+n)}_{MN}$ and κ is the $(n+4)$-dimensional gravitational coupling constant. And the gravitino(s) transforms under supersymmetry as

$$\delta\psi^{\mathbf{A},i}_{\Lambda} = D_\Lambda \varepsilon^{\mathbf{A},i} = (\partial_\Lambda + \frac{1}{4}\Omega_{\Lambda AB}\Gamma^{AB})\varepsilon^{\mathbf{A},i} \quad , \tag{2.2}$$

where $\mathbf{A} = 1, ..., 2^{[\frac{n+4}{2}]}$ is the index for $(4+n)$-dimensional spinors and $i = 1, ..., N$ labels spinors (supersymmetry parameters or gravitinos) of N-extended supergravity. As usual,

$\Gamma^{AB} \equiv \Gamma^{[A}\Gamma^{B]}$, where Γ^A's are gamma matrices satisfying the $SO(3+n,1)$-Clifford algebra. The spin-connection is defined in terms of a Vielbein E^A_Λ: $\Omega_{ABC} \equiv -\tilde{\Omega}_{AB,C}+\tilde{\Omega}_{BC,A}-\tilde{\Omega}_{CA,B}$ where $\tilde{\Omega}_{AB,C} \equiv E^\Lambda_{[A}E^\Pi_{B]}\partial_\Pi E_{\Lambda C}$. Notation $[A \ldots B]$ refers to antisymmetrization of the corresponding indices.

2.1. Kaluza-Klein Compactification of extra Dimensions

We shall now compactify[12] the $(4+n)$-d supergravity down to 4-d. Compactification of extra dimensional part of gravity provides a way of unifying gauge theories with the Einstein-Poincaré gravity theory. The basic idea is that the isometry symmetry of the extra dimensions manifests itself as the gauge symmetry in 4-d after the compactification. In addition, one has to perform a dimensional reduction of the fermionic degrees of freedom as well.

Before we discuss the compactification, we fix our notation. General (space-time) indices running over $(4+n)$-d are denoted by upper-case letters $(A, B, ..., \Lambda, \Pi, ...)$. Lower-case letters $(a, b, ..., \lambda, \pi, ...)$ are for the 4 space-time dimensions, and those with tilde $(\tilde{a}, \tilde{b}, ..., \tilde{\lambda}, \tilde{\pi}, ...)$ are for the n extra spatial dimensions. Latin letters $(A, B, ..., a, b, ...)$ denote flat-tangent space-time indices, and greek letters $(\Lambda, \Pi, ..., \lambda, \pi, ...)$ are reserved for curved space-time indices. The flat Lorenz metric of tangent space is chosen to be $(+ - - \cdots -)$.

The KK Ansätze[13,14] for the Vielbein $E^{(4+n)A}_\Lambda$ and the corresponding metric $g^{(4+n)}_{\Lambda\Pi}$ that give rise to the general dilaton-gauge coupling in the effective 4-d action are of the form:

$$
\begin{aligned}
E^{(4+n)A}_\Lambda &= \begin{bmatrix} \mathrm{e}^{-\frac{1}{2\alpha}\varphi}\mathrm{e}^a_\lambda & \mathrm{e}^{\frac{1}{n\alpha}\varphi}A^{\tilde{\lambda}}_\lambda\Phi^{\tilde{a}}_{\tilde{\lambda}} \\ 0 & \mathrm{e}^{\frac{1}{n\alpha}\varphi}\Phi^{\tilde{a}}_{\tilde{\lambda}} \end{bmatrix} \\
g^{(4+n)}_{\Lambda\Pi} &= \eta_{AB}E^{(4+n)A}_\Lambda E^{(4+n)B}_\Pi = \begin{bmatrix} \mathrm{e}^{-\frac{1}{\alpha}\varphi}g_{\lambda\pi} - \mathrm{e}^{\frac{2\varphi}{n\alpha}}\rho_{\tilde{\lambda}\tilde{\pi}}A^{\tilde{\lambda}}_\lambda A^{\tilde{\pi}}_\pi & -\mathrm{e}^{\frac{2\varphi}{n\alpha}}\rho_{\tilde{\lambda}\tilde{\pi}}A^{\tilde{\lambda}}_\lambda \\ -\mathrm{e}^{\frac{2\varphi}{n\alpha}}\rho_{\tilde{\lambda}\tilde{\pi}}A^{\tilde{\pi}}_\pi & -\mathrm{e}^{\frac{2\varphi}{n\alpha}}\rho_{\tilde{\lambda}\tilde{\pi}} \end{bmatrix} ,
\end{aligned} \tag{2.3}
$$

where $\rho_{\tilde{\lambda}\tilde{\pi}} \equiv \Phi^{\tilde{a}}_{\tilde{\lambda}}\Phi^{\tilde{a}}_{\tilde{\pi}}$ satisfies $\det\rho_{\tilde{\lambda}\tilde{\pi}} = 1$, *i.e.*, $\rho_{\tilde{\lambda}\tilde{\pi}}$ is the unimodular part of the internal metric $g_{\tilde{\lambda}\tilde{\pi}}$, and $\alpha = \sqrt{\frac{n+2}{n}}$.

To obtain the effective 4-d Lagrangian density, one imposes "the right invariance"[14] of the $(4+n)$-d metric $g_{\Lambda\Pi}$ under the action of an isometry of the internal space. These constraints determine the dependence of the metric components on the internal coordinates and the $(4+n)$-d Einstein Lagrangian density (2.1) becomes independent of the internal coordinates, provided proper conditions on the isometry group are imposed. After a trivial integration over the internal coordinates the 4-d Lagrangian density is of the following form (see for example Eq.(8) of Ref.14):

$$\mathcal{L} = -\frac{1}{2}\sqrt{-g}[\mathcal{R} + \mathrm{e}^{-\alpha\varphi}\mathcal{R}_K + \frac{1}{4}\mathrm{e}^{\alpha\varphi}\rho_{\tilde{\alpha}\tilde{\beta}}F^{\tilde{\alpha}}_{\mu\nu}F^{\tilde{\beta}\mu\nu} - \frac{1}{2}\partial_\mu\varphi\partial^\mu\varphi - \frac{1}{4}\rho^{\tilde{\alpha}\tilde{\beta}}\rho^{\tilde{\gamma}\tilde{\delta}}(D_\mu\rho_{\tilde{\alpha}\tilde{\gamma}})(D^\mu\rho_{\tilde{\beta}\tilde{\delta}}) + \lambda(\det\rho_{\tilde{\alpha}\tilde{\beta}} - 1)] \ , \tag{2.4}$$

where $\mathcal{R}_K$ is the Ricci scalar* defined in terms of the unimodular part $\rho_{\tilde{\alpha}\tilde{\beta}}$ of the internal metric, $F^{\tilde{\alpha}}_{\mu\nu} \equiv \partial_\mu A^{\tilde{\alpha}}_\nu - \partial_\nu A^{\tilde{\alpha}}_\mu - g f^{\tilde{\alpha}}_{\tilde{\beta}\tilde{\gamma}} A^{\tilde{\beta}}_\mu A^{\tilde{\gamma}}_\nu$, where $f^{\tilde{\alpha}}_{\tilde{\beta}\tilde{\gamma}}$ is the structure constant for the internal isometry group and g is the gauge coupling constant of the isometry group, is the field strength of the gauge field $A^{\tilde{\alpha}}_\mu$, $D_\mu\rho_{\tilde{\alpha}\tilde{\beta}} = \partial_\mu\rho_{\tilde{\alpha}\tilde{\beta}} - f^{\tilde{\delta}}_{\tilde{\gamma}\tilde{\beta}}A^{\tilde{\gamma}}_\mu\rho_{\tilde{\alpha}\tilde{\delta}}$ is the corresponding gauge covariant derivative and λ is the Lagrangian multiplier. The 4-d gravitational constant κ_4 has been set equal to 1 and $\alpha = \sqrt{\frac{n+2}{n}}$ specifies the coupling constant of the dilaton φ to the gauge fields in the gauge field kinetic energy terms, *i.e.*, $\mathrm{e}^{\alpha\varphi}\rho_{\tilde{\alpha}\tilde{\beta}}F^{\tilde{\alpha}}_{\mu\nu}F^{\tilde{\beta}\mu\nu}$.

The minimum energy configuration for a given topological charge satisfies the Killing spinor equations which are obtained by setting supersymmetric variations of fermionic fields equal to zero. To obtain the Killing spinor equations defined by the Eq.(2.2), one has to perform the dimensional reduction of the fermionic degrees of freedom as well. One has to choose the proper representation for $(4+n)$-d gamma matrices Γ^A, which satisfy the $SO(3+n,1)$ Clifford algebra. They are given in the following form:

$$\Gamma^a = \gamma^a \otimes I \qquad\qquad \Gamma^{\tilde{a}} = \gamma^5 \otimes \gamma^{\tilde{a}} \ , \tag{2.5}$$

where γ^a and $\gamma^{\tilde{a}}$ respectively satisfy $SO(3,1)$ and $SO(n)$ Clifford algebras, I is the $(2^{[\frac{n}{2}]} \times 2^{[\frac{n}{2}]})$-identity matrix and $\gamma^5 \equiv i\gamma^0\gamma^1\gamma^2\gamma^3$. Correspondingly, one has to split[15] the $(4+n)$-d spinor index $\mathbf{A}$ into $\mathbf{A} = (\mathbf{a}, \mathbf{m})$ where $\mathbf{a} = 1, ..., 4$ and $\mathbf{m} = 1, ..., 2^{[\frac{n}{2}]}$, *i.e.*, $\varepsilon^{\mathbf{A}} \equiv \varepsilon^{(\mathbf{a},\mathbf{m})}$, thus corresponding to $2^{[\frac{n}{2}]}$ copies of 4-component spinors in $D = 4$.

Then, the gravitino transformation(s) (2.2) are expressed in terms of 4-d quantities as follows:

$$\begin{aligned}\delta\psi^{\mathbf{m}}_\mu &= \partial_\mu\varepsilon^{\mathbf{m}} + \frac{1}{4}\{\omega_{\mu ab} - \frac{1}{\alpha}\mathrm{e}^c_\mu\eta_{c[a}\mathrm{e}^\nu_{b]}\partial_\nu\varphi + \frac{1}{2}\mathrm{e}^{\alpha\varphi}\rho_{\tilde{\mu}\tilde{\lambda}}A^{\tilde{\mu}}_\mu F^{\tilde{\lambda}}_{ab}\}(\gamma^{ab})\varepsilon^{\mathbf{m}} \\ &+ \frac{1}{2}\mathrm{e}^{\frac{\alpha}{2}\varphi}\sum_{\tilde{b}}\{\frac{1}{2}\mathrm{e}^c_\mu\Phi_{\tilde{\lambda}\tilde{b}}F^{\tilde{\lambda}}_{ca} - \frac{1}{n\alpha}\Phi_{\tilde{\mu}\tilde{b}}A^{\tilde{\mu}}_\mu\partial_a\varphi - \frac{1}{2}A^{\tilde{\mu}}_\mu D_a\Phi_{\tilde{\mu}\tilde{b}} - \frac{1}{2}A^{\tilde{\mu}}_\mu\Phi^{\tilde{c}}_{\tilde{\mu}}\Phi^{\tilde{\lambda}}_{\tilde{b}}D_a\Phi_{\tilde{\lambda}\tilde{c}}\}(\gamma^{a5})\varepsilon^{\mathbf{m}}_{\tilde{b}} \\ &+ \frac{1}{2}(\Phi^{\tilde{\lambda}}_{\tilde{a}}D_\mu\Phi_{\tilde{\lambda}\tilde{b}} - \Phi^{\tilde{\lambda}}_{\tilde{b}}D_\mu\Phi_{\tilde{\lambda}\tilde{a}})\varepsilon^{\mathbf{m}}_{\tilde{a}\tilde{b}}\end{aligned} \tag{2.6}$$

---

* This term describes the self-interactions among scalar fields and vanishes for an Abelian isometry group, *i.e.*, if the internal space is an n-torus.

$$\delta\psi_{\tilde{\mu}}^{\mathbf{m}} = \frac{1}{8}\mathrm{e}^{\alpha\varphi}\rho_{\tilde{\mu}\tilde{\lambda}}F_{ab}^{\tilde{\lambda}}(\gamma^{ab})\varepsilon^{\mathbf{m}} - \frac{1}{2}\mathrm{e}^{\frac{\alpha}{2}\varphi}\{\frac{1}{n\alpha}\Phi_{\tilde{\mu}\tilde{b}}\partial_a\varphi + \frac{1}{2}(D_a\Phi_{\tilde{\mu}\tilde{b}} + \Phi_{\tilde{\mu}}^{\tilde{c}}\Phi_{\tilde{b}}^{\tilde{\lambda}}D_a\Phi_{\tilde{\lambda}\tilde{c}})\}(\gamma^{a5})\tilde{\varepsilon}_{\tilde{b}}^{\mathbf{m}}$$
$$- \frac{1}{2}(f_{\tilde{\beta}\tilde{\gamma}}^{\tilde{\alpha}}\rho_{\tilde{\alpha}\tilde{\mu}}\Phi_{\tilde{a}}^{\tilde{\beta}}\Phi_{\tilde{b}}^{\tilde{\gamma}} - f_{\tilde{\beta}\tilde{\mu}}^{\tilde{\alpha}}\Phi_{\tilde{a}}^{\tilde{\beta}}\Phi_{\tilde{\alpha}\tilde{b}} + f_{\tilde{\beta}\tilde{\mu}}^{\tilde{\alpha}}\Phi_{\tilde{b}}^{\tilde{\beta}}\Phi_{\tilde{\alpha}\tilde{a}})\tilde{\varepsilon}_{\tilde{a}\tilde{b}}^{\mathbf{m}} \quad , \tag{2.7}$$

where the spinors with tilde are defined as $\tilde{\varepsilon}_{\tilde{b}}^{\mathbf{m}} \equiv (\gamma^{\tilde{b}})_{\mathbf{n}}^{\mathbf{m}}\varepsilon^{\mathbf{n}}$, $\tilde{\varepsilon}_{\tilde{a}\tilde{b}}^{\mathbf{m}} \equiv (\gamma^{\tilde{a}\tilde{b}})_{\mathbf{n}}^{\mathbf{m}}\varepsilon^{\mathbf{n}}$, $\Phi_{\tilde{a}}^{\tilde{\pi}} = (\Phi^{-1})_{\tilde{\pi}}^{\tilde{a}}$ and the index $\tilde{a}$ in $\Phi_{\tilde{\mu}}^{\tilde{a}}$ is lowered by $\eta_{\tilde{a}\tilde{b}}$.

2.2. Killing spinor equations and constraints on spinors

We shall now address the Killing spinor constraints for the 4-d Abelian charged black hole Ansätze, *i.e.*, we choose the isometry group of the internal space to be Abelian, *i.e.*, $U(1)^n$. In this case, with a proper choice of gauge one has

$$\rho_{\tilde{\alpha}\tilde{\beta}} = \mathrm{diag}(\rho_1, ..., \rho_{n-1}, \prod_{k=1}^{n-1}\rho_k^{-1}) \ . \tag{2.8}$$

The spherically symmetric Ansatz for the 4-d space-time metric† is given by

$$ds^2 = g_{\mu\nu}dx^\mu dx^\nu = \lambda(r)dt^2 - \lambda^{-1}(r)dr^2 - R(r)(d\theta^2 + \sin^2\theta d\phi^2) \tag{2.9}$$

and the scalar fields φ and ρ_i depend on the radial coordinate r, only. The electric and magnetic fields in spherically symmetric space-time defined by (2.9) take the form:

$$F_{tr}^{\tilde{a}} = E^{\tilde{a}}(r) \ , \qquad F_{\theta\phi}^{\tilde{a}} = P^{\tilde{a}}\sin\theta \ , \ \tilde{a} = 4, ..., (n+3) \ , \tag{2.10}$$

where $E^{\tilde{a}}(r) = \frac{\tilde{Q}^{\tilde{a}}}{R\mathrm{e}^{\alpha\varphi}\rho_i}(i \equiv \tilde{a} - 3 = 1, ..., n)$ and $\tilde{Q}^{\tilde{a}}$ is related to the physical electric charge‡ by $Q^{\tilde{a}} = \mathrm{e}^{-\alpha\varphi_\infty}\rho_{i\infty}^{-1}\tilde{Q}^{\tilde{a}}$. Here the subscript ∞ denotes the asymptotic values of fields at infinity.

---

† The transformation to the flat tangent frame defined by the metric $\eta_{ab} = \mathrm{diag}(1, -1, -1, -1)$ is achieved by the Vierbein components of the form:

$$\mathrm{e}_t^{\hat{t}} = \lambda^{1/2} \qquad \mathrm{e}_\theta^{\hat{\theta}} = R^{1/2} \qquad \mathrm{e}_\phi^{\hat{\phi}} = R^{1/2}\sin\theta \qquad \mathrm{e}_r^{\hat{r}} = \lambda^{-1/2} \quad ,$$

where the hat denotes the tangent space index and the flat space gamma matrices are ordered as $\gamma^{\hat{t}} = \gamma^0, ..., \gamma^{\hat{r}} = \gamma^3$.

‡ The physical electric charge Q is defined in terms of the asymptotic limit of the electric field, *i.e.*, $E \sim \frac{Q}{r^2}$ in the limit $r \to \infty$.

Given the above Ansätze, the Killing spinor equations which are obtained by setting (2.7) equal to zero give rise to the constraints between the lower two-component spinors $\varepsilon_\ell^{\mathbf{m}}$ and the upper two-component spinors $\varepsilon_u^{\mathbf{m}}$ ($(\varepsilon^{\mathbf{m}})^T \equiv (\varepsilon_u^{\mathbf{m}}, \varepsilon_\ell^{\mathbf{m}})^T$)):

$$(\gamma^{\tilde{a}})^{\mathbf{m}}_{\mathbf{n}}\ \varepsilon_\ell^{\mathbf{n}} = \eta_{\tilde{a}} e^{i\theta_{\tilde{a}}(r)} \varepsilon_u^{\mathbf{m}} \ ; \quad \eta_{\tilde{a}} = \pm 1 \quad , \tag{2.11}$$

where the phase $\theta_{\tilde{a}}$ is defined as

$$e^{i\theta_{\tilde{a}}(r)} \equiv \left[(\frac{\tilde{Q}^{\tilde{a}}}{e^{\alpha\varphi}\rho_i} - iP^{\tilde{a}})/(\frac{\tilde{Q}^{\tilde{a}}}{e^{\alpha\varphi}\rho_i} + iP^{\tilde{a}}) \right]^{\frac{1}{2}} \quad . \tag{2.12}$$

The constraints of the type (2.11) can be satisfied if the phase difference for the pair of different $\tilde{a}$'s is $\frac{\pi}{2}$: $\theta_{\tilde{a}} - \theta_{\tilde{b}} = \pm\frac{\pi}{2}$, $\tilde{a} \neq \tilde{b}$. This implies that only two $U(1)$ gauge factors can survive in order to have non-zero Killing spinors.

Without loss of generality, we choose two $U(1)$ factors to be associated with the last two internal coordinates. Then, the Killing spinor equations arising from Eq.(2.6) along with Eq.(2.11) give rise to the following first order differential equations:

$$(\lambda' - \frac{1}{\alpha}\lambda\partial_r\varphi) - e^{-\frac{\alpha}{2}\varphi}\lambda^{\frac{1}{2}}R^{-1}\sum_{\tilde{a}=n+2}^{n+3}\eta_{\tilde{a}}\tilde{Q}^{\tilde{a}}e^{i\theta_{\tilde{a}}(r)}\rho_i^{-\frac{1}{2}} = 0 \ , \tag{2.13}$$

$$2 = \sqrt{\lambda}R(\frac{R'}{R} - \frac{2}{\alpha}\partial_r\varphi) + ie^{\frac{\alpha}{2}\varphi}R^{-\frac{1}{2}}\sum_{\tilde{a}=n+2}^{n+3}\eta_{\tilde{a}}P^{\tilde{a}}e^{i\theta_{\tilde{a}}(r)}\rho_i^{\frac{1}{2}} \ . \tag{2.14}$$

These equations along with (2.12), then, imply that if one $U(1)$ gauge factor has a non-zero electric (or magnetic) charge, then the other $U(1)$ factor should have no electric (or magnetic) charge. Furthermore, the constraint on the phase difference, $\theta_{n+2} - \theta_{n+3} = \pm\frac{\pi}{2}$, for different $U(1)$ factors requires that each $U(1)$ factor should have a different type of monopole charge. *Therefore, supersymmetric spherical solutions choose the vacuum where the isometry group of the internal space is broken down to* $U(1)_E \times U(1)_M$.

One can show[10] that the constraints on spinors of type (2.11) are compatible with the reality and chirality conditions on Dirac spinors in each dimension $5 \leq D \leq 11$.

2.3. Bogomol'nyi Bound

In this section, we derive the Bogomol'nyi bound on the ADM mass[16] of black holes in the Abelian KK theories. For that purpose, one has to introduce the Nester-like two-form[17]:

$$\hat{E}^{\mu\nu} \equiv \frac{1}{2}\bar{\varepsilon}\Gamma^{\mu\nu\rho}\hat{\nabla}_\rho\varepsilon + c.c. \quad . \tag{2.15}$$

The derivation consists of evaluating the surface integral of the Nester's two-form (2.15), which is related through the Stokes theorem to the volume integral of its covariant derivative in the following way:

$$\int_{\Sigma} dS_\mu e\nabla_\nu \hat{E}^{\mu\nu} = \int_{\Sigma} dS_\mu \partial_\nu (e\hat{E}^{\mu\nu}) = \frac{1}{2}\int_{\partial\Sigma} dS_{\mu\nu} e\hat{E}^{\mu\nu} \quad , \tag{2.16}$$

where Σ is a space-like hypersurface with the boundary $\partial\Sigma$ at spatial infinity.

One has to note that $\delta\psi_{\mu 4} \equiv \hat{\nabla}_\rho\varepsilon$ in (2.15) is the supersymmetry transformation for the 4-d "physical gravitino(s)"[18]. The 4-d physical gravitino is defined in terms of the components $\psi^{\mathbf{m}}_\mu$ and $\psi^{\mathbf{m}}_{\tilde{\mu}}$ of the $(4+n)$-d gravitino ψ_Π as $\psi^{\mathbf{m}}_{\mu 4} \equiv \mathrm{e}^{-\frac{1}{2\alpha}\varphi} e^a_\mu [\psi^{\mathbf{m}}_\Lambda E^\Lambda_a + \gamma_a\gamma_5(\gamma^{\tilde{a}})^{\mathbf{m}}_{\mathbf{n}}\psi^{\mathbf{n}}_\Lambda E^\Lambda_{\tilde{a}}]$ and additionally one has to redefine the 4-d fermionic fields (dilatinos) as $\chi^{\mathbf{m}}_{\tilde{a}} \equiv \mathrm{e}^{-\frac{\varphi}{2\alpha}}\psi^{\mathbf{m}}_\Lambda E^\Lambda_{\tilde{a}}$. Then, the surface and volume integrals in (2.16) are evaluated to be

$$\int_{\partial\Sigma} dS_{\mu\nu} e\hat{E}^{\mu\nu} = (\bar{\varepsilon}_\infty\gamma^\mu\varepsilon_\infty)P^{ADM}_\mu + \sum_{\tilde{a}=4}^{n+3} \bar{\varepsilon}_\infty \mathrm{e}^{\frac{\alpha}{2}\varphi_\infty}\rho^{\frac{1}{2}}_{i\infty}(iP^{\tilde{a}} - \gamma_5 Q^{\tilde{a}}) \otimes \gamma^{\tilde{a}}\varepsilon_\infty \quad , \tag{2.17}$$

$$\begin{aligned}\int_{\Sigma} dS_\mu e\nabla_\nu \hat{E}^{\mu\nu} = \int_{\Sigma} dS_\mu e[&-\frac{1}{2}(\overline{\hat{\nabla}_\nu\varepsilon^{\mathbf{m}}})\gamma^{\mu\nu\rho}\hat{\nabla}_\rho\varepsilon^{\mathbf{m}} + \frac{1}{2}(\overline{\delta\chi^{\mathbf{m}}_{\tilde{a}}})[\frac{1}{2}(\gamma^{\tilde{a}}\gamma^{\tilde{b}})^{\mathbf{m}}_{\mathbf{n}} + \eta^{\tilde{a}\tilde{b}}\delta^{\mathbf{m}}_{\mathbf{n}}]\gamma^\mu(\delta\chi^{\mathbf{n}}_{\tilde{b}}) \\ &+ (G^{\mu\nu} - T^{\mu\nu})(\overline{\varepsilon^{\mathbf{m}}}\gamma_\nu\varepsilon^{\mathbf{m}}) + c.c.] \quad ,\end{aligned} \tag{2.18}$$

where P^{ADM}_μ is the ADM 4-momentum and the physical charges of the system are defined as $Q^{\tilde{a}} \equiv \frac{1}{2}\int_{\partial\Sigma} dS_{\mu\nu}F^{\tilde{a}\mu\nu}$ and $P^{\tilde{a}} \equiv \frac{1}{2}\int_{\partial\Sigma} dS_{\mu\nu} e(\mathrm{e}^{-\alpha\varphi}\rho_i^{-1}\tilde{F}^{\tilde{a}\mu\nu})$, where $\tilde{F}^{\tilde{a}\mu\nu} \equiv \frac{1}{2}e^{-1}\mathrm{e}^{\alpha\varphi}\rho_i\varepsilon^{\mu\nu\alpha\beta}$ $\times F^{\tilde{a}}_{\alpha\beta}$ is the dual electromagnetic field strength tensor and $e = \sqrt{-g}$.

The Bogomol'nyi bound on the ADM mass M follows from the fact that the volume integral (2.18) is semi-positive definite provided spinors ε satisfy the (modified) Witten's condition, *i.e.*, $n \cdot \hat{\nabla}\varepsilon = 0$ (n is the 4-vector normal to Σ). In the case of non-supersymmetric $U(1) \times U(1)$ configurations, *i.e.*, $\hat{\nabla}_\mu\varepsilon^{\mathbf{m}} \neq 0$ and $\delta\chi^{\mathbf{m}}_{\tilde{a}} \neq 0$, the surface integral (2.17) is positive if and only if all the eigenvalues of the Hermitian matrix sandwiched between the spinor ε_∞ are positive. This leads to the following inequality*

$$M > e^{\frac{\alpha}{2}\varphi_\infty}\sqrt{(\rho^{\frac{1}{2}}_{(n-1)\infty}|P^{n+2}| + \rho^{\frac{1}{2}}_{n\infty}|Q^{n+3}|)^2 + (\rho^{\frac{1}{2}}_{(n-1)\infty}|Q^{n+2}| + \rho^{\frac{1}{2}}_{n\infty}|P^{n+3}|)^2} \quad . \tag{2.19}$$

For the supersymmetric $U(1)_E \times U(1)_M$ configuration, the constraints (2.11) on the upper- and lower-component spinors select out electric and magnetic charges from different terms in (2.17). In this case the ADM mass of the supersymmetric configuration is of the form

$$M_{ext} = e^{\frac{\alpha}{2}\varphi_\infty}(\rho^{\frac{1}{2}}_{n\infty}|P^{n+2}| + \rho^{\frac{1}{2}}_{(n-1)\infty}|Q^{n+3}|) \quad . \tag{2.20}$$

2.4. Supersymmetric $U(1)_E \times U(1)_M$ Solutions

The supersymmetric solutions satisfy self-dual equations which are obtained from the Killing spinor equations $\delta\psi^{\mathbf{m}}_\mu = 0 = \delta\psi^{\mathbf{m}}_{\tilde{\mu}}$ along with the constraints on spinors $\varepsilon^{\mathbf{m}}$. The angular dependence[19] of the spinors $\varepsilon^{\mathbf{m}}$ is determined from $\mu = \theta, \phi$ components of the Killing spinor equations to be

$$(\varepsilon^{1,\mathbf{m}}_{\ell,u}, \varepsilon^{2,\mathbf{m}}_{\ell,u}) = e^{i\sigma^2\theta/2}e^{i\sigma^3\phi/2}(a^{1,\mathbf{m}}_{\ell,u}(r), a^{2,\mathbf{m}}_{\ell,u}(r)) \quad . \tag{2.21}$$

Choosing the electromagnetic field associated with the second to the last coordinate to be electric and that corresponding to the last coordinate to be magnetic, one obtains the self-dual equations[10] satisfied by the fields and metric components. One may simplify the problem of solving rather complicated coupled first order differential equations by deducing from the equations that metric components and scalar fields are related in the following way:

$$\lambda R = (r - r_H)^2 \quad , \tag{2.22}$$

$$\rho_i = \rho_{i\infty}e^{-\frac{2}{n\alpha}(\varphi-\varphi_\infty)} \quad , \quad i = 1, \dots, n-2 \quad , \tag{2.23}$$

$$\rho_{n-1} = \rho_{(n-1)\infty}\lambda e^{-\frac{2-n}{n\alpha}(\varphi-\varphi_\infty)} \quad , \tag{2.24}$$

$$\lambda = \frac{\eta_e \mathbf{Q} e^{-\frac{1}{\alpha}(\varphi-\varphi_\infty)} + \eta_m \mathbf{P} e^{\frac{1}{\alpha}(\varphi-\varphi_\infty)}}{\eta_e \mathbf{Q} + \eta_m \mathbf{P}} \quad , \tag{2.25}$$

where r_H is the event horizon, $\mathbf{Q} \equiv e^{\frac{\alpha}{2}\varphi_\infty}\rho^{\frac{1}{2}}_{n\infty}Q$ and $\mathbf{P} \equiv e^{\frac{\alpha}{2}\varphi_\infty}\rho^{\frac{1}{2}}_{(n-1)\infty}P$.

---

* It is possible to generalize the Bogomol'nyi bound for non-supersymmetric configurations with $U(1)^n$ ($n > 2$), however, the expressions are cumbersome.

Substituting the above relations (2.22) – (2.25) into the self-dual equations, one obtains an ordinary differential equation for the dilaton field φ, which is solved to be

$$e^{\frac{2}{\alpha}(\varphi-\varphi_\infty)} = \frac{r-|\mathbf{P}|}{r-|\mathbf{Q}|} \ , \tag{2.26}$$

where one has to identify $r_H = \eta_e \mathbf{Q} - \eta_m \mathbf{P}$ and choose the signs of $\eta_{e,m}$ so that $\eta_e \mathbf{Q} = |\mathbf{Q}|$ and $-\eta_m \mathbf{P} = |\mathbf{P}|$. Then, the explicit expressions for the 4-d metric components and electric fields are of the from:

$$\lambda = \frac{r-|\mathbf{Q}|-|\mathbf{P}|}{(r-|\mathbf{Q}|)^{\frac{1}{2}}(r-|\mathbf{P}|)^{\frac{1}{2}}} \ , \tag{2.27}$$

$$R = r^2(1-\frac{|\mathbf{Q}|+|\mathbf{P}|}{r})(1-\frac{|\mathbf{Q}|}{r})^{\frac{1}{2}}(1-\frac{|\mathbf{P}|}{r})^{\frac{1}{2}} \ , \tag{2.28}$$

$$E(r) = \frac{Q}{(r-|\mathbf{P}|)^2} \ . \tag{2.29}$$

In addition, the spinor components are of the form:

$$a_u^{\mathbf{m}}(r) = a_{u\infty}^{\mathbf{m}}\left(\frac{r-|\mathbf{Q}|-|\mathbf{P}|}{r-|\mathbf{P}|}\right)^{\frac{1}{4}} \ . \tag{2.30}$$

One of the surprising results is that the space-time structure described by the metric components λ and R is independent of the dimensionality n of the compactified space. This result can be understood by noting that Eq.(2.23) implies that $n-2$ combinations of scalar fields are constant, *i.e.*, $\rho_i e^{\frac{2}{n\alpha}\varphi} = const.$ $(i = 1, ..., n-2)$. Therefore, one can define new scalar fields $\Phi \equiv \frac{\sqrt{2}}{\alpha}\varphi$ and $\chi_i \equiv \frac{1}{\sqrt{2}}[\ln \rho_i + \frac{2-n}{n\alpha}\varphi]$ $(i = n-1, n)$ in terms of which the Lagrangian density (2.1) reduces effectively to that of 6-d KK theory:

$$\begin{aligned}\tilde{\mathcal{L}} = -\frac{1}{2}\sqrt{-g}[\mathcal{R} &- \frac{1}{2}\partial_\mu\Phi\partial^\mu\Phi - \frac{1}{2}\partial_\mu\chi_{n-1}\partial^\mu\chi_{n-1} - \frac{1}{2}\partial_\mu\chi_n\partial^\mu\chi_n \\ &+ \frac{1}{4}e^{\sqrt{2}(\Phi+\chi_{n-1})}F^{n+2}_{\mu\nu}F^{n+2\ \mu\nu} + \frac{1}{4}e^{\sqrt{2}(\Phi+\chi_n)}F^{n+3}_{\mu\nu}F^{n+3\ \mu\nu}] \ ,\end{aligned} \tag{2.31}$$

with the constraint $\chi_{n-1}+\chi_n = const.$. If either of charges vanishes, *e.g.*, $P = 0$, one linear combination of fields, *e.g.*, $\sqrt{\frac{1}{3}}(\Phi - 2\chi_n)$, becomes constant and the other orthogonal linear combination, *e.g.*, $\sqrt{\frac{2}{3}}(\Phi + \chi_n)$, is the dilaton field of 5-d KK theory[20,21].

3. Non-Supersymmetric Solutions

In the previous Chapter, we showed that supersymmetric 4-d KK black holes can be obtained if and only if the isometry group of the internal space is broken down to $U(1)_E \times U(1)_M$. As an initial attempt to address non-extreme black holes, we shall therefore confine our attention to the $U(1)_E \times U(1)_M$ non-extreme black holes, only.* In this case, the Lagrangian density is effectively that of 6-d KK theory (Eq.(2.31)).

In order to obtain such non-extreme solutions, one has to solve the Einstein equations and Euler-Lagrangian equations for scalar and electric fields arising from the Lagrangian density (2.31). These equations can be written in the following form:

$$(\lambda' R)' + \frac{1}{2}R^{-1}e^{-\sqrt{2}\varphi}\rho\tilde{Q}^2 + \frac{1}{2}R^{-1}e^{\sqrt{2}\varphi}\rho P^2 \tag{3.1}$$

$$(\lambda R)'' = 2 \tag{3.2}$$

$$R^{-1}e^{\sqrt{2}\varphi}\rho P^2 = \frac{1}{\sqrt{2}}(\lambda R\varphi')' + (\lambda R\rho^{-1}\rho')' \tag{3.3}$$

$$R^{-1}e^{-\sqrt{2}\varphi}\rho\tilde{Q}^2 = -\frac{1}{\sqrt{2}}(\lambda R\varphi')' + (\lambda R\rho^{-1}\rho')' \quad , \tag{3.4}$$

where $\rho \equiv e^{\chi_{n-1}}$ and the prime denotes differentiation with respect to r. Recall, $\tilde{Q} = e^{\sqrt{2}\varphi_\infty}\rho_\infty Q$.

The above equations imply the following relations between the metric components and the scalar field ρ:

$$\lambda R = (r - r_+)(r - r_+ + 2\beta) \tag{3.5}$$

$$\lambda = \frac{\rho}{\rho_\infty}\left(\frac{r - r_+}{r - r_+ + 2\beta}\right) \quad , \tag{3.6}$$

where r_+ is defined to be the outermost horizon and $\beta > 0$ is the non-extremality parameter.† Substituting the relations (3.5) and (3.6) into equations (3.3) and (3.4), one obtains ordinary

---

* Configurations with more general charge configurations will be addressed elsewhere[23]. Preliminary results indicate that other more general charge combinations, *e.g.*, configurations with $n \geq 2$ electric charges, only, do not exist.

† Note, the role of the non-extremality parameter β is very similar to the one used in describing non-extreme supergravity walls[9,22].

differential equations each for the following combinations of fields: $X \equiv \sqrt{2}\varphi + 2\ln\lambda - \sqrt{2}\varphi_\infty$ and $Y \equiv -\sqrt{2}\varphi + 2\ln\lambda + \sqrt{2}\varphi_\infty$. Solving the resultant equations and using the relations (3.5) and (3.6), one obtains the following explicit solutions with regular horizons:

$$\lambda = \frac{r - r_+}{(r - r_+ + P')^{1/2}(r - r_+ + Q')^{1/2}} \tag{3.7}$$

$$R = r^2(1 - \frac{r_+ - 2\beta}{r})(1 - \frac{r_+ - P'}{r})^{\frac{1}{2}}(1 - \frac{r_+ - Q'}{r})^{\frac{1}{2}} \tag{3.8}$$

$$e^{\sqrt{2}(\varphi - \varphi_\infty)} = \frac{r - r_+ + Q'}{r - r_+ + P'} \tag{3.9}$$

$$\rho_{n-1} = \rho_{n-1\,\infty}\frac{r - r_+ + 2\beta}{(r - r_+ + P')^{1/2}(r - r_+ + Q')^{1/2}} \quad , \tag{3.10}$$

where $r_+ = \beta + \frac{|\mathbf{P}|\sqrt{\mathbf{P}^2+\beta^2} - |\mathbf{Q}|\sqrt{\mathbf{Q}^2+\beta^2}}{|\mathbf{P}| - |\mathbf{Q}|}$, $P' = \beta + \sqrt{\mathbf{P}^2 + \beta^2}$ and $Q' = \beta + \sqrt{\mathbf{Q}^2 + \beta^2}$, while the ADM mass of the configurations is of the form:

$$M = 2\beta + \sqrt{\mathbf{P}^2 + \beta^2} + \sqrt{\mathbf{Q}^2 + \beta^2} \quad . \tag{3.11}$$

Note that in the limit $\beta \to 0$ the above expressions reduce to those with $r_+ = r_H = |\mathbf{P}| + |\mathbf{Q}|$, $P' = |\mathbf{P}|$, and $Q' = |\mathbf{Q}|$, and thus correspond to the supersymmetric solutions described in the previous Chapter.

4. Global space-time and thermal properties

We first describe the singularity structure of 4-d space-time defined by the metric coefficients (3.7) and (3.8). For the non-extreme solutions, *i.e.*, $\beta > 0$, there is a space-like singularity at $r = r_+ - 2\beta$ which is hidden behind a horizon at $r = r_+$. The global space-time of the non-extreme solutions is that of the Schwarzschild black hole (see Fig. 1). In the limit $\beta \to 0$, *i.e.*, when the supersymmetric (extreme) limit is reached, the singularity becomes null, *i.e.*, it coincides with the horizon (See Fig. 2a). In the limit that either Q or P is zero, the singularity becomes naked (See Fig. 2b).

We now address thermal properties of these solutions. The Hawking temperature[24], which can be calculated by identifying the inverse of the imaginary time period[25] of a functional path integral, turns out to be $T_H = \frac{|\lambda'(r_+)|}{4\pi}$. With the explicit solution (3.7) for λ one obtains:

$$T_H = \frac{1}{4\pi[\beta + (\mathbf{Q}^2 + \beta^2)^{\frac{1}{2}}]^{\frac{1}{2}}[\beta + (\mathbf{P}^2 + \beta^2)^{\frac{1}{2}}]^{\frac{1}{2}}} \quad . \tag{4.1}$$

As β decreases T_H increases and reaches the upper bound $T_{H\,ext} = 1/4\pi|\mathbf{PQ}|$ in the extreme limit $\beta \to 0$. In the extreme limit when either $\mathbf{Q}$ or $\mathbf{P}$ becomes zero, T_H is infinite.

The entropy S of the system, determined as $S = \frac{1}{4}\times$ (the surface area of the event horizon)[26], is of the following form:

$$S = 2\pi\beta[\beta + (\mathbf{Q}^2 + \beta^2)^{\frac{1}{2}}]^{\frac{1}{2}}[\beta + (\mathbf{P}^2 + \beta^2)^{\frac{1}{2}}]^{\frac{1}{2}} \quad . \tag{4.2}$$

where we have used the explicit solution (3.8) for R. For the non-extreme solution, the entropy is finite. However, once the extreme limit is reached the entropy becomes zero.

5. Conclusions

We presented recent developments in the study of 4-d black hole solutions in Kaluza-Klein (KK) theory. In particular, we summarized recent results[10] where all supersymmetric Abelian black holes of such a type have been found. It turns out that spherically symmetric Abelian configurations exist only when the internal isometry group is broken down to $U(1)_M \times U(1)_E$, *i.e.*, such configurations are dyonic black holes with both the electric Q and magnetic P charges necessarily arising from *different* $U(1)$ factors. In the limit when either Q or P becomes zero, such solutions correspond to the supersymmetric black holes of 5-d KK theory, first derived by Gibbons and Perry[3] .

Such black hole configurations present another examples of non-trivial 4-d vacuum configurations of an effective theory with $N \geq 2$ supergravity. Interestingly, such vacuum solutions possess the electric-magnetic duality of a new kind, *i.e.*, the duality associated with different $U(1)$ factors, which is a subject of further study.

We have also presented results for non-extreme $U(1)_E \times U(1)_M$ 4-d KK black holes. Those are configurations with mass $M > M_{ext}$, where M_{ext} is the mass of the corresponding supersymmetric (extreme) configuration. The solutions, parameterized in terms of non-extremality parameter $\beta > 0$, have a global space-time structure of Schwarzschild black holes, with the temperature T_H [entropy S] increasing [decreasing] as β decreases. In the

Figure 1

Figure 1 The Penrose diagram (in the (r, t) plane) for non-extreme $U(1)_E \times U(1)_M$ 4-d Kaluza-Klein black holes. The space-like singularity (jagged line) at $r = r_+ - 2\beta$ (β is the non-extremality parameter) is hidden behind the horizon (dashed line) at $r = r_+$.

Figure 2a

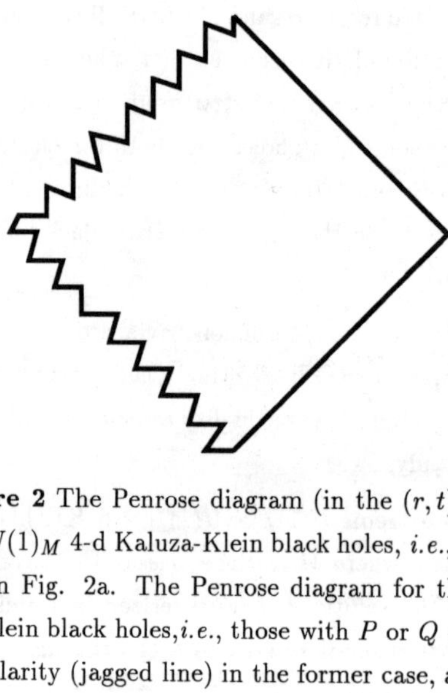

Figure 2b

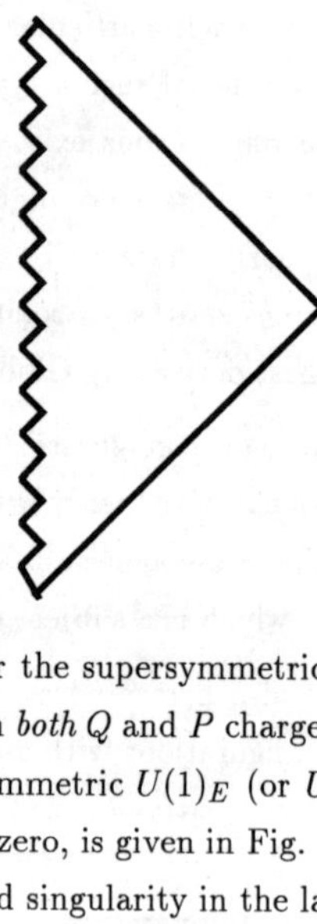

Figure 2 The Penrose diagram (in the (r, t) plane) for the supersymmetric (extreme) $U(1)_E \times U(1)_M$ 4-d Kaluza-Klein black holes, *i.e.*, those with *both* Q and P charges non-zero, is given in Fig. 2a. The Penrose diagram for the supersymmetric $U(1)_E$ (or $U(1)_M$) 4-d Kaluza-Klein black holes, *i.e.*, those with P or Q charge nonzero, is given in Fig. 2b. Note a null singularity (jagged line) in the former case, and a naked singularity in the latter one.

limit $\beta \to 0$, T_H $[S]$ reaches the upper [lower] limit $T_{H\,ext} = 1/(4\pi\sqrt{|PQ|})$ $[S = 0]$ and the space-like singularity becomes null. Notably, the extreme limit is reached smoothly.

Acknowledgments The work is supported by U.S. DOE Grant No. DOE-EY-76-02-3071, and the NATO collaborative research grant CGR 940870.

REFERENCES

1. G.W. Gibbons and C.M. Hull, Phys. Lett. **109B** (1982) 190; G.W. Gibbons, G. Horowitz, S.W. Hawking and M. Perry, Comm. Math. Phys. **88** (1983) 295.
2. R. Kallosh, A. Linde, T. Ortín, A. Peet and A. van Proeyen, Phys. Rev. **D46** (1992) 5278.
3. G.W. Gibbons and M. J. Perry, Nucl. Phys. **B248** (1984) 629.
4. M.J. Duff, R.R. Khuri, R. Minasian, J. Rahmfeld, Nucl. Phys. **B418** (1994) 195; M.J. Duff and J. Rahmfeld, CTP-TAMU-25-94 preprint (1994), hep-th # 9406105.
5. G.W. Gibbons, D. Kastor, L.A.J. London, P.K. Townsend and J. Traschen, Nucl. Phys. **B416** (1994) 880.
6. M. Cvetič, S. Griffies and S.-J. Rey, Nucl. Phys. **B381** (1992) 301; M. Cvetič and S. Griffies, Phys. Lett. **285B** (1992) 27; M. Cvetič, R. Davies, S. Griffies and H.H. Soleng, Phys. Rev. Lett. **70** (1993) 1191.
7. M. Cvetič and D. Youm, UPR-593-T preprint (1993), to be published in Phys. Rev. **D**; M. Cvetič, UPR-596-T preprint (1993), hep-th # 9312184, in the *Proceedings of the International Conference on High Energy Physics*, Marseille, July 21-28, 1993 (World Scientific 1994); UPR-600-T preprint (1994), hep-th # 9402089, to be published in Phys. Lett. **D**.
8. G.W. Gibbons, Nucl. Phys. **B207** (1982) 337; G.W. Gibbons and K. Maeda, Nucl. Phys. **B298** (1988) 741; D. Garfinkle, G. Horowitz and A. Strominger, Phys. Rev. **D43** (1991) 3140; C.F.E. Holzhey and F. Wilczek, Nucl. Phys. **B380** (1992) 447.
9. M. Cvetič and H.H. Soleng, UPR-635-T/CERN-TH.7475/94(1994) preprint, hep-th # 9411170.
10. M. Cvetič and D. Youm, UPR-623-T preprint (1994), hep-th # 9409119.

11. C. Wetterich, Nucl. Phys. **B211** (1983) 177; T. Kugo and P. Townsend, Nucl. Phys. **B221** (1983) 375.

12. T. Kaluza, *Sitz. Preuss. Akad. Wiss.* **K1** (1921) 966; O. Klein, Z. Phys. **37** (1926) 895; T. Appelquist and A. Chodos, Phys. Rev. **D28** (1983) 772; J. Scherk and J.H. Schwarz, Nucl. Phys. **B153** (1979) 61; M.J. Duff, NI-94-015 preprint (1994), hep-th # 9410046, talk given at The Oskar Klein Centenary Symposium, Stockholm, Sweden, September 19-21, 1994.

13. Y.M. Cho and P.G.O. Freund, Phys. Rev. **D12** (1975) 1711; Y.M. Cho and Pong Soo Jang, Phys. Rev. **D12** (1975) 3789; Y.M. Cho, J. Math. Phys. **16** (1975) 2029.

14. Y.M. Cho, Phys. Rev. **D35** (1987) 2628.

15. C. Wetterich, Nucl. Phys. **B222** (1983) 20.

16. R. Arnowitt, S. Deser and C.W. Misner, Phys. Rev. **122** (1961) 997; L.F. Abbott and S. Deser, Nucl. Phys. **B195** (1982) 76.

17. E. Witten, Comm. Math. Phys. **80** (1981) 381; J. Nester, Phys. Lett. **83A** (1981) 241.

18. E. Cremmer and B. Julia, Nucl. Phys. **B159** (1979) 141.

19. M. Cvetič, S. Griffies and S.-J. Rey, Nucl. Phys. **B389** (1993) 3.

20. P. Dobiasch and D. Maison, Gen. Rel. and Grav. **14**, 231 (1982).

21. D. Pollard, J. Phys. **A16** (1983) 565; D.J. Gross and M.J. Perry, Nucl. Phys. **B226** (1983) 29; R.D. Sorkin, Phys. Rev. Lett. **51** (1983) 87; G.W. Gibbons and D.L. Wiltshire, Ann. Phys. **167** (1986) 201.

22. M. Cvetič, S. Griffies and H.H. Soleng, Phys. Rev. Lett. **71** (1993) 670; M. Cvetič, S. Griffies and H.H. Soleng, Phys. Rev. **D48** (1993) 2613.

23. M. Cvetič and D. Youm, work in progress.

24. S. Hawking, Commun. Math. Phys. **43** (1975) 199.

25. G.W. Gibbons and S.W. Hawking, Phys. Rev. **D15** (1977) 2752; G.W. Gibbons and M.J. Perry, Proc. R. Soc. London **A358** (1978) 467.

26. J.D. Bekenstein, Phys. Rev. **D7** (1973) 2333; **D9** (1974) 3292; S.W. Hawking, Phys. Rev. Lett. **26** (1971) 1344; J.M. Bardeen, B. Carter and S.W. Hawking, Comm. Math. Phys. **31** (1973) 161; I. Moss, Phys. Rev. Lett. **69** (1992) 1852; M. Visser, Phys. Rev. **D48** (1993) 583; R. Kallosh, T. Ortín and A. Peet, **D47** (1993) 5400.

Round Table Discussion

U.S. and Polish Involvement in the
Future High Energy Experiments

Stephen Reucroft and Ryszard Sosnowski

After a few brief introductory comments by the mediators, several members of the audience made verbal comments. The most substantial of these have been written down and included in the following. In all cases the name of the contributor is identified.

The usefulness of this session was limited by two factors: as the last session on the last day, many of the workshop participants had left already or were in the process of leaving; the amount of time allotted (1 hour) was a bit short. We had just started to get going when we had to stop. Nevertheless, many interesting points were made, both about experience with past experiments and also about the possibilities for the future.

Reucroft LEP, LHC and TeVatron Timescales

The intention in this short presentation was to give some focus to the subsequent discussion by showing a few relevant parameters of the LEP, LHC and TeVatron programmes from today until $\sim$ 2010. These are as follows. LEP will continue to run at the Z^0 until the end of 1995, by which time each of the four experiments will have accumulated some 6 million Z^0 events. From $\sim$ 1996 until a little before 2000, the machine energy will be above the WW threshold. On a similar timescale, the TeVatron will continue to date data with the existing experiments and with the existing machine configuration until the end of 1995, with the intention of collecting ~ 100 pb^{-1}. After that, there will be a short period of fixed-target running. ¿From $\sim$ 1998 until $\sim$ 2003 or so, the machine will have the Main Injector and an instantaneous luminosity 10 times higher than at present. Both detectors, CDF and DØ , will be upgraded significantly and should collect at least 2000 pb^{-1} each. After that, the TeVatron will move to b-physics, with either new detectors or suitably modified CDF and/or DØ . At the time of the workshop, it was still hoped that LHC turn-on would be $\sim$ 2003, but with a more realistic expectation of 2005. Even 2005 is looking optimistic as I type this. Whatever is the reality, we all hope that LHC is fully operational by the year 2010 and with two experiments (CMS and ATLAS ?) both collecting data at the rate of ~ 100 fb^{-1} per year.

Polish and U.S. physicists are collaborating on experiments at all three of these facilities.

Settles U.S.-Polish Participation in an e^+e^- Linear Collider

It is surely true, as has been stated in this round-table discussion, that the LHC should have a high priority among the future projects involving common participation of Polish and US physicists. Also obtaining approval for the LHC should have the *highest* priority of all particle physicists at the moment, because the LHC will provide a window to new physics which would be disastrous to lose, for everyone involved in this business.

However, I would also like to underline the point that I made in the talk at this workshop on the study of supersymmetry at future colliders, namely that the LHC and a 500 GeV e^+e^- linear collider are truely complementary. Both are needed to map out the SUSY world if it exists: the hadron collider can discover and measure the squarks and gluinos predicted by many of the SUSY models, while the e^+e^- collider is needed for the sleptons, charginos, and neutralinos. If mother nature has not chosen to be supersymmetric, then the complementarity argument still holds, for studies of the top quark and for understanding electroweak-symmetry breaking. This conclusion is the result of numerous workshops held throughout the world over the past few years.

Therefore, in thinking about U.S.-Polish collaboration in common future projects, it would seem that a well-rounded physics program would include participation in the 500 GeV e^+e^- linear collider. There is a dream and a hope that this will be a world machine, and a first step in this direction has been taken with the formation of a World Collaboration to coordinate the R&D efforts underway at the various labs. This collaboration consists of institutions from Asia, the FSU, Europe, and America, and can be joined by any laboratory which is willing to contribute. The easiest way to do that in Europe is at DESY where there are two R&D projects to build prototype linear accelerators: one for a superconding machine, TESLA, and one for a warm machine, S-BAND. Both collaborations are international, with US participation, so that U.S.-Polish contacts can be enriched. Further, detector concepts are now starting to be developed, so that experimentalists interested in building detectors and doing the experiments could make an important contribuition by joining the discussion at this early date.

Let us hope that a way can be found to solve the technical challenge posed by the machine, and this is happening in the enormous R&D push proceeding at the moment, and to solve the unbelievably difficult political challenge posed by proventialism. Polish physicists who share these feelings for the priorities could help out enormously, by stating their mind about the politics, and I am sure they shall.

Janusz A. Zakrzewski
Isntitute of Experimental Physics
University of Warsaw

Contribution to the Round Table discussion

1. Polish research groups from Cracow and Warsaw have participated in the HERA project at DESY, Hamburg, since 1984. Polish phzysicists, engineers and technicians made important contribution both to the construction of the HERA storage ring and that of the detectors, H1 and ZEUS.

2. The Warsaw group constructed major parts of the ZEUS detector, viz. the Backing Calorimeter (BAC) and the Veto Wall The Cracow group built the gas system of the Backing Calorimeter and the luminosity detector. Both groups have been responsible for the running of these parts and for their maintanance and service.

3. Polish groups from Cracow and Warsaw collaborate closely with American groups from Argonne, Brookhaven and several universities, participating in the ZEUS experiments. In particular, American groups are responsible for the uranium Barrel Calorimeter (BCAL).

Heavy Ion Experiments NA35 and NA49
or how to collaborate with Americans doing physics in Europe

Helena Białkowska
Sołtan Institute for Nuclear Studies, Warsaw

In the fall of '86 at CERN, fully stripped oxygen ions were accelerated to 200 GeV/N, with sulphur coming in '87, and lead at 160 GeV/N expected in November '94. There are a dozen experiments with heavy ions at CERN, and I report on a sequence NA35 - NA49, in which a sizable group of polish physicists participates actively. The NA35 - NA49 expts are dominated by two groups: Germans from many institution, but mainly Franfurt University and Munich MPI, and Americans from LBL and Seattle, with large amount of money coming from DoE and German funds. Polish participants in these expts include 8 physicists, 2 PhD s and a few part-time technicians. They take part in the runs at CERN, contribute to apparatus building and testing, work on the software developpment and simulations, and work at the physics analysis, both at home site and abroad. The computing basis consists of IBM RISC 6000 and HP workstations, with telnet facilities to CERN and Frankfurt.

The main idea of both experiments is to study full events of ion collisions, with wide coverage of most of the phase space. In the NA35 set up, a streamer chamber with both visual (film) coverage and electronic (CCD) registration , accompanied by a prototype time projection chamber (TPC) were the main detectors. In the NA49, large TPC's cover most of the phase space, with particle identification by dE/dx measurement.

The main physics resuls from NA35 concern:

- cross sections
- multiplicity distributions
- particle spectra
- space-time dimensions of the production region (interferometry)
- strangeness production.

The NA49 run with 160 GeV/N Pb beam is dedicated to the study of full events with possibly more than one non-trivial collision characteristics, such as large transverse momenta, enhanced strangeness, large K/π ratio, etc.

High Energy Experiments with American and Polish Participation

At CERN:

A Light Universal Detector for the Study of Correlations between Photons and Charged Particles,

- Oak Ridge Nat. Lab., Tennessee University Knoxville
- Warsaw.

Large Acceptance Measurements of Photons and Charged Particles in Heavy Ion Reactions

- Oak Ridge Nat. Lab., Tennessee University Knoxville
- Warsaw

Large Acceptance Hadron Detector for an Investigation of Pb-induced Reactions at the CERN SPS

- LBL, UCLA, University of Washington
- Cracow, Warsaw

Antiproton Induced Fission and Fragmentation of Nuclei

- Florida State
- Warsaw

Measurements of the Spin-Dependent Structure Functions of Proton and Deuteron

- University of Houston, UCLA, Northeastern, Northwestern, UC Santa Cruz, University of Virginia
- Warsaw

Interactions of 180 GeV/Nucleon ^{207}Pb Nuclei in Emulsion Chamber with Copper and Lead Targets

- Louisiana State, University of Minnesota
- Cracow

DELPHI - Collaboration

- Iowa State University ames
- Cracow, Warsaw

At DESY:

ZEUS - Collaboration

- Ten US Universities and Laboratories
- Cracow, Warsaw

H 1 - Collaboration

- UC Davis, SLAC
- Cracow

In USA:

Experiment PHOBOS (nucleus - nucleus collisions at RHIC)

- MIT, BNL, Oak Ridge Nat. Lab., University of Illinois, University of Meryland, Yale University
- Cracow

KLM Collaboration (nucleus - nucleus collisions)

- University of Minnesota, Louisiana State University
- Cracow

Experiment JACEE

- Louisiana State University, University of Washington, University of Alabama
- Cracow

LIST OF PARTICIPANTS

Ignatios Antoniadis
Ecole Polytechnique, 48, Route de Saclay
F-91128 Palaiseau France

Richard Arnowitt
Center for Theoretical Physics, Dept. of Physics,
Texas A & M, College Station, Texas 77843-4242 USA

Constantin Bachas
Ecole Polytechnique, 48, Route de Saclay
F-91128 Palaiseau, France

Howie Baer
Dept. of Physics, Florida State University
Tallahassee, Florida 32306-3016, USA

Jonathan Bagger
Dept. of Physics and Astronomy, John Hopkins University
Charles and 34th Streets, Baltimore, MD 21218, USA

Alfred Bartl
Institut für Theoretische Physik Universität Wien
Boltzmanngasse 5, A-1090 Wien, Austria

Mar Bastero-Gil
INTA-LAEFF, Apdo. de Correos 50.727
28080 Madrid, Spain

Zurab Berezhiani
INFN, Sezione Ferrara,
44100 Ferrara, Italy

Norma Mankoč Borštnik
Dept. of Physics, University of Ljubljana
Jadranska 19, Ljubljana, 61111, Slovenia

Alessandro Bottino
Dipartimento di Fisica Teorica, Turin University,
via P. Giuria 1, 10125 Torino Italy

Gabriel Cardoso
Humboldt University, Institut Physik Elementarteilchen
D-10115 Berlin, Germany

Ali Chamseddine
Institut für Theoretische Physik, ETH
Hönggerberg, CH-8093 Zürich, Switzerland

Piotr Chankowski
Institute of Theoretical Physics,
ul. Hoża 69, 00-681 Warszawa, Poland

Mirjam Cvetič
Dept. of Physics, University of Pennsylvania
Philadelphia 19104-6396, USA

Sally Dawson
Brookhaven National Laboratory,
P.O.Box 5000, Upton, New York 11973-5000, USA

Jean Pierre Derendinger
Physics Institute, University of Neuchatel
Breguet 1, CH-2000 Neuchatel, Switzerland

Sergio Ferrara
Theory Division, CERN
1211 Geneva 23, Switzerland

Edi Gava
SISSA, via Beirut 2/4
I-34014 Trieste, Italy

Richard Grimm	Centre de Physique Theorique, CNRS - Luminy Case 907, F-13288 Marseille Cedex 9, France
Bohdan Grządkowski	Institute of Theoretical Physics, ul. Hoża 69 00-681 Warszawa, Poland
Howard Haber	SCIPP, Natural Sciences II, University of California, Santa Cruz, CA 95064, USA
Katri Huitu	Research Institute for High Energy Physics, P.O.Box 9 Helsinki University, FIN-00014 Helsinki, Finland
Jan Kalinowski	Institute of Theoretical Physics, ul. Hoża 69 00-681 Warszawa, Poland
Ritva Kinnunen	Research Institute for High Energy Physics, Helsinki University Siltavuorenpenger 20 C, FIN-00014 Helsinki, Finland
Ian Kogan	Dept. of Theoretical Physics, University of Oxford Keble Road, Oxford, OX1 3NP, United Kingdom
Jerzy Kowalski-Glikman	Institute for Theoretical Physics, Wrocław Uniwersity Pl. Maxa Borna 9, 50-204 Wrocław, Poland
Lawrence Krauss	Case Western Reserve University, Dept. of Physics Cleveland, OH 44106, USA
Marek Olechowski	Institute of Theoretical Physics, ul. Hoża 69 00-681 Warszawa, Poland
Peter Minkowski	Institut fur Theoretische Physik, Universität Bern Sidlerstrasse 5, CH-3012 Bern, Switzerland
Rabi Mohapatra	Dept. of Physics, University of Maryland Elementary Particles Group, College Park, MD 20742, USA
Thomas Mohaupt	Desy IfH Zeuthen, Platanenallee 6 D-15738 Zeuthen, Germany
Carlos Muñoz	Departamento de Fisica Teorica C-XI, Madrid University E-28049 Cantoblanco Madrid, Spain
Kumar Narain	ICTP, P.O.Box 586 I-34100 Trieste, Italy
Pran Nath	Dept. of Physics, Northeastern University 360 Huntigton Avenue, Boston, MA 02115, USA
Keith Olive	School of Physics and Astronomy, University of Minnesota Minneapolis, MN 55455, USA
Burt A. Ovrut	Dept. of Physics, Pennsylvania University Philadelphia, Pa 19104-6396, USA
Jacek Pawełczyk	Institute of Theoretical Physics, ul. Hoża 69 00-681 Warszawa, Poland
Pascal Perrodo	PPE Division, CERN 1211 Geneva 23, Switzerland

Zbigniew Płuciennik	Institute of Theoretical Physics, ul. Hoża 69 00-681 Warszawa, Poland
Heath Pois	Davis Institute for High Energy Physics, Dept. of Physics University of California, Davis, California 95616, USA
Stefan Pokorski	Institute of Theoretical Physics, ul. Hoża 69 00-681 Warszawa, Poland
Mariano Quiros	TH Division, CERN 1211 Geneva 23, Switzerland
Stuart Raby	Dept. of Physics, Ohio State University 174 W. 18th Ave, Columbus, OH 43210, USA
Lisa Randall	Center for Theoretical Physics, MIT Cambridge, MA 02139, USA
Stephen Reucroft	Dept. of Physics, Northeastern University 360 Huntigton Avenue, Boston, MA 02115, USA
Janusz Rosiek	Institut de Fisica Corpuscular, Universitat de Valencia 46100 Burjassot (Valencia), Spain
Jonathan Rosner	Enrico Fermi Institute, University of Chicago 5640 S. Ellis Avenue, Chicago , IL 60637, USA
Janis Ruza	Nuclear Research Center, Miera str. 31, LV-2169 Salaspis, Latvia
Carlos Savoy	Service de Physique Theorique de Saclay F-91191 Gif-sur-Yvette Cedex, France
Thomas Scott	SCIPP, Natural Sciences II University of California, Santa Cruz, CA 95064, USA
Ron Settles	MPI, Foehringer Ring 6 80805 Munich, Germany
Krzysztof Śliwa	Physics Dept., Tufts University Medford, MA 02155, USA
Andre Sopczak	PPE Division, CERN 1211 Geneva 23, Switzerland
Ryszard Sosnowski	IPJ, ul. Hoża 69 00-681 Warszawa, Poland
Tomasz Taylor	Dept. of Physics, Northeastern University 360 Huntigton Avenue, Boston, MA 02115, USA
Stefan Theisen	Sektion Physik, University of Munich Theresienstr. 37, D-W-8000 Munchen 2, Germany
Jose W. F. Valle	Institut de Fisica Corpuscular, Universitat de Valencia 46100 Burjassot (Valencia), Spain
Andrew White	Dept. of Physics, University of Texas at Arlington Arlington, Texas 76019, USA

FROM BLACK CLOUDS TO BLACK HOLES

(2nd Edition)

by

J V Narlikar

(Inter-Univ. Ctr. for Astronomy and Astrophysics)

This book describes in nontechnical language one of the success stories of modern (twentieth-century) astronomy. It presents us with the physical picture of what constitutes a star, a description of how a star evolves with time, how its shape and brightness change, how it manufactures the chemical elements deep in its interior, what makes it explode... The presentation also includes exotic objects such as supernovae, pulsars, neutron stars and white dwarfs, and of course, black holes. This revised edition brings the discussion up to date with the inclusion of astronomical events like Supernova 1987A and findings from the Hubble Space Telescope as well as other observations. The book is appropriate as supplementary material for an elementary course on astronomy and astrophysics.

Readership: For the general reader.

180pp (approx.) *981-02-2032-4*
Pub. date: Summer 1995 *981-02-2033-2(pbk)*

• *Please write in for price information*

World Scientific
An International Publisher

1060 Main Street, River Edge NJ 07661, USA **Toll-free: 1-800-227-7562** Fax: 1-201-487-9656 Tel: 1-201-487-9655